Wheat Science

Globally, significant work has been done to enhance our current understanding of the nutritional and anti-nutritional properties, processing, storage, bioactivity, and product development of wheat, opening new frontiers for further improvement. *Wheat Science: Nutritional and Anti-Nutritional Properties, Processing, Storage, Bioactivity, and Product Development* addresses the topics associated with the advances in understanding the wheat biochemical, nutritional, and rheological quality. Improving crop varieties by either conventional breeding or transgenic methods to obtain nutritionally enhanced crops has the advantage of making a one-time investment in research and development to have sustainable products.

Features:

- Includes topics associated with the nutritional composition and anti-nutritional properties
- Addresses the effects of different processing technologies on flour yield and end products
- Reviews the effects of storage on nutritional, baking and rheological quality, organoleptic quality, etc.

Processing and storage technologies have impacted the nutritional quality and the bioavailability of nutrients in wheat. Due to its peculiar grain protein composition, especially gluten protein, wheat has extensive usage in making numerous end products, eaten round the clock. Researchers have demonstrated a significant effect of alteration of flour-processing technologies on the rheological quality of end products. This book provides a holistic understanding and covers recent developments of wheat science under one umbrella. Emphasis is placed on current trends and advances in nutritional and anti-nutritional properties, processing, storage, bioactivity, and product development. Additionally, efforts have been made to compile the available information on the application of different ingredients of wheat in the industry and pharma sectors.

Cereals: Science and Processing Technology

Series Editors: Sneh Punia and Manoj Kumar

Maize
Nutritional Composition, Processing, and Industrial Uses
Edited by Sukhvinder Singh Purewal, Pinderpal Kaur, Sneh Punia, Kawaljit Singh Sandhu, Surender Kumar Singh, and Maninder Kaur

Wheat Science
Nutritional and Anti-Nutritional Properties, Processing, Storage, Bioactivity, and Product Development
Edited by Om Prakash Gupta, Sunil Kumar, Anamika Pandey, Mohd. Kamran Khan, Sanjay Kumar Singh, and Gyanendra Pratap Singh

For more information about this series, please visit: www.routledge.com/Cereals/book-series/CSPT

Wheat Science

Nutritional and Anti-Nutritional
Properties, Processing, Storage,
Bioactivity, and Product Development

Edited by Om Prakash Gupta, Sunil Kumar,
Anamika Pandey, Mohd. Kamran Khan,
Sanjay Kumar Singh, and
Gyanendra Pratap Singh

CRC Press
Taylor & Francis Group
Boca Raton London New York

CRC Press is an imprint of the
Taylor & Francis Group, an **informa** business

Designed cover image: © ICAR-Indian Institute of Wheat and Barley Research, Karnal-132001, Haryana, India

First edition published 2024
by CRC Press
6000 Broken Sound Parkway NW, Suite 300, Boca Raton, FL 33487–2742

and by CRC Press
4 Park Square, Milton Park, Abingdon, Oxon, OX14 4RN

CRC Press is an imprint of Taylor & Francis Group, LLC

ISBN: 978-1-032-29374-5 (hbk)
ISBN: 978-1-032-31080-0 (pbk)
ISBN: 978-1-003-30793-8 (ebk)

DOI: 10.1201/9781003307938

Typeset in Times
by Apex CoVantage, LLC

Contents

About the Editors

Dr. Om Prakash Gupta is Senior Scientist (Plant Biochemistry) at ICAR-Indian Institute of Wheat and Barley Research, Karnal, Haryana, India. He obtained his M.Sc. and Ph.D. degrees from ICAR-Indian Agricultural Research Institute, New Delhi. He has been actively engaged in research and teaching in the field of wheat quality and molecular biology since 2008. His major research interests include small RNA's role during various biotic and abiotic stresses, molecular basis of Fe and Zn biofortification, and nutritional and processing quality. Dr. Gupta has authored and coauthored in more than 80 research publications and served as editorial board member of many peer-reviewed journals. Dr. Gupta is the recipient of several academic awards including Jawaharlal Nehru Award for outstanding doctoral thesis by ICAR, University Silver Medal, Aspee Gold Medal Dr. Kirtikar Memorial Gold Medal, and Chowdhary Charan Singh Memorial Award. He is fellow of the Society for Advancement of Wheat and Barley Research.

Dr. Sunil Kumar is currently serving as Principal Scientist (Biochemistry) at Quality and Basic Science Division at ICAR-IIWBR, Karnal. He obtained his M.Sc. and Ph.D. degrees in Biochemistry from CCS Haryana Agricultural University, Hisar, Haryana, India. He has served as Assistant Professor at MPUAT, Udaipur, Rajasthan, India, and as Assistant Scientist at CCSHAU, Hisar, from 2005 to 2011 before joining ICAR as Senior Scientist and contributed toward teaching and research work. His areas of research interest are plant biochemistry, antioxidative system and post-harvest biology. Throughout 16 years of service experience, he has worked/completed various institutional/externally funded research projects. He has also served as I/c PC KVK, Fazilka. Dr. Sunil Kumar has 30 research papers (of international and national repute), one edited book, and more than 50 other publications including technical bulletins, book chapters, conference abstracts, practical manuals, and popular articles to his credit. Currently, he is actively involved in processing and enhancing nutritional quality of wheat using biochemical/molecular approaches, and in wheat-breeding program for high yield and wider adaptability in North Western Plains.

Dr. Anamika Pandey is currently working in the capacity of Assistant Professor at Department of Soil Science and Plant Nutrition, Faculty of Agriculture, Selcuk University, Turkey. She is also handling a 1001 national project funded by The Scientific and Technological Research Council of Turkey (TUBITAK) related to the determination of novel genes regulating boron toxicity tolerance in some of the Poaceae family species. She earned her Ph.D. degree in biotechnology from India. Her current research focuses on identifying the novel genes and transcriptional factors in wild wheat species, which are responsible for making them tolerant toward different biotic and abiotic stress conditions. With a research experience of 14 years in plant molecular biology, she has published several articles in SCI-indexed journals and is in the editorial and reviewer board of different reputed journals such as *PLOS Sustainability and Transformation* and BMC Plant Biology.

Dr. Mohd. Kamran Khan is presently working as Assistant Professor in Department of Soil Science and Plant Nutrition, Faculty of Agriculture, Selcuk University, Konya, Turkey. He obtained his Ph.D. in Biotechnology from Sam Higginbottom University of Agriculture, Technology and Sciences, Allahabad, India. Dr. Khan has around 14 years of research experience in the area of molecular biology and plant biotechnology. His research is evidenced by his publications in journals of international repute like *AoB Plants*, *Frontiers in Plant Science*, *Plants*, *Plos One*, *3 Biotech*, *Genetic Resources*, and *Crop Evolution*. He has more than 40 research publications with more than 250 citations including the chapters in the books published from Elsevier and CRC Press Ltd. He is on editorial and reviewer board of different reputed journals such as *Plos One*, *Journal of Applied Genetics*, and *Scientific Reports*. The research interests of Dr. Khan include biochemical and molecular changes in crop plants, especially wheat under different biotic and abiotic stress conditions. He is also interested in looking for the role of nanoparticles in regulating different abiotic stresses in plants.

Dr. Sanjay Kumar Singh, an ARS, is Principal Scientist (Plant Breeding) at ICAR-Indian Agricultural Research Institute, Pusa, New Delhi, India. He graduated with silver medal and post-graduated with JS Aggarwal Memorial Gold medal from CSA University of Agriculture and Technology, Kanpur, Uttar Pradesh, India, and doctorate from BHU, Varanasi. He has developed 28 wheat varieties for different agro-ecological conditions of the country and 54 trait specific genetic stocks in wheat. His area of expertise is wheat improvement for abiotic stresses, quality traits and hybrid wheat. He successfully handled 35 research projects funded by CIMMYT-Mexico, DFID-UK, CSIRO-Australia, NFSM, DBT, CSIR, DAC&FW, GoI, ICAR (CRP, AP Cess, NATP, Institute), and DIT, MoC&IT, etc. He has accomplished assignments of Head, IIWBR-Regional Station,

Dalang Maidan, Lahaul and Spiti, Himachal Pradesh, India – a national summer nursery facility; Zonal Coordinator, Peninsular zone of the AICRP on Wheat and Barley and Nodal Officer, Seed at IIWBR, Karnal. Dr Singh is Chair, Expert Working Group (EWG) on "Breeding Methods & Strategies" of the G-20 Nations Wheat Initiative. He has more than 380 publications of national/ international repute to his credit. He is recipient of several awards including ICAR-Lal Bahadur Shastri Young Scientist Award, 2007; UP-CST Young Scientist Award, 2008; VS Mathur Memorial Award, 2019; and S Nagarajan Memorial Award, 2020, and Fellow-ISGPB, ISPGR, and SAWBAR. He has international exposure as Visiting Scientist to Mexico, the United States, France, the Netherlands, and Morocco. Dr Singh has very close linkages with KVKs, progressive farmers, and seed growers and has trained more than 1,000 farmers/scientists in latest wheat technologies.

Dr. Gyanendra Pratap Singh has vast experience in wheat research for more than 25 years. He has developed 37 wheat varieties and 34 genetic stocks suited to different agro-climatic zones of the country. Dr. Singh has published more than 160 research and review papers in reputed peer journals. He is recipient of several prestigious awards such as Rafi Ahmed Kidwai award and BP Pal Gold Medal. He is fellow of many scientific academies and societies such as National Academy of Agricultural Sciences, Indian Society of Genetics and Plant Breeding, Society of Advancement of Wheat Research, and Society for Scientific Development in Agriculture and Technology. He has actively organized several workshops and symposia and chaired many scientific sessions. He is actively involved in mobilization of wheat technologies from laboratory to farmer's field.

Contributors

Divya Ambati
ICAR-Indian Institute of Oilseeds
  Research
Rajendranagar, Hyderabad, India

Ankush
ICAR-Indian Institute of Wheat and
  Barley Research
Karnal, Haryana, India

Manju Bala
ICAR-Central Institute of Post-Harvest
  Engineering and Technology,
  P.O. PAU
Ludhiana, Punjab, India

Shweta Bijla
Directorate of Mushroom Research
  (DMR)
Solan, Himachal Pradesh, India

Suma Biradar
University of Agricultural Sciences
Dharwad, Karnataka, India

Jaipraksh Bisen
ICAR-National Rice Research
  Institute
Cuttack, Odisha, India

Hanuman Bobade
Punjab Agricultural University
Ludhiana, Punjab, India

Krishna Bahadur Chhetri
Krishi Vigyan Kendra
Siwan, Bihar, India

R. S. Chhokar
ICAR-Indian Institute of Wheat and
  Barley Research
Karnal, Haryana, India

Anil Kumar Dixit
ICAR-National Dairy Research Institute
Karnal, Haryana, India

Rahul Gajghate
ICAR-Indian Agricultural Research
  Institute
Indore, Madhya Pradesh, India

Monika Garg
National Agri-Food Biotechnology
  Institute
Mohali, Punjab, India

Nitin Kumar Garg
Sri Karan Narendra Agriculture University
Jobner, Rajasthan, India

S. C. Gill
ICAR-Indian Institute of Wheat and
  Barley Research
Karnal, Haryana, India

Velu Govindan
International Maize and Wheat
  Improvement Center
Texcoco, Mexico

Om Prakash Gupta
ICAR-Indian Institute of Wheat and
  Barley Research
Karnal, Haryana, India

Muzaffar Hasan
ICAR-Central Institute of Agricultural
  Engineering
Bhopal, Madhya Pradesh, India

Maria Itria Ibba
International Maize and Wheat
  Improvement Center
Texcoco, Mexico

Singh J. B.
ICAR-Indian Agricultural research
    Institute
Indore, Madhya Pradesh, India

Poonam Jasrotia
ICAR-Indian Institute of Wheat and
    Barley Research
Karnal, Haryana, India

Saurabh Joshi
Agriculture University
Jodhpur, Rajasthan, India

Ankita Kandpal
ICAR-National Institute of Agricultural
    Economics and Policy Research
New Delhi, India

Babita Kathayat
ICAR-National Dairy Research Institute
    Karnal
Karnal, Haryana, India

Harmandeep Kaur
Punjab Agricultural University
Ludhiana, Punjab, India

Satveer Kaur
National Agri-Food Biotechnology
    Institute
Mohali, Punjab, India
Panjab University
Chandigarh, Punjab, India

Hanif Khan
ICAR-Indian Institute of Wheat and
    Barley Research
Karnal, Haryana, India

Rinki Khobra
ICAR-Indian Institute of Wheat and
    Barley Research
Karnal, Haryana, India

Gopalareddy Krishnappa
ICAR-Indian Institute of Wheat and
    Barley Research
Karnal, Haryana, India
ICAR-Sugarcane Breeding Institute
Coimbatore, Tamil Nadu, India

Amresh Kumar
ICAR-National Institute for Plant
    Biotechnology
New Delhi, India

Anuj Kumar
ICAR-Indian Institute of Wheat and
    Barley Research
Karnal, Haryana, India

Neeraj Kumar
ICAR-Indian Institute of Wheat and
    Barley Research
Karnal, Haryana, India

Nitesh Kumar
ICAR-Indian Institute of Wheat and
    Barley Research
Karnal, Haryana, India

Ravindra Kumar
ICAR-Indian Institute of Wheat and
    Barley Research
Karnal, Haryana, India

Sumit Kumar
Dr. Rajendra Prasad Central
    Agricultural University
Pusa, Samastipur, Bihar, India

Sunil Kumar
ICAR-Indian Institute of Wheat and
    Barley Research
Karnal, Haryana, India

Vishnu Kumar
ICAR-Indian Institute of Wheat and
    Barley Research
Karnal, Haryana, India

Arti Kumari
ICAR-Indian Agricultural Research
    Institute
New Delhi, India

Binita Kumari
Rashtriya Kishan (PG) College
Shamli, Uttar Pradesh, India

Sarita Kumari
Dr. Rajendra Prasad Central
    Agricultural University
Pusa, Samastipur, Bihar, India

Kiran Kumara T. M.
ICAR-National Institute of Agricultural
    Economics and Policy Research
New Delhi, India

Chirag Maheshwari
ICAR-Indian Agricultural Research
    Institute
New Delhi, India

Sanjit Maiti
ICAR-National Dairy Research Institute
    Karnal
Karnal, Haryana, India

Gulshan Kumar Malik
Indian Institute of Technology
Kharagpur, West Bengal, India

H. M. Mamrutha
ICAR-Indian Institute of Wheat and
    Barley Research
Karnal, Haryana, India

Sandeep Mann
ICAR-Central Institute
    of Post-Harvest Engineering and
    Technology, P.O. PAU
Ludhiana, Punjab, India

Manas Mathur
Suresh Gyan Vihar University
Jaipur, Rajasthan, India

Nand Lal Meena
ICAR-Indian Agricultural Research
    Institute
New Delhi, India
ICAR-National Bureau of Plant Genetic
    Resources
New Delhi, India

Maninder Meenu
National Agri-Food Biotechnology
    Institute
Mohali, Punjab, India

C. N. Mishra
ICAR-Indian Institute of Wheat and
    Barley Research
Karnal, Haryana, India

Soumya Mohapatra
ICAR-National Dairy Research
    Institute
Karnal, Haryana, India

Vanita Pandey
ICAR-Indian Institute of Wheat and
    Barley Research
Karnal, Haryana, India

Neha Patwa
Kurukshetra University
Kurukshetra, Uttar Pradesh,
    India

Rahul M. Phuke
ICAR-Central Institute for Cotton Research
Nagpur, Maharashtra, India

Rakesh Kumar Prajapat
Suresh Gyan Vihar University
Jaipur, Rajasthan, India

Harsha B. R.
Krishi Vigyan Kendra
Siwan, Bihar, India

Sendhil R.
Pondicherry University
Kalapet, Puducherry, India

Sewa Ram
ICAR-Indian Institute of Wheat and
    Barley Research
Karnal, Haryana, India

N. D. Rathan
ICAR–Indian Agricultural Research
    Institute
New Delhi, India

Krishna Viswanatha Reddy
ICAR–Central Tobacco Research
    Institute
Rajahmundry, Andhra Pradesh, India

Uday G. Reddy
University of Agricultural Sciences
Dharwad, Karnataka, India

Shiv Ram Samota
ICAR-Indian Institute of Wheat and
    Barley Research
Karnal, Haryana, India

Mukesh Saran
Manipal University Jaipur
Jaipur, Rajasthan, India

Biswajit Sen
ICAR-National Dairy Research Institute
    Karnal
Karnal, Haryana, India

Deepak Sharma
Jaipur National University
Jaipur, Rajasthan, India

Megha Sharma
University of Rajasthan
Jaipur, Rajasthan, India

Saloni Sharma
National Agri-Food Biotechnology
    Institute
Mohali, Punjab, India

Ajeet Singh
ICAR-Indian Institute of Wheat and
    Barley Research
Karnal, Haryana, India

Akhlash P. Singh
GGDSD College Chandigarh (Panjab
    University)
Chandigarh, Punjab, India

Anju Mahendru Singh
ICAR-Indian Agricultural Research
    Institute
PUSA, New Delhi, India

Charan Singh
ICAR-Indian Institute of Wheat and
    Barley Research
Karnal, Haryana, India

Gyanendra Singh
ICAR-Indian Institute of Wheat and
    Barley Research
Karnal, Haryana, India

Gyanendra Pratap Singh
ICAR-Indian Institute of Wheat and
    Barley Research
Karnal, Haryana, India

Jyoti Prakash Singh
ICAR-National Bureau of Agriculturally
    Important Microorganisms (NBAIM)
Mau, Uttar Pradesh, India

Pradeep Kumar Singh
ICAR-Indian Agricultural Research
    Institute
New Delhi, India

Ravi P. Singh
International Maize and Wheat
    Improvement Center
Texcoco, Mexico

Sanjay Kumar Singh
ICAR-Indian Agricultural Research
  Institute
PUSA, New Delhi, India

Ramesh Soni
Govt. National P. G. College
Sirsa, Haryana, India

Surya Tushir
ICAR-Central Institute of Post-Harvest
  Engineering and Technology, P.O. PAU
Ludhiana, Punjab, India

Aruna Tyagi
ICAR-Indian Agricultural Research
  Institute,
New Delhi, India

B. S. Tyagi
ICAR-Indian Institute of Wheat and
  Barley Research
Karnal, Haryana, India

Ganesh Upadhyay
CCS Haryana Agricultural
  University
Hisar, Haryana, India

Prathap V.
ICAR-Indian Agricultural Research
  Institute
New Delhi, India

Sai Prasad S. V.
ICAR-Indian Institute of
  Rice Research
Rajendranagar, Hyderabad, India

Deep Narayan Yadav
ICAR-Central Institute of Post-Harvest
  Engineering and Technology
Ludhiana, Punjab, India

K. J. Yashavanthakumar
Agharkar Research Institute
Pune, Maharashtra, India

Preface

Wheat (*Triticum aestivum* L.) is the primary staple food crop providing the bulk of food calories (50%), at least 30% of Fe and Zn intake and 20% dietary energy and protein consumption worldwide; thus, it is essential to improve its nutritional quality. With the prediction that world's population is expected to reach nine billion by 2050, the wheat yield needs to increase by over 60% along with nutritional characteristics with the available arable land. Under this challenging scenario, the emphasis must be toward improvement in productivity and adaptation to environmental challenges. The Consultative Group on International Agricultural Research (CGIAR) has predicted that the average global wheat yields will need to increase to approximately 5 tonnes per ha from the current average of 3.3 tonnes per ha by 2050. This cereal grain can be stored for future use under specified conditions for a prolonged time without impacting the nutritional profile. Though wheat is rich in carbohydrates, fiber, protein, phytochemicals, and antioxidants, it falls short in certain constituents like iron and zinc and essential amino acid like lysine. Such deficiencies, sometimes, are associated with malnutrition, if not supplemented by other sources, resulting in complex health issues and disturbance of socioeconomic balance. Fortification and biofortification-based approaches can extend a fruitful approach in alleviating micronutrient deficiencies. With a range of functionally existent ploidy levels starting from 2n (monococcum), 4n (durum) to 6n (aestivum), wheat has a diverse species range, though majority of today's cultivation is restricted to aestivum with a minor cultivation of durums. Such complexity of genome makes biotechnological and molecular refinement of wheat crop very challenging.

Over the period, significant research works have been done to understand the science of wheat quality involving processing, biochemical, rheological, nutritional, molecular, and biotechnological tools; globally, however, the knowledge is scattered. Therefore, this book on "Wheat Science" presents an effort to encompass the arena of wheat breeding, agronomy, quality, processing, nutritional quality, storage, biotechnology, and health-related aspects of wheat. It entails production practices, trade, consumption, buffer stock, and management of wheat throughout the world. In initial few chapters, we have covered the role of wheat in global food security along with its nutritional and phytochemical profile. In mid-section, we have discussed the effect and impact of storage, milling, and processing on nutritional quality of wheat followed by molecular mechanisms of bioactive compounds on human health. The book also touches upon the role of fortification and biofortification toward micronutrient enrichment in wheat grains and associated processed products and their techno-functional properties. The scope of biotechnological and molecular approaches in improving nutritional qualities along with gluten-related disorders has also been discussed in brief. Therefore, this book shall provide a milestone to break untrodden intricacies of wheat improvement program on multiple aspects of wheat science benefitting different wheat stakeholders, including scholars, students, teachers, researchers, specialists, food scientist, cereal biochemists, policy makers, molecular breeders and biologists, biotechnologists, academicians, and professionals working in the area of wheat quality enhancement.

Editors

Dr. Om Prakash Gupta
Senior Scientist (Plant Biochemistry), Division of Quality and Basic Sciences, ICAR-Indian Institute of Wheat and Barley Research, Karnal, Haryana, India

Dr. Sunil Kumar
Principal Scientist (Plant Biochemistry), Division of Quality and Basic Sciences, ICAR-Indian Institute of Wheat and Barley Research, Karnal, Haryana, India

Dr. Anamika Pandey
Assistant Professor, Department of Soil Science and Plant Nutrition, Selcuk University, Konya, Turkey

Dr. Mohd. Kamran Khan
Assistant Professor, Department of Soil Science and Plant Nutrition, Selcuk University, Konya, Turkey

Dr. Sanjay Kumar Singh
Principal Scientist (Plant Breeding), Division of Genetic, ICAR-Indian Agricultural Research Institute, New Delhi, India

Dr. Gyanendra Pratap Singh
Director, ICAR-National Bureau of Plant Genetic Resources, New Delhi, India

1

Wheat

Origin, History, and Production Practices

**Sanjay Kumar Singh, Anju Mahendru Singh,
Om Prakash Gupta, and Pradeep Kumar Singh**

CONTENTS

1.1 Introduction

Wheat (*Triticum species*) is one of the universal cereals of old world agriculture (Zohary and Hopf, 2000) and the world's foremost crop plant (Feldman et al., 1995; Gustafson et al., 2009), which represents a large part of the history of agriculture itself besides its ancestry (Heun et al., 1997; Lev-Yadun et al., 2000; Salamini et al., 2002; Charmet, 2011; Riehl et al., 2013; Bilgic et al., 2016). In prehistoric times, it was cultivated throughout Europe and was one of the most valuable cereals of ancient Persia, Greece, and Egypt. According to Paleo-botanists and archaeologists, the modern domesticated form of wheat was originated in South-eastern Anatolia, around the region of Diyarbakir Province in present-day Turkey at about 8500 BC. In India, evidence from Mohenjo-daro excavations showed wheat cultivation over 5,000 years ago. Presently, wheat makes up a significant part of the world's diet and provides about one-fifth of the calories and proteins used by humans, making it the second-most important staple, after rice (Dubcovsky and Dvorak, 2007; Shewry, 2009; Hawkesford et al., 2013; Shiferaw et al., 2013; Chakraborty et al., 2020). Wheat grain contains 60–80% carbohydrates mainly as starch, 8–15% proteins, 1.5–2.0% fats, 2–3% crude fibres, 1.5–2.0% minerals, and traces of vitamins with 11% water and provides about 340 calories of energy (Shewry and Hey, 2015). Nearly 55% of carbohydrates intake and 20% of food calories consumed in the world are attributed to wheat (Breiman and Graur, 1995). Asia represents the largest share (58%) of the culinary uses of wheat, followed by Europe (18%), but the per capita consumption is quite low, that is, 63.62 kg per annum, compared to other continents of the world (FAO, 2011).

Wheat is extensively grown on 17% of all crop areas, in the temperate, Mediterranean-type, and subtropical parts of both hemispheres at wide range of altitudes, from 67° N in Norway, Finland, and Russia to 45°S in Argentina (Feldman, 1995). Wheat is the second-most important cereal crops globally and contributes about one-third of grain production, but it remains the largest cultivated and traded cereal in the world among all crops (OECD-FAO, 2020). In 2020, more than 760.9 million tonnes (mt) of wheat were produced on 219 million hectares around the world with a productivity of 3.47 tons/ha, and estimates indicated 0.33% increase in global wheat production in 2020–21 by producing 778.63 million metric tons (Statistica, 2022; USDA, 2022). Regional share of wheat in global production indicated largest contribution of 45.7% and 33.5% by Asia (347.92 mt) and Europe (255.02 mt), respectively, to the global wheat basket (FAOSTAT, 2020). Although wheat is produced in over 123 countries, but the EU, China, India, Russia, the United States, Australia, Ukraine, Pakistan, Canada, and Argentina, the top ten wheat-producing countries (Figure 1.1), contribute over 82% of total wheat production (FAOSTAT, 2020).

Production trends indicated roughly equal split of wheat production between the developing and developed world where the developing regions including China and Central Asia account for roughly 53% of the total harvested area and 50% of the

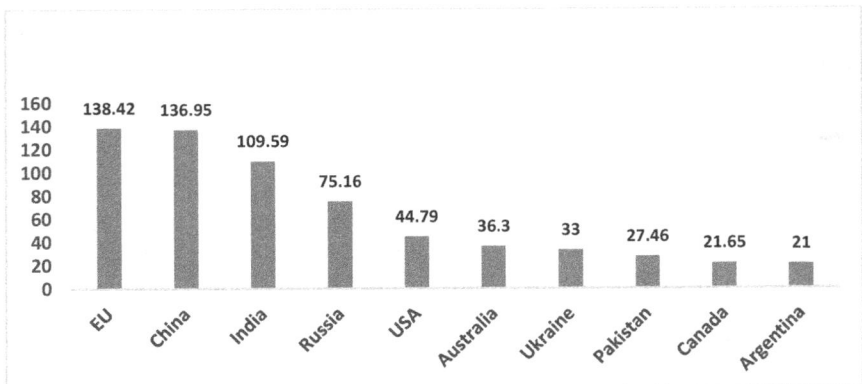

FIGURE 1.1   Top ten wheat-producing countries during 2020–21.

global production (Shiferaw et al., 2013). China and India together contribute more than 70% of total wheat production in Asia. Current estimates of wheat production showed a dip of approximately 1.06% of global wheat production of 770.3 mt during 2021–22 (FAO, 2021). India produced 109.59 million tons of wheat from 31.1 m ha area during 2020–21 and contributed about 34% to the total food grain production in India (DAC&FW, 2021). The major wheat-growing states in India are Uttar Pradesh, Punjab, Haryana, Madhya Pradesh, Rajasthan, and Bihar which contribute approximately 91.5% of India's wheat produce.

The average yield trends for wheat indicated higher yield levels in Europe (4.13 tons/ha) than the global average of 3.47 tonnes/ha (t/ha), whereas yield levels in North America (3.45 t/ha) and Asia (3.42 t/ha) are close to global wheat average yield (FAOSTAT, 2020). The country-wise yield levels showed over 8.0 tons/ha yield levels in the top yielding countries New Zealand (8.96 t/ha), the Netherlands (8.61 t/ha), Belgium (8.49 t/ha), and Ireland (8.37 t/ha). However, the average yield of 3.44 t/ha in India is close to the world average yield. The trend of area, production, and productivity showed a significant quantum jump in wheat in post-independence period in India where increments of 2.23 times in area under wheat (+21.91 mha: 222.50%), 15.49 times in wheat production (+102.15 mt: 1548.73%), and 4.11 times in wheat productivity (+2754 kg/ha: 411.23%) were recorded (Singh et al., 2022). This gargantuan jump in production, post-independence, is attributed to the increased crop productivity followed by area. Precisely, the impact was more evident since the inception of the All India Coordinated Research Project leading to semi-dwarf wheat-based Green Revolution. Despite of recent risks of climate change, decade-wise analysis indicated increasing trends in area, production, and productivity in recent decades.

The wheat is widely adopted by mankind owing to its high environmental adaptability because of its allopolyploid nature and, thereby, genomic plasticity in addition to its excellent food/feed qualities regarding carbohydrates, proteins, vitamin content, and unique elastic property of its gluten, which make it suitable for more diverse use of its flour (Dubcovsky and Dvorak, 2007). Approximately 95% of wheat cultivated is hexaploid with the remaining 5% being durum wheat (*T. turgidium* L.) and few other less important types (Shewry, 2009). Global production is struck with the primary challenge to produce more food for the burgeoning population due to productivity going towards stagnation in most of the wheat-growing areas. The demand for wheat is expected to increase by 60% by 2050 from the present level for which at least 1.6% per annum increase in wheat yield is required (GCARD, 2012; Wheat Initiative, 2013; Fischer et al., 2014; FAO, 2017; Poudel and Poudel, 2020). During the Green Revolution, the gains in crop productivity were driven by an increased availability of hybrid seeds and fertilizer. The major concern nowadays is declining crop productivity in some countries, which is as high as 25% per year (Rajaram, 2012). For wheat, in spite of the declining growth rate in yield, the overall yields are closer to 1.5% per annum and are able to meet the demand of the world's growing population (Mittal, 2022).

Modernization of agriculture and high external input use in wheat cultivation made total factor productivity (TFP) a more reliable indicator of crop productivity than crop yields. Studies have shown plateaued technical efficiency over the years in predominated rice–wheat cropping system particularly in South Asia where high input usage, resource degradation and decelerating TFP being the norm for wheat cultivation (Mittal and Kumar, 2000; Kumar et al., 2004; Kumar and Mittal, 2006; Karim and Talukder, 2008; Bhushan, 2016; Ali et al., 2017), the degradation of resource base and progressive intensification of grain cultivation attributed to the deceleration of TFP growth rates which have been further driven down due to climate change in recent years (Ali and Byerlee, 2002; Byerlee et al., 2003; Hughes et al., 2017). The shrinking resources and environmental conditions in present climate change era necessitate focused efforts to increase yields, resistance to biotic and abiotic stresses, and improved climate-smart agronomic practices for high-input use efficiency and thereby sustainable crop production (Pingali, 1999; Mittal and Lal, 2001; Kumar et al., 2008).

1.2 Origin

As a globally cultivated allopolyploid crop, bread wheat has been shaped for success by multiple steps of domestication and polyploidization (Pont and Salse, 2017; Glemin et al., 2018). Taxonomically, bread wheat (*Triticum aestivum* L.), also known as common wheat, is an annual, predominantly autogamous species belonging to the Dinkel section in genus *Triticum L.* of the tribe *Triticeae* of the family Gramineae. The tribe *Triticeae* of the family *Gramineae* is economically the most important of the grass family, as it contains numerous important crop and forage species (wheat, barley, rye, and others). Wild wheats are commonly divided into two sub-groups: goat grasses in the genus *Aegilops L.* and wild wheats in the genus *Triticum L.* All cultivated wheats are included in the genus *Triticum*. Further, wild species in *Aegilops* and *Triticum* occur in a polyploid series, based on a haploid number of n = x = 7 chromosomes. Diploid (2x), tetraploid (4x) and hexaploid (6x) species of *Aegilops* occur naturally, whereas wild species of *Triticum* occur only at the diploid and tetraploid levels. Cultivated wheat exists at the diploid, tetraploid and hexaploid levels. Percival (1921) categorised into 11 species (Table 1.1), and later Feldman and Sears (1981) have listed 13 diploid, 12 tetraploid, and 5 hexaploid species of *Triticum*. Recently, taxonomists have recognized 24 wild species of *Aegilops* including weedy *A. mutica* Boiss and four wild and two domesticated species, for a total of six in the genus *Triticum* (Slageren, 1994).

TABLE 1.1

Different Species of Genus *Triticum*

Species	Common Name
T. monococcum L.	Einkorn wheat
T. dicoccum Schubl.	Emmer wheat
T. orientale mihi.	Khorasan wheat
T. durum Desf.	Macaroni wheat
T. polonicum l.	Polish wheat
T. turgidum L.	Rivet or Cone wheat
T. pyramidale mihi.	Egyptian cone wheat
T. vulgare host.	Bread wheat
T. compactum Host.	Club wheat
T. sphaerococcum mihi.	Indian dwarf wheat
T. spelta L.	Large spelt or Dinkel wheat

Source: Percival (1921)

TABLE 1.2

Centres of Origin of *Triticum Species*

Species	Common Name	Centre of Origin
T. aestivum	Bread wheat	Central Asia, The Near East
T. compactum	Club wheat	Central Asia
T. sphaerococcum	Shot wheat	Central Asia
T. dicoccum	Emmer wheat	Abyssinia
T. durum	Durum wheat	The Near East, Mediterranean, Abyssinia
T. turgidum	Rivet or cone wheat	The Near East, Mediterranean, Abyssinia
T. monococcum	Einkorn wheat	The Near East

Source: Zeven and Zhukovskyii (1975), Hawkes (1982)

According to the earliest historic records, wheat was an important cultivated cereal in South Western Asia, its geographical centre of origin. Many wild species of *Triticum* are found in Lebanon, Syria, northern Israel, and eastern Turkey. Wheat was cultivated in ancient Greece and Egypt in prehistoric times. The Central Asia, Near East, Mediterranean, and Abyssinia regions are the world centres of diversity for wheat and its related species (Table 1.2).

Bread wheat is an allohexaploid species, composed of 21 chromosome pairs organized in three subgenomes, A, B, and D, Genome BBAADD, 2n = 6x = 42 (Sears, 1952). The three genomes of bread wheat (A, B, and D) are derived from three diploid wild ancestors (Blake et al., 1999; Petersen et al., 2006; Marcussen et al., 2014; El-Baidouri et al., 2017), *Triticum urartu* (AuAu, 2n = 2x = 14), an unknown member of the Sitopsis section (closely related to *Aegilops speltoides*, BB, 2n = 2x = 14), and *Aegilops tauschii* (DD, 2n = 2x = 14). Apparently, the diploid, tetraploid, and hexaploid wheat species had all undergone domestication before they were widely cultivated. These three genomes were combined through two sequential polyploidization events (Feldman et al., 1995; Huang et al., 2002) as represented in Figure 1.2. The first polyploidization occurred between *T. urartu* (contributing the A genome) and the Sitopsis species (source of the B genome) about 500,000–150,000 years ago, resulting in the appearance of tetraploid wild emmer wheat (*Triticum turgidum* ssp. dicoccoides) belonging to the emmer wheat section (Chalupska et al., 2008; IWGSC, 2014; IWGSC, 2018). The egg donor in this initial hybridization event was the donor of the B genome, and the pollen donor contributed the A genome. A domestication of wild tetraploid emmer wheat resulted in the sequential appearance of hulled domesticated emmer wheat (*Triticum turgidum* ssp. dicoccum) and large-seeded, free-threshing durum wheat (*Triticum turgidum* ssp. durum) (Dubcovsky and Dvorak, 2007; Avni et al., 2017). Because of its favourable large-seed and free-threshing traits, tetraploid durum wheat had become the major wheat crop by approximately 3,000 years ago. About 10,000 years ago, a second polyploidization event followed the natural hybridization between the domesticated tetraploid emmer wheats (AABB) and *Ae. tauschii* (DD), finally giving rise to modern allohexaploid bread wheat *T. aestivum* (Peng et al., 2011; Dvorak et al., 2012; Wang et al., 2013; Pont et al., 2020; Yao et al., 2020) probably in the Fertile Crescent region that nowadays comprises Northern Iran (Zohary et al., 2012). It is believed that this event of second polyploidization occurred when the

Aegilops speltoids
$2n = 2x = 14$ SS (BB)   X   *Triticum urartu*
$2n = 2x = 14$ AA

↓

1st polyploidization

BBAA
T. turgidum ssp. dicoccoides
$2n = 4x = 28$ Wild
Fragile rachis, Hard glumes

↓

BBAA Dicoccum wheat
T. turgidum ssp dicoccum
$2n = 4x = 28$ Cultivated
Tough rachis, Soft glumes

↓

BBAA Durum wheat
T. turgidum ssp durum
$2n = 4x = 28$ Cultivated
Tough rachis, Soft glumes

Dicoccum & Durum Wheat

Triticum turgidum ssp. *dicoccum*
BBAA   Cultivated   X   *Aegilops tauschii*
$2n = 4x = 28$ DD   Wild
$2n = 2x = 14$

↓

2nd Polyploidization

BBAADD
T. aestivum ssp. *spelta*
$2n = 6x = 42$ Cultivated
Tough rachis, Hard glumes

↓

BBAADD
Bread wheat
T. aestivum ssp. *aestivum*
$2n = 6x = 42$ Cultivated
Tough rachis, Soft glumes

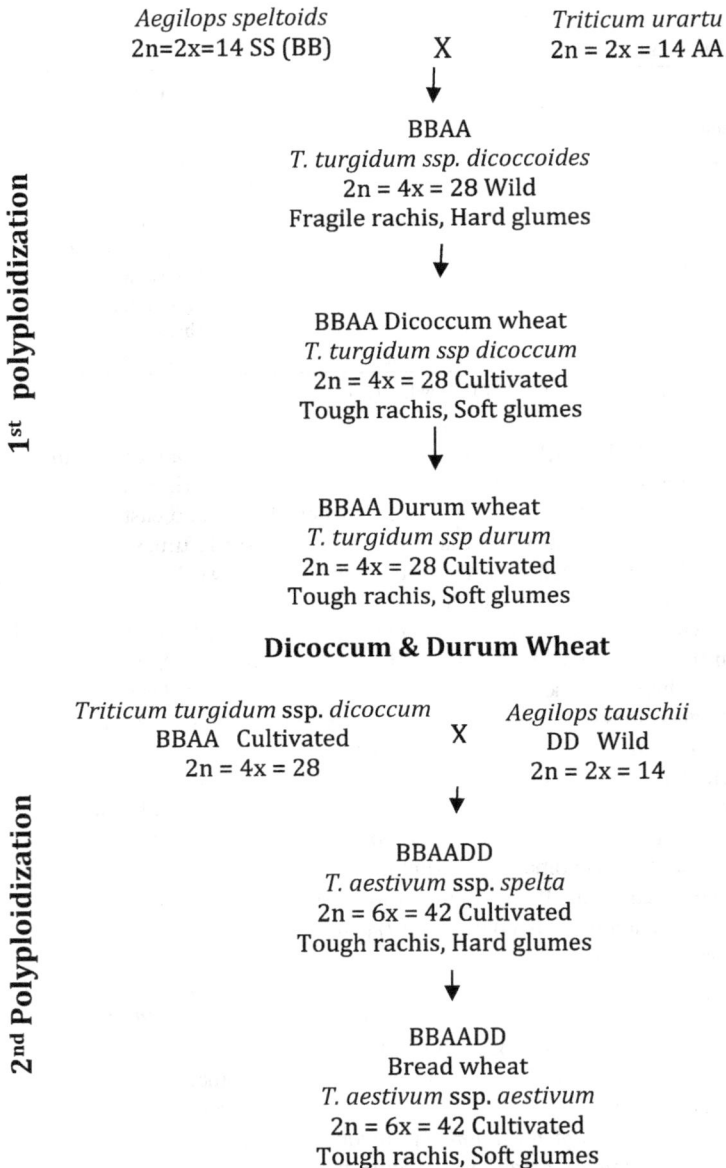

FIGURE 1.2   Origin of bread wheat.

cultivation of domesticated emmer wheat spread into the natural range of *Ae. tauschii* (Salamini et al., 2002). From its origins in the Near East, farming expanded throughout Europe, Asia, and Africa, together with various domesticated plants and animals. After domestication and centuries of cultivation and selection, this *T. aestivum* exhibited broader adaptability and many other favourable agronomic traits, which enabled it to expand further than durum to become the most important cultivated wheat crop worldwide (Kihara, 1944; McFadden and Sears, 1946; Dubcovsky and Dvorak, 2007; Salse et al., 2008; El Baidouri et al., 2017; Pont and Salse, 2017).

1.3 Wheat Research in India

Wheat grown in India prior to 20th century was mostly consisted of mixtures of various botanical forms usually referred to as 'sorts' among which Sharbati, Dara, Safed Pissi, Chandausi, Karachi, Choice White, Hard Red Calcutta, Lal Kanak, Lal Pissi Jaipur Local, Kharchia Local, Mondhya 417, Muzaffar Nagar White, Buxar White, etc., were some prominent local bread wheat sorts utilized nationally and internationally. Sir Albert Howard and his wife Lady G.L.C. Howard initiated systematic bread wheat improvement work in 1905 at Imperial (now Indian) Agricultural Research Institute (IARI) at Pusa in Bihar. Later, wheat improvement was also undertaken at government agricultural colleges at Lyallpur (now in Pakistan), Kanpur, Sabour, Pawarkheda (Madhya Pradesh), Nagpur, Akola, and Niphad (Maharashtra). In the era of pre-semi dwarf bread wheat varieties (1905–62), selection from the mixtures resulted in some important purelines like NP4, NP6, NP12, NP22, Type9, Type11, Pb type8A, C13, C46, AO13, AO49, AO68, AO69, AO85, AO88, and AO90. Some of these varieties earned international recognition due to their excellent grain appearance and are still considered to be valuable genetic resource. Hybridization among these varieties and the exotic wheats was utilized in wheat breeding from the early years of the 20th century which resulted in varieties like NP52, NP80–5, NP120, NP165, NP710, NP770, NP783, NP824, PbC518, PbC591, Niphad4, and AO113. Exotic sources such as Ridley, Padova I, and Padova II gained acceptance, and Federation 41, Kononso, Thatcher, Kenya C 10854, Kenya 48F, Democrat, Spalding's, Ridit, Reliance, prolific, Gabo, Timstein, Bobin, Gaza, and Regent contributed significantly in the Indian wheat breeding programme as donor lines for incorporating rust resistance led to development of disease-resistant cultivars like K53, K54, RS31–1 and Kenphad25. The varieties developed until early 1960s were tall in stature and prone to lodging and, therefore, non-responsive to high fertilizer usage, for example, K 65, K 68, and Hyb 65. Similar to those of bread wheat improvement programme, the durum improvement was initiated by H.M. Chibber in 1918–19 in the erstwhile Bombay Presidency by making single plant selections out of the cultivated landraces like Bansi, Kathia, Gangajali, Haura, Jalalia, Jamli, Khandwa, and Malvi. These selections yielded superior lines like Bansi 103, Bansi 162, Bansi 168, Bansi 224, and Baxi 228–18. The durum wheat evolved in India is good source of genes for drought and heat tolerance, and it is mostly used as chapati or *dalia* due to their unique quality characteristics. Subsequent efforts using exotic germplasm can lead to the development of improved durums like A 206, A 624, Amrut, N 59, NP 404, A-9-30-1, N 5749, Hybrid 23, Ekdania 69, Narasingarh 111, Bijaga yellow, and Bijaga red (Singh et al., 2006b).

The real breakthrough in productivity occurred from the introduction of the Norin (having *Rht* dwarfing gene)-based semi-dwarf Mexican wheat varieties, viz, Lerma Rojo 64-A, Sonora 63, Sonora 64, and Mayo 64 in the early 1960s, possessing a unique set of high productivity traits such as non-lodging habit, high fertilizer responsiveness, and resistance to rust and foliar diseases. Among these, Lerma Rojo 64-A and Sonora

64 carrying the genes for dwarfism and resistance to rusts were released by the Central Varietal Release Committee in 1965 which laid the foundation for increased wheat production. At the same time, the establishment of the All India Coordinated Wheat Improvement Project (AICWIP) in 1965 was an important milestone for systematic developments in wheat research resulting in a major breakthrough in wheat production and productivity. Later on, the advanced lines from CIMMYT provided the base material for development and commercial release of important amber-coloured varieties, namely, S227 (Kalyansona), S307, S308 (Sonalika), S331, Chhoti Lerma, and Safed Lerma which became the harbingers of the 'Green Revolution' in India. During the Green Revolution, yields increased dramatically year on year following the introduction of reduced height alleles which not only increased the harvest index but also allowed the application of more nitrogen without crop lodging (Hawkesford et al., 2013). The unprecedented growth in the productivity of wheat achieved through the Green Revolution in many developing regions of Asia, Africa, and Latin America has helped overcome famines and has saved millions of lives, and, because of this, wheat is often described as the 'Miracle Crop' of the 20th century. This growth in productivity was made possible through technological achievements in developing the semi-dwarf, high yielding varieties and favourable policy and institutional support in ensuring farmer access to new seeds, fertilizers, markets, and irrigation infrastructure (Datt and Ravallion, 1998; Fan and Hazell, 2001; Evenson and Rosegrant, 2003; Renkow and Byerlee, 2010). After the initial impact of the Green Revolution in high production zones through the exploitation of *Rht-B1* and *Rht-D1* dwarfing genes in conjunction with disease resistance (Reynolds and Borlaug, 2006), hybridizations between Mexican and Indian varieties led to the release of WL711, HD2009, WH147, HD1982, UP262, HUW12, HD2189, NI5439, and many others which took the wheat revolution to newer heights. Varieties such Lok1, HUW234, HD2285, HD2329, HD2189, and some derivatives of *Veery* of CIMMYT, viz., HUW206 and HS207, were very popular among the farmers during 1980s. The development of wheat varieties CPAN3004, WH542, PBW343, and PBW373 having 1B/1R translocation through the utilization of winter wheat gene pool provided a quantum jump in wheat productivity by way of resistance to diseases, enhanced morpho-physiological traits, and wider adaptability. Later, HD2687, UP2338, RAJ3765, K9107, NW1014, HP1744, GW273, GW322, GW366, and MACS2496 covered significant wheat acreage in different parts of the country. During early 2000s, PBW 343 was the dominant wheat variety, and, after its susceptibility to yellow rust, DBW17 caught the farmers' attention as a suitable replacement for PBW343 in NWPZ. The release of landmark varieties PBW343, DBW17, HD2967, and HD3086 brought out a revolution in the high productive environments of northern India. The robust and tall plant types with thick stem has led to a change in the plant type of the varieties with the newer genotypes being more efficient and adaptive. Recent releases DBW187, DBW222, and DBW303 have revolutionized the wheat cultivation in the breadbasket of India with a yield potential of more than 8.0 tons/ha.

1.4 Production Zones and Conditions

Wheat in India is cultivated in almost every state except Kerala, thus representing diverse crop growing conditions and situations. Wheat cultivation in India extends from 9°N (Palni hills) to above 35°N (Srinagar valley of Jammu and Kashmir); thus, the wheat crop is exposed to a wide range of agro-climatic changes such as humidity, temperature, and photoperiod during crop season, soil types, altitudes, latitudes, and cropping systems. Based on the production conditions, India is divided into five mega wheat-producing zones, namely Northern Hills Zone (0.82 mha), North Western Plain Zone (12.33 mha), North Eastern Plain Zone (8.85 mha), Central Zone (6.84 mha), and Peninsular Zone (0.71 mha). NWPZ, NEPZ, and Central Zones are the main contributors to wheat production where approximately 91.5% of the total wheat production comes from states of Uttar Pradesh, Punjab, Haryana, Madhya Pradesh, Rajasthan, and Bihar. Food security of India largely depends on the highest wheat-producing zone of the country NWPZ, comprising Punjab, Haryana, Delhi, western Uttar Pradesh, Tarai regions of Uttarakhand, Jammu and Kathua districts of Jammu and Kashmir, and Una district and Paonta valley of Himachal Pradesh. North Eastern Plain Zone (NEPZ) includes Eastern Uttar Pradesh, Bihar, Jharkhand, Odisha, West Bengal, Assam, and Plains of North Eastern states. There is a large possibility to increase wheat production in this zone which is referred as 'the sleeping giant'; therefore, there is a need that farmers of this zone adopt improved wheat production technologies to have better harvest. In the central zone, wheat is cultivated in 6.84 mha area of Madhya Pradesh, Chhattisgarh, Gujarat, Kota Udaipur region of Rajasthan, and Bundelkhand region of Uttar Pradesh. Although the average productivity of central zone is very less as compared to Punjab and Haryana mainly due to heat stress and less water availability, but irrigated areas have productivity at par to that of the states of Haryana and Punjab. The quality of wheat is excellent in this region, and this zone is known for the production of durum wheat suitable for making products like pasta, noodle, macaroni, suji, and *dalia*. Wheat grown in this region has got higher demand in domestic as well as international market and also fetches good price. Good-quality Karnal bunt free wheat is also grown in the Peninsular zone comprising Maharashtra, Karnataka, Goa, and Palni hills of Tamil Nadu. This region is also home to *Triticum dicoccum*. In Northern hill zones, wheat is cultivated in very small area, mostly under rainfed conditions. The growing period of wheat is variable from one agro-climatic zone to another that affects the vegetative and grain-filling duration leading to differences in attainable yield. The maximum wheat-growing duration is in Northern Hill Zone, and the minimum is in Peninsular Zone.

Wheat is cultivated during winter season from mid-October to April (except in higher hills of north India where the harvesting of wheat is done in the month of May). Sowings of wheat are initiated when the average of day–night temperatures is equal to 23°C. The months of December and January remain to be coldest, followed by comparatively warmer and higher temperatures in the months of March to April coinciding to later grown stages of the crop till maturity. Wheat is mainly grown under three production conditions, viz., timely sown, irrigated; late sown, irrigated; and timely sown, restricted irrigation. Nearly 89% of the wheat area in the country is irrigated, and most of it lies in north India. The central, peninsular, and hilly areas have comparatively lower coverage of area under irrigation and grow mostly rainfed. In recent years, a new situation of timely sown, restricted/limited irrigation has emerged in some of the areas of the central and peninsular parts where water for irrigation is not available in sufficient quantity, and, thus, the wheat crop is grown with one to two irrigations only (Singh and Kumar, 2021).

1.5 Varietal Spectrum

The agro-climatic conditions, local preferences and wheat-based food habits, prevalence of diseases and pests, wheat-based cropping systems, availability of irrigation, and related input factors have a direct bearing on the types of wheat varieties to be developed for commercial cultivation in the country. The farmers of the country have been provided with a choice of varieties during last 60 years since inception of All India Coordinated Wheat and Barley Improvement Programme in 1965. The Indian wheat improvement programme has significantly contributed to the release of 501 wheat varieties including *Triticum aestivum, T. durum, T. dicoccum*, and triticale through Central Variety Release Committee (CVRC) or State Variety Release Committee (SVRC) for different agro-climatic zones along with relevant production technology (Table 1.3). This included 422 bread wheat, 68 durum, 07 dicoccum, and 04 triticale varieties.

Some varieties released in recent past have contributed significantly to enhance wheat production, and, among them, DBW17, PBW550, HD2967, HD3086, WH1105, K0307, GW 322, GW 366, HI 1544, and MACS 6222 are notable in different areas. The recently released cultivars are presented in Table 1.4, which are becoming popular among the farmers. Most of the farmers are still growing old varieties having low productivity like Sonalika, Raj 3765, UP 262, WH 147, and HUW 234. In this situation, farmers have to grow recommended new varieties. A list of improved varieties recommended for different production condition is given in Table 1.4.

1.6 Wheat Production Technology

The adoption of recommended production technology is a key component for realizing higher yields and reducing the gap between experimental and farmers' fields. The optimized production technology for different production conditions in India is summarized as follows (Singh et al., 2014; Singh et al., 2016).

1.6.1 Soil Testing

It gives information about soil fertility of the particular field so that one can apply fertilizers as per soil requirement/recommendations. It is done usually in October or

TABLE 1.3

Released Wheat Varieties in India (1965–2021)

Crop Species	CVRC	SVRC	Total
Bread wheat (*T. aestivum*)	267	155	422
Durum wheat (*T. durum*)	45	23	68
Dicoccum wheat (*T. dicoccum*)	6	1	7
Triticale	4	-	4
Total	**322**	**179**	**501**

TABLE 1.4

Wheat Varieties for Diverse Agro-Climatic and Production Conditions

Production Condition	Improved Cultivars
Early sown, irrigated	DBW 327, DBW 332, DBW 187, DBW 303, WH 1270
Timely sown, irrigated	DBW 222, HD 3226, WB 02, HPBW 01, PBW 723, DBW 88, HD 3086, WH 1105
Late sown, irrigated	PBW 771, PBW 752, DBW 173, DBW 90, WH 1124, DBW 71, HD 3059, PBW 590, PBW 757, HI 1621, HD 3271
Timely sown, restricted irrigation	DBW 296, HUW 838, HI 1628, NIAW 3170, HD 3237, HI 1620, WH 1142, HD 3043, PBW 660, PBW 644, WH 1080
Sodic soils/others	KRL 19, KRL 210, KRL 213
Timely sown, irrigated	DBW 222, HD 3249, HD 3086, DBW 187, K 1006, HD 2967, NW 5054, DBW 39, RAJ 4120, K 0307
Late sown, irrigated	HI 1621, DBW 107, HD 3118, HD 2985, HI 1563, HD 3271
Timely sown, restricted irrigation	DBW 252, HI 1612, K 1317, HD 3171, HD 3293
Sodic soils/others	KRL 19, KRL 210, KRL 213
Timely sown, irrigated	GW 513, HI 1636, HI 1544, GW 366, GW 322, HI 8759 (d), HD 4728 (d), HI 8737 (d), HI 8713 (d), MPO 1215 (d)
Late sown, irrigated	HI 1634, CG 1029, RAJ 4238, MP 3336, MP 1203
Timely sown, restricted irrigation	DBW 110, HI 1500, MP 3173, MP 3288, DDW 47(d), UAS 466 (d), HI 8627 (d), HI 8823 (d)
Sodic soils/others	KRL 19, KRL 210, KRL 213
Timely sown, irrigated	DBW 168, MACS 6478, UAS 304, MACS 6222, DDW 48 (d), MACS 3949 (d), WHD 948 (d), UAS 428 (d), UAS 415 (d), HW 1098 (dic), MACS 2971 (dic), DDK 1029 (dic), DDK 1025 (dic)
Late sown, irrigated	HI 1633, HD 3090, AKAW 4627, HD 2932, Raj 4083
Timely sown, restricted irrigation	NIAW 3170, MP 1358, HI 1605, DBW 93, NIAW1415, PBW 596, UAS 375, UAS 347, MACS 4028 (d), HI8777 (d), UAS 446 (d), NIDW 1149 (d), GW 1346 (d), HI 8802 (d), HI 8805 (d), MACS 4058 (d)
Timely sown, irrigated	HS 562, HPW 349, VL 907, HS 507
Late sown, irrigated/rainfed	HS 490, VL 892
Timely sown, rainfed	HS 542, HPW 251, VL 829
For high-altitude areas	VL 832

Abbreviations: (d) – durum wheat; (dic) – dicoccum wheat

after the harvesting process of crops. For soil testing, the samples are taken from all corners and middle of the field, mixed well and from it, 500 g soil sample is taken for testing in nearby soil-testing laboratory.

1.6.2 Field Preparation

Good field preparation helps in saving the irrigation water by uniform irrigation and weed control. Pre-sowing irrigation followed by ploughing with disc harrow, tiller, and leveller at field capacity is needed for optimum field conditions. Field preparation is done from October, based on crop production condition by good ploughing, followed by planking for moisture conservation. Field should be well prepared by disc harrow and cultivator as per the requirement of the soil. Field should be well levelled so the use of laser land leveller is quite beneficial. Three-to-four field preparation operations should be done, followed by levelling. To conserve soil moisture, there should not be deep/heavy cultivation. If sowing is to be done by zero tillage, there is no need to prepare field.

1.6.3 Seed Rate and Treatment

Seed rate is an important factor to get maximum yield levels. The recommended seed rate for early sown irrigated, timely sown irrigated, and timely sown restricted irrigation conditions is 100 kg/ha whereas it is 125 kg/ha for late sown irrigated and timely sown rainfed conditions. Seed treatment gives protection to soil and seed-borne diseases and promotes better germination resulting in healthier plants. It should be done one day before sowing. Raxil @1.0 g/kg seed or Vitavax @ 2.0 g/kg seed or Thiram + Vitavax (1:1) may be used for seed treatment for the control of smut diseases. Seed treatment drum may be used for seed treatment only.

1.6.4 Sowing Time and Method

Wheat sowing starts from mid-October to January on the basis of production conditions. Irrigated timely sown crop is sown during November 1–15, whereas irrigated late sown crop is sown during December 5–15. The rainfed crop in some areas, specially NHZ, is sown during 15–31 October. Crop under early sown irrigated and timely sown restricted irrigation condition is sown from 25th October to 5th November. The suitable temperature required for sowing of wheat is 21–25°C. Sowing is to be done by drilling seed and fertilizer drill. If there is loose straw in the field use of rotary disc drill, Happy Seeder is recommended. Crop residues control weeds in field, and residue retention helps in conserving moisture and increases soil organic matter. Thus, crop residues should not be burnt as these kill beneficial insects. The depth of sowing should be around 5 ± 2 cm, and row spacing for timely sown should be 20 cm. Under late sown condition, row-to-row spacing should be 18 cm. For rainfed areas, line spacing should be 23 cm depending upon soil moisture condition.

1.6.5 Fertilizer Application

All essential nutrients are made available through balanced and integrated nutrient management for a healthy crop with higher productivity. Usually, the application of chemical fertilisers in the form of N:P:K is given, which vary from zone and production condition. Deficiency of one element may affect the availability of other element. Other micronutrients may be applied as per requirement and deficiency in the soil on the basis of soil-testing reports. The recommended fertiliser dose (RFD) for timely sown irrigated condition is 150:60:40 NPK/ha in NWPZ/NEPZ and 120:60:40 NPK/ha in NHZ/CZ/PZ, whereas RFD for late sown irrigated condition is 120:60:40 NPK/ha in NWPZ/NEPZ and 90:60:40 NPK/ha in CZ/PZ. For irrigated conditions, 1/3rd N and full P and K should be applied at sowing and remaining 2/3rd N should be applied in two equal splits at first and second irrigation processes. The RFD under timely sown restricted irrigation condition is 90:60:40 NPK/ha, and timely sown rainfed condition is 60:30:20 NPK/ha where full dose should be applied at sowing. The seed and fertilizer should not be placed at the same place in the soil else they will affect the germination adversely and productivity will be less. It is suggested to place fertilizers deeper than seed if sown with seed-cum-fertilizer drill.

1.6.6 Irrigation Management

Irrigation can be applied depending upon availability of water and requirement of crop. Generally, pre-sowing irrigation is given in October followed by three to six irrigations as per requirement and the availability of irrigation water. Bunds may be prepared in and around the field, and field should be divided in equal parts so that irrigation may be uniform, easy, and quick. In the event of availability of six irrigations, the crop stages for irrigation are crown root initiation (CRI) at 21 days after sowing (DAS), tiller completion at 40–45 DAS, booting/late jointing at 60–65 DAS, heading/flowering at 80–85 DAS, milk stage at 100–105 DAS, and dough stage at 115–120DAS. CRI and heading are the two most sensitive stages where moisture stress causes maximum damage to this crop. Irrigation scheduling can be managed for five (21, 45, 65, 85, and 105 DAS), four (21, 45, 85, and 105 DAS), and three (21, 65, and 105 DAS) irrigation availability. In zero-tillage sown wheat crop also, irrigation needs to be managed similar to the conventional method of wheat cultivation. In raised bed sown crop, if moisture is less at sowing time, then light irrigation should be applied immediately after sowing for better germination and crop stand. Under restricted irrigation condition, one pre-sowing irrigation and another at 45 DAS are recommended whereas rainfed crop may be given pre-sowing irrigation if water is available.

1.6.7 Weed Management

The infestation of both grassy and broad leaf weed is one of the major biotic constraints in wheat production. The major weeds in wheat crop are *Phalaris minor (mandusi/kanki)*, *Avena ludoviciana (jangli jai)*, *Chenopodium album (bathua)*, *C. murale (kharbathua,)*, *Cyperus rotundus (montha)*, *Medicago sativa (maina)*, *Melilotus alba (senji)*, *Malwa parviflora (chughra)*, *Convolvulus arvensis (hirankhuri)*, *Cirsium arvense (kandai)*, *Anagallis arvensis (krishnanil)*, *Argemone mexicana (satyanashi)*, *Rumex dentatus (jangali palak)*, etc. Manual weeding with hand hoe is beneficial for weed control after first irrigation at optimum moisture, but the farmers prefer chemical weed control due to cost and time effectiveness. The use of weed free seed, early sowing, and reduced spacing also helps in weed control. Depending on the weed flora infesting the field, herbicide should be selected. For the control of broad-leaved weeds, the spray of Metsulfuron (Algrip) @ 4 g or Carfentrazone (Affinity) @ 20 g or 2,4 D @ 500 g per hectare dissolved in 250–300 litre of water at 30–35 DAS is recommended, whereas spray of Isoproturon @ 1000 g or Clodinafop (Topik/Point/Jhatka) @ 60 g or Pinoxaden (Axial 5 EC) @ 40 g or Fenoxaprop (Puma Power) @100 g or Sulfosulfuron @ 25 g per hectare dissolved in 250–300 litre water at 30–35 DAS is recommended for the control of narrow-leaved weeds. For the control of both narrow- and broad-leaved weeds, spray of Isoproturon (Isogaurd 75 WP) @ 500 g or Sulfosulfuron (Leader/SF10/Safal) @ 13 g or Sulfosulfuron + Metsulfuron (Total) @ 16 g or Mesosulfuron+ Iodosulfuron (Atlantis) @ 160 g or Fenoxaprop+ Metribuzin (Accord plus) @ 500–600 ml dissolved in 250–300 litre of water in a hectare at 30–35 DAS is recommended. Pendimethalin (Stomp) @ 1000–1500 ml dissolved in 500 litre of water can also be used in a hectare within 3 days of sowing for weed control.

1.6.8 Crop-Protection Measures

Rusts, foliar blight, Karnal bunt, powdery mildew, smut diseases, and termites are major biotic stresses impacting wheat crops. The use of resistant cultivars is the best option to avoid these biotic stresses. However, chemical control measures are also recommended to control the occurrence of these diseases and pests. The spray of propiconazole (Tilt 25 EC @ 0.1%) or tebuconazole (Folicur 250EC @ 0.1%) or triademefon (Bayleton 25WP @ 0.1%) is recommended to control rust, powdery mildew, and Karnal bunt diseases. Seed treatment with Carboxin (75 WP @ 2.5 g/kg seed) or Carbendazim (50 WP @ 2.5 g/kg seed) or Tebuconazole (2DS @ 1.25 g/kg seed) is recommended to control smut diseases. Seed treatment with chlorpyriphos @ 0.9g a.i./kg seed or thiamethoxam 70WS (Cruiser 70WS) @ 0.7 g a.i./kg seed or Fipronil (Regent 5FS @ 0.3 g a.i./kg seed) is very effective for termite control.

1.6.9 Harvesting, Threshing, and Storage

Generally, wheat is manually harvested, but for quick harvesting, combine harvester should be used to avoid losses in grain yield due to shattering and lodging. When the moisture level of grain is 20%, it is the proper time for manual harvesting in which bundles are made and dried for 3–4 days and threshed by thresher. Wheat should be harvested 4–5 days before it is dead ripe to avoid shattering losses. Combine harvesting is convenient due to synchronized maturity of wheat crop, and it is used when the grain moisture is below 14%. Grain quality will be better at the optimum time of harvesting. Before storage, grains should be dried by spreading on tarpaulin plastic sheets in bright sunlight to keep a moisture level below 12%. Bins and silos made of GI sheets are used for storage. Nowadays, aluminium bins, puma bins, silos, and polylined bags are available for storage. Farmers can store wheat grains in their traditional storage as well. To protect from storage insects or pests, it is necessary to fumigate the produce with FDB @ 5 g/tons or aluminium phosphide @ 3 g/ton and keeping room sealed for 24 hours. Storage at high-moisture condition should be avoided to maintain seed/grain viability and minimize storage pest damage.

1.7 Emerging Challenges

Wheat improvement programme has significantly contributed in yield enhancement at global and national levels, and the occurrence of Green Revolution in South Asia especially in India is a landmark success story. Increasing production and productivity in a sustainable basis in economic, social and environmental terms, while considering the diversity of agricultural conditions, is one of the most important challenges that the world faces today (G20 meeting, 2012). The demand by 2050 is predicted to increase by 50% from today's levels. To meet this demand, global annual yield has to increase by almost 50% from the present level (Giraldo et al., 2019) through increased adoption of modern varieties and improved productivity. Although the global food security has been ensured till date, but there are numerous emerging challenges to sustain future wheat production targets. Climate change and variations (Aryal et al., 2016; Mehar et al., 2016; Mittal and Hariharan, 2018), wide yield gap between the farmer's field and experimental farms (Ghimire et al., 2012; Pavithra et al., 2017), unavailability of improved seeds (Lantican et al., 2016), use of self-saved old seeds (Abeyo et al., 2020), change in biotic and abiotic stress spectrum (Pingali, 1999; Lantican et al., 2016), cropping pattern shifts (Islam et al., 2020), depleting soil quality due to over extraction of nutrients, micronutrient deficiencies, declining water tables, salinity, sodicity, etc., are the biggest threats as these impact wheat yield potential and productivity.

1.8 Priorities in Wheat Improvement

There is a need of multidisciplinary collaborations for effective planning, implementation, and evaluation of research priorities. Most priorities in wheat breeding remain the same and for which faster development and the accumulation of knowledge from different fields are needed to provide new strategies and paths to reach these goals. Increasing yield potential has been prioritized as the main objective of wheat breeding over many decades in order to meet the food requirements of an ever-increasing population (Borlaug, 2007; Pingali, 2012). Disease resistance is the next most important trait since the first breeding attempts by Knight in 1787 until today in different countries (Lupton, 1987). The old diseases like rusts are still a potential threat for wheat cultivation, but the most recent diseases such as wheat blast are also posing a concern for wheat cultivation worldwide (Wulff and Dhugga, 2018). Tolerance to abiotic stresses (especially drought, heat, cold, acid soils, and lodging), dual-purpose (forage and grain), improved nutrient use, various quality traits and grain bio-fortification efficiency, and increasing photosynthetic capacity were other traits of importance for wheat improvement. These traits are prioritised within each environmental region and over time. The most important steps for wheat improvement were already reviewed for different traits, that is yield potential (Evenson and Gollin, 2003; Reynolds and Borlaug, 2006), rust resistance (Singh et al., 2006a), drought tolerance (Mwadzingeni et al., 2016), physiological efficiency (Reynolds et al., 2011; Rangan et al., 2016), bio-fortification (Xu et al., 2011; Umar et al., 2019), low gluten content (Green and Jabri, 2003; Spaenij-Dekking et al., 2005; Catassi et al., 2013; Rosell et al., 2014) which are becoming potentially important targets for wheat breeding in recent years. Additionally, the identification of resource use, adaptation and mitigation strategies for targeted environments, bio-risk analysis, development of disease prediction modules, integrated pest management strategies, crop modelling, tailoring genotypes in cropping system perspective, public and private partnerships, (Hobbs and Morris, 1996; Wheat initiative, 2013; IIWBR, 2015) and strategies for reduced yield gaps including participatory approach (Joshi and Witcombe, 1996; Witcombe et al., 1996; Dixon et al., 2006; Joshi et al., 2017; Abeyo et al., 2020) are key components in achieving future targets.

1.9 Approaches in Wheat Improvement

Wheat is a self-pollinated species, and, therefore, the conventional structure of its breeding programmes includes the use of artificial hybridizations between selected genotypes, something already performed for more than two centuries, and different forms of selection within segregating populations (Lupton, 1987; Scheeren et al., 2011). It is recognized that these conventional procedures were, and will continue to be, the main approach for the development of wheat cultivars worldwide. However, new tools and techniques are assisting this conventional process, increasing its success rate and reducing costs, time, and labour. Improving wheat becomes more difficult due to needs to "match" quantity and quality, allying yield with grain and flour quality (Shewry, 2009). Also, restricted genetic variability in wheat and its genome size, complexity, and polyploid nature constitute a challenge when applying advanced breeding techniques.

Wheat is probably the one among all crop species in which the use of wild and cultivated relatives was extensively exploited as a source of variability for its improvement since long ago, as early as plant breeding itself (Bedo and Láng, 2015). Despite the restricted variability within wheat germplasm, there is an immeasurable richness in variation existing in related species belonging to its secondary and tertiary gene pools (Schneider et al., 2008; Dempewolf et al., 2017; Zhang et al., 2017) that makes gene introgression more prominent (King et al., 2017). Chromosomal translocation of 1RS-1BL between wheat and rye (*Secale cereale* L.) by E.R. Sears is the most important introgression to date in wheat, leading to increased yield potential and resistance/tolerance to biotic and abiotic stresses (Schlegel and Korzun, 1997; Rabinovich, 1998; Crespo-Herrera et al., 2017). The development of synthetic wheat (Liu et al., 2006; Mujeeb-Kazi et al., 2008) and pre-breeding (Molnár-Láng et al., 2015; Singh et al., 2018; Singh et al., 2021a) for trait improvement are some of the remarkable achievements in wheat improvement. Mutation induction through chemical or physical mutagens has been widely used in wheat (Parry et al., 2009) for the induction of variability for various traits including resistance to herbicides (Pozniak and Hucl, 2004), increased amylose content, and starch resistance (Slade et al., 2012). The detection of mutants is very difficult (Parry et al., 2009) due to the polyploid nature of wheat, but tilling methods (Slade et al., 2005; Uauy et al., 2009) and high-resolution melting analysis (Dong et al., 2009) have proven to be efficient for the detection of mutations in different genomes of hexaploid wheat. Mutation breeding led to the development of 265 wheat cultivars in different countries (FAO/IAEA, 2022).

The use of molecular markers for QTL mapping and marker-assisted selection has been growing (Buerstmayr et al., 2009; Gupta et al., 2017), and different meta-analyses indicated that AFLP, RFLP, and SSR were the most used markers (Chao et al., 1989; Nagaoka and Ogihara, 1997; Song et al., 2005; Hospital, 2009; Gupta et al., 2010) during 1990s to 2000s. Recent revolution using the high-throughput genotyping platforms using DArT (Akbari et al., 2006) and SNP (Cavanagh et al., 2013; Wang et al., 2014a; Winfield et al., 2016) markers, genotyping by sequencing (Poland et al., 2012a; Saintenac et al., 2013), and genome-wide association studies (GWAS) facilitated the generation of 20 to 450 K loci maps in wheat and allowed capture of a larger genetic diversity (Cabrera-Bosquet et al., 2012; Poland et al., 2012b; Kollers et al., 2013; Arruda et al., 2016; Liu et al., 2016a; Guo et al., 2017). The limitation of MAS to aid the selection for a few genes or alleles at a time has been overcome by a recent approach of genomic selection (GS) to handle numerous genes of quantitative nature (Meuwissen et al., 2001) which perform selection and prediction of breeding values based only on genotyping (Jannink et al., 2010) with greater accuracy (Heffner et al., 2009; Poland et al., 2012b; Arruda et al., 2015; Bassi et al., 2016). The International Wheat Genome Sequencing Consortium (IWGSC) facilitated the generation of a reference genome of wheat based on hexaploid wheat cultivar Chinese Spring (IWGSC, 2014; IWGSC, 2018) which has been considered a step-change for more effective use of genomics in breeding. The development of transgenic plants (Shewry, 2009; Hilbeck et al., 2015; Lynas, 2012) for various traits of resistance to biotic (Zhao et al., 2006; Janni et al., 2008; Li et al., 2015) and abiotic stresses (Gao et al., 2009), baking quality (Rakszegi et al., 2005), and cisgenic plants (Tester and Langridge, 2010) was an effective tool, but did not get suitable acceptance for large-scale commercial cultivation (Vasil et al., 1992; Vasil, 2007; Wulff and Dhugga, 2018). The most recent and promising innovation of genome or gene editing (Sander and Joung, 2014; Bortesi and Fischer, 2015; Brozynska et al., 2016; King et al., 2017), which accurately targets segments of the genome for modification, by deletion, insertion, or substitution of nucleotides (Sander and Joung, 2014), has proven to be successful (Shan et al., 2013; Upadhyay et al., 2013; Shan et al., 2014; Wang et al., 2014b; Zhang et al., 2016) despite the great complexity of its extensive, redundant, and polyploid genome. The precise gene replacement in gene editing can also mitigate linkage drag making it a potential tool for gene introgression from wild relatives into wheat background (Wang et al., 2018).

The development and cultivation of hybrid cultivars are one of the innovative approaches for breaking yield barriers (Singh et al., 2010: Longin et al., 2013) which indicate the future importance in changing climatic conditions (Mühleisen et al., 2014) due to higher yield stability. Hybrids provide protection against various biotic (Longin et al., 2013; Miedaner et al., 2017; Beukert et al., 2020) and abiotic stresses (Longin et al., 2013; Gomaa et al., 2014) and exhibit improved grain quality, enhanced fertilizer response, better root growth, and high rate of grain filling (Hallauer et al., 1988; Kindred and Gooding, 2005; Xu et al., 2011; Ahmad et al., 2016). However, several difficulties associated with hybrid development and seed production limited its cultivation in less than 1% of the global area under wheat (Longin et al., 2012; Whitford et al., 2013; Kempe et al., 2014; Würschum et al., 2018a; Würschum et al., 2018b; Thorwarth et al., 2018; Miedaner and Laidig, 2019). Now research in the development and cultivation of hybrids seems to be a priority in wheat breeding due to a huge accumulation of knowledge and new technologies (Zhao et al., 2015; Liu et al., 2016b; Gaynor et al., 2017; Watson et al., 2018; Gupta et al., 2019; Singh et al., 2021b).

The speed breeding, based on photoperiod, light, and temperature manipulation (artificially) in growth chambers and glasshouses, is a recent tool to shorten breeding cycle of wheat cultivars (Watson et al., 2018) and expedite phenotyping for various traits (Ghosh et al., 2018; Singh et al., 2020). The use of high-throughput phenotyping also aims to evaluate several traits in a large number of plants over a short period of time (de Souza, 2010; Cabrera-Bosquet et al., 2012; Araus and Cairns, 2014). This technique comprises several highly optimized and automated steps (Kipp et al., 2014; Fahlgren et al., 2015) under controlled conditions (Fahlgren et al., 2015; Richard et al., 2015) as well as in-field conditions (Comar et al., 2012; Kipp et al., 2014; Haghighattalab et al., 2016; Tattaris et al., 2016). These recent techniques become effective components for faster cultivar development programmes.

1.10 Conclusion

Wheat, being a nutri-rich grain, has a significant role in ensuring food and nutritional security for establishing zero hunger as committed under the Sustainable Development Goals (SDGs). Modern wheat cultivars usually refer to two species, namely, hexaploid bread wheat and tetraploid durum wheat. Wheat is a superb model organism for the evolutionary theory of allopolyploid speciation, adaptation, and domestication in plants (Gustafson et al., 2009). During agricultural development, early domesticates were gradually replaced first by landraces and traditional varieties and later by genetically less diverse modern cultivars. This has resulted in genetic bottlenecks in modern breeding processes of improved germplasm and caused substantial genetic erosion (Tanksley and McCouch, 1997; Nevo, 2004; Fu and Somers, 2009) and thus increased susceptibility and vulnerability to environmental stresses, pests, and diseases (Fu and Somers, 2009; Nevo 2009, 2011). Although the narrow gene pool available for the development of superior varieties is of major concern heightened by increasing global population predictions, breeding programmes had considerable success in producing higher yielding varieties in the past with the limited available variation. There is growing evidence that wheat yields are plateauing due to the exhaustion of the available genetic variation compounded by environmental change (Brisson et al., 2010; Charmet, 2011; Ray et al., 2013). Thus, there is an urgent need to identify new sources of genetic variation that can be used in future genetic improvement to develop superior wheat varieties for feeding the ever-increasing human population. Though experiencing diversity bottlenecks, wheat has strong adaptability to diverse environments and end uses (Nevo et al., 2002) as it compensates these bottlenecks by capturing part of the genetic diversity of its progenitors and by generating new diversity at a relatively fast pace (Dubcovsky and Dvorak, 2007). Therefore, the best strategy for wheat improvement is to utilize the adaptive genetic resources of the wild progenitors, wild emmer *T. dicoccoides*, and other wheat relatives for which germplasm collections are essential to conserve biodiversity and efficient utilization for future wheat improvement programmes (Feldman and Sears, 1981; Nevo and Beiles, 1989; Nevo, 2007; Johnson, 2008; Xie and Nevo, 2008; Nevo and Chen, 2010).

The current challenges of adversaries of climate change, restricted availability of arable land and water, diversification in cropping pattern, constant evolution of pathogens, and quality requirements are major obstacles in sustaining the pace of wheat productivity and production which posed a significant pressure on wheat-breeding programmes to maintain productivity along with improved resistance to biotic and abiotic stresses. These challenges also linked issues of varietal replacement, sustainable input utilization, and changing socio-economics of the growing population. It is important to increase the yield levels in a sustainable manner and reduce the yield gaps through the use of improved cultivars, efficient utilization of inputs for better soil health, adoption of recommended production technologies for weed and pest management, and improved extension services in participatory mode.

Breeding has been responsible for increasing wheat yields and improving many other traits, such as resistance to biotic and abiotic stresses and grain quality. In order to meet its global demand, efforts in the development and implementation of improved strategies must continuously take place in wheat-breeding programs. Classical methods of wheat breeding, which are largely based on crosses and phenotypic selection, have been the most used plant breeding methods around the globe for more than one century and significantly contributed to develop elite breeding lines and release of the largest number of cultivars (Prohens, 2011). This approach seems to be main strategy for future wheat improvement programmes especially in developing countries. Plant breeding has experienced innovations and revolutions throughout its existence, and wheat has been witness to most of these transformations and probably will continue as an ally of the transformations to come. The availability of fast-track improved genomic approaches has made it possible to mobilize gene-bank variation to the breeding pipelines (Singh et al., 2021a) and discover a large number of novel genes, which can be easily incorporated into inferior genotypes using marker-aided interventions. The application of next-generation sequencing (NGS) platforms and bioinformatics tools has revolutionized the pre-breeding activities (Singh et al., 2018) and wheat genomics to identify marker–trait associations and direct insertion of a gene of interest through gene editing and phenotypic selection by genomic selection for faster wheat improvement. The combined approaches of conventional methodologies supported by modern tools and techniques will probably predominate in future breeding programs.

REFERENCES

Abeyo B, Badebo A, Gebre D and Listman M. 2020. Achievements in fast-track variety testing, seed multiplication and scaling of rust resistant varieties: lessons from the wheat seed scaling project. Ethiopia. CDMX, Mexico, and Addis Ababa, Ethiopia: The International Maize and Wheat Improvement Center (CIMMYT) and the Ethiopian Institute of Agricultural Research (EIAR).

Ahmad E, Kamar A and Jaiswal JP. 2016. Identifying heterotic combinations for yield and quality traits in bread wheat (*Triticum aestivum* L.). Electron J Plant Breed 7(2):352–61.

Akbari M, Wenzl P, Caig V, Carling J, Xia L, Yang S, Uszynski G, Mohler V, Lehmensiek A, Kuchel H, Hayden MJ, Howes N, Sharp P, Vaughan P, Rathmell B, Huttner E and Kilian A. 2006. Diversity arrays technology (DArT) for high-throughput profiling of the hexaploid wheat genome. Theor Appl Genet. 113:14 09–20. https://doi.org/10.1007/s00122-006-0365-4.

Ali A, Hussain M, Hassan S and Nadeem N. 2017. Estimation of total factor productivity growth of major grain crops in Pakistan: 1972–2013. Int Rev Soc Sci 54:237–45.

Ali M and Byerlee D. 2002. Productivity growth and resource degradation in Pakistan's Punjab: a decomposition analysis. Econ Dev Cult Chang 504:839–63.

Araus JL and Cairns JE. 2014. Field high-throughput phenotyping: the new crop breeding frontier. Trends Plant Sci 19:52– 61. https://doi.org/10.1016/j.tplants.2013.09.008.

Arruda MP, Brown PJ, Brown-Guedira G, Krill AM, Thurber C, Merrill KR, Foresman BJ, Frederic L and Kolb FL. 2016. Genome-wide association mapping of *Fusarium* head blight resistance in wheat using genotyping-by-sequencing. Plant Genome 9(1) :1–14. https://doi.org/10.3835/plantgenome.2015.04.0028.

Arruda MP, Brown PJ, Lipka AE, Krill AM, Thurber C and Kolb FL. 2015. Genomic selection for predicting *Fusarium* head blight resistance in a wheat breeding program. The Plant Genome 8(3) :1–12. https://doi.org/10.3835/plantgenome.2015.01.0003.

Aryal JP, Sapkota TB, Jat ML, Jat HS, Rai M, Mittal S, Stirling CM and Sutaliya JM. 2016. Conservation agriculture-based wheat production better copes with extreme climate events than conventional tillage-based systems: a case of untimely excess rainfall in Haryana, India. Agric Ecosyst Environ 233:325–35.

Avni R, Nave M, Barad O, Baruch K, Twardziok SO, Gundlach H, Hale I, Mascher M, Spannagl M, Wiebe K, Jordan KW, Golan G, Deek J, Ben-Zvi B, Ben-Zvi G, Himmelbach A, MacLachlan RP, Sharpe AG, Fritz A, Ben-David R, Budak H, Fahima T, Korol A, Faris JD, Hernandez A, Mikel MA, Levy AA, Steffenson B, Maccaferri M, Tuberosa R, Cattivelli L, Faccioli P, Ceriotti A, Kashkush K, Pourkheirandish M, Komatsuda T, Eilam T, Sela H, Sharon A, Ohad N, Chamovitz

DA, Mayer KFX, Stein N, Ronen G, Peleg Z, Pozniak CJ, Akhunov ED and Distelfeld A. 2017. Wild emmer genome architecture and diversity elucidate wheat evolution and domestication. Science 357:93–7.

Bassi FM, Bentley AR, Charmet G, Ortiz R and Crossa J. 2016. Breeding schemes for the implementation of genomic selection in wheat (*Triticum* spp.). Plant Sci 242 :23–36 https://doi.org/10.1016/j.plantsci.2015.08.021.

Bedo Z and Láng L. 2015. Wheat breeding: current status and bottlenecks. In: Molnar M, Ceoloni C and Doležel J (Eds) Alien introgression in Wheat: cytogenetics, molecular biology, and genomics, 1st edn. New York: Springer International Publishing, pp. 203–62.

Beukert U, Liu G, Thorwarth P, Boeven PHG, Friedrich C, Longin H, Zhao Y, Ganal M, Serfling A, Ordon F and Reif JC. 2020. The potential of hybrid breeding to enhance leaf rust and stripe rust resistance in wheat. Theor Appl Genet 133:2171–81.

Bhushan S. 2016. TFP growth of wheat and paddy in post-green revolution era in India: parametric and non-parametric analysis. Agric Econ Res Rev 291:27–40.

Bilgic H, Hakki EE, Pandey A, Khan MK and Akkaya MS. 2016. Ancient DNA from 8400 year-old Çatalhöyük Wheat: implications for the origin of Neolithic agriculture. PLoS One 11:e01 51974. https://doi.org/10.1371/journal.pone.0151974.

Blake NK, Lehfeldt BR, Lavin M and Talbert LE. 1999. Phylogenetic reconstruction based on low copy DNA sequence data in an allopolyploid: The B genome of wheat. Genome 42:351–60.

Borlaug NE. 2007. Sixty-two years of fighting hunger: personal recollections. Euphytica 157:287– 97. https://doi.org/10.1007/s10681-007-9480-9.

Bortesi L and Fischer R. 2015. The CRISPR/Cas9 system for plant genome editing and beyond. Biotechnol Adv 33:41– 52. https://doi.org/10.1016/j.biotechadv.2014.12.006.

Breiman A and Graur D. 1995. Wheat evolution. ISR J Plant Sci 43(2):85–98.

Brisson N, Gate P, Gouache D, Charmet G, Oury FX and Huard F. 2010. Why are wheat yields stagnating in Europe? A comprehensive data analysis for France. Field Crops Res 119:201–12.

Brozynska M, Furtado A and Henry RJ. 2016. Genomics of crop wild relatives: expanding the gene pool for crop improvement. Plant Biotechnol J 14:10 70–85. https://doi.org/10.1111/pbi.12454.

Buerstmayr H, Ban T and Anderson JA. 2009. QTL mapping and marker-assisted selection for Fusarium head blight resistance in wheat: a review. Plant Breed 128 :1–26. https://doi.org/10.1111/j.1439-0523.2008.01550.x.

Byerlee D, Ali M and Siddiq A. 2003. Sustainability of the rice–wheat system in Pakistan's Punjab: how large is the problem? In: *Improving the productivity and sustainability of rice-wheat systems: issues and impacts*, vol. 65, pp. 77–95.

Cabrera-Bosquet L, Crossa J, von Zitzewitz J, Serret MD and Araus JL. 2012. High-throughput phenotyping and genomic selection: the Frontiers of crop Breeding converge. J Integr Plant Biol 54:3 12–20. https://doi.org/10.1111/j.1744-7909.2012.01116.x.

Catassi C, Bai JC, Bonaz B, Bouma G, Calabrò A, Carroccio A, Castillejo G, Ciacci C, Cristofori F, Dolinsek J, Francavilla R, Elli L, Green P, Holtmeier W, Koehler P, Koletzko S, Meinhold C, Sanders D, Schumann M, Schuppan D, Ullrich R, Vécsei A, Volta U, Zevallos V, Sapone A and Fasano A. 2013. Non-celiac gluten sensitivity: the new frontier of gluten related disorders. Nutrients 5:3 839–53 https://doi.org/10.3390/nu5103839.

Cavanagh CR, Chao S, Wang S, Huang BE, Stephen S, Kiani S, Forrest K, Saintenac C, Brown-Guedira GL, Akhunova A, See D, Bai GH, Pumphrey M, Tomar L, Wong D, Kong S, Reynolds M, Lopez M, da Silva, Bockelman H, Talbert L, Anderson

JA, Dreisigacker S, Baenziger S, Carter A, Korzun V, Morrell PL, Dubcovsky J, Morell MK, Sorrells ME, Hayden MJ and Eduard Akhunov E. 2013. Genome-wide comparative diversity uncovers multiple targets of selection for improvement in hexaploid wheat landraces and cultivars. Proc Natl Acad Sci 110:8057–62. https://doi. org/10.1073/pnas.1217133110.

Chakraborty M, Mahmud NU, Gupta DR, Tareq FS, Shin HJ and Islam T. 2020. Inhibitory effects of linear lipopeptides from a marine *Bacillus subtilis* on the wheat blast fungus *Magnaporthe oryzae* Triticum. Front Microbiol 1 1:665. https://doi.org/10.3389/fmicb.2020.00665.

Chalupska D, Lee HY, Faris JD, Evrard A, Chalhoub B, Haselkorn R and Gornicki P. 2008. Acc homoeo loci and the evolution of wheat genomes. Proc Natl Acad Sci USA 105:9691–96.

Chao S, Sharp PJ, Worland AJ, Warham EJ, Koebner RMD and Gale MD. 1989. RFLP-based genetic maps of wheat homoeologous group 7 chromosomes. Theor Appl Genet 78(4):49 5–504. https://doi.org/10.1007/BF00290833.

Charmet G. 2011. Wheat domestication: lessons for the future. C R Biol 334:2 12–20. https://doi.org/10.1016/j.crvi.2010.12.013.

Comar A, Burger P, de Solan B, Baret F, Daumard F and Hanocq JF. 2012. A semi-automatic system for high throughput phenotyping wheat cultivars in-field conditions: description and first results. Funct Plant Biol 39:914 https://doi.org/10.1071/FP12065.

Crespo-Herrera L, Garkava-Gustavsson L and Åhman I. 2017. A systematic review of rye (*Secale cereale* L.) as a source of resistance to pathogens and pests in wheat (*Triticum aestivum* L.). Hereditas 154(14) :1–9. https://doi.org/10.1186/s41065-017-0033-5.

DAC&FW. 2021. Latest APY state data. *Directorate of Economics and Statistics, Department of Agricultural Cooperation & Farmers Welfare, Govt of India.* https://eands.dacnet.nic.in/APY_96_ To_06. htm. Accessed on 12.7.2022.

Datt G and Ravallion M. 1998. Farm productivity and rural poverty in India. J Develop Stud 34, 62–85.

de Souza N. 2010. High-throughput phenotyping. Nat Methods 7(1): 36. https://doi.org/10.1038/nmeth.f.289.

Dempewolf H, Baute G, Anderson J, Kilian B, Smith C and Guarino L. 2017. Past and future use of wild relatives in crop Breeding. Crop Sci. 10 70–82. https://doi.org/10.2135/cropsci2016.10.0885.

Dixon J, Nalley L, Kosina P, LaRovere R, Hellin J and Aquino P. 2006. Adoption and economic impact of improved wheat varieties in the developing world. J Agric Sci 144:489–502.

Dong C, Vincent K and Sharp P. 2009. Simultaneous mutation detection of three homoeologous genes in wheat by high resolution melting analysis and mutation surveyor. BMC Plant Biol 9:143 https://doi.org/10.1186/1471-2229-9-143.

Dubcovsky J and Dvorak J. 2007. Genome plasticity a key factor in the success of polyploid wheat under domestication. Science 316:18 62–66. https://doi.org/10.1126/science.1143986.

Dvorak J, Deal KR, Luo M-C, You FM, von Borstel K and Dehghani H. 2012. The origin of spelt and free-threshing hexaploid wheat. J Hered 103:426–41.

El Baidouri M, Murat F, Veyssiere M, Molinier M, Flores R, Burlot L, Alaux M, Quesneville H, Pont C and Salse J. 2017. Reconciling the evolutionary origin of bread wheat (Triticum aestivum). New Phytol 213:1477– 86. https://doi.org/10.1111/nph.14113.

Evenson RE and Gollin D. 2003. Assessing the impact of the green revolution, 1960 to 2000. Science 300: 758–62 https://doi.org/10.1126/science.1078710.

Evenson RE and Rosegrant MW. 2003. The economic consequences of crop genetic improvement programmes. In Evenson R and Gollin D (Eds) Crop variety improvement and its effect on productivity: the impact of international agricultural research. Wallingford: CABI Publishing.

Fahlgren N, Gehan MA and Baxter I. 2015. Lights, camera, action: high-throughput plant phenotyping is ready for a close-up. Curr Opin Plant Biol 24 :93–9. https://doi.org/10.1016/j.pbi.2015.02.006.

Fan S and Hazell P. 2001. Returns to public investments in the less favoured areas of India and China. Am J Agri Econ 83(5):1217–22.

FAO. 2011. Food and agriculture organiz ation. www.fao.org/statistics/en

FAO. 2017. The future of food and agriculture—trends and challenges. Rome: FAO.

FAO. 2020. Data: production. Online database Crop Prod Harvest Area. www.fao.org/faostat/en/#data/QC. Accessed on 26.4.2020.

FAO. 2021. World food situation. Cereal Supply and Demand Brief. www.fao.org/worldfood situation/csdb/en/. Accessed on 12.7.2022.

FAO/IAEA. 2022. Mutant variety database. https://nucleus.iaea.org/sites/mvd/SitePages/Home.aspx. Accessed on 7.7.2022.

Feldman M. 1995. Wheats. In: Smartt J and Simmonds NW (orgs) Evolution of crop plants. Harlow: Longman Scientific and Technical, pp. 185–92.

Feldman M, Lupton FGH and Miller TE. 1995. Wheats. In: Smartt J and Simmonds NW (Eds) Evolution of crop plants, 2nd edn. Harlow: Longman Scientific & Technical, pp. 184–92.

Feldman M and Sears ER. 1981. The wild gene resources of wheat. Sci Am 244:102–13.

Fischer RA, Byerlee D and Edmeades G. 2014. Crop yields and global food security: will yield increase continue to feed the world? (ACIAR monograph no. 158). Canberra: Australian Centre for International Agricultural Research, p. 634.

Fu YB and Somers DJ. 2009. Genome-wide reduction of genetic diversity in wheat breeding. Crop Sci 49:161–8.

G20 Meeting. 2012. Agriculture vice-ministers and deputies meeting report. www. mofa.go.jp/policy/economy/g20_summit/2012/pdfs/am_report_e.pdf

Gao SQ, Chen M, Xia LQ, Xiu HJ, Xu ZS, Li LC, Zhao CP, Cheng XG and Ma YZ. 2009. A cotton (Gossypium hirsutum) DRE-binding transcription factor gene, GhDREB, confers enhanced tolerance to drought, high salt, and freezing stresses in transgenic wheat. Plant Cell Rep 28:3 01–11. https://doi.org/10.1007/s00299-008-0623-9.

Gaynor RC, Gorjanc G, Bentley AR, Ober ES, Howell P, Jackson R, Mackay IJ and Hickey JM. 2017. A two-part strategy for using genomic selection to develop inbred lines. Crop Sci 57(5):2372–86.

GCARD. 2012. Breakout session P1.1 National Food Security – The wheat initiative – An international research initiative for wheat improvement context – the problems being addressed. In: Proceedings of the GCARD—Second global conference on agricultural research for development. Uruguay: GCARD.

Ghimire S, Mehar M and Mittal S. 2012. Influence of sources of seed on varietal adoption behaviour of wheat farmers in Indo-Gangetic Plains of India. Agric Econ Res Rev 25:399–408.

Ghosh S, Watson A, Gonzalez-Navarro OE, Ramirez-Gonzalez RH, Yanes L, Mendoza-Suárez M, Simmonds J, Wells R, Rayner T, Green P, Hafeez A, Hayta S, Melton RE, Steed A, Sarkar A, Carter J, Perkins L, Lord J, Tester M, Osbourn A, Moscou MJ, Nicholson P, Harwood W, Martin C, Domoney C, Uauy C, Hazard B, Wulff BBH and Hickey LT. 2018. Speed breeding in growth chambers and glasshouses for crop breeding and model plant research. Nat Protoc 13:29 44–63. https://doi.org/10.1038/s41596-018-0072-z.

Giraldo P, Benavente E, Manzano-Agugliaro F and Gimenez E. 2019. Worldwide research trends on wheat and barley: a bibliometric comparative analysis. Agronomy 9:352. https://doi.org/10.3390/agronomy9070352.

Glemin S, Scornavacca C, Dainat J, Burgarella C, Viader V, Ardisson M, Sarah G, Santoni S, David J and Ranwez V. 2018. Pervasive hybridizations in the history of wheat relatives. bioRxiv, 300848. https://doi.org/10.1101/300848.

Gomaa MA, El-Banna MNM and Gadalla A. 2014. Heterosis, combining ability and drought susceptibility index in some crosses of bread wheat (*Triticum aestivum* L.) under water stress conditions. Middle East J Agric Res 3:338–45.

Green PHR and Jabri B. 2003. Coeliac disease. Lancet 362:3 83–91. https://doi.org/10.1016/S0140-6736(03)14027-5.

Guo Z, Chen D, Alqudah AM, Röder MS, Ganal MW and Schnurbusch T. 2017. Genome-wide association analyses of 54 traits identified multiple loci for the determination of floret fertility in wheat. New Phytol 214:2 57–70. https://doi.org/10.1111/nph.14342.

Gupta PK, Balyan HS and Gahlaut V. 2017. QTL analysis for drought tolerance in wheat: present status and future possibilities. Agronomy 7:5. https://doi.org/10.3390/agronomy7010005.

Gupta PK, Balyan HS, Gahlaut V, Saripalli G, Pal B, Basnet BR and Joshi AK. 2019. Hybrid wheat: past, present and future. Theor Appl Genet 132(9):2463–83. https://doi.org/10.1007/s00122-019-03397-y.

Gupta PK, Langridge P and Mir RR. 2010. Marker-assisted wheat breeding: present status and future possibilities. Mol Breed 26:1 45–61. https://doi.org/10.1007/s11032-009-9359-7.

Gustafson P, Raskina O, Ma X and Nevo E. 2009. Wheat evolution, domestication, and improvement. In: Carver BF (Ed) Wheat: science and trade. Danvers: Wiley, pp. 5–30.

Haghighattalab A, González Pérez L, Mondal S, Singh D, Schinstock D, Rutkoski J, Ortiz-Monasterio I, Singh RP, Goodin D and Poland J. 2016. Application of unmanned aerial systems for high throughput phenotyping of large wheat breeding nurseries. Plant Methods 12(35) :1–15. https://doi.org/10.1186/s13007-016-0134-6.

Hallauer AR, Russell WA and Lamkey KR. 1988. Corn breeding. In: Sprague GF and Dudley JW (Eds) Corn and corn improvement, 3rd edn. Madison, WI: American Society Agronomy, pp. 463–564.

Hawkes, JG. 1982. Genetic conservation of recalcitrant species – an overview. In: Withers LA and Williams JT (Eds) Crop genetic resources – The conservation of difficult material (Proc. Int. Workshop). Reading: University of Reading, pp. 83–92.

Hawkesford MJ, Araus J-L, Park R, Calderini D, Miralles D, Shen T and Zhang JP. 2013. Prospects of doubling global wheat yields. Food Energy Secur 2(1): 34–48. https://doi.org/10.1002/fes3.15.

Heffner EL, Sorrells ME and Jannink J-L. 2009. Genomic selection for crop improvement. Crop Sci 49(1): 12. https://doi.org/10.2135/cropsci2008.08.0512.

Heun M, Schäfer-Pregl R, Klawan D, Castagna R, Accerbi M, Borghi B and Salamini F. 1997. Site of einkorn wheat domestication identified by DNA fingerprinting. Science. 278(5341):1 312–4. https://doi.org/10.1126/science.278.5341.1312.

Hilbeck A, Binimelis R, Defarge N, Steinbrecher R, Székács A, Wickson F, Antoniou M, Bereano PL, Clark EA, Hansen M, Novotny E, Heinemann J, Meyer H, Shiva V and Wynne B. 2015. No scientific consensus on GMO safety. Environ Sci Eur 27(4) :1–6. https://doi.org/10.1186/s12302-014-0034-1.

Hobbs P and Morris M. 1996. Meeting South Asia's future food requirements from rice-wheat cropping systems: priority issues facing researchers in the post-green revolution era (NRG Paper 96–01), Mexico, DF: CIMMYT, Mexico.

Hospital F. 2009. Challenges for effective marker-assisted selection in plants. Genetica 136:3 03–10. https://doi.org/10.1007/s10709-008-9307-1.

Huang S, Sirikhachornkit A, Su X, Faris J, Gill B, Haselkorn R and Gornicki P. 2002. Genes encoding plastid acetyl-CoA carboxylase and 3-phosphoglycerate kinase of the Triticum/Aegilops complex and the evolutionary history of polyploid wheat. Proc Nat Acad Sci 99:8133–8.

Hughes N, Lawson K and Valle H. 2017. Farm performance and climate: climate-adjusted productivity for broad-acre cropping farms (ABARES Research Report). Canberra: ABARES.

IIWBR. 2015. Vision-2050. Karnal: Indian Institute of Wheat & Barley Research, p. 34.

International Wheat Genome Sequencing Consortium. 2014. A chromosome-based draft sequence of the hexaploid bread wheat (Triticum aestivum) genome. Science, 345(6194):12 51788. https://doi.org/10.1126/science.

International Wheat Genome Sequencing Consortium. 2018. Shifting the limits in wheat research and breeding using a fully annotated reference genome. Science, 361(6403): 7191 https://doi.org/10.1126/science.aar.

Islam MT, Gupta DR, Hossain A, Roy KK, He X, Kabir MR, Singh PK, Khan MAR, Rahman M and Wang GL. 2020. Wheat blast: A new threat to food security. Phytopathol Res 2(28): 1–13. https://doi.org/10.1186/s42483-020-00067-6.

Janni M, Sella L, Favaron F, Blechl AE, Lorenzo GD and D'Ovidio R. 2008. The expression of a bean PGIP in transgenic wheat confers increased resistance to the fungal pathogen Bipolaris sorokiniana. Mol Plant-Microbe Interact 21:1 71–77. https://doi.org/10.1094/MPMI-21-2-0171.

Jannink J-L, Lorenz AJ and Iwata H. 2010. Genomic selection in plant breeding: From theory to practice. Brief Funct Genomics 9:1 66–77. https://doi.org/10.1093/bfgp/elq001.

Johnson RC. 2008. Gene banks pay big dividends to agriculture, the environment, and human welfare. PLoS Biol 6(6):e148. https://doi.org/10.1371/journal.pbio.0060148.

Joshi A and Witcombe JR. 1996. Farmer participatory crop improvement. II. Participatory varietal selection: a case study in India. Exp Agric 32:461–77.

Joshi KD, Rehman AU, Ullah G, Nazir MF, Zahara M, Akhtar J, Khan M, Baloch A, Khokhar J, Ellahi E, Khan A, Suleman M and Imtiaz M. 2017. Acceptance and competitiveness of new improved wheat varieties by smallholder farmers. J Crop Improve 31(4):608–27. https://doi.org/10.1080/15427528.2017.1325808.

Karim M and Talukder RK. 2008. Analysis of total factor productivity of wheat in Bangladesh. Bangladesh J Agric Econ 31:1–17.

Kempe K, Rubtsova M and Gils M. 2014. Split-gene system for hybrid wheat seed production. Proc Natl Acad Sci 111:909 7–102. https://doi.org/10.1073/pnas.1402836111.

Kihara H. 1944. Discovery of the DD-analyser, one of the ancestors of vulgare wheat. Agric Hortic 19:889–90.

Kindred DR and Gooding MJ. 2005. Heterosis for yield and its physiological determinants in wheat. Euphytica 142:149–59.

King J, Grewal S, Yang C, Hubbart S, Scholefield D, Ashling S, Edwards KJ, Allen AM, Burridge A, Bloor C, Davassi A, da Silva GJ, Chalmers K and King IP. 2017. A step change in the transfer of interspecific variation into wheat from Amblyopyrum muticum. Plant Biotechnol J 15:217– 26. https://doi.org/10.1111/pbi.12606.

Kipp S, Mistele B, Baresel P and Schmidhalter U. 2014. High-throughput phenotyping early plant vigour of winter wheat. Eur J Agron 52:2 71–78. https://doi.org/10.1016/j.eja.2013.08.009.

Kollers S, Rodemann B, Ling J, Korzun V, Ebmeyer E, Argillier O, Hinze M, Plieske J, Kulosa D, Ganal MW and Röder MS. 2013. Whole genome association mapping of Fusarium head blight resistance in European winter Wheat (Triticum aestivum L.). PLoS One 8:e57500. https://doi.org/10.1371/journal.pone.

Kumar P, Kumar A and Mittal S. 2004. Total factor productivity of crop sector in the Indo-Gangetic Plain of India: sustainability issues revisited. Ind Econ Rev 39(1):169–201.

Kumar P and Mittal S. 2006. Agricultural productivity trends in India: sustainability issues. Agric Econ Res Rev 50:71–88.

Kumar P, Mittal S and Hossain M. 2008. Agricultural growth accounting and total factor productivity in South Asia: a review and policy implications. Agric Econ Res Rev 21:145–72.

Lantican MA, Braun HJ, Payne TS, Singh RP, Sonder K, Baum M, van Ginkel M and Erenstein O. 2016. Impacts of international wheat improvement research, 1994–2014. Mexico: CIMMYT.

Lev-Yadun S, Gopher A and Abbo S. 2000. The cradle of agriculture. Science 288:1 602–3. https://doi.org/10.1126/science.288.5471.1602.

Li X, Shin S, Heinen S, Dill-Macky R, Berthiller F, Nersesian N, Clemente T, McCormick S and Muehlbauer GJ. 2015. Transgenic wheat expressing a barley UDP-glucosyltransferase detoxifies deoxy-nivalenol and provides high levels of resistance to Fusarium graminearum. Mol Plant-Microbe Interact 28:1237– 46. https://doi.org/10.1094/MPMI-03-15-0062-R.

Liu B, Asseng S, Liu L, Tang L, Cao W and Zhu Y. 2016a. Testing the responses of four wheat crop models to heat stress at anthesis and grain filling. Glob Chang Biol 22:1890–903. https://doi.org/10.1111/gcb.13212.

Liu G, Zhao Y, Gowda M, Longin CFH, Reif JC and Mette MF. 2016b. Predicting hybrid performances for quality traits through genomic-assisted approaches in central European wheat. PLoS One 11: e0158635. https://doi.org/10.1371/journal.pone.

Liu S, Zhou R, Dong Y, Li P and Jia J. 2006. Development, utilization of introgression lines using a synthetic wheat as donor. Theor Appl Genet 112:1360– 73. https://doi.org/10.1007/s00122-006-0238-x.

Longin CFH, Gowda M, Mühleisen J, Ebmeyer E, Kazman E, Schachschneider R, Schacht J, Kirchhoff M, Zhao Y and Reif JC. 2013. Hybrid wheat: quantitative genetic parameters and consequences for the design of breeding programs. Theor Appl Genet 126:2791– 801. https://doi.org/10.1007/s00122-013-2172-z.

Longin CFH, Mühleisen J, Maurer HP, Zhang H, Gowda M and Reif JC. 2012. Hybrid breeding in autogamous cereals. Theor Appl Genet 125:1087– 96. https://doi.org/10.1007/s00122-012-1967-7.

Lupton FGH. 1987. Wheat breeding. Dordrecht: Springer.

Lynas M. 2012. Rothamsted's aphid-resistant wheat – a turning point for GMOs? Agric Food Secur 1(17) :1–3. https://doi.org/10.1186/2048-7010-1-17.

Marcussen T, Sandve SR, Heier L, Spannagl M, Pfeifer M, IWGSC, Jakobsen KS, Wulff BBH, Steuernagel B, Mayer KFX and Olsen OA. 2014. Ancient hybridizations among the ancestral genomes of bread wheat. Science 345(6194):1250092. https://doi.org/10.1126/science.1250092.

McFadden ES and Sears ER. 1946. The origin of *Triticum spelta* and its free-threshing hexaploid relatives. J Hered 37:1 07–16. https://doi.org/10.1093/oxfordjournals.jhered.a105590.

Mehar M, Mittal S and Prasad N. 2016. Farmers coping strategies for climate shock: is it differentiated by gender? J Rural Stud 44:1 23–31. https://doi.org/10.1016/j.jrurstud.2016.01.001.

Meuwissen THE, Hayes BJ and Goddard ME. 2001. Prediction of total genetic value using genome-wide dense marker maps. Genetics 157:1819–29.

Miedaner T and Laidig F. 2019. Hybrid breeding in rye (Secale cereal L.). In: Al-Khayri JM, et al (Eds) Advances in plant breeding strategies, vol. 5. Cereals and legumes. Cham: Springer, pp. 1–31.

Miedaner T, Schulthess AW, Gowda M, Rief JC and Longin CFH. 2017. High accuracy of predicting hybrid performance of Fusarium head blight resistance by midparent values in wheat. Theor Appl Genet 130:4 61–70. https://doi.org/10.1007/s00122-016-2826-8.

Mittal S. 2022. Wheat and barley production trends and research priorities: a global perspective. In: Kashyap PL, Gupta V, Gupta OP, Sendhil R, Gopalareddy K, Jasrotia P and Singh GP (Eds) New horizons in wheat and barley research – global trends, breeding and quality enhancement. Singapore: Springer Nature, pp. 3–18. https://doi.org/10.1007/978-981-16-4449-8.

Mittal S and Hariharan VK. 2018. Mobile based climate services impact on farmers' risk management ability in India. Clim Risk Manag 22: 42–51. https://doi.org/10.1016/j.crm.2018.08.003.

Mittal S and Kumar P. 2000. Literacy, technology adoption, factor demand and productivity: an econometric analysis. Indian J Agric Econ 55(3):490–9.

Mittal S and Lal RC. 2001. Productivity and sources of growth for wheat in India. Agric Econ Res Rev 14(2):109–20.

Molnár-Láng M, Ceoloni C and Doležel J. 2015. Alien introgression in wheat: cytogenetics, molecular biology, and genomics, 1st edn. New York: Springer International Publishing.

Mühleisen J, Piepho HP and Maurer HP. 2014. Yield stability of hybrids versus lines in wheat, barley, and triticale. Theor Appl Genet 127:309–16.

Mujeeb-Kazi A, Gul A, Farooq M, Rizwan S and Ahmad I. 2008. Rebirth of synthetic hexaploids with global implications for wheat improvement. Aust J Agric Res 59(5):3 91–95. https://doi.org/10.1071/AR07226.

Mwadzingeni L, Shimelis H, Dube E, Laing MD and Tsilo TJ. 2016. Breeding wheat for drought tolerance: progress and technologies. J Integr Agric 15:9 35–43. https://doi.org/10.1016/S2095-3119(15)61102-9.

Nagaoka T and Ogihara Y. 1997. Applicability of inter-simple sequence repeat polymorphisms in wheat for use as DNA markers in comparison to RFLP and RAPD markers. Theor Appl Genet 94:59 7–602. https://doi.org/10.1007/s001220050456.

Nevo E. 2004. Genomic diversity in nature and domestication. In: Henry R (Ed) Diversity and evolution of plants. Genotypic and phenotypic variation in higher plants. Wallingford: CABI Publishing CAB International, pp. 287–315.

Nevo E. 2007. Evolution of wild wheat and barley and crop improvement: studies at the Institute of Evolution. ISR J Plant Sci 55:251–63.

Nevo E. 2009. Ecological genomics of natural plant populations: The Israeli perspective. In Plant Genomics. Berlin: Springer, pp. 321–44.

Nevo E. 2011. Triticum. In Wild crop relatives: Genomic and breeding resources. Berlin: Springer, pp. 407–56.

Nevo E and Beiles A. 1989. Genetic diversity of wild emmer wheat in Israel and Turkey. Theor Appl Genet 77:421–55.

Nevo E and Chen G. 2010. Drought and salt tolerances in wild relatives for wheat and barley improvement. Plant Cell Environ 33:670–85. https://doi.org/10.1111/j.1365-3040.2009.02107.x.

Nevo E, Korol AB, Beiles A and Fahima T. 2002. Evolution of wild emmer and wheat improvement: Population genetics, genetic resources and genome organization of wheat's progenitor, Triticum dicoccoides. Berlin: Springer, p. 64.

OECD-FAO. 2020. *Agricultural outlook 2020–2029*. Rome: FAO . https://doi.org/10.1787/112c23b-en.

Parry MAJ, Madgwick PJ, Bayon C, Tearall K, Hernandez-Lopez A, Baudo M, Rakszegi M, Hamada W, Al-Yassin A, Ouabbou H, Labhilili M and Phillips AL. 2009. Mutation discovery for crop improvement. J Exp Bot 60:28 17–25. https://doi.org/10.1093/jxb/erp189.

Pavithra S, Mittal S, Bhat SA, Birthal PS, Shah SA and Hariharan VK. 2017. Spatial and temporal diversity in adoption of modern wheat varieties in India. Agric Econ Res Rev 30(1):57–72.

Peng Z, Li X, Yang Z and Liao M. 2011. A new reduced height gene found in the tetraploid semi-dwarf wheat landrace Aiganfanmai. Genet Mol Res 10:2349–57.

Percival J. 1921. The wheat plant – a monograph. London: Duckworth and Co., p. 154.

Petersen G, Seberg O, Yde M and Berthelsen K. 2006. Phylogenetic relationships of Triticum and Aegilops and evidence for the origin of the A, B, and D genomes of common wheat (Triticum aestivum). Mol Phylogenet Evol 39: 70–82. https://doi.org/10.1016/j.ympev.2006.01.023.

Pingali PL. 1999. CIMMYT 1998–99: world wheat facts and trends. Global wheat research in a changing world: challenges and achievements. Mexico: CIMMYT.

Pingali PL. 2012. Green revolution: impacts, limits, and the path ahead. Proc Natl Acad Sci 109:12 302–8. https://doi.org/10.1073/pnas.0912953109.

Poland JA, Brown PJ, Sorrells ME and Jannink J-L. 2012a. Development of high-density genetic maps for barley and wheat using a novel two-enzyme genotyping-by-sequencing approach. PLoS One 7:e 32253. https://doi.org/10.1371/journal.pone.0032253.

Poland JA, Endelman J, Dawson J, Rutkoski J, Wu S, Manes Y, Dreisigacker S, Crossa J, Sánchez-Villeda H, Sorrells ME and Jannink J-L. 2012b. Genomic selection in wheat breeding using genotyping-by-sequencing. Plant Genome J 5:1 03–13. https://doi.org/10.3835/plantgenome2012.06.0006.

Pont C, Leroy T, Seidel M, Tondelli A, Duchemin W, Armisen D, Lang D, Bustos-Korts D, Goué N, Balfourier F, Molnar-Lang M, Lage J, Kilian B, Ozkan H, Waite D, Dyer S, Letellier T, Alaux M, Russell J and Salse J. 2020. Tracing the ancestry of modern bread wheats. Nat. Genet 51:1–7.

Pont C and Salse J. 2017. Wheat paleohistory created asymmetrical genomic evolution. Curr Opin Plant Biol 36: 29–37. https://doi.org/10.1016/j.pbi.2017.01.001.

Poudel P and Poudel MR. 2020. Heat stress effects and tolerance in wheat: a review. J Biol Today World 9:217. https://doi.org/10.35248/2322-3308.20.09.217.

Pozniak CJ and Hucl PJ. 2004. Genetic analysis of Imidazolinone resistance in mutation-derived lines of common wheat. Crop Sci 44: 23. https://doi.org/10.2135/cropsci2004.2300.

Prohens J. 2011. Plant breeding: a success story to be continued thanks to the advances in genomics. Front Plant Sci 2:51.

Rabinovich SV. 1998. Importance of wheat-rye translocations for breeding modern cultivar of *Triticum aestivum L.* Euphytica 100:3 23–40. https://doi.org/10.1023/A:1018361819215.

Rajaram S. 2012. Strategy for increasing wheat productivity. In Proceedings of the Regional Consultation on Improving Wheat Productivity in Asia, Bangkok, Thailand; 26–27 April, 2012. 224 p.

Rakszegi M, Békés F, Láng L, Tamás L, Shewry PR and Bedo Z. 2005. Technological quality of transgenic wheat expressing an increased amount of a HMW glutenin subunit. J Cereal Sci 42: 15–23. https://doi.org/10.1016/j.jcs.2005.02.006.

Rangan P, Furtado A and Henry RJ. 2016. New evidence for grain specific C4 photosynthesis in wheat. Sci Rep 6: 31721. https://doi.org/10.1038/srep31721.

Ray DK, Mueller ND, West PC and Foley JA. 2013. Yield trends are insufficient to double global crop production by 2050. PLoS ONE 8:e66428.

Renkow M and Byerlee D. 2010. The impacts of CGIAR research: A review of recent evidence. Food Policy 35:391–402.

Reynolds M, Bonnett D, Chapman SC, Furbank RT, Mane's Y, Mather DE and Parry MAJ. 2011. Raising yield potential of wheat. Overview of a consortium approach and breeding strategies. J Exp Bot 62:4 39–52. https://doi.org/10.1093/jxb/erq311.

Reynolds MP and Borlaug NE. 2006. Impacts of breeding on international collaborative wheat improvement. J Agric Sci 144(1) :3–17. https://doi.org/10.1017/S0021859606005867.

Richard C, Hickey LT, Fletcher S, Jennings R, Chenu K and Christopher JT. 2015. High-throughput phenotyping of seminal root traits in wheat. Plant Methods 11:13. https://doi.org/10.1186/s13007-015-0055-9.

Riehl S, Zeidi M and Conard NJ. 2013. Emergence of agriculture in the foothills of the Zagros Mountains of Iran. Science 341:6 5–7. https://doi.org/10.1126/science.1236743.

Rosell CM, Barro F, Sousa C and Mena MC. 2014. Cereals for developing gluten-free products and analytical tools for gluten detection. J Cereal Sci 59:3 54–64. https://doi.org/10.1016/j.jcs.2013.10.001.

Saintenac C, Jiang D, Wang S and Akhunov E. 2013. Sequence-based mapping of the Polyploid Wheat genome. G3: Genes Genom Genet 3:11 05–14. https://doi.org/10.1534/g3.113.005819.

Salamini F, Özkan H, Brandolini A, Schäfer-Pregl R and Martin W. 2002. Genetics and geography of wild cereal domestication in the near east. Nat Rev Genet 3:4 29–41. https://doi.org/10.1038/nrg817.

Salse J, Chagué V, Bolot S, Magdelenat G, Huneau C, Pont C, Belcram H, Couloux A, Gardais S, Evrard A, Segurens B, Charles M, Ravel C, Samain S, Charmet G, Boudet N and Chalhoub B. 2008. New insights into the origin of the B genome of hexaploid wheat: evolutionary relationships at the SPA genomic region with the S genome of the diploid relative *Aegilops speltoides*. BMC Genom 9:555. https://doi.org/10.1186/1471-2164-9-555.

Sander JD and Joung JK. 2014. CRISPR-Cas systems for editing, regulating and targeting genomes. Nat Biotechnol 32:3 47–55. https://doi.org/10.1038/nbt.2842.

Scheeren PL, Caierão E, Silva MS and Bonow S. 2011. Melhoramento no Brasil de trigo. In: Trigo no Brasil: Embrapa, p. 488.

Schlegel R and Korzun V. 1997. About the origin of 1RS.1BL wheat-rye chromosome translocations from Germany. Plant Breed 116:5 37–40. https://doi.org/10.1111/j.1439-0523.1997.tb02186.x.

Schneider A, Molnár I and Molnár-Láng M. 2008. Utilisation of Aegilops (goatgrass) species to widen the genetic diversity of cultivated wheat. Euphytica 163 :1–19. https://doi.org/10.1007/s10681-007-9624-y.

Sears ER. 1952. Homoeologous chromosomes in Triticum aestivum. Genetics 37:624.

Shan Q, Wang Y, Li J and Gao C. 2014. Genome editing in rice and wheat using the CRISPR/CAS system. Nat Protoc 9:239 5–410. https://doi.org/10.1038/nprot.2014.157.

Shan Q, Wang Y, Li J, Zhang Y, Chen K, Liang Z, Zhang K, Liu J, Xi JJ, Qiu JL and Caixia Gao C. 2013. Targeted genome modification of crop plants using a CRISPR-Cas system. Nat Biotechnol 31(8):6 86–88. https://doi.org/10.1038/nbt.2650.

Shewry PR. 2009. Wheat. J Exp Bot 60:15 37–53. https://doi.org/10.1093/jxb/erp058.

Shewry PR and Hey SJ. 2015. The contribution of wheat to human diet and health. Food Energy Secur 4(3):178–202.

Shiferaw B, Smale M, Braun HJ, Duveiller E, Reynolds M and Muricho G. 2013. Crops that feed the world 10. Past successes and future challenges to the role played by wheat in global food security. Food Secur 5(3):29 1–317. https://doi.org/10.1007/s12571-013-0263-y.

Singh G, Tyagi BS, Gupta A, Kumar V, Tiwari V, Chatrath R, Tiwari R, Singh SK, Saharan MS, Sharma RK, Gupta RK and Sharma I. 2016. Wheat: A guide on special features of varieties for different production conditions in India (India Research Bulletin no. 36). Karnal: Indian Institute of Wheat & Barley Research, p. 68.

Singh RP, Hodson DP, Jin Y, Huerta-Espino J, Kinyua MG, Wanyera R, Njau P and Ward RW. 2006a. Current status, likely migration and strategies to mitigate the threat to wheat production from race Ug99 (TTKS) of stem rust pathogen. CAB Rev Perspect Agric Vet Sci Nutr Nat Resour 1(54): 1–13. https://doi.org/10.1079/PAVSNNR20061054.

Singh RP, Kumar A, Gupta RK, Sharma RK, Saharan MS, Chokkar RS, Chandra S, Singh SK, Chand R & Sharma I. 2014. Bharat mein gehoon ki unnat kheti (Extension Bulletin no. 48). Karnal: Directorate of Wheat Research, p. 44.

Singh S, Jighly A, Sehgal D, Burgueño J, Joukhadar R, Singh SK, Sharma A, Vikram P, Sansaloni CP, Govindan V, Bhavani S, Randhawa M, Solis-Moya E, Singh S, Pardo N, Arif MAR, Laghari KA, Basandrai D, Shokat S, Chaudhary HK, Saeed NA, Basandrai AK, Ledesma-Ramírez L, Sohu VS, Imtiaz M, Sial MA, Wenzl P, Singh GP and Bains NS. 2021a. Direct introgression of untapped diversity into elite wheat lines. Nat Food 2:819–27.

Singh S, Vikram P, Sehgal D, Burgueño J, Sharma A, Singh SK, Sansaloni CP, Joynson R, Brabbs T, Ortiz C, Solis-Moya E, Govindan V, Gupta N, Sidhu HS, Basandrai AK, Basandrai D, Ledesma-Ramires L, Suaste-Franco MP, Fuentes-Dávila G, Moreno JI, Sonder K, Singh VK, Singh S, Shokat S, Arif MAR, Laghari KA, Srivastava P, Bhavani S, Kumar S, Pal D, Jaiswal JP, Kumar U, Chaudhary HK, Crossa J, Payne TS, Imtiaz M, Sohu VS, Singh GP, Bains NS, Hall A and Pixley KV. 2018. Harnessing genetic potential of wheat germplasm banks through impact-oriented-pre-breeding for future food and nutritional security. Sci Rep 8:12527. https://doi.org/10.1038/s41598-018-30667-4.

Singh SK, Chatrath R and Mishra B. 2010. Perspective of hybrid wheat research: a review. Ind J Agric Sci 80(12):1013–27.

Singh SK, Dubey S and Desai SA. 2021b. Hybrid wheat: prospects and future perspectives. In Bishnoi SK, Khan H, Singh SK and Singh GP (Eds) Breeding frontiers in wheat. Jodhpur: Agrobios, pp. 331–65.

Singh SK and Kumar S. 2021. Advances in wheat and barley production technologies. In: Yadav AS, Kumar N, Arora S, Srivastava DS and Pant H (Eds) Advances in crop production and climate change. New Delhi: New India Publishing Agency, pp. 27–59.

Singh SK, Kumar S, Kashyap PL, Sendhil R and Gupta OP. 2022. Wheat. In: Ghosh PK, Das A, Saxena R, Kar G and Vijay D (Eds) Trajectory of 75 years of Indian agriculture after independence. Singapore: Springer Nature Pte Ltd.

Singh SK, Kundu S, Kumar D, Srinivasan K, Mohan D and Nagarajan S. 2006b. Wheat. In Dhillon BS, Saxena S, Agrawal PK and Tyagi RK (Eds) Plant genetic resources: Foodgrain crops. New Delhi: Narosa Publishing House, pp. 58–89.

Singh SK, Singh S and Singh GP. 2020. Speed breeding for development of new crop varieties. In NAHEP-CAAST Training manual on pre-breeding and molecular breeding approach – Two important pillars for the vegetable and crop improvement, Feb 17–March 1, New Delhi: NIPB, pp. 20–5.

Slade AJ, Fuerstenberg SI, Loeffler D, Steine MN and Facciotti D. 2005. A reverse genetic, nontransgenic approach to wheat crop improvement by TILLING. Nat Biotechnol 23: 75–81. https://doi.org/10.1038/nbt1043.

Slade AJ, McGuire C, Loeffler D, Mullenberg J, Skinner W, Fazio G, Holm A, Brandt KM, Steine MN, Goodstal JF and Knauf VC. 2012. Development of high amylose wheat through TILLING. BMC Plant Biol 12: 69. https://doi.org/10.1186/1471-2229-12-69.

Song QJ, Shi JR, Singh S, Fickus EW, Costa JM, Lewis J, Gill BS, Ward R and Cregan PB. 2005. Development and mapping of microsatellite (SSR) markers in wheat. Theor Appl Genet 110:5 50–60. https://doi.org/10.1007/s00122-004-1871-x.

Spaenij-Dekking L, Kooy-Winkelaar Y, van Veelen P, Drijfhout JW, Jonker H, Soest LV, Smulders MJM, Bosch D, Gilissen LJWJ and Koning F. 2005. Natural variation in toxicity of Wheat: potential for selection of nontoxic varieties for celiac disease patients. Gastroenterology 129:797–806. https://doi.org/10.1053/j.gastro.2005.06.017.

Statista, 2022. www.statista.com/statistics/263977/world-grain-production-by-type/2021–22. Accessed on 28.6.2022.

Tanksley SD and McCouch SR. 1997. Seed banks and molecular maps: unlocking genetic potential from the wild. Science 277:1063–66.

Tattaris M, Reynolds MP and Chapman SC. 2016. A direct comparison of remote sensing approaches for high-throughput phenotyping in plant Breeding. Front Plant Sci 7:1–9. https://doi.org/10.3389/fpls. 2016.01131.

Tester M and Langridge P. 2010. Breeding technologies to increase crop production in a changing world. Science 327: 818–22 https://doi.org/10.1126/science.1183700.

Thorwarth P, Piepho HP, Zhao Y, Ebmeyer E, Schacht J, Schachschneider R, Kazman E, Reif JC, Würschum T, Friedrich C and Longin H. 2018. Higher gain yield and higher grain protein deviation underline the potential of hybrid wheat for a sustainable agriculture. Plant Breed 137:326–37.

Uauy C, Paraiso F, Colasuonno P, Tran RK, Tsai H, Berardi S, Comai L and Dubcovsky J. 2009. A modified TILLING approach to detect induced mutations in tetraploid and hexaploid wheat. BMC Plant Biol 9:115. https://doi.org/10.1186/1471-2229-9-115.

Umar M, Nawaz R, Sher A, Ali A, Hussain R and Khalid MW. 2019. Current status and future perspectives of biofortification in wheat. Asian J Res Crop Sci 4(4) :1–14. https://doi.org/10.9734/ajrcs/2019/v4i430079.

Upadhyay SK, Kumar J, Alok A and Tuli R. 2013. RNA-guided genome editing for target gene mutations in Wheat. G3: Genes Genom Genet 3:2 233–8. https://doi.org/10.1534/g3.113.008847.

USDA. 2022. www.worldagriculturalproduction.com/crops/wheat.aspx. Accessed on 28.6.2022

van Slageren MW. 1994. Wild wheats: a monograph of Aegilops L and Amblypyrum (Jaub, and Spach) Eig (Poaceae). Wageningen: Wageningen Agricultural University Papers 1994.

Vasil IK. 2007. Molecular genetic improvement of cereals: transgenic wheat (*Triticum aestivum* L.). Plant Cell Rep 26:1 133–54 https://doi.org/10.1007/s00299-007-0338-3.

Vasil V, Castillo AM, Fromm ME and Vasil IK. 1992. Herbicide resistant fertile transgenic Wheat plants obtained by micro-projectile bombardment of regenerable embryogenic callus. Nat Biotechnol 10:6 67–74. https://doi.org/10.1038/nbt0692-667.

Wang J, Luo MC, Chen Z, You FM, Wei Y, Zheng Y and Dvorak J. 2013. *Aegilops tauschii* single nucleotide polymorphisms shed light on the origins of wheat D-genome genetic diversity and pinpoint the geographic origin of hexaploid wheat. New Phytol 198:925–37.

Wang S, Wong D, Forrest K, Allen A, Chao S, Huang BE, Maccaferri M, Salvi S, Milner SG, Cattivelli L, Mastrangelo AM, Whan A, Stephen S, Barker G, Wieseke R, Plieske J, IWGSC, Lillemo M, Mather D, Appels R, Dolferus R, Brown-Guedira G, Korol A, Akhunova AR, Feuillet C, Salse J, Morgante M, Pozniak C, Luo MC, Dvorak J, Morell M, Dubcovsky J, Ganal M, Tuberosa R, Lawley C, Mikoulitch I, Cavanagh C, Edwards KJ, Hayden M and Akhunov E. 2014a. Characterization of polyploid wheat genomic diversity using a high-density 90,000 single nucleotide polymorphism array. Plant Biotechnol J 12:787–96. https://doi.org/10.1111/pbi. 12183.

Wang Y, Cheng X, Shan Q, Zhang Y, Liu J, Gao C and Qiu JL. 2014b. Simultaneous editing of three homoeoalleles in hexaploid bread wheat confers heritable resistance to powdery mildew. Nat Biotechnol 32:9 47–51. https://doi.org/10.1038/nbt.2969.

Wang Z, Huang L, Wu B, Hu J, Jiang Z, Qi P, Zheng Y and Liu D. 2018. Characterization of an integrated active Glu-1Ay allele in common wheat from wild emmer and its potential role in flour improvement. Int J Mol Sci 19:923.

Watson A, Ghosh S, Williams MJ, Cuddy WS, Simmonds J, Rey MD, Hatta MAM, Hinchliffe A, Steed A, Reynolds D, Adamski NM, Breakspear A, Korolev A, Rayner T, Dixon LE, Riaz A, Martin W, Merrill Ryan M, Edwards D, Batley J, Raman R, Carter J, Rogers C, Domoney C, Moore G, Harwood W, Nicholson P, Dieters MJ, DeLacy IH, Zhou J, Uauy G, Boden SA, Park RF, Wulff BBH and Hickey LT. 2018. Speed breeding is a powerful tool to accelerate crop research and breeding. Nat Plants 4(1): 23–29. https://doi.org/10.1038/s41477-017-0083-8.

Wheat Initiative. 2013. An international vision for wheat improvement. www.wheat initiative. org/vision-paper.

Whitford R, Fleury D, Reif JC, Garcia M, Okada T, Korzun V and Langridge P. 2013. Hybrid breeding in wheat: technologies to improve hybrid wheat seed production. J Exp Bot 64:54 11–28. https://doi.org/10.1093/jxb/ert333.

Winfield MO, Allen AM, Burridge AJ, Barker GLA, Benbow HR, Wilkinson PA, Coghill J, Waterfall C, Davassi A, Scopes G, Pirani A, Webster T, Brew F, Bloor C, King J, West C, Griffiths S, King I, Bentley AR, Edwards KJ. 2016. High-density SNP genotyping array for hexaploid wheat and its secondary and tertiary gene pool. Plant Biotechnol J 14(5):119 5–206. https://doi.org/10.1111/pbi.12485.

Witcombe JR, Joshi A, Joshi KD and Sthapit BR. 1996. Farmers participatory crop improvement. I. Varietal selection and breeding methods and their impacts on biodiversity. Exp Agric 32:445–60.

Wulff BBH and Dhugga KS. 2018. Wheat—the cereal abandoned by GM. Science 361: 451–2. https://doi.org/10.1126/science.aat5119.

Würschum T, Langer SM, Longin CFH, Tucker MR and Leiser WL. 2018a. A three component system incorporating Ppd D1, copy number variation at Ppd B1, and numerous small effect quantitative trait loci facilitates adaptation of heading time in winter wheat cultivars of worldwide origin. Plant Cell Environ 41(6):1407–16.

Würschum T, Liu G, Boeven PHG, Friedrich C, Longin H, Mirdita V, Kazman E, Zhao Y and Reif JC. 2018b. Exploiting the RHT portfolio for hybrid wheat breeding. Theor Appl Genet 131:1433–42.

Xie W and Nevo E. 2008. Wild emmer: genetic resources, gene mapping and potential for wheat improvement. Euphytica, 164:603–14.

Xu Y, An D, Li H and Xu H. 2011. Review: Breeding wheat for enhanced micronutrients. Can J Plant Sci 91: 231–7. https://doi.org/10.4141/CJPS10117.

Yao Z, Zhao X, Li Y, Xu J, Bi A, Kang L, Xu D, Chen H, Wang Y, Wang YG, Liu S, Jiao C, Lu H, Wang J, Yin C, Jiao Y and Lu F. 2020. *Triticum* population sequencing provides insights into wheat adaptation. Nat Genet 52:1412–22.

Zeven AC and Zhukovsky PM. 1975. Dictionary of cultivated plants and their centers of diversity. Wageningen: PUDOC, p. 219.

Zhang H, Mittal N, Leamy LJ, Barazani O and Song BH. 2017. Back into the wild—apply untapped genetic diversity of wild relatives for crop improvement. Evol Appl 10(1):5–24. https://doi.org/10.1111/eva.12434.

Zhang Y, Liang Z, Zong Y, Wang Y, Liu J, Chen K, Qiu JL, and Gao C. 2016. Efficient and transgene-free genome editing in wheat through transient expression of CRISPR/Cas9 DNA or RNA. Nat Commun 7:12 617. https://doi.org/10.1038/ncomms12617.

Zhao TJ, Zhao SY, Chen HM, Zhao QZ, Hu ZM, Hou BK and Xia GM. 2006. Transgenic wheat progeny resistant to powdery mildew generated by Agrobacterium inoculum to the basal portion of wheat seedling. Plant Cell Rep 25:1199–204. https://doi.org/10.1007/s00299-006-0184-8.

Zhao Y, Li Z, Liu G, Jiang Y, Maurer HP, Würschum T, Mock H-P, Matros A, Ebmeyer E, Schachschneider R, Kazman E, Schacht J, Gowda M, Friedric C, Longin H and Jochen C. Reif JC. 2015. Genome-based establishment of a high-yielding heterotic pattern for hybrid wheat breeding. Proc Natl Acad Sci 112(51):156 24–29. https://doi.org/10.1073/pnas.1514547112.

Zohary D and Hopf M. 2000. Domestication of plants in the old world: The origin and spread of cultivated plants in West Asia; Europe, and the Nile Valle. New York: Oxford University Press.

Zohary D, Hopf M and Weiss E. 2012. Domestication of plants in the old world: The origin and spread of domesticated plants in Southwest Asia, Europe, and the Mediterranean Basin. Oxford: Oxford University Press.

2

Wheat Production, Trade, Consumption, and Stocks

Global Trends and Prospects

**Sendhil R., Kiran Kumara T. M., Ankita Kandpal,
Binita Kumari, and Soumya Mohapatra**

CONTENTS

2.1 Introduction

Wheat is the most worldwide grown cereal grain (Nelson, 1985) serving as the staple food of the global population. Wheat is not only rich in starch but also rich in protein, vitamins, minerals, dietary fibres, and phytochemicals (Shewry and Hey, 2015). About 13–17% of the wheat grain is bran, of which 53% is fibre content (Sramkova et al., 2009). More than 800 million population worldwide suffers from chronic hunger with a maximum population hailing from sub-Saharan Africa and South Asia (Sharma et al., 2016). As per the FAO statistics, 2021, the percent of undernourishment is 9.8% worldwide while 11.7% of the world's population suffers from severe food insecurity leading to undernutrition. Studies suggest that malnutrition can be successfully addressed if the diet of the people consists of balanced proportions of proteins, carbohydrates, fat, and micronutrients (Collins, 2007). Wheat is rich in all these nutrients and provides nearly 55% of carbohydrates and 20% of calories consumed around the world (Breiman and Graur, 1995; Opine et al., 2015). Thus, it is evident that wheat plays an important role in alleviating hunger and malnutrition in the world.

DOI: 10.1201/9781003307938-2

Owing to the importance of wheat in alleviating global hunger, the production of wheat in the world must increase to 840 million tonnes by 2050 (Sharma et al., 2015). However, the shrinking of land and other essential resources is one of the major challenges to achieving the required wheat production. Wheat is placed first among the cereals in terms of consumption due to its nutritive value and relative ease in its harvest, storage, transport, and processing as compared to other grains (Posner, 2000). The demand for wheat has been growing, and it is predicted that the global demand would increase by about 60% by the year 2050 (Rosegrant and Agcaoili, 2010). Wheat production faces several biotic and abiotic stresses, which pose threat to achieving the production targets. Also, the productivity of the crop is declining every year. Ever-increasing population combined with declining farm sizes also poses a serious threat to fulfilling the increasing demand for grain. Thus, there is a need to have a deep insight into the production, consumption, and trade of the crop in the world.

In the present chapter, we analyzed the production, trade, consumption, and stock scenario across major wheat-growing countries/regions in terms of growth trends using the data sourced from the DES (2022), FAOSTAT (2022), and USDA (2022). The chapter has also highlighted the constraints which hamper the production and trade in the wheat-growing regions, and various policies which can help to overcome those constraints have been suggested.

2.2 Global Trends in the Area, Production, and Productivity of Wheat

Data on area, production, and productivity of wheat were collected and compiled for the period of 1970 triennium ending (TE) to 2020 TE from the FAOSTAT, 2022 database for required analysis and policy formulations. The inference can be drawn from Figure 2.1 that at the global level, the acreage of wheat has oscillated between 200 and 240 Mha between 1970 and 2020 TE, with a peak growth rate during 1982 TE (240 Mha) and gradually slowed down toward the current acreage of 220 Mha. By 2020, wheat was the most widely grown crop in the world, with a global area of 220 Mha, showing an increasing trend in the area of about 10% since 2019 TE. The area under wheat has been more or less stagnant after the 1990 TE which can be attributed to the continuous increasing and declining trends in area in the case of Asian and European countries, respectively. In terms of production, global production of wheat amounted to over 765 million metric tonnes, which is an increase of over 30 million tonnes as compared to 2019 TE. Despite the oscillating pattern of wheat acreage, there has been a continuous increasing trend in global wheat production since 1996 TE (+36%) due to a substantive rise in yield levels (+38%).

Asian and European countries are the major contributors to the global production of wheat with a steady growth pattern in the respective production levels. Asian countries occupy the largest share in global wheat production (45%, 2020 TE), followed by European (35%, 2020 TE) and American (15%, 2020 TE) countries with small shares by Africa and Oceania (3–4%). The relative production share of all the countries has remained more or less similar since the last quarter century. Interestingly, in the United States, 5–6% decline in wheat production was observed mainly due to the changing cropping pattern and production technologies. The global yield level of wheat has shown a steady growth of about 1.5 to 3.5 tonnes/ha, quadrupling the production

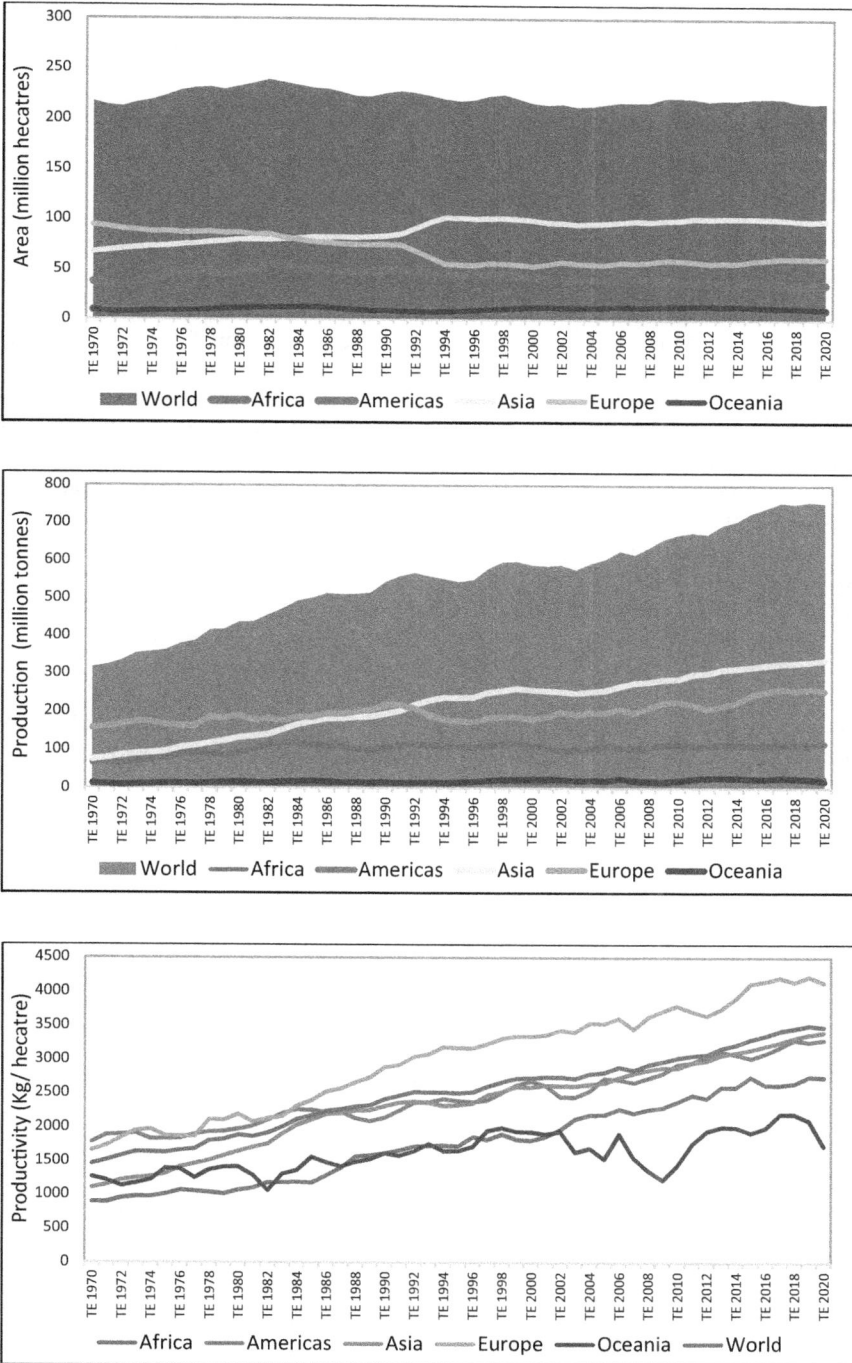

FIGURE 2.1 Global trends in the (a) area, (b) production, and (c) productivity for wheat-growing regions.

level over the period. All the regions have registered increasing yield patterns over the period of time, except Oceania. With the relative stability in global wheat acreage over the past half century, the yield level of wheat has shown a consistent rise which can be explained by a modest increase in the production level (Erenstein et al., 2022).

2.3 Regional Scenario in the Area, Production, and Productivity of Wheat

Although wheat is grown all over the world, five regions, that is, Asia, America, Europe, Africa, and Oceania, are considered for the present study (Table 2.1). Asia contributes to the maximum acreage (45.79%) under wheat crop, followed by Europe (28.46%) and America (16.36%); whereas, Africa and Oceania have a smaller share of the global wheat area (4–5% each). All the five regions have exhibited a positive growth rate during 2020 TE. Asia has shown the highest growth rate in the area during the study period (47.06%), followed by Oceania (15.24%) and Africa (9.02%); however, Europe and America have shown a continuous declining trend in the area. In the case of the Asian region, South Asia has registered the highest growth rate in acreage (66.85%), followed by Central Asia, but, East Asia, Southeast Asia, and West Asia have shown a declining trend in 2020 TE. The maximum decrease in the wheat area was observed in South Africa (1.38 Mha, −72.38%), followed by South Europe (6.69 Mha, −56.75%) and the Middle African region (0.02 Mha, −48.07%).

Similarly, the increasing trend in wheat acreage was highest in the case of West Africa (0.06 Mha, +399.56%), followed by North Europe (0.43 Mha, +173.76%). Among the sub-regions, South Asia occupies the largest share of global wheat area in 2020 TE (22.92%), followed by East Europe (20.05%) and North America (11.65%). The continuous rise in the wheat area may be attributed due to the Green Revolution which combined high-yielding varieties of wheat with input and irrigation

TABLE 2.1

Trends in Wheat Area for Major Global Regions

Region	Area (Mha)						Global Share (TE 2020, %)	TE 1970 to TE 2020 (% Change)
	TE 1970	TE 1980	TE 1990	TE 2000	TE 2010	TE 2020		
Africa	**9.05**	**8.31**	**8.26**	**8.91**	**9.38**	**9.86**	**4.56**	**9.02**
East Africa	1.27	0.75	0.94	1.31	1.88	2.13	0.99	68.49
Middle Africa	0.03	0.02	0.01	0.01	0.01	0.01	0.01	−48.07
North Africa	5.84	5.68	5.41	6.71	6.76	7.12	3.29	22.05
South Africa	1.91	1.85	1.85	0.82	0.67	0.53	0.24	−72.38
West Africa	0.01	0.01	0.05	0.06	0.06	0.07	0.03	399.56

TABLE 2.1 *(Continued)*

Region	Area (Mha)						Global Share (TE 2020, %)	TE 1970 to TE 2020 (% Change)
	TE 1970	TE 1980	TE 1990	TE 2000	TE 2010	TE 2020		
Americas	**37.42**	**46.47**	**49.00**	**41.88**	**39.11**	**35.39**	**16.36**	**−5.42**
Central America	0.87	0.73	1.02	0.72	0.77	0.57	0.26	−35.08
North America	28.64	36.41	38.47	33.02	29.99	25.20	11.65	−12.04
South America	7.91	9.33	9.51	8.15	8.34	9.63	4.45	21.85
Asia	**67.35**	**79.58**	**83.68**	**99.21**	**99.56**	**99.04**	**45.79**	**47.06**
Central Asia	–	–	–	12.11	16.46	14.21	6.57	17.30
East Asia	25.94	29.90	30.67	28.91	24.56	24.39	11.28	−5.98
South Asia	29.72	37.18	40.34	44.50	46.62	49.58	22.92	66.85
Southeast Asia	0.07	0.09	0.12	0.09	0.10	0.06	0.03	−10.01
West Asia	11.62	12.41	12.55	13.59	11.82	10.79	4.99	−7.09
Europe	**94.00**	**86.41**	**75.30**	**53.87**	**59.56**	**61.55**	**28.46**	**−34.52**
East Europe	73.85	67.93	56.15	34.45	40.22	43.37	20.05	−41.27
North Europe	1.60	1.94	2.97	3.85	4.28	4.39	2.03	173.76
South Europe	11.79	9.23	8.00	6.88	5.68	5.10	2.36	−56.75
West Europe	6.75	7.32	8.18	8.69	9.37	8.69	4.02	28.62
Oceania	**9.06**	**10.98**	**9.06**	**12.06**	**13.38**	**10.44**	**4.83**	**15.24**
Australia and New Zealand	9.06	10.98	9.06	12.06	13.38	10.44	4.83	15.24
Melanesia	0.00	0.00	0.00	0.00	0.00	0.00	0.00	131.96
World	**216.87**	**231.76**	**225.31**	**215.93**	**220.98**	**216.28**	**100.00**	**−0.27**

Data source: FAOSTAT (2022)

Note: The bold values indicate the data for the whole region. The subsequent rows indicate the data for the sub-regions.

management practices. Expansion and intensification of the wheat crop are mostly dominated by South Asian and some European countries thereby necessitating the scope to divert the improved production practices toward other global regions.

Table 2.2 presents the global production trends of wheat. The production level of wheat had witnessed a rising trend over a period of time. Global wheat production has increased twice from 315.41 Mt during 1970 TE to 752.68 Mt during 2020 TE, corresponding to a substantial growth of 138.64%. In terms of the global share of wheat production, Asia holds the largest proportion (44.75%), followed by Europe (33.81%) and America (15.46%); together, they contribute about 94% to global wheat production. All five regions showed a positive growth rate in wheat production during the study period.

TABLE 2.2

Trends in Wheat Production for Major Global Regions

Region	Production (Mt)						Global Share (TE 2020, %)	TE 1970 to TE 2020 (% Change)
	TE 1970	TE 1980	TE 1990	TE 2000	TE 2010	TE 2020		
Africa	**8.05**	**8.83**	**13.25**	**16.15**	**22.33**	**26.95**	**3.58**	**234.67**
East Africa	1.12	0.98	1.51	1.86	3.43	5.99	0.80	434.92
Middle Africa	0.03	0.02	0.01	0.02	0.02	0.02	0.00	−44.09
North Africa	5.51	6.01	9.19	12.13	16.91	19.00	2.52	245.07
South Africa	1.38	1.79	2.48	2.04	1.86	1.85	0.25	34.45
West Africa	0.02	0.03	0.06	0.11	0.10	0.10	0.01	333.73
Americas	**66.17**	**91.53**	**105.21**	**112.30**	**114.29**	**116.34**	**15.46**	**75.83**
Central America	2.40	2.68	4.03	3.26	3.94	3.06	0.41	27.75
North America	54.46	76.28	83.95	90.04	89.10	84.59	11.24	55.32
South America	9.31	12.57	17.23	19.00	21.25	28.69	3.81	208.22
Asia	**74.02**	**130.57**	**193.39**	**256.76**	**286.11**	**336.86**	**44.75**	**355.08**
Central Asia	–	–	–	15.06	22.68	21.81	2.90	44.87
East Asia	29.22	58.17	93.25	108.64	115.47	134.53	17.87	360.36
South Asia	32.08	52.26	74.52	105.34	119.63	150.76	20.03	369.92
Southeast Asia	0.04	0.08	0.14	0.10	0.18	0.11	0.02	209.79
West Asia	12.68	20.07	25.47	27.63	28.15	29.65	3.94	133.77
Europe	**155.69**	**189.39**	**216.96**	**180.13**	**226.18**	**254.45**	**33.81**	**63.43**
East Europe	103.50	121.04	123.81	71.18	110.01	144.45	19.19	39.57
North Europe	6.01	9.80	19.26	25.37	28.07	27.88	3.70	363.82
South Europe	21.22	22.57	22.97	20.43	19.14	19.99	2.66	−5.81
West Europe	24.96	35.97	50.92	63.15	68.96	62.13	8.25	148.94
Oceania	**11.48**	**15.35**	**14.58**	**23.31**	**19.34**	**18.08**	**2.40**	**57.55**
Australia and New Zealand	11.48	15.35	14.58	23.31	19.34	18.08	2.40	57.55
Melanesia	0.00	0.00	0.00	0.00	0.00	0.00	0.00	−38.83
World	**315.41**	**435.68**	**543.40**	**588.65**	**668.25**	**752.68**	**100.00**	**138.64**

Data source: FAOSTAT (2022)

Note: The bold values indicate the data for the whole region. The subsequent rows indicate the data for the sub-regions.

The highest growth rate was observed in the case of Asia (+262.84 Mt, 336.86%), followed by Africa (+18.9 Mt, 234.67%) and America (+50.17 Mt, 75.83%). Positive growth in production was observed in almost all the sub-regions, except Middle Africa (−44.09%), Melanesia (−38.83%), and Southern Europe (−5.81%) during the study period. Within the sub-regions, the highest growth rate in production was registered in East Africa (+4.87 Mt, 434.92%, South Asia (+118.68 Mt, 369.92%), North Europe (+21.87 Mt, 363.82%), and West Africa (+0.08 Mt, 333.73%). Although acreage under wheat was continuously fluctuating over the period, production levels in all the regions have witnessed a positive trend due to high-yielding and climate-resilient wheat varieties and improved management techniques.

The global productivity level of wheat witnessed an increasing trend from 1,455 kg/ha during 1970 TE to 3,480 kg/ha during 2020 TE (Table 2.3), with a growth rate of 139.18%. There was a positive change observed in each region during the study period; Asia witnessed the highest growth (+2,302 kg/ha, 209.47%), followed by

TABLE 2.3

Trends in Wheat Productivity for Major Global Regions

Region	Productivity (kg ha⁻¹)						TE 1970 – TE 2020 (% Change)	Rank (TE 2020)
	TE 1970	**TE 1980**	**TE 1990**	**TE 2000**	**TE 2010**	**TE 2020**		
Africa	**891**	**1,064**	**1,610**	**1,809**	**2,373**	**2,733**	**206.74**	**17**
East Africa	883	1,312	1,606	1,428	1,825	2,803	217.28	15
Middle Africa	1,095	754	938	1,434	1,327	1,185	8.26	24
North Africa	943	1,059	1,705	1,801	2,492	2,666	182.67	18
South Africa	724	969	1,320	2,475	2,775	3,530	387.83	7
West Africa	1,673	1,993	1,141	1,856	1,716	1,456	−12.98	23
Americas	**1,781**	**1,969**	**2,135**	**2,681**	**2,925**	**3,288**	**84.60**	**12**
Central America	2,742	3,664	3,972	4,561	5,119	5,408	97.23	4
North America	1,918	2,094	2,162	2,728	2,972	3,359	75.15	10
South America	1,182	1,350	1,809	2,328	2,552	2,982	152.23	14
Asia	**1,099**	**1,641**	**2,310**	**2,589**	**2,873**	**3,401**	**209.47**	**9**
Central Asia	–	–	–	1,246	1,373	1,535	23.13	22
East Asia	1,126	1,945	3,038	3,758	4,702	5,517	389.82	3
South Asia	1,079	1,405	1,847	2,367	2,565	3,040	181.82	13
Southeast Asia	533	879	1,110	1,063	1,730	1,862	249.17	19

(Continued)

TABLE 2.3 (*Continued*)

Region	Productivity (kg ha⁻¹)						TE 1970 – TE 2020 (% Change)	Rank (TE 2020)
	TE 1970	TE 1980	TE 1990	TE 2000	TE 2010	TE 2020		
West Asia	1,092	1,618	2,027	2,033	2,381	2,742	151.07	16
Europe	**1,656**	**2,188**	**2,881**	**3,344**	**3,791**	**4,133**	**149.51**	**5**
East Europe	1,402	1,777	2,205	2,067	2,723	3,329	137.53	11
North Europe	3,749	5,056	6,461	6,594	6,565	6,323	68.64	2
South Europe	1,800	2,446	2,873	2,973	3,360	3,920	117.81	6
West Europe	3,692	4,915	6,222	7,267	7,362	7,149	93.65	1
Oceania	**1,259**	**1,410**	**1,609**	**1,932**	**1,436**	**1,725**	**37.04**	**21**
Australia and New Zealand	1,259	1,410	1,609	1,932	1,436	1,725	37.04	20
Melanesia	2,815	1,498	1,061	1,863	–	719	-74.44	25
World	**1,455**	**1,880**	**2,409**	**2,726**	**3,023**	**3,480**	**139.18**	**8**

Data source: FAOSTAT (2022)

Note: The bold values indicate the data for the whole region. The subsequent rows indicate the data for the sub-regions.

Africa (+1,842 kg/ha, 206.74%) and Europe (2,477 kg/ha, 149.51%). Among the sub-regions, all witnessed an increasing pattern in yield, except Melanesia (−74.44%) and West Africa (−12.98%). East Asia (+4,391 kg/ha, 389.82%), South Africa (+2,806 kg/ha, 387.83%), Southeast Asia (+ 1,329 kg/ha, 249.17%), and East Africa (+1,920 kg/ha, 217.28%) are some of the sub-regions with highest growth rate over the period of time. The positive growth trend in yield levels of wheat can be attributed to the adoption of good agricultural practices, varietal improvement in various national and international institutions, and the adoption of modern cultivation practices at the farmers' level (Joshi et al., 2007).

The top 20 countries contributing highest to the global area, production, and productivity of wheat in the year 2020 TE are presented in Figure 2.2. India occupies the first position in terms of area (39.21%), followed by Russian Federation (33%), China (31%), the United States (21%), and Kazakhstan (15.68%). However, in terms of production, China takes the top slot with 135 M tonnes (17.64% of total production), followed by India (110 Mt) and Russian Federation (78 Mt).

The high production level of wheat in China can be attributed to the adoption of resource conservation technologies as well as the improved agricultural environment in the country (Wang et al., 2009). The average productivity of wheat in 2020 TE was highest in New Zealand (9,100 kg/ha), followed by Belgium (8,900 kg/ha) and the Netherlands (8,100 kg/ha). These countries accomplished high productivity levels despite occupying very less area under wheat. The main reasons for higher wheat

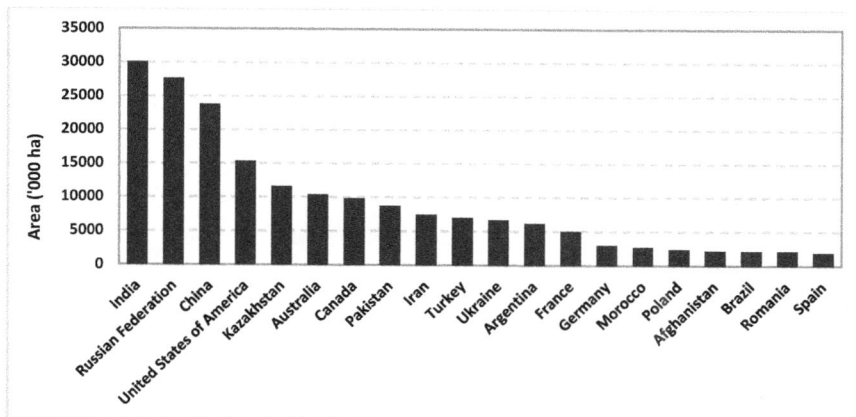

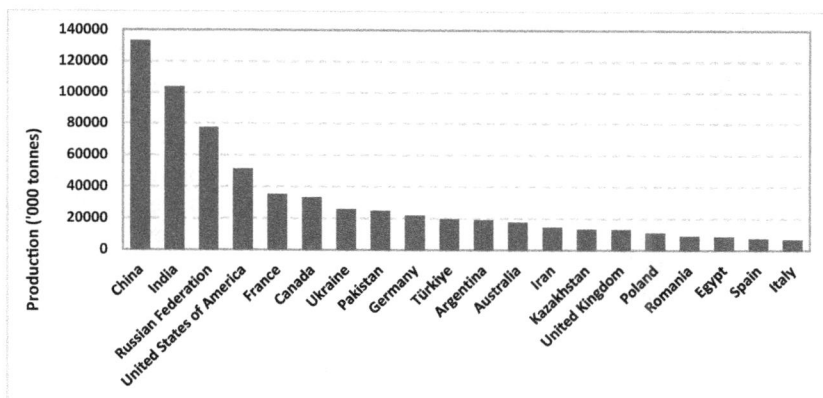

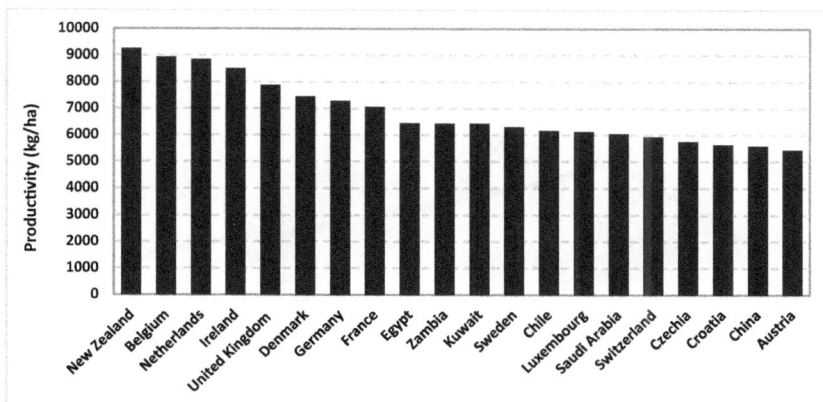

FIGURE 2.2 Top 20 countries in the (a) wheat area, (b) production, and (c) productivity (TE 2020).

Data source: FAOSTAT (2022)

yield in New Zealand can be better cultivar parameters, canopy structure, higher bio-mass production, and ultimately higher yield (Senapati and Semenov, 2019). India, with the first rank in the area and second in production, failed to bag a position among the top 20 countries in terms of productivity. Low wheat yield levels in India could be due to increasing heat and other climatic stress, improper irrigation management as well as a threat from several forms of diseases (wheat rust), and adverse selection of crop varieties (Joshi et al., 2007).

2.4 Productivity Level Vis-a-Vis Efficiency Gap

Figure 2.3 depicts the comparison between the productivity and efficiency gap for all the wheat-growing countries, as productivity is considered one of the major determi-nants of efficiency level. Efficiency gap (EG) is a ratio that explains the difference in the crop yield level of a particular region and the yield level of the region with the highest productivity, that is, benchmark yield level (Sendhil et al., 2022; Singh et al., 2017). EG is computed by using the following formula: EG = 1 − (Actual yield ÷ Benchmark yield) × 100

It can be inferred from Figure 2.3 that New Zealand is considered the benchmark country for EG analysis with the highest productivity of 9,100 kg/ha (2020 TE) owing to the wide cultivation of winter wheat which is highly productive (having a long duration) in comparison to the spring wheat (have a short duration and cul-tivated in regions like India). EG is maximum for Somalia (more than 99%) and minimum for the Netherlands. China, the largest producer of wheat, registered an EG of more than 60% owing to the type of wheat under cultivation. The range of EG among countries calls for region-specific research interventions for augment-ing productivity and minimizing the yield gap as well as the realization of higher yield potential.

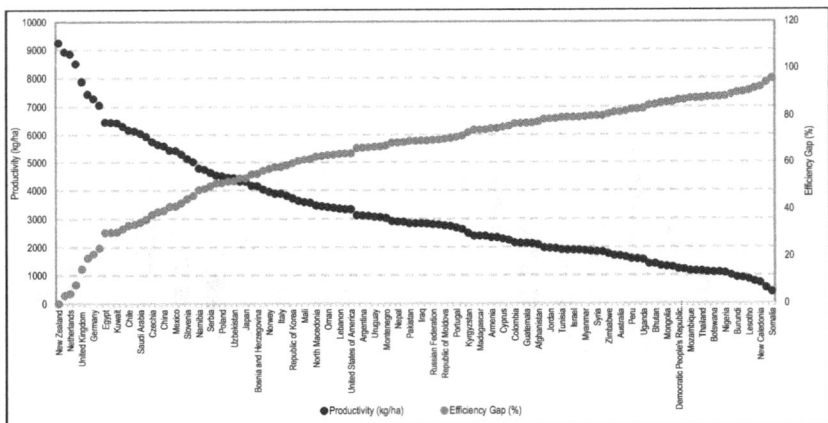

FIGURE 2.3   Productivity levels (kg ha^{-1}) vis-à-vis efficiency gap (%).

Data source: FAOSTAT (2022)

2.5 Trends in Wheat Trade

2.5.1 Export Scenario

The world's biggest wheat producers have attained self-sufficiency in wheat production, which requires an active global trade linking the areas with surplus production with the deficit areas. Table 2.4 presents the pattern of wheat exports across

TABLE 2.4

Trends in Wheat Exports for Major Global Regions

Region	Exports ('000 tonnes)					
	TE 1970	TE 1980	TE 1990	TE 2000	TE 2010	TE 2020
Africa	**52.25**	**137.08**	**585.31**	**187.08**	**144.83**	**149.68**
East Africa	40.77	6.29	0.06	59.55	37.83	11.48
Middle Africa	0.00	0.19	0.08	0.00	0.42	0.00
North Africa	2.71	0.17	0.82	2.08	5.92	6.89
South Africa	8.56	128.94	582.14	105.04	84.89	125.91
West Africa	0.20	1.49	2.21	20.41	15.76	5.40
Americas	**26,595.57**	**52,270.60**	**56,240.99**	**55,856.66**	**53,942.87**	**61,637.45**
Caribbean	0.00	0.00	2.05	0.34	0.15	0.39
Central America	99.08	17.64	134.08	338.70	992.49	612.04
North America	24,114.59	48,792.29	51,313.85	45,303.68	44,373.30	49,163.08
South America	2,381.91	3,460.67	4,791.01	10,213.94	8,576.93	11,861.94
Asia	**186.44**	**1435.44**	**2,644.58**	**5,661.96**	**5,799.80**	**7,130.01**
Central Asia	–	–	–	3,523.21	4487.75	5,712.30
East Asia	6.43	1.66	29.90	3.97	44.86	5.37
South Asia	142.73	422.81	56.49	282.74	306.83	1,104.37
Southeast Asia	15.82	16.72	43.70	5.04	33.21	38.65
West Asia	21.45	994.24	2,514.49	1,847.00	927.15	269.33
Europe	**13,667.72**	**14,567.11**	**31,317.77**	**35,499.16**	**68,344.59**	**110,166.13**
East Europe	6,255.56	3,340.07	3,123.81	6,372.60	30,775.40	72,153.89
North Europe	370.47	1,047.90	4,801.47	5,102.73	6,954.15	7881.67
South Europe	731.39	224.05	2,238.34	823.41	1,858.24	2231.90
West Europe	6,310.30	9,955.09	21,154.16	23,200.42	28,756.80	27,898.68
Oceania	**6,124.09**	**10,883.08**	**11,381.34**	**16,498.93**	**13,058.64**	**10,783.39**
Australia and New Zealand	6,124.09	10,883.08	11,381.34	16,498.93	13,058.59	10,781.84
Melanesia	0.00	0.00	0.00	0.00	0.05	1.55
World	**46,626.07**	**79,293.31**	**102,170.00**	**113,703.79**	**141,290.73**	**189,866.67**

Data source: FAOSTAT (2022)

Note: The bold values indicate the data for the whole region. The subsequent rows indicate the data for the sub-regions.

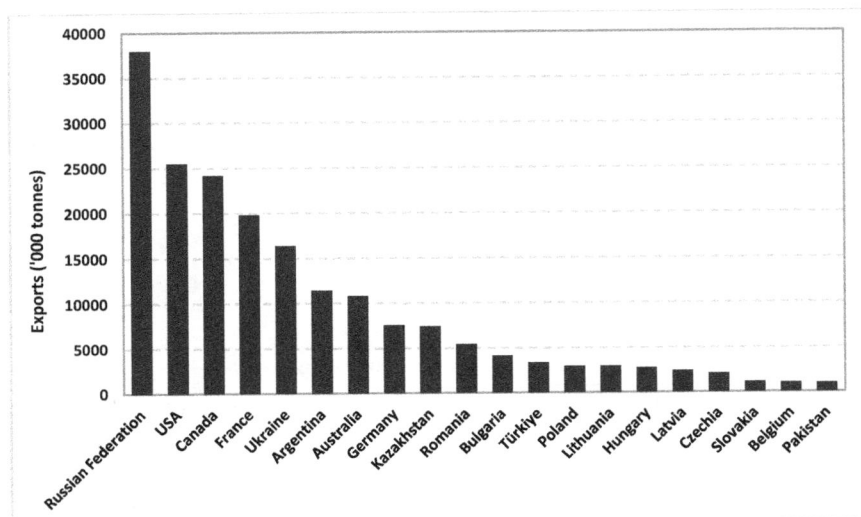

FIGURE 2.4 Top 20 countries in wheat exports (TE 2020).

major global regions and subregions. Total world exports of wheat in 2020 TE were around 190 Mt, out of which Europe registered the highest quantity of exports (110 Mt), followed by America (61.63 Mt) and Asia (7.13 Mt). All the regions increased their exports during the study period (1970 TE to 2020 TE), except Oceania, thereby raising the global exports from 141.29 Mt in 2019 TE to 189.86 Mt in 2020 TE. Among the subregions, East Europe registered the highest level of wheat export, followed by North America, West Europe, and South America. Though Asian countries have been ranked top in terms of wheat area and production, the consumption demand has increased and is attributed to the rising population and economic growth. Therefore, Asian countries are not able to realize their full export potential (Sendhil et al., 2022).

Among the major wheat-exporting countries, Russian Federation tops the list with an export quantity of 38.76 Mt (23.92% of global export) during 2020 TE (Figure 2.4), followed by the United States (25.49 Mt), and Canada (24.87 Mt). In order to bridge the gap between consumption demand and availability of wheat, trade plays an important role to diversify the dynamic production and consumption pattern.

2.5.2 Import Scenario

Considering the dynamism of wheat processing and consumption pattern, substantial heterogeneity was found underlying wheat trade in each region. Table 2.5 indicates that the global import of wheat has registered positive growth (+140.65 Mt, 304.89%) during the study period, that is, 1970 TE to 2020 TE. All the regions and subregions covered in the study have witnessed an increased import trend over the past half-century, which can be attributed to the high levels of spatial variations between production level and consumption preference of wheat.

TABLE 2.5

Trends in Wheat Imports for Major Global Regions

Region	Imports ('000 tonnes)					
	TE 1970	**TE 1980**	**TE 1990**	**TE 2000**	**TE 2010**	**TE 2020**
Africa	**3319.24**	**10556.35**	**14728.04**	**20970.35**	**35686.14**	**46516.92**
East Africa	260.10	613.96	1,165.95	2,376.13	4,774.50	5,670.83
Middle Africa	121.66	308.10	301.42	448.61	867.09	1,792.40
North Africa	2,485.22	8,114.54	11,951.83	14,414.52	23,036.77	27,848.24
South Africa	54.85	44.33	324.85	761.19	1,586.36	2,332.33
West Africa	397.40	1475.41	983.99	2,969.89	5,421.42	8873.12
Americas	**5,413.64**	**10,331.87**	**8,042.57**	**19,297.70**	**20,550.91**	**23,988.26**
Caribbean	555.46	1,225.91	1,727.68	1,609.07	1,690.61	1,756.11
Central America	308.65	1,301.17	1,304.31	3,786.70	4,532.37	6,231.16
North America	31.40	3.82	479.31	2,076.96	2,606.85	2,452.11
South America	4,518.13	7,800.97	4,531.26	11,824.97	11,721.08	13,548.88
Asia	**19,439.81**	**28,684.46**	**44,617.02**	**44,095.68**	**50,750.29**	**78,165.78**
Central Asia	–	–	–	1005.68	2,118.33	4,157.05
East Asia	10,957.04	17,852.60	23,990.80	12,416.45	10,989.72	15,678.41
South Asia	5,509.37	3,983.22	8,662.48	11,647.28	9,142.63	8,442.05
Southeast Asia	1,193.62	3,171.92	4,291.58	7,801.51	10,826.45	25,753.14
West Asia	1,779.78	3,676.72	7,672.16	11,224.77	17,673.16	24,135.14
Europe	**17,937.35**	**28,721.39**	**33,886.55**	**27,169.30**	**32,940.02**	**36,884.10**
East Europe	3,960.79	14,873.45	19,207.66	4,594.22	1,803.10	2,346.53
North Europe	5,220.35	3,516.37	1,803.55	2,502.10	2,992.06	3,831.54
South Europe	1,856.15	4,815.95	6,493.21	12,271.95	14,895.18	16,163.99
West Europe	6,900.06	5,515.62	6,382.13	7,801.03	13,249.68	14,542.04
Oceania	**24.90**	**102.41**	**288.14**	**436.04**	**576.08**	**1,242.22**
Australia and New Zealand	22.61	28.55	148.26	184.17	283.79	803.73
Melanesia	1.73	73.19	139.12	251.67	291.29	437.17
Micronesia	0.01	0.00	0.00	0.00	0.00	0.00
Polynesia	0.55	0.67	0.76	0.20	0.99	1.32
World	**46,134.94**	**78,396.47**	**101,562.32**	**111,969.07**	**140,503.43**	**186,797.28**

Data source: FAOSTAT (2022)

Note: The bold values indicate the data for the whole region. The subsequent rows indicate the data for the sub-regions.

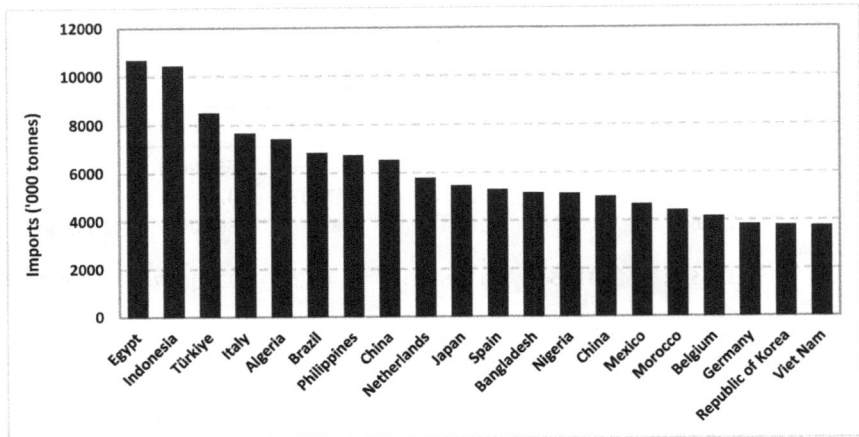

FIGURE 2.5   Top 20 countries in wheat imports (TE 2020).

Asia is the major importer of wheat (41.84%) in 2020 TE, followed by Africa (24.89%), Europe (19.74%), and America (12.83%). Asia, being the largest producer of wheat, surprisingly tops the list of major wheat-importing regions due to the increase in income, urbanization, change in food preferences and lifestyles as well as the inclusion of high-value foods in the dietary patterns (Drewnowski et al., 2009). Among the subregions, North Africa is the major importer of wheat (27.84 Mt) owing to the region's strong population growth, climate-constrained production, and shift in the dietary pattern which have forced the country to heavily rely on imports (Nigatu and Motamed, 2015). Egypt, Indonesia, Turkey, and Italy are some of the major wheat-importing countries (Figure 2.5) during 2020 TE and account for about 40% of global wheat imports. Though these countries are self-sufficient in terms of wheat production, still they are the major importers of wheat grain, and, by using the advanced processing infrastructures, they become the leading exporter of wheat flour in the world.

2.6 Trends in Wheat Stocks

Stocking the surplus quantity of food grains in the central pool aids one country to meet the operational requirements in times of food crisis or increased price variability, especially in the case of the importing countries. It can be inferred from Table 2.6 that the pattern of wheat-opening stock for all the global regions has increased over the period of time (1970 TE to 2020 TE) due to which the global opening stock has registered a growth of about 135.61% (161.85 Mt in 1970 TE to 381.35 Mt in 2020 TE). Asia contributed the highest to the global opening stock of wheat (212.98 Mt, 55.85%), followed by Europe (87.9 Mt, 23.04%) and America (50.56 Mt, 13.25%). Among the subregions, East Asia ranks at the top in terms of wheat opening stock (135.12 Mt, 35.45%), which could be due to the strong governmental policies as well as new-fangled buffer stock operations (Lin et al., 2022).

TABLE 2.6

Trends in Wheat Opening Stocks for Major Global Regions

Region	Opening Stocks ('000 tonnes)							
	TE 2012	**TE 2013**	**TE 2014**	**TE 2015**	**TE 2016**	**TE 2017**	**TE 2018**	**TE 2019**
Africa	**20,064.0**	**20,358.9**	**21,189.7**	**21,006.3**	**22,922.6**	**23,851.6**	**24,729.4**	**24,886.5**
East Africa	3,703.0	3,831.8	4,024.0	3,977.2	3,955.4	3,883.4	3,704.5	3,346.9
Middle Africa	196.2	212.0	235.6	277.9	354.1	441.8	556.5	652.1
North Africa	12,654.6	13,027.4	14,041.5	14,414.0	16,721.3	17,861.6	18,730.3	18,896.3
South Africa	497.9	686.4	899.5	933.8	654.7	509.5	342.9	478.9
West Africa	3,012.4	2,601.3	1,989.1	1,403.4	1,237.1	1,155.3	1,395.2	1,512.3
Americas	**12,899.1**	**10,628.0**	**18,809.1**	**28,179.5**	**40,124.1**	**45,829.0**	**48,584.9**	**50,566.5**
Caribbean	181.1	176.5	164.5	193.1	248.8	263.9	257.1	193.6
Central America	1,053.4	961.5	959.2	945.6	1,021.6	1,172.6	1,203.9	1,224.0
North America	1,894.8	2,263.2	12,154.8	22,305.7	32,967.9	38,388.4	42,162.3	45,298.0
South America	9,769.8	7,226.7	5,530.6	4,735.1	5885.8	6,004.0	4,961.6	3,850.9
Asia	**112,248.8**	**11,5112.4**	**12,3764.4**	**13,7314.9**	**15,7540.4**	**17,7601.3**	**19,7728.0**	**212,985.8**
Central Asia	6458.6	5056.2	5896.0	4947.0	7299.6	9895.5	12803.4	14104.0
East Asia	63108.0	62943.9	65564.5	73210.9	87210.0	103214.5	120753.8	135127.1
South Asia	28139.3	29593.7	29168.7	31569.4	31546.6	32289.5	31934.1	32675.1
Southeast Asia	3607.4	3886.7	3895.3	3643.5	3309.9	3505.2	3743.8	4446.1
West Asia	10935.5	13632.0	19239.9	23944.0	28174.3	28696.6	28492.7	26633.5
Europe	**12288.7**	**11888.1**	**32724.6**	**56204.9**	**81925.2**	**85373.8**	**88931.5**	**87901.2**
East Europe	5817.3	5642.7	13075.9	20867.5	28730.5	29614.0	29824.7	28123.2
North Europe	4552.1	4152.7	4259.0	5776.8	8002.1	8703.1	8474.7	6847.1
South Europe	617.3	712.2	4251.3	7960.7	11631.5	13551.9	15209.2	17498.6
West Europe	1302.0	1380.6	11138.5	21599.9	33561.2	33504.7	35423.0	35432.3
Oceania	**4355.5**	**5144.9**	**5945.1**	**5433.5**	**5040.3**	**4735.0**	**5089.9**	**5015.6**
Australia and New Zealand	4176.9	4976.6	5771.6	5255.9	4873.3	4584.0	4981.3	4937.7
Melanesia	178.3	167.9	173.2	177.3	166.6	150.6	108.3	77.7
Polynesia	0.3	0.3	0.3	0.3	0.3	0.3	0.3	0.3
World	**161856.0**	**163132.3**	**202432.9**	**248139.1**	**307552.6**	**337390.7**	**365063.7**	**381355.7**

Data source: FAOSTAT (2022)

Note: The bold values indicate the data for the whole region. The subsequent rows indicate the data for the sub-regions.

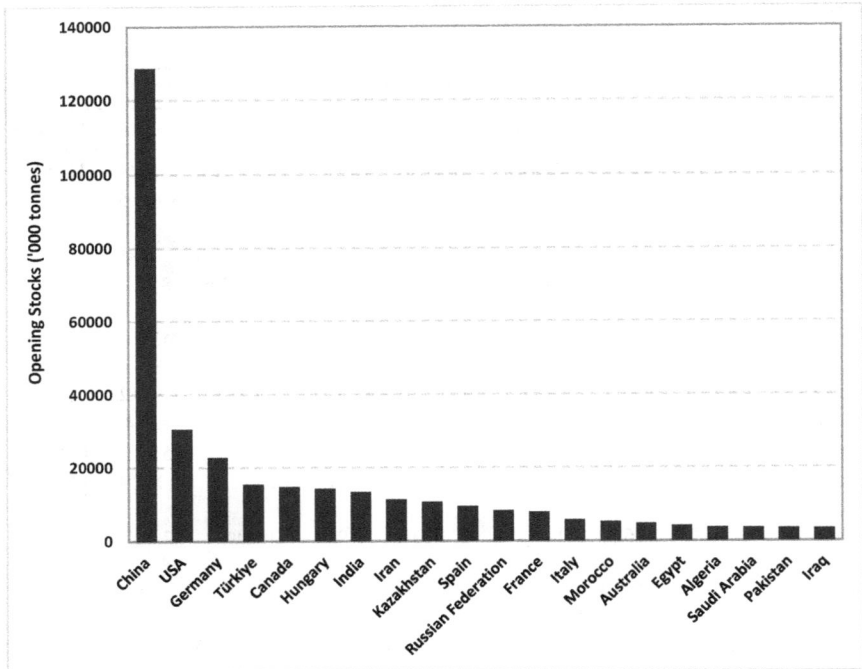

FIGURE 2.6   Top 20 countries in wheat opening stocks (TE 2019).

Figure 2.6 depicts the top 20 countries in wheat opening stocks in 2019 TE, and it can be understood that China alone contributes more than 85% to the global wheat opening stock, followed by the United States, Germany, and Turkey. The highest opening stock of China could be due to the population explosion and to meet their consumption requirement as well as the largest foreign reserve of the country (Li et al., 2014). However, India, being the second largest producer of wheat, registered an opening stock of less than 200 Mt, which sheds light on the need for an efficient management of the wheat supply chain as well as strengthening the existing buffer stock operations (Saini and Kozicka, 2014).

2.7   Trends in Wheat Consumption and Closing Stocks

Globally, wheat is one of the cheapest sources of all the required nutrients and micronutrients; therefore, it is the staple food for a majority of the regions in the world. Asian regions like South Asia and East Asia top the list of the consumption levels of wheat (Table 2.7), followed by the European Union and the Soviet Union. China has been ranked first in terms of wheat consumption (148 Mt) in 2021/22, followed by the European Union (109.75 Mt) and India (107.91 Mt), which could be due to various factors like population growth, surplus production and access to the public, change in consumer preference, and rising income level. (Sendhil et al., 2020;

TABLE 2.7

Trends in Wheat Consumption and Ending Stocks ('000 Tonnes)

Region	2018/19	2019/20	2020/21	2021/22
Consumption				
North America	46,606	47,586	46,809	47,433
South America	28,695	29,465	29,685	29,430
European Union	106,300	107,250	104,750	109,750
Other Europe	20,697	20,421	18,725	20,565
Former Soviet Union – 12	76,435	75,560	78,645	78,110
Middle East	59,919	62,860	64,322	63,640
North Africa	46,100	46,250	46,720	47,020
Sub-Saharan Africa	30,319	32,975	34,583	34,335
East Asia	137,750	138,638	162,368	161,151
South Asia	138,996	139,880	148,436	155,592
Southeast Asia	26,165	26,355	26,030	26,385
Oceania	10,610	9,410	9,435	9,955
Others	3,614	3,694	3,818	3,565
Ending Stocks				
North America	36,030	33,869	28,929	21,823
South America	5,876	6,786	6,651	4,921
European Union	15,798	13,110	10,693	14,361
Other Europe	3,224	3,850	2,497	2,869
Former Soviet Union –12	14,883	13,141	18,139	21,693
Middle East	14,412	17,849	16,851	14,185
North Africa	15,537	14,357	11,142	11,253
Sub-Saharan Africa	2,926	4,220	4,486	4,039
East Asia	141,028	153,011	147,013	144,955
South Asia	21,466	28,203	34,242	28,966
Southeast Asia	6,186	6,240	5,728	5,401
Oceania	4,769	3,029	4,640	5,161
Others	408	548	614	474
Consumption				
Algeria	10,750	10,950	11,150	11,370
Brazil	11,900	11,900	11,800	11,750
Canada	9,120	9,750	9,133	9,500
China	125,000	126,000	150,000	148,000
Egypt	20,100	20,300	20,600	20,500
The European Union	106,300	107,250	104,750	109,750
India	95,629	95,403	102,217	107,911
Indonesia	10,600	10,300	10,300	10,600
Iran	16,100	17,200	17,400	18,200

(Continued)

TABLE 2.7 (*Continued*)

Region	2018/19	2019/20	2020/21	2021/22
Morocco	10,700	10,400	10,400	10,600
Pakistan	25,400	25,500	26,300	27,700
Russia	40,500	40,000	42,500	41,750
Turkey	18,800	20,000	20,600	20,200
Ukraine	8,800	8,300	8,700	10,000
The United Kingdom	15,417	15,196	13,455	15,300
The United States	29,986	30,436	30,476	30,533
Others	177,104	181,459	184,545	183,267
Ending Stocks				
Algeria	5,219	5,358	4,992	3,917
China	138,088	150,015	144,120	141,916
The European Union	15,798	13,110	10,693	14,361
India	16,992	24,700	27,800	21,467
Iran	4,936	4,786	4,336	4,786
Russia	7,778	7,228	11,380	12,088
Ukraine	1,555	1,504	1,505	5,842
The United States	29,386	27,985	23,001	17,962
Others	62,791	63,527	63,798	57,762

Data source: USDA (2022)

Svanidze et al., 2019; Nasurudeen et al., 2006). East Asia occupies the top position in terms of global closing stock of wheat (144.96 Mt) out of which China alone contributes 141.92 Mt (98%). The overall analysis between 2018/19 and 2021/22 indicates that the consumption of wheat across regions witnessed an increase. However, wheat closing stocks show a mixed pattern which is attributed to the change in production and consumption.

2.8 Production and Trade-Related Constraints in Wheat

Agriculture production is significantly affected by several agronomic and socio-economic challenges such as shrinking land base, declining land quality, inefficient resource use, low water use efficiency, and heat intolerance, and pest and disease infestation has made the task of increasing the production even more complex. Although global wheat production has shown an increasing trend over the years, there are still significant gaps existing between the demand and supply of wheat. The adoption of intensive agricultural practices such as indiscriminate use of fertilizer, pesticides, and other inputs has threatened the sustainable production of wheat. In addition, detrimental impacts of climate change have further worsened the situation and made wheat farming uneconomical, especially for small and marginal holders. Delayed sowing in most of the wheat-growing regions, lack of access and availability of quality seeds due to weak institutional linkages, price volatility in the global wheat market, shrinking

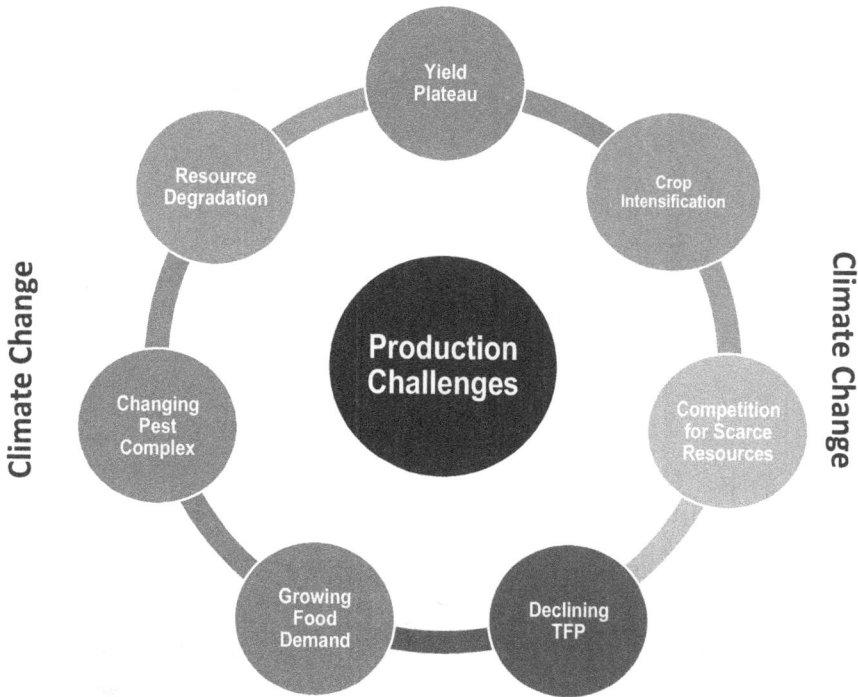

FIGURE 2.7 Challenges in wheat production.

Source: Adapted from Ramadas et al. (2019).

farm holdings, and low resource use efficiency are major challenges and or constraints for wheat production. Further, total factor productivity in technologically advanced wheat growing is declining due to low technical and economic efficiency in resource use (Kandpal et al., 2022). Ramadas et al. (2019) highlighted the major production challenges in wheat in the Indian agricultural landscape (Figure 2.7).

A plethora of empirical studies highlighted that trade can be a powerful engine for the growth and development of the economy (Hallaert and Munro, 2009). Countries that have sufficient infrastructure, efficient transport structure, minimal documentation, and other trade-related requirements often enjoy enormous benefits from the trade (Llanto, 2011). However, it is often challenging for developing nations to realize the maximum benefits from the trade due to various trade-related constraints. Roy and Banerjee (2010) highlighted the key factors which are obstacles to efficient trade and categorized them as *'Gateway issues'* and *'Behind-the-border issues'* (Figure 2.8).

The volume of trade is primarily driven by behind-the-border issues, which are critical for higher growth in trade. In addition to this, Wilson et al. (2004) also identified that port efficiency, custom, regulatory environment, and lack of infrastructure are the key factors that act as impediments to regional trade in agriculture. Further, by employing a similar framework, Llanto (2011) identified the key constraints to trade in the South Asian region to be (i) high tariffs, (ii) inefficient logistics and port operation,

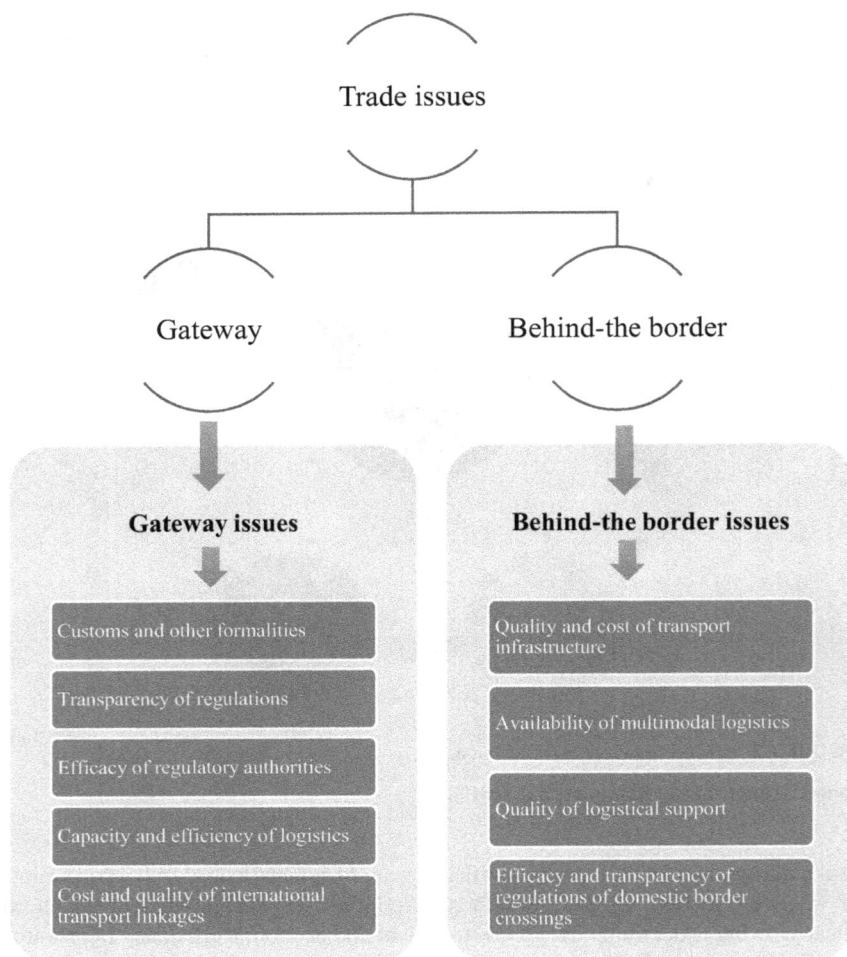

FIGURE 2.8   Key issues in trade.

(iii) the presence of an unfavorable regulatory environment, and (iv) geo-political conflicts. Furthermore, high transaction costs and price volatility in the global wheat
market are other major factors that significantly determine the wheat trade.

2.9 Conclusion and Policy Imperatives

Sustainable intensification through the adoption of conservation practices as well as
ecofriendly technologies in agriculture ensures food security and adaptation to climate change. The global trend from TE 1970 to TE 2020 indicated that the area under
wheat has increased by 10%. Similarly, production (+142%) and productivity (+135%)

in wheat have also shown an increasing trend of more than twofold. Regional-level analysis shows that Asia and European countries are the largest contributors to global wheat production with a steady growth in the respective production levels. India occupies the first position in terms of area, whereas China and New Zealand rank first in terms of production and productivity, respectively. Although the performance of wheat production among different countries is largely satisfactory, yet significant efficiency gaps are observed among major wheat-producing regions. This necessitates specific research intervention for augmenting the productivity level and minimizing the yield gap as well as the realization of higher yield potential. Further, the global wheat export and import have also significantly increased and registered positive growth from TE 1970 to TE 2020. However, although Asia is a major wheat producer, it has not realized the full export potential compared to other regions. It is evident that wheat is a staple food for billions of households with all the required nutrients, and the wheat stock and consumption level have significantly increased all over the regions. To increase wheat production, attention needs to be given to the development of improved varieties which can withstand all situations and make it to reach the farmers' field through extensive outreach programs. Further, policies need to be developed and implemented like incentivizing farmers to promote the adoption of climate-smart farm practices to produce more wheat with fewer resources in the context of climate change. To gain maximum benefits from trade and to increase export competitiveness, domestic support should be given without distorting the trade. Finally, there is a need to develop a framework to create enabling environment for easy access to quality seeds (for producers), affordable nutri-rich cereals (for consumers through an efficient distribution system), investment in R&D (for researchers to realize the genetic gain), and infrastructure at all levels, which strengthens the complete value chain.

REFERENCES

Breiman, A., & Graur, D. (1995). Wheat evolution. *Indian Journal of Plant Sciences*, 43(2), 85–98.

Collins, S. (2007). Treating severe acute malnutrition seriously. *Archives of Disease in Childhood*, 92(5), 453–461.

DES. (2022). Directorate of Economics and Statistics, Department of Agriculture and Farmers Welfare, Ministry of Agriculture and Farmers Welfare, Go vernment of India. https://eands.dacnet.nic.in/

Drewnowski, A., & Popkin, B.M. (2009). The nutrition transition: New trends in the global diet. *Nutrition R eviews*, 55, 31–43. https://doi.org/10.1111/j.1753-4887.1997.tb01593.x.

Erenstein, O., Jaleta, M., Mottaleb, K.A., Sonder, K., Donovan, J., & Braun, H. (2022). Global trends in wheat production, consumption and trade. In M.P. Reynolds & H.-J. Braun (ed s.), *Wheat Improvement*, 47–68. https://doi.org/10.1007/978-3-030-90673-3_4.

FAOSTAT. (2022). *Food and Agriculture Organization of the United Nations*. Rome. www.fao.org/faostat/en/

Hallaert, J.J., & Munro, L. (2009). *Binding constraints to trade expansion: Aid for trade objectives and diagnostics tools*. OECD Trade Policy Working Papers, No. 94, OECD Publishing, France. https://doi.org/10.1787/5kmlbl6glf5d-en

Joshi, A.K., Mishra, B., & Chatrath, R. (2007). Wheat improvement in India: present status, emerging challenges and future prosp ects. *Euphytica*, 157, 431–446. https://doi.org/10.1007/s10681-007-9385-7.

Kandpal, A., Kumara, K., Sendhil, R., & Balaji, S.J. (2022). Technical efficiency in Indian wheat production: regional trends and way forward. In *New Horizons in Wheat and Barley Research*, 475–490. Singapore: Springer. https://doi.org/10.1007/978-981-16-4134-3_17.

Li, Y., Zhang, W., Ma, L., Wu, L., Shen, J., Davies, W.J., Oenema, O., Zhang, F., & Dou, Z. (2014). An analysis of China's grain production: looking back and looking forward. *Food and Energy Security*, 3(1), 19–32.

Lin, H.I., Yu, Y.Y., Wen, F.I., & Liu, P.T. (2022). Status of food security in East and Southeast Asia and challenges of climate change. *Climate*, 10, 40–49. https://doi.org/10.3390/cli10030040.

Llanto, G.M. (2011). *Binding constraints to regional cooperation and integration in South Asia*. Asian Development Bank, Mandaluyong, Philippines. https://www.adb.org/sites/default/files/linked-documents/rcs-south-asia-2011-2015-oth-03.pdf

Nasurudeen, P., Kuruvila, Anil, Sendhil, R., & Chandrasekar, V. (2006). The dynamics and inequality of nutrient consumption in India. *Indian Journal of Agricultural Economics*, 61(3), 362–370.

Nelson, J.H. (1985). Wheat: it's processing and utilization. *American Journal of Clinical Nutrition*, 41(5), 1070–1076.

Nigatu, G., & Motamed, M. (2015). Middle East and North Africa region: an important driver of world agricultural trade. *A Report from the Economic Research Service, USDA, AES-88*. www.ers.usda.gov

Opine, O.O., Jideani, A.I.O., & Beswa, D. (2015). Composition and functionality of wheat bran and its application in some cereal food products. *International Journal of Food Science and Technology*, 50(12), 2509–2518.

Posner, E.S. (2000). Wheat. In K. Kulp, G. Joseph, & J. Ponto (eds.), *Handbook of Cereal Science and Technology*. London: CRC Press.

Ramadas, S., Kumar, T.K., & Singh, G.P. (2019). Wheat production in India: trends and prospects. In *Recent Advances in Grain Crops Research*. London: IntechOpen. https://doi.org/10.5772/intechopen.86341.

Rosegrant, M.W., & Agcaoili, M. (2010). *Global Food Demand, Supply and Price Prospects to 2010*. Washington, DC: International Food Policy Research Institute.

Roy, J., & Banerjee, P. (2010). Connecting South Asia: The centrality of trade facilitation for regional economic integration. *Promoting Economic Cooperation in South Asia: Beyond SAFTA*, pp. 110–138.

Saini, S., & Kozicka, M. (2014). Evolution and critique of buffer stocking policy of India (Working Paper 283). New Delhi: Indian Council for Research on International Economic Relations (ICRIER).

Senapati, N., & Semenov, M.A. (2019). Assessing yield gap in high productive countries by designing wheat ideotypes. *Sci Rep*, 9(1), 5516. https://doi.org/10.1038/s41598-019-40981-0.

Sendhil, R., Kumara, T.M.K., Ramasundaram P., Sinha, M., & Kharkwal, S. (2020). Nutrition status in India: dynamics and determinants. *Glo bal Food Security*, 26, 100455. https://doi.org/10.1016/j.gfs.2020.100455.

Sendhil, R., Kumari, B., Khandoker, S., Jalali, S., Acharya, K.K., Gopalreddy, K., Sign, G.P., & Joshi, A.K. (2022). Wheat in Asia: trends, challenges and research priorities. In *New Horizons in Wheat and Barley Research*, 33–61. Singapore: Springer. https://doi.org/10.1007/978-981-16-4449-8_3.

Sharma, I., Tyagi, B.S., Singh, G., Venkatesh, K., & Gupta, O.P. (2015). Enhancing wheat production – A global perspective. *Indian Journal of Agricultural Sciences*, 85(1), 3–13.

Sharma, P., Dwivedi, S., & Singh, D. (2016). Global poverty, hunger, and malnutrition: A situational analysis. In U. Singh, C. Praharaj, S. Singh, & N. Singh (eds.), *Biofortification of Food Crops*. New Delhi: Springer.

Shewry, P.R., & Hey, S.J. (2015). The contribution of wheat to human diet and health. *Food and Energy Security*, 4(3), 178–202.

Singh, S., Chand, R., Sendhil, R., & Singh, R. (2017). Tracking the performance of Indian agriculture. *Indian Journal of Agricultural Sciences*, 87(12), 1619–1626.

Sramkova, Z., Gregova, E., & Sturdik, E. (2009). Chemical composition and nutritional quality of wheat grain. *Acta Chimica Slovaca*, 2(1), 115–138.

Svanidze, M., Götz, L., Djuric, I., & Glauben, T. (2019). Food security and the functioning of wheat markets in Eurasia: a comparative price transmission analysis for the countries of Central Asia and the South Caucasus . *Food Security*, 11, 733–752. https://doi.org/10.1007/s12571-019-00933-y.

USDA. (2022). United States Department of Agriculture. Washington, DC. https://www.fas.usda.gov/data/

Wang, F., He, Z., Sayre, K., Shengdong, L., Jisheng, S., Feng, B., & Kong, L. (2009). Wheat cropping systems and technologies in China. *Field Crops Re search*, 111(3), 181–188. https://doi.org/10.1016/j.fcr.2008.12.004.

Wilson, J.S., Mann, C.L., & Otsuki, T. (2004). Assessing the potential benefit of trade facilitation: A global perspective (Policy Research Working Paper no. 3224). Washington, DC: World Bank.

3

Pre- and Post-Harvest Management of Wheat for Improving the Productivity, Quality, and Resource Utilization Efficiency

Neeraj Kumar, Ganesh Upadhyay, Krishna Bahadur Chhetri,
Harsha B. R., Gulshan Kumar Malik, Ravindra Kumar, Poonam Jasrotia,
Shiv Ram Samota, Nitesh Kumar, R. S. Chhokar and S. C. Gill

CONTENTS

DOI: 10.1201/9781003307938-3

3.1 Introduction

Rice and wheat are the staple foods which hold the major share in Indian food basket. Presently, the production of rice and wheat in India is estimated to be 130.29 million and 106.84 million tonnes (Mt), respectively, which represent 41.26% and 33.84% share of these crops in total food grains production (DES, 2022). The provision of minimum support price, suitable agro-climatic conditions, and increased mechanization have been the prime reasons for intensive cultivation of rice and wheat crops, especially in northern Indian plains. However, the intensive adoption of rice–wheat system, particularly in Punjab and Haryana states, caused a destructive effect on the soil quality, groundwater table, nutrient balance, and crop productivity in addition to the challenge of residue management (Chauhan et al., 2012; Bhatt et al., 2016; Kumar et al., 2022, 2023). In fact, the critical window period of 20–25 days between the harvest of paddy and sowing of wheat crop, timely unavailability of suitable residue-handling machines, their high cost, and inaptness of rice residue as fodder have been the prime factors responsible for burning of rice residue as wheat sowing beyond 30th November in a year can reduce the wheat yield by 26.8 kg day⁻¹ ha⁻¹ (Tripathi et al.,

2005). The burning of rice residue has been associated with a destructive effect on natural resources, loss of soil microorganisms, reduced efficacy of certain herbicides, and loss of straw retained nutrients. The alternative crop production methods like direct drilling of wheat under residue-free or residue-covered conditions using zero-till drill or conservation agriculture (CA)-based machinery have gained attention due to their ability of timely seeding of crop and cost reduction on the elimination of seed-bed preparation. Further, the integration of zero-tillage with surface retention of crop residue provides multifarious benefits on moisture conservation, weed suppression, reduced soil erosion, improved soil organic carbon, higher water-holding capacity, and reduced soil bulk density (Mondal et al., 2019; Kumar et al., 2020; Saurabh et al., 2021). Jat et al. (2019) found 36% increase in system productivity, 43% augmentation in net profit, and 31.2% rise in system protein yield along with improved soil quality and 33% savings of irrigation water under CA-based rice–wheat–mung bean system over conventional rice–wheat system. It is clear that recycling crop residue back to soil as surface mulch adds multiple benefits to natural resources as well as growers. However, crop residue also serves as food and shelter for many insect-pests, requiring timely and adequate measures for their proper control, failing which a penalty on yield and grain quality may be observed. The yield losses in wheat due to animal pest, pathogens and viruses could be as more as 8.7, 15.6 and 2.5%, respectively, depending upon the severity (Oerke, 2006). Moreover, the options of weed control by tillage operation are left out under CA-based cultivation, and it requires proper weed-control strategies to realize the similar crop yield and proper grain quality as with conventional tillage practice. The yield losses in wheat due to weeds can be as high as 80% depending upon weed infestation (Chhokar et al., 2012).

The other management practices on nutrients, water, and harvesting and threshing also impart significant role in yield, wheat quality, and profitability (Sarkar et al., 2013). In fact, the adoption of recommended pre-harvest management practices is prerequisite to obtain healthy grains, making them suitable for further processing and value addition. Like pre-harvest, post-harvest management also plays a crucial role in post-harvest losses, supply chain, quality of wheat product, and overall profitability (Grover and Singh, 2013; Amentae et al., 2017). Kalsa et al. (2019) reported that loss due to insects in wheat stored in farmers' storage can be as high as 14%. In India, inadequate management of food grains causes post-harvest losses of 7–10% of total production from farm to market level and 4–5% loss at market and distribution levels (World Bank study, 1999). It can be understood that both pre- and post-harvest management practices have decisive roles in ensuring the supply of quality wheat product to the consumers. In this chapter, an attempt has been made to provide the highlights on pre- and post-harvest management of wheat for improving productivity, quality, and resource utilization efficiency.

3.2 Pre-Harvest Management in Wheat

3.2.1 Seasonal Pattern of Wheat

Wheat is a *Rabi* crop, and its growing span varies according to agro-climatic conditions. The variation in agro-climatic conditions affects vegetative and reproductive time of wheat, which produces a significant effect on potential yield. The sowing and harvesting time of wheat in different zones are presented in Table 3.1.

TABLE 3.1

Sowing and Harvesting Time of Wheat (*Triticum aestivum*) in Different Zones

Zone	Jan	Feb	Mar	April	May	June	July	Aug	Sep	Oct	Nov	Dec
NHZ*											(05-11 Nov) (26 Nov-02 Dec)	(17-23 Dec)
NWPZ#	(01-07 Jan)										(05-11 Nov)	(10-16 Dec)
NEPZ$	(01-07 Jan)										(12-18 Nov)	(10-16 Dec)
CZ^											(12-18 Nov)	(03-09 Dec) (24-31 Dec)
PZ**											(05-11 Nov) (26 Nov-02 Dec)	(17-23 Dec)

Legend: Timely Sowing | Late Sowing | Very Late Sowing | Harvesting

Notes: *Northern Hills Zone, #North Western Plain Zone, $North Eastern Plain Zone, ^Central Zone, **Peninsular Zone

3.2.2 Selection of Suitable Cultivars

The selection of appropriate wheat cultivar according to sowing, irrigation, and climatic conditions is a preliminary step to realize the higher grain yield. The cultivation of wheat cultivar in a different zone than recommended results in declined yield due to alteration in vegetative and reproductive period of wheat. The new high-yielding cultivars of wheat require additional fertilizer quantity above the recommended dose and spray of growth retardant for the maximization of the yield with reduced risk of lodging. The performance of a particular wheat cultivar is influenced by various pre-harvest management practices, which also bridge the gap between actual and potential yield. The major pre-harvest management practices pertaining to wheat are discussed in the following sections.

3.2.3 Land Preparation

The feasibility and judiciousness in crop production can be ensured with the increase in crop rotation intensities, timely preparation of seedbeds, decrease in drudgery, and effective utilization of various resource inputs and fuel expenditure through energy-efficient machinery. The use of resource-saving machinery, for instance, laser land levellers, rotary tillage implements (rotavator, power harrow, and powered discs), and the combinations of active–passive and passive–passive configured implements, is very useful for the timely preparation of seedbed, reducing the harmful effects of multiple vehicular passes on soil structure and minimizing the fuel expenditure costs, thereby resulting in more profit to the farmers. The concept of properly matching the tractor horsepower with the implement size and operating it at the optimal setting aids in lowering the time, manpower, and fuel costs in comparison to the traditional practices of seedbed preparation (Alam, 2000; Mehta et al., 2011; Mrema et al., 2014; Petrov et al., 2020; Nataraj et al., 2021). Laser-guided land levelling is one such modern sustainable practice, which has made a remarkable contribution in conserving irrigation water. The traditional leveling implements leave some areas of the field with high elevation, causing these regions to suffer from water stress conditions, while some areas of the field having low elevation undergo surplus water conditions. It causes the variability in water and nutrients applied to crops, resulting in uneven crop stand and yield. These problems can be addressed by the laser land leveler which removes the soil from areas having higher elevation and fills it in the area of lower elevation, thereby precisely leveling the farm within ±20 mm of average farm elevation. The adoption of laser land leveling helps in reducing irrigation water demand (20–30%), abiotic stresses, and energy inputs in addition to raising agricultural productivity (10–20%) and profitability (Jat et al., 2006; Kaur et al., 2012; Shahani et al., 2016). Chen et al. (2022) witnessed 23.2% higher yield of winter wheat along with 30–40% increase in net-return through an integrated approach of precision land leveling along with precision seeding over conventional farmers' practice. Amidst climate change and declining resources, the adoption of laser land leveling followed by resource-conserving implements for seed-bed preparation and sowing would be beneficial to farmers to reduce input costs and improve net-profitability.

The PTO-operated rotary tillage machinery, for instance, rotavator/rotary tiller, power harrow (vertical axis), or powered-discs, are widely used by farmers throughout the world as they enable fine soil pulverization and proper mixing of manures and fertilizers. Multiple tillage passes given in the traditional seedbed preparation, namely ploughing, disking, cultivating, and planking can be substituted by the reduced passes of rotary tillage implements (Marenya and du Plessis, 2006; Upadhyay and Raheman, 2020a; Balsari et al., 2021; Hensh et al., 2021). Rotavator is reviewed to be more fuel-efficient and time-saving in contrast to the passive-type machinery with the additional benefit of its ability to carry out the puddling operation in wetlands (Destain and Houmy, 1990; Prasad, 1996; Kankal et al., 2016; Choudhary et al., 2021). The PTO-operated disc-type implements outperform the passive disc-type implements with higher residue incorporating ability, pulverizing index, operating depth, and fuel-energy utilization along with less draft, slippage, and required field trips (Islam et al., 1994; Nalavade et al., 2010; Upadhyay and Raheman, 2018, 2019).

Several scientists worked on the different configurations of active–passive and passive–passive tools/implements for seedbed preparation and found promising results (Sarkar et al., 2021; Bovas et al., 2022). Integrating at least two tillage operations together can lower the operating costs for seedbed preparation and saves time. As the combined or integrated machinery could establish seedbeds in less field trips, this would further assist in reducing the subsoil compaction issue caused by tractor–tire traffic. The passive–passive configurations outperform the traditional practices with respect to energy consumption, time, and operating costs (Watts and Patterson, 1984; Sahoo, 2005; Raheman and Roul, 2013; Alkhafaji, 2020). Moreover, with active–passive combinations, the negative draft generated by the powered tools/implements and tractor-engine power is efficiently utilized. These integrations are energy-efficient and demand fewer tillage field trips than traditional implements to attain the proper seedbed quality (Shinners et al., 1990; Anpat and Raheman, 2017; Upadhyay and Raheman, 2018, Usaborisut and Prasertkan, 2019). Kailappan et al. (2001) reported cost savings of 44–55% and time-saving of 50–55% in preparing seedbeds with an integrated machinery. The active–passive combination machinery by Usaborisut and Prasertkan (2019) reduced the draft and energy demand of subsoiling by 4.4–11.3% and 10.5–15.3% compared to the subsoiler.

3.2.4 Sowing/Planting

In the Indian Indo-Gangetic Plains (IGP) region, the rice–wheat cropping system is extended over a wide region stretching from Punjab to West Bengal, and it has performed a crucial part in the Indian food security. In IGP, rice is generally grown in the warm, humid season (May/June to October/November) and wheat in the cool, dry season (November/December to February/March) (Gupta, 2003). The time window between paddy harvesting and wheat sowing in northwest India is about 7 to 10 and 15 to 20 days for Basmati and coarse rice grain varieties, respectively (IARI, 2012; Lohan et al., 2018). Combine harvesters cut the paddy crop at a particular height over field surface, thus generating two distinctive crop residue components i.e. anchored stubble and heaps of loose straw. An immense load of loose straw about 7.5 tonnes per hectare is left on the surface. Due to lack of availability of effective machinery and techniques for crop residue handling and for establishing the subsequent crop, farmers favour its *in-situ* burning on the farm. Furthermore, repeated puddling operations for the sowing of paddy crop result in the development of a sub-surface hardpan having an increased penetration resistance and reduced hydraulic conductivity, which impede the transfer of water/air and further hamper the establishment of subsequent crop roots (Aggarwal et al., 1995; Hassan and Gregory, 1999; Timsina and Connor, 2001; Kukal and Aggarwal, 2003). Subsoil compaction reduced water and nutrients use efficiency of wheat followed after rice by 38% and 22%, respectively, resulting in reduced root length density chiefly because of a decrease in nutrient uptake (12 to 35% for nitrogen, 17 to 27% for phosphorous, and up to 24% for potassium) (Sur et al., 1980; Ishaq et al., 2001).

Many scientists have examined the impact of late sowing on phenology and wheat crop production. Sowing in mid-to-late November results in lower wheat yield because of elevated temperatures at the time of the grain-filling phase. Timsina et al. (2008) reported that, in general, sowing wheat beyond 15 November in Punjab, India, reduces the grain yield between 0.25% to 0.75%, 0.5% to 1.0%, and 0.8% to 1.0% for varieties having yield estimation under 4 t ha^{-1}, 4 to 6 t ha^{-1}, and greater than 6 t ha^{-1}, respectively. Singh et al. (2016) carried out a simulation study to analyse the effects of mulching and sowing date on wheat grain yield in clay-loam and sandy-loam soils at Ludhiana, Punjab. Their findings suggested that the optimal sowing date for wheat grain is affected by both soil type and the presence/absence of mulch. Further, the management of residue after rice harvest plays a prominent role in timely seeding of wheat. The simultaneous task of paddy straw handling and direct wheat seeding can address both challenges of residue management and timely seeding.

Roto seed drill is an important implement for an effective handling of paddy crop residues and timely seeding of wheat. Sharma et al. (2008) and Dixit et al. (2014) reported that the Roto seed drill is beneficial for sowing wheat after paddy harvesting saving time and fuel costs as compared to traditional operation. However, it operates successfully in fields free from loose straw and leaves only standing stubble or little quantity of loose straw. The necessity of partial elimination of loose straw from paddy farms before wheat sowing is the major constraint of this machine (Singh, 2016). In recent times, resource-conserving machines such as Turbo Happy Seeder (THS), super seeder, and strip-till drill have been used for residue management and timely seeding of wheat and other crops.

In THS, the cutting and shredding of straw are accomplished with inverted 'γ' type serrated flails fitted on a rotor (1,000 to 1,300 rev min⁻¹) of the residue management drum. These flails shear standing stubbles near the ground and also throw the residue over saw-edged blades fitted on the interior section of the management drum to achieve fine chopping and shredding. Sidhu et al. (2015) observed an improved yield of wheat sown in paddy stubble fields with the THS as compared to the conventional practice of straw burning followed by tillage and sowing operations. The use of THS also decreased fuel expenditure and ensured the optimum sowing window and reduction in the irrigation requirement. Despite the aforementioned merits, the farmer's acceptance towards THS is slow. Sharma et al. (2008) and Chhokar et al. (2018) identified the inadequacy of THS in wet residue, more energy demand, less time availability, and incompatibility in direct seeding of wheat on sugarcane-trashed fields as the primary obstructions in its adoption under wheat–sugarcane cropping systems.

Super Seeder is tillage-cum-sowing machinery equipped with a rotavator fitted with 'LJF' blades at the front, followed by sowing and compaction units at the rear. It simultaneously prepares the seedbed and sows the wheat crop. Additionally, it can be used in residue-covered field, thereby incorporating the crop residue into soil followed by seeding operation and compaction of the soil. The shape of 'LJF'-type blades gives the benefit of gradual increment in bite-width as opposed to constant bite-width in traditional 'L'-type blades during their impact on the ground. The 'LJF' blade shape lessens the chances of choking of blades with paddy straw and does its effective incorporation in the soil. But, for its efficient operation, tractors having engine power greater than 60 hp are required making it uneconomical for small and medium farmers compared to the THS and strip-till drill.

Strip-till drill is a tractor's PTO-operated machinery for the direct sowing of various crops. It comprises 'J' type blades, which provide tillage in narrow strips required for proper seeds and soil contact and more favourable germination. It is an intermediate option between traditional and no-till practices and provides the advantage of improved seed germination as with normal tillage operation in direct seeding of crops without complete coverage of soil. Gangwar et al. (2006) witnessed a greater mean wheat yield with reduced (strip) tillage over conventional and zero tillage options under sandy loam soil. The strip-till drill permits the farmers for advanced sowing of wheat in paddy-harvested fields with comparable or greater yield as compared to traditional practice (Chaudhary and Singh 2002; Shukla et al., 2008; Hossain et al., 2012).

3.2.5 Nutrient Management

In the past few decades, excessive or blind application of fertilizers has synced the food production with human consumption level, but it has depleted the resources and also caused the environmental problems. Over application of nutrients has increased the yield but with decreasing rate which has reduced the nutrient-use efficiency (NUE). Decrease in NUE has exploited the resources to the next level. Soil-testing-based precise approach of nutrient application with the right amount at the right place in the right way at the right time is necessary to enhance NUE and yield of wheat. Quality of wheat can be increased by selecting a variety which is rich in qualitative parameters *viz.*, protein and micronutrients (Fe and Zn). Supplying the deficient nutrients to the crops is also a prerequisite to overcome the problem of nutrient deficiency in the food chain at the human level. The nutrient management of wheat should be done by following steps for improving productivity, quality, and resource utilization efficiency.

3.2.5.1 Soil Testing

It is the basic and primary requirement for proper application of fertilizers. Primary nutrients, *viz.*, nitrogen, phosphorus and potassium, must be applied on the basis of soil testing. Other deficient nutrients could be applied either by soil testing analysis or based on visual plant deficiency symptoms.

3.2.5.2 Source and Amount of Fertilizers

Nutrient-use efficiency of conventional fertilizers remains within 30–35%, 18–20%, and 35–40% for N, P, and K, respectively. So the use of slow-release fertilizers, smart fertilizers, nanofertilizers, and customized fertilizers could bring up the nutrient-use efficiency. Slow-release fertilizers like neem-coated urea have more NUE as they slowly release in the soil and meet the demand of crop for a longer period. Smart fertilizers have an emphasis on controlled-release systems to synchronize nutrient availability with the plant demands, thus reduce losses to the environment by enhancing the nutrient-use efficiency. Nanofertilizers have a role in improving the nutrient-use efficiency by exploiting unique properties of nanoparticles within the range of nano-dimension from 1 to 100 nm. Customized fertilizers make it easier to apply a whole spectrum of required plant nutrients in the proper proportions for different stages of crop growth and development. Multi-nutrient carriers including macro, secondary, and/or micronutrients from both inorganic and organic sources are referred to as customized fertilizers.

Other than recommended dose of fertilizer (150 kg nitrogen, 60 kg phosphorus, and 60 kg potassium per hectare) approach, soil test crop response (STCR) and nutrient expert approach can be used to calculate the required amount of fertilizers in wheat. This approach not only provides appropriate amount of fertilizers to the wheat but also gives the maximum yield (Table 3.2).

TABLE 3.2

Different Tools for Real-Time Nitrogen Management in Wheat

Tools	Critical Values	% Saving of N	Effect on Grain Yield	References
Leaf colour chart (LCC)	≤4	30%	No significant difference	Reena et al. (2017)
SPAD meter	42	18.8%	13.4% yield increment	Ghosh et al. (2020)
GreenSeeker	0.66	20%	No significant difference	Gupta (2006)
Soil test crop response (STCR)	FN = 5.31 T − 0.51 SN (New Delhi)	-	-	AICRP on STCR
Nutrient expert (NE)	As per crop advisor (NE)	-	16% yield increment	Pampolino et al. (2012)

3.2.5.3 Method of Application of Fertilizers

Nitrogen is the most limiting nutrient of the wheat. Whole nitrogen should not be applied as a basal dose because of its high mobility in the soil which makes it deficient in later stages. It should be applied in two to three splits. Nutrients, *viz.*, P (essential for root development and tillering) and K (quality nutrient), should be applied as a basal dose as per the requirement of the crop based on soil testing. There is no need to supply sufficient nutrients like calcium, magnesium, and some micronutrients. Sulphur can be applied as basal dose and foliar spray if there is drop in N:S below 15–16:1. On the soil testing or deficiency symptoms of the crop, micronutrients can be applied either by mixing with macronutrients or by foliar spray.

3.2.5.3.1 Fertigation

It is the process of the application of fertilizers with irrigation water, which allows for the application of small amounts of fertilizer per fertigation event, allowing for greater fertigation frequency flexibility. In fertigation, fertilizer solution is evenly distributed in irrigation. The availability of nutrients is abundant to the crop plants, so efficiency is increased. In fertigation, liquid and water-soluble fertilizers are used. Through fertigation, nutrients and water are delivered near the active root zone, resulting in a greater crop absorption. Because fertigation distributes water and fertilizer evenly across all crops, an increase in 25–50% yield is possible. The fertilizer application through fertigation technique demonstrates 80–90% fertilizer use efficiency, resulting in the saving of more than 25% nutrients. Along with saving water and fertilizers, time, labour, and energy are all significantly reduced as a result of this method. In case of nitrogen fertigation, urea is an excellent choice for use in micro irrigation systems. It's very soluble and dissolves in a non-ionic form, so it won't react with other compounds in the water. Urea also has no effect on precipitation. In drip fertigation, nitrogenous fertilizers such as urea, ammonium nitrate, ammonium sulphate, calcium ammonium sulphate, and calcium ammonium nitrate are employed. In P fertigation, phosphate salts may precipitate when phosphorus is added to irrigation water. Fertigation tends to be better with phosphoric acid and mono ammonium phosphate. Potash application through fertigation does not cause any kind of precipitation if used in the forms of KNO_3, KCl, K_2SO_4, and mono potassium phosphate. Micronutrients, *viz.*, Fe, Mn, Zn, Cu, B, and Mo, can also be used in drip fertigation.

3.2.5.4 Time of Application

All primary nutrients which are required in higher quantities can be applied as basal dose except nitrogen. Though nitrogen, being the most limiting and highly mobile nutrient in the soil, is required appropriately at different growth stages of crop for maximizing the yield and enhancing its use efficiency, so, the most efficient use of nitrogen may be realized by matching N supply with crop need, which means applying the proper quantity of N at the optimal physiological stage of nutrient requirement. A potential approach is to use leaf colour chart (LCC), chlorophyll metre (or SPAD metre), GreenSeeker (NDVI meter), and others to assess the plant's N demands and then adjust the time of N administration in wheat (Table 3.2).

3.2.5.4.1 Leaf Colour Chart

The leaf colour chart (LCC) is an inexpensive tool for determining the nitrogen demand of crop in real time. LCC indicates the nitrogen deficiency based on visual observation of differences in colour intensity of stripes and plant leaves. LCC contains six shades of green colour ranging from yellowish green to dark green. The LCC score increases with the intensification of green colour. If the mean score of at least six leaves falls below critical value (LCC of 4), a dose of 20–25 kg N ha^{-1} is applied. Then, LCC reading at 7 days' interval is repeated till 50 days after transplanting. The application of nitrogen @50 kg ha^{-1} as basal and 30 kg ha^{-1} as top dressing in wheat with LCC reading less than 5 recorded 17.7 and 51.5% higher grain yield over recommended dose of nitrogen (100 kg ha^{-1}) and zero nitrogen, respectively (Dineshkumar et al., 2013). Singh et al. (2014) reported that LCC-based nitrogen fertilizer application produced equivalent or more grain yield than the blanket N recommendation with the average saving of 29 kg N ha^{-1} in wheat. LCC optimizes the nitrogen use in wheat with similar or high yield levels irrespective of the source of nitrogen (Parihar et al., 2018).

3.2.5.4.2 SPAD Metre

The SPAD chlorophyll metre is excellent for quickly determining the chlorophyll content and nitrogen status of crops. Soil Plant Analysis Development (SPAD, Minolta Camera Co., Osaka, Japan) chlorophyll meter is a fast and non-destructive method which allows *in-situ* measurement of chlorophyll content (N status). SPAD meter measures the leaf colour as a representation for leaf N using green colour intensity as an indicator. The readings from SPAD are taken at 7- to 10-days intervals, starting from 21 days after sowing for wheat continues up to the first (10%) flowering. Readings should be taken on one point midway between the leaf base and tip. The readings thus obtained are utilized for nitrogen application on the basis of two approaches.

4.   i. Sufficiency index approach: If the ratio of SPAD meter reading value in the question plot to well fertilized reference plot lies below 0.9, then the nitrogen is applied in the question plot (Hussain et al., 2000).

   ii. Threshold value approach: In this approach, a dose of nitrogen is applied when SPAD value falls below threshold value [critical SPAD value = 42 for wheat as suggested by Bijay-Singh et al. (2002)].

In a study, Bijay-Singh et al. (2002) observed the maximum tillering and 20% higher wheat yield on applying 30 kg N ha^{-1} based on SPAD meter reading over conventional practice. Khurana et al. (2008) used a critical SPAD value of 42 to show the significant gains in agronomic (63%) and apparent recovery (59%) in N efficiency as compared to farmers' practice.

3.2.5.4.3 GreenSeeker

GreenSeeker is a hand-held optical sensor which works on the basis of the reflectance of infrared light projected on a specified crop area of 0.61 m × 0.61 m when placed 0.6 to 1 meter above the target area (Bijay-Singh et al., 2011). It is well correlated to leaf area index so it measures the photosynthetic efficiency of plants by sensing. The values of NDVI (normalised difference vegetation index) over crop canopy in

GreenSeeker range from 0 to +1.0. This tool is also used for real-time nitrogen application in wheat. GreenSeeker-based strategy nitrogen fertilizer application in wheat saves 82% of nitrogen with statistically similar yield over farmers' practice (Li et al., 2009). GreenSeeker-based strategy has achieved the highest agronomic efficiency 111% for winter wheat over other nitrogen management practices (Cao et al., 2017).

3.2.6 Water Management

Being a *Rabi* season crop, wheat requires comparatively less water than the *Kharif* season cereals. The water requirement of wheat is 45–65 cm and varies depending on soil and climatic factors. The rainfall received during wheat growing period (long-term average = 12 cm) does not fulfill total water requirement of crops; hence, irrigation is necessary to fulfil the remaining needs of water (Singh et al., 2016). Excessive and improper uses of water resources in the past decades have depleted the groundwater and other resources. There is a need of agronomic practices which will not only enhance the water-use efficiency of crop but also sustain the water resources from over-exploitation. These include conjunctive use of water, reduction of water losses through evapotranspiration (ET) and percolation, proper irrigation scheduling, use of efficient irrigation methods and weather-forecasting tools for proper water management.

3.2.6.1 Conjunctive Use of Water

In some countries, there is scarcity of fresh water, so conjunctive use of saline water and fresh water with proper management practices enhances the water productivity without compromising the yield of wheat (Mojid and Hossain, 2013).

3.2.6.2 Methods to Reduce Water Losses

In arid and semi-arid regions, water losses takes place mainly by two ways, viz., evapotranspiration (out of total evapotranspiration, transpiration causes 74 to 76%, while evaporation causes 24 to 26% loss as reported by Tfwala et al. (2021)) and percolation (excessive application of water by traditional flooding in water-sufficient areas).

Evapotranspiration losses could be minimized by using the water-use-efficient wheat variety and mulching. The selection of a high water-use-efficient variety could be helpful in proper utilization of water and also enhances the grain yield. Straw mulching, a possible practice with resource conservation tools, could provide many benefits to wheat by conserving soil moisture, maintaining soil temperature, enhancing nutrient status in soil, preventing soil and water loss, improving soil microbial activities, and also by reducing weed flora. Mulching of wheat boosts grain yield and water efficiency as compared to non-mulched wheat due to improved soil physical and chemical properties, as well as increased soil micro fauna activity by increasing soil moisture (Chen et al., 2015).

Percolation loss is a result of flood irrigation which is a conventional method of irrigation. Flood irrigation method causes deep drainage below root zone causes water use inefficiency and temporary waterlogging leads to aeration stress in wheat (Kukal and Aggarwal, 2003). Excess water in the field leaches down the precious nutrients and bring anaerobic situation for a short period of time. To maximize water-use-efficiency, careful irrigation scheduling is necessary with appropriate amount of water, as flood irrigation reduces water use efficiency and sometimes productivity also (Qiu et al., 2008). The appropriate irrigation scheduling improves crop yield and water conservation and helps in the consumption of applied inputs properly. The first approach relies on irrigating the crop at critical growth stages of crop, while second focuses on irrigating the crop on the basis of soil moisture depletion in the field, regardless of growth stages. Irrigation requirements varies from soil to soil which may be six to eight irrigations in sandy loam soil while three to four irrigations in heavy clay soil sown wheat (Kumar, 2009). The stages like crown root initiation (21–25 days after sowing), tillering (40–45 DAS), jointing (60–65 DAS), flowering (80–85 DAS), milking (105–110 DAS), and dough (120–125 DAS) are the critical growth stages for irrigation scheduling. Application of 45 mm of water rather than 60 mm at every stage could also save the irrigation water without any significant loss in wheat yield (Meena et al., 2019). Apart from the time of irrigation, the methods of irrigation application and crop cultivation also help in conserving the water along with improved yield. The cultivation methods like furrow irrigated raised bed system and micro-irrigation systems such as sprinkler and drip irrigation are beneficial for enhancing water-use efficiency and productivity of wheat as discussed later in the chapter.

3.2.6.2.1 Furrow Irrigated Raised Bed System (FIRBS)

FIRBS method reduces seed and nitrogen requirements by 25% and irrigation water usage by up to 40%. Flood irrigation is avoided in this approach so the water is applied in the furrows which save water thus increasing the water-use efficiency, and fertilizer is applied exclusively to the raised beds.

3.2.6.2.2 Sprinkler Irrigation

Flood irrigation generates uneven water distribution, water loss through seepage and deep percolation, excessive weed growth, salinization, and waterlogging, all of which have an impact on land and crop output. Sprinkler irrigation can achieve great irrigation efficiency, which is not always attainable with surface irrigation systems. This approach has a high irrigation efficiency, making it suitable to hilly terrain and light soil, and can save up to 60% of water (Shankar et al., 2015). Liu et al. (2013) found that irrigating wheat through sprinkler method reduced evapotranspiration by 4 to 23% and recorded more crop water productivity (18–57%) and irrigation water productivity (21–81 %) over surface-irrigated wheat. Sprinkler also recorded 4.0 to 22.1% more dry matter accumulation over surface irrigation. Mini-sprinkler irrigation under zero tillage with residue saved 43.3 and 25.0% of water over surface-irrigated wheat under zero tillage with residue and conventional tillage without residue (farmers' practice), respectively (Singh et al., 2021). Mini-sprinkler irrigation under zero tillage also recorded similar yield with farmers' practice.

3.2.6.2.3 Surface Drip Irrigation

Drip irrigation, also known as trickle irrigation, is one of the most effective irrigation methods. It comprises a number of hollow plastic pipes of small diameter having emitters or drippers on the surface through which water is dropped into soil at very slow rate (3–15 litres h^{-1}). Drip irrigation has enhanced the yield and productivity of some crops (particularly spaced crops), labour and cost savings, energy savings, less pumping hours and therefore, simpler irrigation, better crop development, and improved soil health, in addition to water savings. Drip irrigation has more overall irrigation efficiency (90–95%) than sprinkler irrigation (75%). Drip irrigation could save more than 20% of irrigation water as compared to surface-irrigated wheat (Kharrou et al., 2011). Chouhan et al. (2014) reported that drip-irrigated wheat recorded 12.14 and 21.76% more grain yield and water productivity than sprinkler-irrigated wheat, respectively.

3.2.6.2.4 Sub-Surface Drip Irrigation

Sub-surface drip irrigation is the regular discharge of small amounts of water below the soil surface from discrete emission points or line sources. Sub-surface perforated pipes leak water as it flows by gravity, and irrigation water is delivered to the crops via soil capillary. Sub-surface drip irrigation minimizes evaporation losses, allowing better water and nutrients' application to the root zone, which leads to more enhanced fertilizer and water-use efficiency, reduces weed growth, and lowers labour costs. It also makes cultural farming activities easier in sub-surface drip irrigation than in surface drip irrigation (Ayars et al., 1999). Due to lower evaporation losses, sub-surface drip irrigation has lowered ET by 26% compared to flood irrigation and 15% compared to surface drip irrigation, resulting in an increased grain yield and biomass (Umair et al., 2019). Sidhu et al. (2019) reported that use of irrigation water reduced by 42–53% in sub-surface drip irrigation (laterals spacing of 67.5 cm at 15-cm depth) under conservation agriculture based wheat cultivation as compared to a flood irrigation in the same crop establishment method. Sub-surface drip irrigation has improved crop and irrigation water productivity by 25 and 20% over flood irrigation, respectively.

Other than time and application of irrigation, judicious and precise application of irrigation could be achieved by using weather-forecasting tools. By knowing upcoming rainfall, smart application of irrigation could be done which will not only save the water but also increase the grain yield by creating less waterlogged situations in the field.

3.2.7 Weed Management Strategies for Wheat Crop

Among various pests, weeds are considered as major biotic limiting factor in crop production. Weeds, besides reducing the yield and the quality of the produce, also hinder in harvesting operations. Globally, weeds cause about 18 to 29% yield reduction in wheat, subjected to weed population, type of weed flora, and its infestation duration (Oerke, 2006). Wheat, in general, is infested with both grass and broadleaved weeds. The most common and economically troublesome monocot and dicot weeds in wheat-based cropping systems include *Phalaris spp., Avena spp., Lolium spp., Alopecurus spp., Bromus spp., Chenopodium album* L, *Raphanus raphanistrum* L., and *Kochia scoparia* L. (Collavo et al., 2011; Chhokar et al., 2012). Weeds create competition for the resources, and the lodging nature of some weeds like *Phalaris minor* wanes yield as well as quality of wheat. *Phalaris minor* competes more fiercely for light because it has more and larger leaves than wheat (Malik and Singh, 1993). Lodging also creates difficulty in harvesting operations and lowers production and quality by shrivelling the grains. Mixing of weed seeds in wheat grains also decreases their export to foreign countries as a quarantine measure. Therefore, integrated weed management practices are essential for proper and timely control of weeds in wheat. The weed management approach is generally classified into three categories, viz., cultural, mechanical, and chemical weed control as discussed next.

3.2.7.1 Cultural Methods of Weed Management

In order to effectively manage the weeds in wheat, a multi-layered hierarchy of cultural, mechanical, and chemical techniques is used. Different and frequently inconsistent methods of cultural weed management exist, which are detailed in Table 3.3.

TABLE 3.3

Effect of Different Cultural Practices on Weed Management

Practice	Weeds Controlled	Details	References
Stale seedbed technique	Most of weed flora	Stale seed bed technique controlled 28.6% of weeds (*Avena fatua, Phalaris minor, Chenopodium album*) and enhance the wheat production by 5%.	Safdar et al. (2011)
Zero tillage	*Phalaris minor*	Zero tillage practice reduced the *P. minor* by 40–45% and contributed to 25–30% in yield increment.	
Competitive verities	A significant range of weed flora	More competitive varieties having fast germination, more leaf area index, taller stature, improved root spread, and higher allelopathic effect suppress the weeds more efficiently.	Jabran and Farooq (2013), Blackshaw et al. (2002)
Sowing time adjustments	*Phalaris minor Avena ludoviciana*	Early sowing of wheat reduced *P. minor* by 15–20% and increased the grain yield by 10–15%, while late sowing reduced the infestation of *Avena ludoviciana*.	Singh et al. (1995, 1999)
Mulching	*Phalaris minor, Poa annua, Capsella bursa-pastoris, Stellaria media*	Mulching in wheat has potentially reduced the weed population by 38–80%, which contributed to yield increment by 12–32%.	Batish et al. (2007)
Crop rotation	A significant range of weed flora, *Avena fatua* L.	*A. fatua* was significantly lowered while shifting from rice–wheat cropping system (67%) to sugarcane–vegetables–wheat (9%) and cotton–pigeon pea–wheat (16%).	Malik and Singh (1995)
Allelopathy effect	*Avena fatua* L., *Phalaris minor*	Mulberry extracts resulted in a complete inhibition of the wild oat and *Phalaris minor* germination.	*Jabran* et al. (2010)
Irrigation time	*Chenopodium album* L., *Chenopodium murale* L.	Pre-sowing irrigation lowered the dry weight of *Chenopodium album* and *Chenopodium murale* by 21 and 25%, respectively.	Singh and Singh (2004)
Fertilizer management	*Avena fatua, Sonchus arvensis, Setaria viridis C. album*	Low-nitrogen fertilized plots had less occurrence of weeds and enhanced the grain yield by 10%.	Blackshaw et al. (2004)

3.2.7.2 Mechanical Weed Management

It consists of the uprooting and removal of weeds using a variety of tools and instruments. Manual weeding requires a lot of manpower and time. Its profitability is very poor because labour is scarce and expensive. Moreover, weeds that morphologically resemble crops, such as *P. minor* and *Avena ludoviciana* before flowering, make mechanical weeding challenging. However, when wheat is seeded in FIRBS or in lines in a flatbed system, mechanical control can be used efficiently.

3.2.7.3 Chemical Weed Control

Chemical weed control is recommended because it is more effective, less expensive, and takes less time. Additionally, it avoids the crop's mechanical harm that results from mechanical weeding. Furthermore, the control is more effective because during mechanical control, the weeds that invariably escape due to morphological similarity to crop are killed even within the rows. Choosing the right herbicides for the type of weed flora that is infesting the crop and applying the herbicides at the right amount and timing are both essential for effective weed control (Table 3.4). However, the adoption of zero tillage agronomic practice and repeated dependence on the same group of herbicides for weed management evolved herbicide-resistant weeds in wheat. The report of herbicide-resistant weed species given in International Survey of Herbicide Resistant Weeds by Heap in 2022 ranked wheat first in the number of herbicide-resistant weed species (a total of 77 weed species herbicide-resistance cases globally). Dependency on new herbicides is also limited in some countries due to unawareness, government policies, and higher costs which divert out the focus towards integrated approach of weed management. Integrated weed management considers all relevant control

TABLE 3.4

Different Wheat Herbicides, Their Optimum Doses, Time of Application, and Target Weeds

Herbicide	Dose (a.i.) g ha^{-1}	Grasses	Broad Leaf	Application Time
Clodinafop	60	√		Post-emergence
Fenoxaprop-ethyl	100–120	√		Post-emergence
Pinoxaden	35–40	√		Post-emergence
Sulfosulfuron	25	√	√	Post-emergence
Mesosulfuron + iodosulfuron	12 + 2.4	√	√	Post-emergence
Sulfosulfuron + metsulfuron	30 + 2	√	√	Post-emergence
Pyroxasulfone	127.5	√	√	Pre-emergence and early post-emergence
Pendimethalin	1,500	√	√	Pre-emergence
Metsulfuron + carfentrazone	25 (5 + 20)		√	Post-emergence
Metsulfuron	4		√	Post-emergence
2,4-D	500		√	Post-emergence

Source: Chhokar et al. (2012)

measures and methods available locally and should be opted for effective and eco-friendly weed management in wheat. Integrated weed management approach in wheat is a multilayer hierarchy which includes cultural, mechanical, and chemical methods.

3.2.8 Disease Management

Wheat crop is attacked by a number of diseases (Table 3.5) which significantly affect the crop performance. In Indian north-western plain zone, stripe rust or yellow rust (*Puccinia striiformis* f. sp. *tritici*), brown rust or leaf rust (*Puccinia recondita*), Karnal bunt (*Tilletia indica*), powdery mildew (*Blumeria graminis* f. sp. *tritici*), and loose smut (*Ustilago segetum* var. *tritici*) diseases are very important, while some diseases like FHB and powdery mildew are emerging in NWPZ of India. Symptoms-based

TABLE 3.5

Some Diseases of Wheat Incited by Different Phytopathogens

Crop	Type of Pathogens	Disease Name	Causal Organism/Pathogen
Wheat (*Triticum aestivum* L.)	Fungi	Yellow rust or stripe rust	*Puccinia striiformis* f.sp. *tritici*
		Brown rust or leaf rust	*Puccinia recondita*
		Black rust or stem rust	*Puccinia graminis* f. sp. *tritici*
		Karnal bunt	*Tilletia indica*
		Loose smut	*Ustilago segetum* var. *tritici*
		Flag smut	*Urocystis agropyri*
		Spot blotch	*Bipolaris sorokiniana*
		Septoria tritici blotch	*Zymoseptoria tritici* (*Septoria tritici*)
		Septoria nodorum blotch	*Parastagonospora nodorum*
		Tan spot	*Pyrenophora tritici-repentis*
		Powdery mildew	*Blumeria graminis* f. sp. *tritici*
		Downy mildew (crazy top)	*Sclerophthora macrospora*
		Fusarium head blight	*Fusarium graminearum*
		Fusarium leaf blotch	*Monographella nivalis* (Anamorph *Microdochium nivale*)
		Wheat blast	*Magnaporthe oryzae* pathotype *Triticum*
		Alternaria leaf blight	*Alternaria triticina*
		Common root rot	*Cochliobolus sativus, Fusarium* spp., and *Pythium* spp.
		Take-all	*Gaeumannomyces graminis tritici*
	Bacteria	Bacterial spike blight (gummosis)	*Rathayibacter tritici; Clavibacter iranicus*
		Basal glume rot and bacterial leaf blight	*Pseudomonas syringae* pv. *atrofaciens; P. syringae* pv. *syringae*
		Bacterial stripe (black chaff)	*Xanthomonas campestris* (pv.) *translucens*
	Nematode	Ear cockle	*Anguina tritici*

diagnosis and management of these diseases are crucial to avoid significant yield losses in the crop and described next.

3.2.8.1 Stripe rust disease

Stripe rust or yellow rust disease is the most serious economic concern for wheat as it can spike yield losses up to 100% in disease-inclined varieties (Chen, 2005). It is incited by *Puccinia striiformis* Westend. f. sp. *tritici* (Pst), the fungus pervasive in cool (10 to 16°C temperature) and wet weather conditions (Chen et al., 2014). In India also, stripe rust is a major problem of wheat production especially in NWPZ. The symptoms of the disease appear as bright yellow-orange pustules or sori occurring in stripes on the leaves producing yellow-orange uredospores (Figure 3.1a) that break through the epidermis, predominantly on the upper leaf surfaces (Prescott et al., 1986). The disease spreads from one season to another season through uredospores.

FIGURE 3.1 Some major diseases of wheat: (a) Stripe or yellow rust, (b) leaf rust, uredosori with ruptured host epidermis (inset), (c) Karnal bunt: infected ear, (d) disease grains, (e) loose smut, (f) fusarium head blight, (g) powdery mildew, and (h) spot blotch.

3.2.8.2 Leaf rust disease

Leaf rust or brown rust is incited by *Puccinia triticina* Eriks. (Pt), most frequent with vast spread among wheat rusts (Bolton et al., 2008; Huerta-Espino et al., 2011). The fungal incitant is prevalent in areas with mild temperatures (15–20°C) and moist conditions (rain or dew). The disease manifests initially as very small, round, orange pustules/sori, which are scattered over the leaves but scarcely on the leaf sheath and stem. These sori become brown with maturity (Figure 3.1b). Towards the end of the season, the telia may be formed in the same pustule. These telia are small, oval to linear, black, and covered by the epidermis. Severe incidence of leaf rust inflicts heavy yield losses.

> *Management of stripe and leaf rusts:* Cultural practices like timely sowing help in avoiding the inoculum build up due to coinciding of conducive weather conditions and susceptible crop stages. Clean cultivation such as the removal of volunteer plants and alternate hosts is most recommended cultural practice to contain the pathogen (Figueroa et al., 2018). The use of genetic resistance is a traditional choice for rust management as toxicants' application are expensive, weather-dependent, and raise environmental and health concerns (Figueroa et al., 2018). Recently developed resistant cultivars effective against stripe rust include DBW 327, DBW 332, DBW 303, WH 1270, DBW 187, HD 3226, WB 2, PBW 723, HPBW 01, DBW 88, HD 3086, DPW 621–50, PBW 752, DBW 90, DBW 296, HUW 838, HD 3298, HI 1621, HD 3271, PBW 757, DBW 173, WH 1124, etc. (for NWPZ) and HS 562, HS 542, HPW 349, HS 507, VL 907, VL 804, HPW 349, HS 507, VL 907, etc. (for NHZ). For leaf rust, DBW 327, DBW 332, WH 1270, DBW 187, DBW 222, HD 3226, PBW 723, DBW 88, HD 3086, DBW 90, HUW 838, HD 3298, HI 1621, HD 3271, HI 1636, GW 513, HI 1634, CG 1029, DBW 110, HI 1633, UAS 375, NIAW 317, HI 1605, HS 562, etc., cultivars may be deployed following zone-specific recommendation. Although resistance-breeding programme against rusts of wheat is very strong particularly in India (ICAR-IIWBR, 2021), the breakdown of resistance in newly bred cultivars opens the scope of use of safer fungicides (Figueroa et al., 2018). On the appearance of the rust diseases, foliar spray with 0.1% Propiconazole 25% EC or Tebuconazole 25.9% E.C. or Triadimefon 25% W.P is suggested. The spray may be repeated after 15–20 days' interval, keeping disease severity and spread in mind (ICAR-IIWBR, 2021).

3.2.8.3 Karnal bunt

The Karnal bunt (KB) of wheat was first reported by Mitra in 1930 from Regional Station of IARI at Karnal, Haryana (Mitra, 1931). This disease is caused by *Neovossia indica* (formerly *Tilletia indica*). Moderate temperatures (19–23 °C), high humidity (>70%), and cloudiness or rainfall during anthesis are favourable environmental conditions for the disease development. It is very tough to diagnose the infection in field as only few kernels in an ear-head are infected, and maximum of the infected kernels do not exhibit symptoms before maturity. The disease produces a dark colour and a fishy smell on infected kernels. Usually, only the germ end of the kernel shows symptoms, but, occasionally, the entire kernel may appear diseased. The darkening of the kernel is a result of the kernel tissue being converted in a teliospore mass by the fungus (Figure 3.1 c, d) (Gupta and Kumar, 2020). In severe cases, grain may appear as black shiny sack of teliospores. The embryo and endosperm are not colonized. The pericarp ruptures during threshing, and teliospores are deposited in soil and remain adhered to the surface of the seed.

Management of Karnal bunt: The disease management can be categorized into the following two segments:

Pre-harvest management strategies: Since the KB pathogen perpetuates through soil and seed, the adoption of disease-free seed for sowing and rotating the cycle with non-host crops can help in managing this disease (Bashyal et al., 2020). The resistant sources against Karnal bunt disease are scarce because KB resistance is polygenic and partially dominant (Bashyal et al., 2020). The available resistant/tolerant varieties recommended for the particular zone should be employed to contain the disease. Biocontrol agents having antagonistic potential against this pathogen include *Trichoderma viride, T. harzianum,* and *Gliocladium deliquescens* (Bashyal et al., 2020). Out of the various fungicides assessed for foliar application, propiconazole (Tilt 250 EC) at the heading stage provided 71.4–100% disease control (Aujla et al., 1989; Singh et al., 1989). Gupta et al. (2020) found propiconazole effective with stage-specific applications combined with soil amendment with *Trichoderma viride.* Some other fungicides such as bitertanol (baycor), tebuconazole (folicur), and cyproconazole (SAN 619F) have also the potential for utilization at times of dire need (Bashyal et al., 2020).

Postharvest management strategies: Post-harvest handling is very crucial to maintain seed quality. Poor handling may result in mechanical damage which is a major cause of seed deterioration, and it can facilitate the introduction and easy establishment of pathogens, tending the seed prone to fungal invasion and limiting storage perspective (Shelar, 2008). Postharvest control of diseases employing mechanical processing is a very crucial and most reasonable newer method for attaining bunt-free grains/seeds in wheat since it is tough to detect the infection in the field, and KB-infected grain exhibits no symptoms before maturity. The seed/grain must be threshed and examined (Gupta and Kumar, 2020). Kumar et al. (2015) found that adjustments in pre-cleaner and screen grader of the processing line assembly can eradicate up to 74.1% bunted seeds of the total KB-infected seeds present in the seed lot. They noticed that tip-infected seeds are hard to remove from healthy seeds as the density of such seeds remains at par with healthy seeds. Dayal et al. (2021) supported their study and improved seed quality in different wheat cultivars by reducing the Karnal-bunt-infected seeds through mechanical seed processing.

3.2.8.4 Loose smut disease

This is an internal seed-borne disease, incited by a fungus *Ustilago segetum* var. *tritici* (Kumar and Gupta, 2020). The disease symptoms appear post ear emergence. In this disease, complete inflorescence (except rachis) turns into black powdery mass of smut teliospores (Figure 3.1e). Initially, this powdery mass is covered with a delicate grey membrane which soon dissolves, and a large number of teliospores are released in the environment. Only the rachis remains intact on tillers, and the infected tillers may remain slightly shorter than healthy tillers (Bashyal et al., 2020). Wind and moderate rain, as well as cool temperatures (16–22°C) are ideal for the disease spread. The teliospores from the infected ear-heads get dispersed by air to the open flowers of the healthy plants, and the disease cycle continues.

Management of loose smut:

The effective cultural practices include the use of disease-free seeds for sowing. In case, the crop is being raised for seed production, proper monitoring at the time of ear emergence is required. The infected tillers should be rouged out after covering them with long paper bags, and these infected tillers should be destroyed carefully (Kumar and Gupta, 2020). Hot water treatment by saturating the seed in cool normal water for 5–6 hours and then immersing it for 2 min in hot water at a temperature of 50°C followed by drying the seed eliminates the resting mycelium of *Ustilago segetum tritici* (Gupta and Kumar, 2020). Solar treatment of wheat seed includes the soaking of the seed for 4 hours in cold water in the month of May–June and followed by sun drying in hot sunshiny days. After soaking, the mycelium in seed becomes active and get killed. Such seed fails to produce infected ears after sowing in the next crop seasons (Gupta and Kumar, 2020). The successful management of loose smut can be achieved efficiently by seed treatment with systemic fungicides like tebuconazole (Raxil 2 DS) at 1.5 g kg^{-1} of seed and carboxin (Vitavax) and carbendazim (Bavistin 50 WP) at 2.5 g/kg of seed. Bread wheat varieties like HS 277, VL 829, PBW 34, Halna and durum wheat varieties like PDW 233, WH 896, HI 8498 and RAJ 1555 can be employed as these cultivars have exhibited resistance against loose smut disease (Bashyal et al., 2020).

3.2.8.5 Fusarium head blight of wheat

Wheat Fusarium head blight (FHB), which is also referred as wheat head scab or ear blight, is principally incited by Ascomycete fungus *Fusarium graminearum* (Figueroa et al., 2018). The initial symptoms manifest as minute water-soaked spots at the base or middle of glumes, rachis or on the first floret to flower which finally may result in premature senescence of the ear-head (Figure 3.1f). In warm and moist conditions, the pathogen sporulates as white-pink salmon mass on infected glumes and spikelets. Co-infections of various cereal-threatening *Fusarium* species including many locally unique species may lead to severe FHB epidemics (Brown and Proctor, 2013). Though FHB is the most serious and hazardous floral disease of wheat causing frequent epidemics in United States, China, the EU, the United Kingdom, Africa, Brazil, etc. (Figueroa et al., 2018), the disease is of minor importance in India, and this has been reported from some areas of Punjab and Himanchal Pradesh (Saharan et al., 2004; Teli et al., 2016). Frequent rains before and during the crop anthesis make the wheat prone against the FHB pathogen, and the prevalence of elevated temperature with the humid conditions during this crop growth stage favours the disease development (Figueroa et al., 2018). The disease causes multifaceted impact on wheat: Yield and quality of grain; the accumulation of trichothecene mycotoxins (Teli et al., 2020);

and subsequently presenting health hazards to humans, cattle, poultry, and natural ecosystem (Figueroa et al., 2018).

Management of FHB: The principle of field sanitation based upon tillage to bury infected crop residues to prevent the next generation of the fungal propagules should be followed for reduction in disease incidence. Inclusion of non-host crops, for example, soybean, can help in disease management and reduction in deoxynivalenol (DON) accumulation (Dill-Macky and Jones, 2000). Disease-free seed should be used for sowing. The use of FHB-resistant cultivars is the most economical control measure (Wegulo et al., 2015) but is demonstrated to be slow and challenging to attain in elite wheats. FHB resistance is governed by multiple major and minor QTLs which are usually linked with a fitness cost or yield penalty (Gilbert and Haber, 2013).

Fungicides are an important FHB disease-control measure. Seed treatment with Carboxin 75% WP or Carbendazim 50% WP @ 2.5 g kg^{-1} seed or Tebuconazole 2DS @1.25 g kg^{-1} seed is helpful in managing the disease. Though triazole (DMIs) mixtures sprayed during the optimum flowering stage could provide partial FHB control (McMullen et al., 2012) due to intrinsic resistance of *F. graminearum* to triazoles. Unfavourable weather situations like rains reduce the efficacy of fungicides due to delayed sprays. Additionally, the use of single spray in managing the FHB in both the early- and late-flowering tillers encounters certain technical hurdles (Figueroa et al., 2018).

3.2.8.6 Powdery mildew

The disease remains prevalent in cooler areas. In India, it is found in hilly regions, foothills and plains of north-western India, and the southern hills (Nilgiris) (Kashyap et al., 2018). The disease symptoms manifest as superficial small patches of greyish white cottony growth on leaves, sheath, stem, and even on ear-heads (Figure 3.1g). The symptoms appear initially on upper surface of leaves and, later, on both surfaces. With an increase in disease severity, the white cottony growth can spread over entire leaf and other aerial parts of the host (Kashyap et al., 2018). At the later phase of the disease, cleistothecia are formed in late crop season, and colour of the patches turns into brownish to dull tan.

Management of powdery mildew: The use of recommended seed rate is an important cultural practice as it helps to avoid dense plant stand, which favours the disease development. Deployment of disease-resistant/tolerant cultivars helps to manage the disease. On the appearance of the disease symptoms, triadimefon 25% WP @0.1% can help in disease management (Kashyap et al., 2018).

3.2.8.7 Spot Blotch Disease

The fungal incitant of spot blotch, *Bipolaris sorokiniana*, can produce disease symptoms on all plant parts, *viz.*, internodes, stem, nodes, leaves, awn, glumes, and seed. Damping off (pre- and post-emergence), foot rot, seedling blight, leaf spot, and spike blight are main diseases which are caused by this pathogen at different growth stages of the wheat plant. In spot blotch, the successful interaction of susceptible host and a virulent *Bipolaris sorokiniana* results in the appearance of small pinheads like dark brown to blackish spots with or without yellowing around it. As the disease progresses, these spots enlarge in size (Figure 3.1h), and several lesions coalesce turning into bigger necrotic regions of up to several centimetres long. In the advanced stage of this disease, the symptoms may be visible on sheath, ear, and awns of the affected plants, and spikelet infection results in grain shrivelling and black point. Warm and humid climate favours the fast growth of pathogen, especially after anthesis in host plant.

Management of spot blotch disease: Agronomic practices such as crop rotation, balanced use of recommended fertilizers, clean cultivation, use of efficient biological control agents besides deployment of resistant/tolerant cultivars, and judicious fungicidal sprays serve as IDM practices for reducing the spot blotch disease levels (Singh et al., 1998). Early sowing of the wheat crop circumvents the coincidence of crop growth stage favourable for pathogen development under hot and humid period. Sources of resistance against *Bipolaris sorokiniana* are scanty among commercial wheat varieties in India. Nevertheless, for an effective management of the disease, the cultivation of recommended (tolerant) varieties, like HI 1612, HD 3171, HD 2985, HI 1563, DBW 39, CBW 38, NW 1014, NW 2036, K 9107, HD 2733, DBW 14, HD 2888, K 0307, DBW 39, and HUW 468, should be encouraged (ICAR-IIWBR, 2021). Fungicides applied in seed treatments include captan, mancozeb, maneb, thiram, carboxin, iprodione, and triadimefon (Stack and McMullen, 1988; Mehta, 1993). Foliar sprays of triademinol, fentinaacetate, propiconazole, Dithane Z-78, iprodione, imazalil, and kresoxim-methyl provide efficacious check on the disease. Though the economic viability of these fungicides in managing high disease pressure under conducive environment may be an issue. Further, Singh et al. (2008) concluded that the foliar sprays with propiconazole at 0.1% (Tilt 25 EC) on first appearance of disease and later at 15-day intervals thrice were most effective in managing the incidence of leaf blight complex incited by *Bipolaris sorokiniana* [*Cochliobolus sativus*] and *Alternaria triticina*). Kavita et al. (2017) evaluated 15 fungicides in vitro against *B. sorokiniana* of barley and found that propiconazole at 0.1% and 0.05% was the most effective in controlling mycelial growth of pathogen. Magar et al. (2020), checked *in vitro* efficacy of eight fungicides against *B. sorokiniana*, using poisoned food technique. Among the fungicides evaluated, propiconazole 25% EC (Tilt) was most effective fungicide which completely inhibited the mycelial growth (Magar et al., 2020). Whereas, Raj (2021) concluded that Tebuconazole 50% + Trifloxystrobin 25% WG (Nativo) and Propiconazole 25% EC (Zerox) at both 75 and 100 ppm were most effective in completely inhibiting the radial growth of pathogen among the tested seven fungicides. Several fungal biocontrol agents including *Trichoderma viride*, *Trichoderma harzianum* and *Trichoderma virens*, *Trichoderma reesei*, and *Chaetomium globosum* (Singh et al., 2018; Darshan et al., 2020) exhibited strong antagonistic capacities against *B. sorokiniana* and demonstrated the potential to be utilized for the spot blotch disease management.

3.2.9 Harvesting and Threshing Management

The time of harvesting and moisture content of the crop along with the harvesting technique used (i.e. manual or mechanical) are the crucial factors that decide the quality and yield of the crop. Early harvest at higher moisture increases drying costs, vulnerability to mould growth, and disease infestation and leads to more broken seeds (Khan and Khan, 2010; Benaseer et al., 2018). However, a delay in sowing the subsequent crop because of the scarcity of labours in the peak season, unforeseen natural calamities, and weather change are the obstructions in the timely harvesting of wheat crop and cause considerable loss to the farmers due to higher shattering losses. The losses could be as high as 2 to 7%, if there is a delay in the harvesting of wheat (Iqbal et al., 1980; Benaseer et al., 2018). The optimum moisture of the grain should be 18–23% at the time of harvest and 12–20% at the time of threshing. According to Ladha et al. (2007), approximately 25 to 50% of the grain value, which basically governs the market price, is lost due to poor scheduling of harvesting and threshing operations, inefficient control of the grain moisture at different phases of processing, and poor handling and milling. Basavaraja et al. (2007) have estimated the storage losses in wheat to be about 33.5% of the total losses, while harvesting and threshing operations together contribute about 17% of total post-harvest losses.

Grover and Singh (2013) found an increase in wheat harvesting losses from 1.5% to 2.5% due to high shattering losses when harvesting was delayed in Punjab, India. The advancement in mechanical harvesting techniques enables timely harvest of the crop. The mechanical harvesting techniques encompass (i) simultaneous harvesting and threshing with self-propelling combines, (ii) reaping and windrowing with self-propelling/tractor-operated reapers, (iii) stripping of grains from standing crop with strippers, and (iv) mechanical reaping and binding with reaper-binders. Reapers enable harvesting of crops mostly at the ground level. The working capacity of reapers may vary from 0.25 to 0.40 ha h^{-1} for operating width of 1.20–2.20 m (Murumkar et al., 2014; Nadeem et al., 2015). Parida (2008) witnessed the cost of operation and manpower requirement of harvesting by a tractor-mounted reaper as 63 and 25% of traditional harvesting with a sickle, respectively. Similarly, the cost of operation and manpower requirement of harvesting by a self-propelled reaper was found to be 51 and 26% of traditional harvesting, respectively. Kumar et al. (2019) also observed cost savings of 32.4% and a benefit–cost ratio of 2.08 during the harvesting of wheat crop with a reaper-binder machine having 1.22 m cutting width as compared to the manual harvest with labours.

During the harvesting of a crop by combine harvesters, both the crop parameters (type, variety, moisture content, grain–straw ratio, and lodging of the crop) and machine parameters (reel index, position of reel, height of cutter bar, forward speed, cylinder rpm, cylinder-concave clearance, blower rpm, and sieves' size) affect the grain losses (Srivastava et al., 1990; Singh, 2016). Lashgari et al. (2008) found the values of cylinder speed 800 rpm, forward speed 1.8 km h^{-1}, and concave clearance 25 mm to be the best setting for a combine harvester (John Deere 955) based on the wheat kernel breakage and seed germination parameters. Kumar et al. (2017) reported less unthreshed grains percentage and kernel breakage with a decrease in operating speed from 4.05 to 3.25 km h^{-1} and an increase in crop moisture content from 16 to 20%. In another study, Sattar et al. (2015) reported total grain losses to be 4.28, 3.85, and 2.92% of the gross yield in manual harvesting and threshing, mechanical reaping and threshing, and combine harvesting operations, respectively. Pawar et al. (2008) found that the total field losses were lesser in wheat crop harvested with combine (4.20%) as compared to harvesting done with a self-propelled vertical conveyor reaper (VCR) and thresher (10.57%). It was found that the use of a combine harvester is more profitable in larger fields, whereas the combination of self-propelling VCR and thresher is a better economical option in smaller farms. Ahuja et al. (2007) reported a 44.9% cost saving and 2.4% increase in the yield during the combine harvesting of wheat crop as compared to manual harvesting followed by power threshing operation. Patel and Varshney (2014) found a forward velocity 1.5 km h^{-1} and crop moisture content 9.16% as the best suitable parameters for harvesting wheat crop with a plot combine harvester.

For the threshing of wheat crop, spike tooth type threshing cylinders are generally preferred by the farmers due to their easy design, low price, less energy consumption, and grain breakage along with their capability to produce good-quality *bhusa*. The requirement of good-quality wheat bhusa suitable for animal consumption led to constraints in the adoption of combine harvesters and the improvement in the design of power threshers that could effectively thresh the grains besides delivering 10 to 20 mm long wheat straw (Tiwari et al., 2019). As reported by Tiwari et al. (2018), the majority of wheat threshers in India have spike tooth-type cylinders powered with a 3–4 kW prime mover. The parameters such as feeding chute angle, spike tooth characteristics (shape, dimensions, and number), cylinder parameters (type, diameter, and speed), and concave parameters (shape, size, and clearance) affect threshing efficiency and threshed grain's quality. Tiwari and Chauhan (2018) reported that round-spiked cylinders having a tip diameter of 600 mm and thickness of 6 mm offered greatest threshing efficiency and good straw quality with least specific power consumption and grain loss in the threshing of wheat crop. Multi-crop threshers based on the axial flow principle are also used for wheat, paddy, oilseed, and pulse crop threshing. These threshers have provisions for independent adjustments of the threshing cylinder and blower speeds to help in lowering the grain breakage and improving the cleaning efficiency. The fed crop advances spirally in-between the cylinder and concave for multiple rotations and gets threshed for a larger time by the repeated blows/impacts of the spike tooth (Khan, 1990; Harrison, 1991, 1992; Sessiz and Ülger, 2003). The CIAE multi-crop thresher suitable for both wheat and paddy consists of a spiked cylinder, straw thrower, aspirator type blower, and cleaning sieves. For paddy crops, the spiral louvers provided in the semi-hexagonal top cover help to advance the crop axially from the feed side to the discharge side. The lengthy straw is thrown with the help of straw throwers. For other crops, the semi-hexagonal cover is substituted with a semi-circular cover, and a semi-circular disc is fitted between the cylinder and straw thrower. It helps to save 26–39% labour and time requirement along with 22% cost of operation compared to manual threshing operation (Pandey et al., 1997; Singh, 2016).

3.3 Post-Harvest Management of Wheat

After successfully harvesting of the wheat, post-harvest losses may occur due to spillage, shattering or rodents, birds, insects, and damage during harvesting in field or storage. Harvesting of wheat with high-moisture content attracts mould infestation. Post-harvest losses can be reduced using modern and available technologies.

3.3.1 Losses During Post Harvest

The study estimates that around 8% post-harvest losses occur in wheat. Post-harvest losses in different activities are presented in Table 3.6. The adequate measures required to reduce the post-harvest losses of wheat are described here:

- Ensure uniform drying of wet wheat grain soon after harvest.
- Using suitable threshing and winnowing processes to avoid losses.
- Use of effective methods to prevent grain contamination, as well as good storage and transportation packing.
- Keeping the right amount of moisture in the air and using insect-control techniques before and during storage.
- Providing aeration and moving stocked grains on a regular basis, as well as stirring grain bulk on occasion.
- Modern transportation infrastructure to reduce losses on the farm and in the market.

Further harvesting of wheat at proper time and moisture remains a prerequisite to diminish losses during post-harvest management activities. The following things should be adopted during the harvesting stage of wheat:

- When the grains of wheat grow hard, the crop should be harvested.
- Harvesting before maturity reduces grain recovery and increases the percentage of immature seeds, broken seeds, etc.
- Harvesting delays result in grain breaking and spillage. Harvesting should be done during the dry summer season to avoid exposure to rodents, birds, and insect and pest infestation. Using the correct harvesting equipment and methods, harvested wheat should be stored separately to avoid cross-pollination.

TABLE 3.6

Post-Harvest Losses of Wheat in Different Activities

Sl. No.	Process	Losses during Different Processes (%)
1	Threshing	1.0
2	Moisture	0.5
3	Rodents	2.5
4	Insects	3.0
5	Birds	0.5
6	Transport	0.5

Source: Report of the Committee on Post-Harvest Losses of Food grains in India, Ministry of Food and Agriculture, Govt. of India, 1971

- Excessive drying and direct sun drying should not be done.
- Proper threshing and winnowing must be carried out. To reduce losses during shipment, the grains must be packaged in clean and appropriate gunny bags.

3.3.2 Grading

The bold-size grains of high quality attract higher prices. With greater purchasing power, quality products are in high demand. Heterogeneous quality is unavoidable because wheat is grown under a variety of agro-climatic settings. As a result, having a single national language for defining quality features is required to ease selling without physical inspection. The following marketing advantages are provided by grading:

- Low transportation and storage costs, as well as the knowledge of current prices and the right markets
- Financial aid is simple, and future trading is simple.
- Increases the market share.
- High-quality products are available to consumers at affordable price.
- Encourages competition.

3.3.2.1 Grade Specifications

Depending on the end usage, different authorities grade it according to different criteria. The majority of the commodity may be classed as hard, semi-hard, or soft on the basis of kernel texture with their colour as white, amber, or red. Physical characteristics such as impurity or refraction, as well as the grain's overall look, are mostly considered by Indian merchants. The various grading equipment used are as follows:

- Wheat sample – 50 grams
- Sample divider by sampler – Tube or scoop
- Machine for cleaning and grading
- Plant for dust collection
- Air separation using screens
- Clean – graders
- De stoners
- Gravity separators
- Silo storage system and pre-cleaning
- CFTRI – Afflation detection kit

3.3.3 Contaminants and Remedies

Soil, water, equipment and storage structures and materials are all sources of grain contamination. To reduce the contamination, all precautions including proper aeration should be taken. The adulterants in wheat grains produce negative impact on consumer's health (Table 3.7). Aflatoxin, a class of mycotoxins produced by fungi, can be contaminated in wheat grains during any process ranging from the field to storage, as long as the conditions are favourable for fungus growth. Some methods to control aflatoxins are described here:

- Wheat should be stored in a moisture-free, dry, secure location.
- It should be ensured that grains are properly dried to prevent the formation of fungus.
- A scientific and proper storing method should be used.
- By using a preventive/curative chemical treatment, one can avoid insect infestation and fungus contamination.
- Grain that has been infested should be segregated.

The adulterants in wheat grains can be examined by visual and other specific tests as presented in Table 3.8.

TABLE 3.7

Health Effects Associated With Different Adulterants in Wheat Grains

Sl. No.	Adulterant Type	Health Hazards
1	Admixtures: Soil, mud, stones, foreign matters	Abrasive impact in the gastrointestinal tract
2	Chemicals: Residues such as zinc, mercury, tin, copper and others as well as pesticide residues above the prescribed limit	May cause damage to the liver and carcinogenic metal toxicity, as well as paralysis
3	Fungal: Salmonella toxins, stem rust, fusarium, loose smut, aspergillus hill bunt, etc.	Vomiting, diarrhoea, paralysis, muscular weakness, liver, kidney, and brain damage
4	Viral: Machupo virus (rodent urine)	Black typhus or Ordog fever
5	Natural	Affects organs

Source: https://agmarknet.gov.in/Others/profile_wheat.pdf

TABLE 3.8

Some Simple Tests for the Detection of Wheat Adulterants

SI. No.	Common Adulterants	Detection Technique
1	Soil, stone, grit in grains	Using grading tools and ocular inspection.
2	Hidden insect infestation in grains	Filter paper should be folded first. Then grains with the help of a hammer must be crushed after putting some grains on Ninhydrin (1% in alcohol)-soaked filter paper. A bluish purple patch denotes the presence of a hidden insect infestation.

Source: https://agmarknet.gov.in/Others/profile_wheat.pdf

3.3.4 Packaging

Food packaging is an important step in ensuring a longer shelf life and maintaining product quality, as well as providing protection against degradation and damage during transportation and storage. Labelling and branding are inextricably linked to packaging. Consumers nowadays demand products in unit bundles. Therefore, the availability of good, sanitized, unadulterated food items is important. Wheat must be packaged with greater care for export. In today's consumerist society, the box not only protects the contents but also draws the customer. The following features must be included in good packaging:

- It must keep wheat in good condition over a longer duration.
- It must be clean and easy to handle and transport.
- It must be recognizable and appealing to customers.
- It must be able to withstand spoilage.
- It must contain relevant information as per the norms.

3.3.4.1 Method of Packing and Packaging Material

To pack the graded wheat, one should use new, dry jute bags or any suitable packaging materials as per the norms laid by concern authority. The insects, fungal infestation, or objectionable odour should not be there in the containers. Each package should be firmly sealed. Each shipment should only contain one type of wheat.

3.3.5 Storage and Transportation

3.3.5.1 Storage

Wheat is the main staple meal for the bulk of the population, therefore it must be stored safely. The shrinking of land holding combined with the rapid increase in population necessitates the development of better storage facilities in order to minimize losses. Furthermore, by extending the marketing period, storage allows for greater price realization. Factors impacting wheat quality during storage and transportation are:

- Temperature
- Moisture and humidity
- Rodents and insects
- Grain quality prior to storage
- Storage structure used
- Sanitation and fumigants used
- Damage to the grain prior to storage
- Ambient factors, etc.

Harvested grains typically have a moisture level of 20%, but a moisture percentage of roughly 12% is recommended for storage. Natural or mechanical resources are used to dry the material. Wheat that has a moisture level of more than 13% at temperatures of 30 to 40°C is sensitive to moulds, which cause musty odours, discoloration, and reduced flour output. At 70% relative humidity, wheat has an equilibrium moisture content of 13.5% (relative humidity). A moisture content of 13 to 14% is appropriate for short-term storage, but 11 to 12% is recommended for longer-term storage of up to five years (Dixit et al., 2015).

3.3.5.2 Storage Structures

Food grains are stored in villages in traditional buildings of various shapes and sizes with varying capacities.

a) Underground storage structures: In states of Bihar, Karnataka, Maharashtra, Andhra Pradesh, and Madhya Pradesh, underground storage is popular particularly in dry portions. Although the lack of oxygen in underground storage buildings reduces infestation, moisture seepage can lead to mould growth and grain damage from rodents.

b) Aboveground storage structures: There are two types of storage structures, viz. indoor and outdoor. These are categorized into:

(i) Conventional storage structures such as mud bins, gunny jute bags, bamboo-reed bins, and metal drums.

(ii) Better storage structures such as enhanced bins, silos, brick-built godowns, and cover and plinth (CAP) storage.

3.3.5.3 Storage Facility

(i) Farmers/producers' storage facilities: The wheat grain should be cleaned as much as possible before storing it, according to the code of practices pre-scribed at the farmer level. The grain should be dried until the moisture content is below 12 to 14% depending on the storage period. Before filling the grain, cleaning and disinfecting the storage structure should be done. During filling of grains in the structure, the grains should be stirred with a large stick for proper settling. Before sealing the structure's cover, pesticides should be applied in the required dosages to the grain. The inlet and outlet lids should be closed as soon as possible. Mud or other binding materials to gaps and crevices should be applied as needed. Weevils and bugs would be unable to access the grain or structure and cause damage. To deter rats, the area around the structure must be cleansed. Also, spilled grain should be cleaned up around the structure. After emptying, proper cleaning and disinfecting of the building should be done. When it comes to the usage of chemicals and enhanced grain storage procedures, it is always a good idea to speak with local officials.

(ii) Rural and mandi godowns: Farmers in rural areas keep their produce in their own houses or in various facilities. It is well known that farmers of marginal and small land holdings are unable to afford the modern storage structures to keep their produce until market prices are favourable for selling. Rural godowns are promoted by the government. Farmers bring their harvest to the *mandi* after the harvesting of crop. It is carried in bulk in bags commonly for storage or during the conveyance. This storage facility needs to be strengthened in order to accommodate proper storage and reduce losses.

(iii) CWC, SWC, and FCI warehouses:

    (a) Central Warehousing Corporation (CWC) – In 1957, the CWC was founded as largest public warehousing company. Food grains, fertilisers, and other items are stored in the godowns.

    (b) State Warehousing Corporation (SWC) – Each state has established its own warehouse. The SWC's primary operating area is in the state's district locations.

    (c) Food Corporation of India (FCI) – FCI has a large storage capacity for food grains and plays an important role in Public Distribution System under National Food Security Act.

3.3.6 Major Stored Grain Pests and Their Control Measures

Insects have long been associated with humankind, causing significant food losses by feeding on grains, contaminating foodstuff with their urine, hairs, and excreta; leaving behind skin casts; creating disease issues, etc. (Gupta et al., 2022). The problem of insect pests in destroying food is particularly more under storage conditions, and this is because of the abundance of food provided by bulk storage, reduced movement of insects in search of food, and protection from the extreme of climate instability. (Walter et al., 2016; Singano et al., 2019). The presence of insects in storage compromises with both quality and the marketing value of grain (Ogendo et al., 2004), and, in some countries like Britain, the grain is being discarded even when there are just minor insect pest infestations. About 10% of the total production of food grains in India is lost due to post-harvest losses brought on by unscientific storage, insects, rodents, microbes, etc. (Kumar, 2017). According to reports, around 500 species of insects have been found to be associated with stored grain products. Among these, about 100 species of insect pests of stored products cause economic losses. Grains crops which suffer from insect damage during storage include cereals (rice, wheat, maize, etc.); pulses (chickpea, mung beans, black gram, green gram, cowpea, etc.); and oil seeds (soyabean, sunflower, linseed, etc.). Stored grain insect pests can be categorised into two groups, viz. primary and secondary pests. Primary pests are those which cause damage to whole grains, while secondary pests damage broken or already damaged grains. Based on where they attack, primary pests are further classified as internal and external feeders. The primary and secondary storage insect pests of stored grain are presented in Tables 3.9 and 3.10, respectively.

3.3.6.1 Cultural Methods

    (i) Sanitation: Cleaning and maintaining the storage structure properly on a regular basis are essential for preventing insect pest infestations (Kerstin and Charity, 2011; Mobolade et al., 2019). Before harvest, all grain-handling equipment is maintained and repaired, weeds are removed from around the bins, and major good hygiene steps like cleaning harvest and transportation equipment are taken. In addition, advanced protective measures including floor treatment with residual insecticide can be done, four to six weeks prior to harvest. In turn, this will eradicate those insects that were not removed during cleaning and migrated into the bin. The top four storage

TABLE 3.9

Primary Storage Insect Pests of Stored Grain

Common Name	Scientific Name	Family	Order
Internal feeders			
Rice weevil	*Sitophilus oryzae, S. zeamais, S. granarius*	Curculionidae	Coleoptera
Lesser grain borer	*Rhyzopertha dominica*	Bostrichidae	Coleoptera
Angoumois grain moth	*Sitotroga cerealella*	Gelechiidae	Lepidoptera
Pulse beetle	*Callosobruchus chinensis, C. maculatus*	Bruchidae	Coleoptera
Drug store beetle	*Stegobium paniceum*	Anobiidae	Coleoptera
Tamarind/peanut beetle	*Pachymeres gonagra*	Bruchidae	Coleoptera
Sweet potato weevil	*Cylas formicarius*	Apionidae	Coleoptera
Potato tuber moth	*Phthorimaea operculella*	Gelechiidae	Lepidoptera
Areca nut beetle	*Araecerus fasciculatus*	Anthribidae	Coleoptera
External feeders			
Red flour beetle	*Tribolium castaneum, Tribolium confusum*	Tenebrionidae	Coleoptera
Khapra beetle	*Trogoderma granarium*	Dermestidae	Coleoptera
Indian meal moth	*Plodia interpunctella*	Phycitidae	Lepidoptera
Fig moth or almond moth	*Ephestia cautella*	Phycitidae	Lepidoptera
Rice moth	*Corcyra cephalonica*	Galleriidae	Lepidoptera

TABLE 3.10

Secondary Storage Insect Pests of Stored Grain

Common Name	Scientific Name	Family	Order
Saw-toothed grain beetle	*Oryzaephilus surinamensis*	Silvanidae	Coleoptera
Long-headed flour beetle	*Latheticus oryzae*	Tenebrionidae	Coleoptera
Flat grain beetle	*Cryptolestus minutas*	Cucujidae	Coleoptera
Grain lice	*Liposcelis divinitorius*	Liposcelidae	Psocoptera

insects whose control decreased by 9–34-folds under poorer sanitation conditions include *C. ferrugineus*, *P. truncatus*, *C. cautella*, and *R. dominica* (Morrison et al., 2019).

(ii) Solar drying: Sun drying is the most well-known practice followed before the storage of any food grains started (Belmain and Stevenson, 2001). This procedure limits the storage misfortunes by moulds, staining, respiration, and insect attack. A legitimate grain cleaning before storage guarantees uniform air circulation inside the storage structure. Air circulation assists with lessening the chances of the development of contagious fungal growth. In particular, the cleaning of grains upgrades the viability of the management tactics like fumigation. Moumouni et al. (2014) reported that the solar drying of cowpeas at temperatures between 16.7 and 37.8° was an efficient way to minimize the damage of *C. maculatus* during storage.

(iii) Frequent monitoring: Regular monitoring of stored grains conditions is necessary in order to detect any potential issues associated with storage moisture, temperature, and pest infestation. Smaller size of insects acts as a limiting factor, and it often becomes difficult or ineffective to identify pest range, unless frequent monitoring is taken into account. For the identification and monitoring of stored grain insect pests, traps baited with synthetic aggregation pheromones have been designed. Pheromone-baited traps are the best monitoring devices in developing nations due to their low cost and species specificity.

3.3.6.2 Mechanical Methods

(i) Hermetic technology: Hermetic technology has received a lot of support for the long-duration storage of grains (Likhayo et al., 2016; Baoua et al., 2016; García-Lara et al., 2020). It involves reducing the use of refrigeration for pest control and food protection without the use of synthetic chemicals. It operates on the principle of deoxygenation. As a result, the respiratory systems of insects and the grain itself increase carbon dioxide levels and decrease oxygen levels to the point where aerobic respiration of pests is minimised. Airtight storage containers like super grain bags made from polyethylene and metal silos can help farmers to reduce post-harvest losses (Kumar and Kalita, 2017). De Groote et al. (2013) proposed that larger grain borer (*Prostephanus truncatus*) can be controlled without insecticides by using hermetic storage, in the form of either metal silos or super bags.

(ii) Monitoring technologies: Stored grain insect damage can be minimized only if the infestation is early detected followed by the implementation of appropriate control measures. Various techniques are being used for the detection and decision-making in storage pest management such as grain probe traps, sticky traps; pheromones light traps and visual lures, berlese funnel and acoustical methods, electrical conductance, machine vision, detection of parasitized stored products, diagnosis of early grain spoilage sensing technology, environmental sensing, acoustic sensing, image sensing, ionizing irradiation, and ozonation.

(iii) Nanotechnology: The use of nanotechnology offers a significant potential for the management of insect pests and pathogens, particularly in stored grains, by delivering pesticides in a targeted and controlled manner (Gharsan et al., 2022). A nanosystem unit consists of two core components: an active ingredient and a carrier. In addition to it, based on their chemical composition, nanoformulations can also be categorised into three main groups including (i) inorganic-based, solid, and non-biodegradable nanoparticles (gold, silver, copper, iron, and silica-based nanoparticles), (ii) organic-based biodegradable nanoparticles (liposomes, solid lipid, and polymeric nanoparticles), and (iii) hybrid (combination of both inorganic and organic components) nanoparticles. For instance the plant-mediated green silver nanoparticles using plant *Euphorbia prostrata* (Zahir et al., 2012) and *Avicennia marina* (Sankar and Abideen, 2015) resulted in higher mortality of *S. oryzae* followed by *T. castaneum* and *R. dominica*. Furthermore, guidelines for the discharge of nanomaterials were framed by the Australian Pesticides and Veterinary Medicines Authority (APVMA) (Bowman and Hodge, 2009; Walker et al., 2017). Due to the widespread use of nanopesticide formulations, there is also rising concern in the scientific community regarding their toxicity and effects on the ecosystem, which calls for additional study and research in these areas (Jasrotia et al., 2022). In light of this, additional efforts are needed for the development of safer and more efficient nanoformulations.

3.3.6.3 Chemical Control Methods

Most of the non-chemical techniques may not provide satisfactory results in insect management when used as sole method or in combination. Thus, in order to get satisfactory results, the use of chemicals has become necessary.

3.3.6.3.1 Surface Treatment

New gunny bags or clean old ones should be used by dipping them in 0.0125% Fenvalerate 20 EC or cypermethrin 25 EC for 10 min before filling them with grains. Malathion emulsion (0.05%) should be sprayed over the floor, walls, and ceiling of empty godowns or containers to disinfect them.

3.3.6.3.2 Seed Treatment

Mixing of malathion 5% @ 250 g per quintal of seed is recommended. Alternatively, by dilution in 500 ml of water, the grains may also be treated with 1.5 ml of cypermethrin 25 EC or 25 ml of malathion 50 EC or 2 ml of Fenvalerate 20 EC or 14 ml of deltamethrin 2.8 EC per quintal of seed. To protect pulses from pulse beetle attack, covering the pulses stored in bulk with 7 cm layer of sand or sawdust or dung ash is suggested.

3.3.6.3.3 Fumigation

Small quantities of grain can be disinfected using metallic drums or wooden boxes. In India, ethylene dichloride and carbon tetrachloride mixture have been advised for the fumigation of food grains in storage at the farm level (Mohapatra et al., 2015), and hydrogen phosphide in the form of aluminium phosphide or methyl bromide for protection in warehouses, godowns, and silos. Use of ethylene dichloride and carbon tetrachloride mixture with an exposure time of 4 days, at the pace of 1 litre for 20 quintals of grain or 35 litres per 100 m^3, is recommended. Methyl bromide is utilized at a rate of 3.5 kg per 100 m^3 of room with 10–12 hours of exposure. The fumigant, hydrogen phosphide (aluminium phosphide), is available in tablet form and can be used at the rate of one tablet (3 g) per metric tonne or 25 tablets per 100 m^3 of space with an exposure period of 7 days. For instance Rajendran and Sriranjini (2008) proposed that the use of essential oils (mainly belonging to *Apiaceae, Lamiaceae, Lauracea e*, and *Myrtaceae*) and their components (cyanohydrins, monoterpenoids, sulphur compounds, thiocyanates, and others) along with CO_2 or ethyl formate effectively control beetle pests such as *Tribolium castaneum, Rhyzopertha dominica, Sitophilus oryzae* and *Sitophilus zeamais* via fumigant toxicity.

3.3.6.3.4 Less Harmful Insecticides

Numerous stored-grain-pests have evolved resistance against commonly used insecticides. Resistance in *Rhyzopertha dominica, T. castaneum*, and *Sitophilus oryzae* against phosphine and *Tribolium castaneum* against lindane and malathion are some of the known cases. On the other hand, many active, less toxic chemical products are currently available for stored product usage, and their use may increase soon in near future. These includes harmless juvenile hormone analogs, diatomaceous earth, and spinosyns (Gupta et al., 2022). Research studies are being conducted to determine the efficacy of insect growth regulator (novaluron), a novel spinosyn (spinetoram), and neonicotinoids (imidacloprid and thiamethoxam) for the management of stored grain pests.

3.3.6.3.5 Biological Control Methods

There is a long list of biocontrol agents enlisted in various insect families such as *Braconidae, Ichneumonidae*, and *Bethylidae* against insect pests of stored commodities. The typical examples of biological control agents that have been used for stored insects include anthocorid bug, *Xylocoris flavipes* (Reuter) for the control of stored commodities pests such as *Tribolium castaneum* and *T. confusum* (Rahman et al., 2009) and the use of egg parasitoid *Trichogramma pretiosum* in peanut warehouses for the suppression of Indian meal moth and so on.

3.3.7 Supply Chain of Wheat

3.3.7.1 Processing and Value Addition

Processing of wheat enhances the value and also lowers storage and handling expenses. Wheat flour is manufactured in 5 to 10 hp burr mills, while *suji* and *maida* and by-products of 13% bran and 3% germ are produced in roller mills. The different processes involved in wheat milling are shown in Figure 3.2.

The following steps are involved in wheat milling:

Cleaning: The cleaning process involves the removal of contaminants and foreign materials, which hamper quality of wheat and rinsing thereafter.

Conditioning: To maintain the gluten quality, the temperature of wheat is maintained below or at 47°C. Because both moistening and heating are done at the same time, hydrothermal treatment is employed for conditioning.

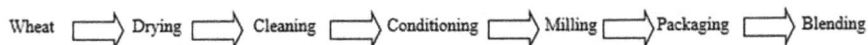

Wheat ⇒ Drying ⇒ Cleaning ⇒ Conditioning ⇒ Milling ⇒ Packaging ⇒ Blending

FIGURE 3.2 Different processes involved in wheat milling.

Grinding: Roller mills, reduction roll systems, and scratch systems in general perform this task.

Packaging: The finished goods are sealed in waterproof bags. They should be stored in a dry and cool environment.

Blending: Due to an increase in health consciousness among customers, some flours are being blended, such as soybean flour, and flour is being fortified with riboflavin, vitamins A and D, and thiamine.

3.3.7.2 Transportation

The mode and cost of transportation play a crucial impact in the transportation of wheat. The transportation cost is a fundamental element in the wide price discrepancies that occur between surplus and deficit locations. From the field to the market, wheat is transported in bulk and bags. Internal markets are typically served by road and rail, with waterways serving as the primary mode of export. However, highways are also utilized to transport people between adjacent countries that are linked by road.

The following factors should be considered while choosing a form of transportation:

- It will be more economical than the other options.
- Its purpose is to save wheat from adverse weather.
- The delivery to the consignee should be made on time.
- It must be insured.
- It should be producer-friendly in terms of transportation payment.

3.3.8 Some of the Profitable Flour Mill Business Ideas

Wheat germ oil: It is derived from the germ of the wheat kernel, which accounts for only 2–3% of the kernel's weight. It's refined vegetable oil with a lot of natural vitamin E, which is a natural antioxidant, as well as a lot of unsaponifiable fraction. Its fatty acid composition has a significant amount of essential fatty acids, which aid in cell renewal. Wheat germ oil contains a high concentration of octacosanol, a long-chain saturated primary alcohol found in a variety of vegetable waxes.

Wheat starch and wheat gluten: Starch is a major carbohydrate found in wheat. Gluten, the primary protein in wheat, is used to make wheat gluten. It's made by scrubbing wheat flour dough with water until all the starch granules are gone, leaving a sticky, insoluble gluten mass that's baked before being eaten.

Atta, maida, suji, and wheat bran: Wheat is ground into a variety of flours, including coarse flour, flour, semolina, bran, and wheat germ. It is also utilised in compound feeds, starch synthesis, and as a feedstock in the manufacturing of ethanol.

Bakery plant: Bread has become one of the important food items in the modern human diet. It is the most widely consumed wheat-based bakery product, such as breads, biscuits, and other pastries. Bread requires wheat flour, yeast, sugar, salt, water, and a shortening agent as raw components.

3.3.9 Market Practices and Constraints

Marketing channels are a collection of interconnected intermediaries that sell produce from producer to consumers. The following are most prevalent wheat distribution channels:

(4)   (i) Private channel: It includes private sector organisations.

      (ii) Institutional channel: It includes both public and cooperative sector organisations. It is vital to the purchase and distribution of wheat.

3.3.10 Cost and Margin

Marketing cost: Marketing expenses are the entire costs incurred during the sale and purchase of the produce until they reach the final consumer. These are some of them:

- Handling fees.
- Storage and transportation fees.
- Handling fees levied by wholesalers and retailers for handling.
- Additional costs for supplementary services such as finance, risk, and market information.

Marketing margin: The profits of the many market functionaries involved in marketing the product from its initial point of manufacturing until it reaches the final user are referred to as marketing margins. By improving the efficiency of the marketing system, marketing costs can be decreased. Handling large quantities of product at once lowers marketing costs and improves efficiency. Marketing costs are reduced thanks to improved handling, packing, and labour efficiency. Adopting tried-and-true management strategies lowers marketing expenditures. Selling value-added items lowers the marketing expenses.

Agricultural commodities have larger marketing margins due to the inherent risk at various phases of the selling process. The following precautions may help to lessen the risk:

- Hedging activities, effective market news services, grading,
- Increased competitiveness in the selling of farm products are all being implemented.

The overall marketing margin's absolute value fluctuates based on the following factors:

Market fee: It is calculated on the basis of the weight of the produce or the value of the product. The buyers are usually the ones who collect it. The market charge varies from one state to the next.

Commission: In most cases, commission payments are made in cash. It varies depending on the market.

Taxes: Tolls, terminal taxes, sales taxes, octroi, and other taxes are levied in different markets. Typically, the seller is responsible for these taxes.

Miscellaneous charges: In addition, there are various other fees that must be paid. Handling, weighing, loading, unloading, cleaning, and monetary and in-kind charitable contributions are just a few examples. These fees are paid by either the vendor or the buyer.

Farmers use marketing data to plan their production and marketing of their products. It is also vital for market participants to make the best trading selections possible.

3.3.11 Information Technology Applications

Information technology has penetrated many aspects of human life and has become an indispensable information tool. To speed agricultural research and development, latest innovation in information technology should be adopted. Selected APMC (Agricultural Produce & Livestock Market Committee) are given computers, peripherals, and software in order to broadcast daily market information such as arrivals and prices.

In Madhya Pradesh, ITC's e-chaupals have done commendable work on soybean production and raw material availability for processing. The establishment of "Agriclinic or Agribusiness Centre and Kisan Cell" will allow farmers to get immediate help with their difficulties over the phone. Kisan call centres are already operational and offering assistance to struggling farmers.

National Agriculture Market (e-NAM) – National Agriculture Market (e-NAM) is expanding to ease farmers. The pan-India trading portal is helping realize the vision of "One Nation, One Market" for agri-produce. National Agriculture Market, or e-NAM, was an innovative agricultural marketing initiative that aimed to improve farmers' digital access to multiple markets and buyers, as well as bring transparency to trade transactions, with the goal of improving price discovery mechanisms and quality-commensurate price realisation and developing the concept of "One Nation One Market" for agricultural produce. e-NAM improved market linkage by integrating 1,000 markets across 18 states and three UTs. On the e-NAM platform, more than 1.69 crore farmers and 1.55 lakh traders have registered so far. Farmers are increasingly trading on the e-NAM platform due to the online and transparent bidding procedure. e-NAM is expanding into "Platforms of Platforms" to create a digital ecosystem that uses individual platforms' expertise across various aspects of the agricultural value chain, such as building and integrating service platforms with e-NaM (QC services, transportation, and delivery). e-NAM is more than a programme – it's a journey that aspires to assist last-mile farmers and improve the industry. Our farmers will profit greatly from this intervention in terms of increasing their revenue by allowing them to obtain competitive and remunerative pricing in a transparent manner without incurring additional costs (*Anonymous, 2021*).

3.3.12 Constraints in Wheat Supply Chain

There are constraints at various stages of wheat supply chain as described here.

- Insufficient storage facilities.
- Lack of grading of wheat in the market despite the premium price of graded wheat.

- Due to high truck transportation costs and a market oversupply, the majority of wheat producers sell their crop at the village level to village traders, itinerant merchants, and other small businesses. Small/marginal farmers are the ones that are most affected by the transportation issue.
- The markets are overwhelmed with wheat arrivals during the peak post-harvest period, producing a slew of issues such as storage, selling space, and traffic congestion. Distress sales are a typical occurrence as a result of excessive supply and the resulting in surplus wheat in the market.
- Farmer financial problems have a crucial part in the adoption of new technology, resulting in increased losses and deterioration of quality.
- Farmers are frequently unaware of market information such as supply, demand, current market pricing, market charges, and so on, which are critical for making timely decisions.

3.4 Conclusion

The rigorous adoption of conventional rice–wheat system in the north western and eastern plain zones instigated detrimental effects on soil quality, nutrient balance, irrigation water budget, crop response, quality, and profitability. The substitution of exhaustive tillage practices with resource-conserving tillage system (zero, strip, and reduced tillage) with or without recycling of crop residue at least in the wheat crop mitigated these issues to some extent. The quality, suitability of wheat for value chain, and profitability can be assured only after the adoption of the latest technologies and proper management of entire wheat chain on every front. The timely and proper management of sowing/planting, nutrient application, water application, weeds control, disease control, harvesting, and threshing in the recommended way remains prerequisite for improving the productivity, resource utilization efficiency, profitability, quality, and suitability of wheat for value addition in addition to reduced losses. The post-harvest management activities take a strategic role in the vertical intensification of harvested wheat grain to wheat-based products for consumers. The selling of value-added wheat products attracts the attention of consumer and ensures premium market price in contrast to the bulk selling of harvested grains at a local level. However, proper and adequate storage facility remains an essential need to reduce the losses due to grain pests and to have high shelf life of value-added wheat products. The awareness of farmers about various pre-harvest management activities concerning yield, quality and profitability, various parameters and regulations for suppling the raw material in wheat industry, and the importance of value addition for attracting a premium market price would be helpful for wheat supply chain.

REFERENCES

Aggarwal, G. C., Sidhu, A. S., Sekhon, N. K., Sandhu, K. S., & Sur, H. S. (1995). Puddling and N management effects on crop response in a rice-wheat cropping system. *Soil and Tillage Research*, 36(3–4), 129–139.
Ahuja, S. S., Rikhi, P., & Dogra, B. (2007). Economics of harvesting and threshing of wheat and paddy in northern India. *Journal of Agricultural Engineering*, 44(2), 14–19.

AICRP on STCR [AICRP on Soil Test Crop Response Correlation]. Four decades of STCR research – crop wise recommendations. *ICAR-Indian Institute of Soil Science, Bhopal.* https://iiss.icar.gov.in/downloads/stcr%20Crop%20wise%20Recommendations.pdf (Accessed 25 April 2023)

Alam, A. (2000). Farm mechanization: Rising energy intensity. *The Hindu Survey of Indian Agriculture*, pp. 181–191.

Alkhafaji, A. J. (2020). Designing and testing of triple combination tillage implement. *Plant Archives*, 20(1), 2363–2366.

Amentae, T. K., Hamo, T. K., Gebresenbet, G., & Ljungberg, D. (2017). Exploring wheat value chain focusing on market performance, post-harvest loss, and supply chain management in Ethiopia: The case of Arsi to Finfinnee market chain. *Journal of Agricultural Science*, 9(8), 22.

Anonymous (2021). National Agriculture Market (e-NAM) is expanding to ease farmers. *Ministry of Agriculture & Farmers Welfare, Government of India.* https://pib.gov.in/PressReleasePage.aspx?PRID=1695193 (Accessed 26 Aug 2022).

Anpat, R. M., & Raheman, H. (2017). Investigations on power requirement of active-passive combination tillage implement. *Engineering in Agriculture, Environment and Food*, 10(1), 4–13.

Aujla, S. S., Sharma, I., Singh, P., Singh, G., Dhaliwal, H. S., & Gill, K. S. (1989). Propiconazole-a promising fungicide against Karnal bunt of wheat. *Pesticides*, 23, 35–38.

Ayars, J. E., Phene, C. J., Hutmacher, R. B., Davis, K. R., Schoneman, R. A., Vail, S. S., & Mead, R. M. (1999). Subsurface drip irrigation of row crops: a review of 15 years of research at the Water Management Research Laboratory. *Agricultural Water Management*, 42(1), 1–27.

Balsari, P., Biglia, A., Comba, L., Sacco, D., Alcatrao, L. E., Varani, M., Mattetti, M., Barge, P., Tortia, C., Manzone, M., & Gay, P. (2021). Performance analysis of a tractor-power harrow system under different working conditions. *Biosystems Engineering*, 202, 28–41.

Baoua, I. B., Amadou, L., Bakoye, O., Baributsa, D., & Murdock, L. L. (2016). Triple bagging hermetic technology for post-harvest preservation of paddy rice *Oryza sativa* L. in the Sahel of West Africa. *Journal of Stored Products Research*, 68, 73–79.

Basavaraja, H., Mahajanashetti, S. B., & Udagatti, N. C. (2007). Economic analysis of post-harvest losses in food grains in India: A case study of Karnataka. *Agricultural Economics Research Review*, 20(1), 117–126.

Bashyal, B. M., Rawat, K., Sharma, S., Gogoi, R., & Aggarwal, R. (2020). Major seed-borne diseases in important cereals: Symptomatology, aetiology and economic importance. In *Seed-Borne Diseases of Agricultural Crops: Detection, Diagnosis & Management* (pp. 371–426). Singapore: Springer.

Batish, D. R., Kaur, M., Singh, H. P., & Kohli, R. K. (2007). Phytotoxicity of a medicinal plant, Anisomeles indica, against Phalaris minor and its potential use as natural herbicide in wheat fields. *Crop Protection*, 26(7), 948–952.

Belmain, S., & Stevenson, P. (2001). Ethnobotanicals in Ghana: reviving and modernising age-old farmer practice. *Pesticide Outlook*, 12(6), 233–238.

Benaseer, S., Masilamani, P., Albert, V. A., Govindaraj, M., Selvaraju, P., & Bhaskaran, M. (2018). Impact of harvesting and threshing methods on seed quality-A review. *Agricultural Reviews*, 39(3), 183–192.

Bhatt, R., Kukal, S. S., Busari, M. A., Arora, S., & Yadav, M. (2016). Sustainability issues on rice–wheat cropping system. *International Soil and Water Conservation Research*, 4(1), 64–74.

Bijay-Singh, Sharma, R. K., Jat, M. L., Martin, K. L., Chandna, P., Choudhary, O. P., Gupta, R. K., . . . & Gupta, R. (2011). Assessment of the nitrogen management strategy using an optical sensor for irrigated wheat. *Agronomy for Sustainable Development*, 31(3), 589–603.

Bijay-Singh, Singh, Y., Ladha, J. K., Bronson, K. F., Balasubramanian, V., Singh, J., & Khind, C. S. (2002). Chlorophyll meter–and leaf color chart–based nitrogen management for rice and wheat in Northwestern India. *Agronomy Journal*, 94(4), 821–829.

Blackshaw, R. E., Molnar, L. J., & Janzen, H. H. (2004). Nitrogen fertilizer timing and application method affect weed growth and competition with spring wheat. *Weed Science*, 52(4), 614–622.

Blackshaw, R. E., O'Donovan, J. T., Harker, K. N., & Li, X. (2002, September). Beyond herbicides: new approaches to managing weeds. In *Proceedings of the International Conference on Environmentally Sustainable Agriculture for Dry Areas* (pp. 305–312), 15–19 September 2002, Shijiazhuang, People's Republic of China.

Bolton, M. D., Kolmer, J. A., & Garvin, D. F. (2008). Wheat leaf rust caused by Puccinia triticina. *Molecular plant pathology*, 9(5), 563–575.

Bovas, J. J. L., Udhayakumar, R., James, P. S., Muthiah, A., Khatawkar, D. S., & James, A. (2022). Combined and multifunctional implements: A promising approach for modern farm mechanization. *Biological Forum—An International Journal*, 14(1), 1376–1383.

Bowman, D., & Hodge, G. (2009). Nanotechnology products in Australia: chemicals, cosmetics and regulatory character. In G. Hodge, D. Bowman, & K. Ludlow (Eds.), *New Global Frontiers in Regulation: The Age of Nanotechnology*. Edward Elgar Publisher, pp. 239–264.

Brown, D. W., & Proctor, R. H. (2013). *Fusarium: Genomics, Molecular and Cellular Biology*. Norfolk: Caister Academic Press.

Cao, Q., Miao, Y., Feng, G., Gao, X., Liu, B., Liu, Y., Li, F., Khosla, R., Mulla, D. J., & Zhang, F. (2017). Improving nitrogen use efficiency with minimal environmental risks using an active canopy sensor in a wheat-maize cropping system. *Field Crops Research*, 214, 365–372.

Chaudhary, V. P., & Singh, B. (2002). Effect of zero, strip and conventional till system on performance of wheat. *Journal of Agricultural Engineering*, 39(2), 27–31.

Chauhan, B. S., Mahajan, G., Sardana, V., Timsina, J., & Jat, M. L. (2012). Productivity and sustainability of the rice–wheat cropping system in the Indo-Gangetic Plains of the Indian subcontinent: problems, opportunities, and strategies. *Advances in agronomy*, 117, 315–369.

Chen, J., Zhao, C., Jones, G., Yang, H., Li, Z., Yang, G., Chen, L., & Wu, Y. (2022). Effect and economic benefit of precision seeding and laser land leveling for winter wheat in the middle of China. *Artificial Intelligence in Agriculture*, 6, 1–9.

Chen, W., Wellings, C., Chen, X., Kang, Z., & Liu, T. (2014). Wheat stripe (yellow) rust caused by *Puccinia striiformis* f. sp. tritici. *Molecular Plant Pathology*, 15(5), 433–446.

Chen, X. M. (2005). Epidemiology and control of stripe rust [*Puccinia striiformis* f. sp. tritici] on wheat. *Canadian Journal of Plant Pathology*, 27(3), 314–337.

Chen, Y., Liu, T., Tian, X., Wang, X., Li, M., Wang, S., & Wang, Z. (2015). Effects of plastic film combined with straw mulch on grain yield and water use efficiency of winter wheat in Loess Plateau. *Field Crops Research*, 172, 53–58.

Chhokar, R. S., Sharma, R. K., Gill, S. C., Singh, R. K., Joon, V., Kajla, M., & Chaudhary, A. (2018). Suitable wheat cultivars and seeding machines for conservations agriculture in rice-wheat and sugarcane-wheat cropping system. *Wheat and Barley Research*, 10(2), 78–88.

Chhokar, R. S., Sharma, R. K., & Sharma, I. (2012). Weed management strategies in wheat-A review. *Journal of Wheat Research*, 4(2), 1–21.

Choudhary, S., Upadhyay, G., Patel, B., & Jain, M. (2021). Energy requirements and tillage performance under different active tillage treatments in sandy loam soil. *Journal of Biosystems Engineering*, 46(4), 353–364.

Chouhan, S. S., Awasthi, M. K., & Nema, R. K. (2014). Maximizing water productivity and yields of wheat based on drip irrigation systems in clay loam soil. *International Journal of Engineering Research & Technology*, 3, 533–535.

Collavo, A., Panozzo, S., Lucchesi, G., Scarabel, L., & Sattin, M. (2011). Characterisation and management of Phalaris paradoxa resistant to ACCase-inhibitors. *Crop Protection*, 30(3), 293–299.

Darshan, K., Aggarwal, R., Bashyal, B. M., Singh, J., Shanmugam, V., Gurjar, M. S., & Solanke, A. U. (2020). Transcriptome profiling provides insights into potential antagonistic mechanisms involved in *Chaetomium globosum* against *Bipolaris sorokiniana*. *Frontiers in Microbiology*, 11, 578115.

Dayal, G., Sharma, A. K., Mishra, C. N., Kamble, U. R., Kumar, R., & Gaurav, S. S. (2021). Mechanical seed processing improves the seed quality and reduces Karnal Bunt incidence in seed lots of varied wheat cultivars. *Journal of Cereal Research*, 13(S1), 51–56.

De Groote, H., Kimenju, S. C., Likhayo, P., Kanampiu, F., Tefera, T., & Hellin, J. (2013). Effectiveness of hermetic systems in controlling maize storage pests in Kenya. *Journal of Stored Products Research*, 53, 27–36.

DES [Directorate of Economics and Statistics] (2022). Fourth advance estimates of production of foodgrains for 2021–22. *Department of Agriculture and Farmers Welfare, Ministry of Agriculture and Farmers Welfare, Government of India*. https://eands. dacnet.nic.in/Advance_Estimate/Time%20Series%204%20AE%202021-22%20 (English).pdf (Accessed 23 Aug 2022).

Destain, M. F., & Houmy, K. (1990). Effects of design and kinematic parameters of rotary cultivators on soil structure. *Soil and Tillage Research*, 17(3–4), 291–301.

Dill-Macky, R., & Jones, R. K. (2000). The effect of previous crop residues and tillage on Fusarium head blight of wheat. *Plant Disease*, 84(1), 71–76.

Dineshkumar, S. P., Patil, B. N., Hiremath, S. M., Koti, R. V., Angadi, V. V., & Basavaraj, B. (2013). Nitrogen management through leaf colour chart (LCC) on growth, yield and yield attributes of emmer wheat [*Triticum dicoccum* (Schrank.) Schulb.] under irrigated condition. *Karnataka Journal of Agricultural Sciences*, 26(3), 350–355.

Dixit, A., Manes, G. S., Cheetu, G. D., Prakash, A., & Panwar, P. (2014). Adoption status of roto seed drill in Punjab. *Agricultural Engineering Today*, 38(3), 24–27.

Dixit, A. K., Jha, S. N., & Kudos, S. K. A. (2015). Four Decades: R&D of All India Coordinated Research Project on Post-Harvest Engineering and Technology. Ludhiana: ICAR-Central Institute of Post-Harvest Engineering and Technology.

Figueroa, M., Hammond-Kosack, K. E., & Solomon, P. S. (2018). A review of wheat diseases—a field perspective. *Molecular Plant Pathology*, 19(6), 1523–1536.

Gangwar, K. S., Singh, K. K., Sharma, S. K., & Tomar, O. K. (2006). Alternative tillage and crop residue management in wheat after rice in sandy loam soils of Indo-Gangetic plains. *Soil and Tillage Research*, 88(1–2), 242–252.

García-Lara, S., García-Jaimes, E., & Ortíz-Islas, S. (2020). Field effectiveness of improved hermetic storage technologies on maize grain quality in Central Mexico. *Journal of Stored Products Research*, 87, 101585.

Gharsan, F. N., Kamel, W. M., Alghamdi, T. S., Alghamdi, A. A., Althagafi, A. O., Aljassim, F. J., & Al-Ghamdi, S. N. (2022). Toxicity of citronella essential oil and its nanoemulsion against the sawtoothed grain beetle Oryzaephilus surinamensis (Coleoptera: Silvanidae). *Industrial Crops and Products*, 184, 115024.

Ghosh, M., Swain, D. K., Jha, M. K., Tewari, V. K., & Bohra, A. (2020). Optimizing chloro-phyll meter (SPAD) reading to allow efficient nitrogen use in rice and wheat under rice-wheat cropping system in eastern India. *Plant Production Science*, 23(3), 270–285.

Gilbert, J., & Haber, S. (2013). Overview of some recent research developments in Fusarium head blight of wheat. *Canadian Journal of Plant Pathology*, 35(2), 149–174.

Grover, D. K., & Singh, J. M. (2013). Post-harvest losses in wheat crop in Punjab: past and present. *Agricultural Economics Research Review*, 26(2), 293–297.

Gupta, A., Kumar, A., Kumar, R., & Verma, K. (2020). Evaluation of foliar applications at different growth stages and soil amendment for management of Karnal bunt disease in wheat. *Seed Research*, 48(1), 25–28.

Gupta, A., & Kumar, R. (2020). Management of seed-borne diseases: an integrated approach. In *Seed-Borne Diseases of Agricultural Crops: Detection, Diagnosis & Management* (pp. 717–745). Singapore: Springer.

Gupta, R. K. (2003). Production constraints of the rice-wheat system In: *Addressing Resource Conservation Issues in Rice-Wheat Systems of South Asia: A Resource Book*. Rice Wheat Consortium for Indo-Gangetic Plains – International Maize and Wheat Improvement Centre, New Delhi, India, pp 32–33.

Gupta, R. K. (2006). Crop canopy sensors for efficient nitrogen management in the Indo-Gangetic plains (Progress report. USDA funded project). In *The Rice-Wheat Consortium, New Delhi International Maize and Wheat Improvement Center (CIMMYT)* (p. 28). Mexico: CIMMYT.

Gupta, R. K., Guroo, M. A., Gani, M., Bali, K., & Kour, R. (2022). Technological innova-tions for the management of insect-pests in stored grains. In *New Horizons in Wheat and Barley Research* (pp. 309–319). Singapore: Springer.

Harrison, H. P. (1991). Rotor power and losses of an axial-flow combine. *Transactions of the ASAE*, 34(1), 60–64.

Harrison, H. P. (1992). Grain separation and damage of an axial-flow. *Canadian Agricultural Engineering*, 34(1), 49.

Hassan, M. M., & Gregory, P. J. (1999). Water transmission properties as affected by crop-ping and tillage systems. *Pakistan Journal of Soil Science*, 16, 29–38.

Hensh, S., Tewari, V. K., & Upadhyay, G. (2021). An instrumentation system to mea-sure the loads acting on the tractor PTO bearing during rotary tillage. *Journal of Terramechanics*, 96, 1–10.

Hossain, M. I., Hossain, I., Mamun, M. A. A., Siddiquie, N. A., Rahman, M. M., & Rahman, M. S. (2012). Two-wheel tractor operated strip tillage seeding equipment for dry land farming. *International Journal of Energy Machinery*, 5(1), 35–41.

Huerta-Espino, J., Singh, R. P., German, S., McCallum, B. D., Park, R. F., Chen, W. Q., . . . & Goyeau, H. (2011). Global status of wheat leaf rust caused by Puccinia tri-ticina. *Euphytica*, 179(1), 143–160.

Hussain, F., Bronson, K. F., Yadvinder, S., Singh, B., & Peng, S. (2000). Use of chlo-rophyll meter sufficiency indices for nitrogen management of irrigated rice in Asia. *Agronomy Journal*, 92(5), 875–879.

IARI. (2012). Crop Residues Management With Conservation Agriculture: Potential, Constraints and Policy Needs. New Delhi: Indian Agriculture Research Institute.

ICAR-IIWBR (2021). Wheat crop health newsletter. In: S. Kumar, P. Jasrotia, P. L. Kashyap, R. Kumar and G. P. Singh (Eds.), *ICAR-Indian Institute of Wheat and Barley Research* (vol. 27, p. 6). Karnal: ICAR.

Iqbal, M., Sheikh, G. S., & Sial, J. K. (1980). Harvesting and threshing losses of wheat with mechanical and conventional methods. *Agricultural Mechanization in Asia (Japan)*, 11(3), 66–70.

Ishaq, M., Ibrahim, M., Hassan, A., Saeed, M., & Lal, R. (2001). Subsoil compaction effects on crops in Punjab, Pakistan. *Soil and Tillage Research*, 60(3–4), 153–161.

Islam, M. S., Salokhe, V. M., Gupta, C. P., & Hoki, M. (1994). Effects of PTO-powered disk tilling on some physical properties of Bangkok clay soil. *Soil and Tillage Research*, 32(2–3), 93–104.

Jabran, K., & Farooq, M. (2013). Implications of potential allelopathic crops in agricultural systems. In *Allelopathy* (pp. 349–385). Berlin and Heidelberg: Springer.

Jabran, K., Farooq, M., Hussain, M., & Ali, M. (2010). Wild oat (*Avena fatua* L.) and canary grass (Phalaris minor Ritz.) management through allelopathy. *Journal of Plant Protection Research*, 50(1), 41–44.

Jasrotia, P., Nagpal, M., Mishra, C. N., Sharma, A. K., Kumar, S., Kamble, U., Bhardwaj, A. K., Kshyap, P. L., Kumar, S., & Singh, G. P. (2022). Nanomaterials for postharvest management of insect pests: current state and future perspectives. *Frontiers in Nanotechnology*, 3, 811056.

Jat, H. S., Kumar, P., Sutaliya, J. M., Kumar, S., Choudhary, M., Singh, Y., & Jat, M. L. (2019). Conservation agriculture based sustainable intensification of basmati rice-wheat system in North-West India. *Archives of Agronomy and Soil Science*, 65(10), 1370–1386.

Jat, M. L., Chandna, P., Gupta, R., Sharma, S. K., & Gill, M. A. (2006). *Laser Land Leveling: A Precursor Technology for Resource Conservation*. Rice-Wheat Consortium Technical Bulletin Series 7, p. 48. New Delhi: Rice-Wheat Consortium for the Indo-Gangetic Plains.

Kailappan, R., Manian, R., Amuthan, G. N., Vijayaraghavan, C., Duraisamy, G. (2001). Combination tillage tool I (design and development of combination tillage tool). *Agricultural Mechanization in Asia, Africa and Latin America*, 32(3), 19–22.

Kalsa, K. K., Subramanyam, B., Demissie, G., Worku, A. F., & Habtu, N. G. (2019). Major insect pests and their associated losses in quantity and quality of farm-stored wheat seed. *Ethiopian Journal of Agricultural Sciences*, 29(2), 71–82.

Kankal, U. S., Karale, D. S., Thakare, S. H., & Khamballkar, V. P. (2016). Performance evaluation of tractor operated rotavator in dry land and wet land field condition. *International Journal of Agricultural Science and Research*, 6(1), 137–146.

Kashyap, P. L., Jasrotia, P., Kumar, S., Singh, D. P., Singh, G. P. (2018). Identification guide for major diseases and insect-pests of wheat. *Technical Bulletin* no. 18, p. 38.

Kaur, B., Singh, S., Garg, B. R., Singh, J. M., & Singh, J. (2012). Enhancing water productivity through on-farm resource conservation technology in Punjab agriculture. *Agricultural Economics Research Review*, 25(347–2016–16901), 79–85.

Kavita, P. S. K., Pande, K. S. K., & Yadav, J. K. (2017). In vitro evaluation of fungicides against Bipolaris sorokiniana causing spot blotch of barley (*Hordeum vulgare* L.). *International Journal of Current Microbiology and Applied Sciences*, 6, 4734–4739.

Kerstin, H., & Charity, M. (2011). Aflatoxin control and prevention strategies in key crops of Sub-Saharan Africa. *African Journal of Microbiology Research*, 5(5), 459–466.

Khan, A. U. (1990). Dual-mode all-crop thresher for Egyptian conditions. *Agricultural Mechanization in Asia, Africa and Latin America*, 21(4), 11–14.

Khan, M. A., & Khan, S. L. (2010). *Trade Development Authority of Pakistan Report on Post Harvest Losses of Rice*. Karachi: Trade Development Authority of Pakistan.

Kharrou, M. H., Er-Raki, S., Chehbouni, A., Duchemin, B., Simonneaux, V., Le Page, M., . . . & Jarlan, L. (2011). Water use efficiency and yield of winter wheat under different irrigation regimes in a semi-arid region. *Agricultural Sciences in China*, 2(3), 273–282.

Khurana, H. S., Phillips, S. B., Alley, M. M., Dobermann, A., Sidhu, A. S., & Peng, S. (2008). Agronomic and economic evaluation of site-specific nutrient management for irrigated wheat in northwest India. *Nutrient Cycling in Agroecosystems*, 82(1), 15–31.

Kukal, S. S., & Aggarwal, G. C. (2003). Puddling depth and intensity effects in rice–wheat system on a sandy loam soil: I. Development of subsurface compaction. *Soil and Tillage Research,* 72(1), 1–8.

Kumar, A., Gupta, A., Atwal, S. S., Maheshwari, V. K., & Singh, C. B. (2015). Post harvest management of Karnal Bunt, a quarantine disease in wheat. *Plant Pathology Journal (Faisalabad)*, 14(1), 23–30.

Kumar, A., Kumar, A., Khan, K., & Kumar, D. (2017). Performance evaluation of harvesting and threshing methods for wheat crop. *International Journal of Pure and Applied Bioscience*, 5(2), 604–611.

Kumar, A., Kumar, A., Kumar, S., & Chandra, S. (2019). Performance evaluation and economics of the reaper-cum-binder machine for the mechanized harvesting of wheat crop at Madhubani district of Bihar. *Journal of Pharmacognosy and Phytochemistry*, 8(4), 63–68.

Kumar, D., & Kalita, P. (2017). Reducing postharvest losses during storage of grain crops to strengthen food security in developing countries. *Foods*, 6(1), 8.

Kumar, N., Chaudhary, A., Ahlawat, O. P., Naorem, A., Upadhyay, G., Chhokar, R. S., Gill, S. C., Khippal, A., Tripathi, S. C., & Singh, G. P. (2023). Crop residue management challenges, opportunities and way forward for sustainable food-energy security in India: A review. *Soil and Tillage Research*, 228, 105641.

Kumar, N., Chhokar, R. S., Meena, R. P., Kharub, A. S., Gill, S. C., Tripathi, S. C., ... & Singh, G. P. (2022). Challenges and opportunities in productivity and sustainability of rice cultivation system: a critical review in Indian perspective. *Cereal Research Communications*, 50, 573–601.

Kumar, N., Chhokar, R. S., Tripathi, S. C., Sharma, R. K., Gill, S. C., & Kumar, M. (2020). Role of conservation agriculture in sustainable food production and challenges. *Findings in Agricultural Research and Management (FARM) Journal*, 4(2), 5–11.

Kumar, R. (2009). *Irrigate Wheat in a Proper Way*. Agropedia, India. http://agropedia.iitk.ac.in/content/irrigate-wheat-proper-way (Accessed 25 April 2023)

Kumar, R. (2017). *Insect Pests of Stored Grain: Biology, Behavior, and Management Strategies* (1st ed.). New York: Apple Academic Press.

Kumar, R., & Gupta, A. (Eds.). (2020). *Seed-Borne Diseases of Agricultural Crops: Detection, Diagnosis & Management* (pp. 1–871). Singapore: Springer.

Ladha, J. K., Pathak, H., & Gupta, R. K. (2007). Sustainability of the rice-wheat cropping system: issues, constraints, and remedial options. *Journal of Crop Improvement,* 19(1–2), 125–136.

Lashgari, M., Mobli, H., Omid, M., Alimardani, R., & Mohtasebi, S. S. (2008). Qualitative analysis of wheat grain damage during harvesting with John Deere combine harvester. *International Journal of Agriculture and Biology*, 10(2), 201–204.

Li, F., Miao, Y., Zhang, F., Cui, Z., Li, R., Chen, X., Xhang, H., Schroder, J., Raun, W. R., & Jia, L. (2009). In-season optical sensing improves nitrogen-use efficiency for winter wheat. *Soil Science Society of America Journal*, 73(5), 1566–1574.

Likhayo, P., Bruce, A. Y., Mutambuki, K., Tefera, T., & Mueke, J. (2016). On-farm evaluation of hermetic technology against maize storage pests in Kenya. *Journal of Economic Entomology*, 109(4), 1943–1950.

Liu, H. J., Kang, Y., Yao, S. M., Sun, Z. Q., Liu, S. P., & Wang, Q. G. (2013). Field evaluation on water productivity of winter wheat under sprinkler or surface irrigation in the North China plain. *Irrigation and Drainage*, 62(1), 37–49.

Lohan, S. K., Jat, H. S., Yadav, A. K., Sidhu, H. S., Jat, M. L., Choudhary, M., Peter, J. K., & Sharma, P. C. (2018). Burning issues of paddy residue management in north-west states of India. *Renewable and Sustainable Energy Reviews*, 81, 693–706.

Magar, P. B., Baidya, S., Koju, R., & Adhikary, S. (2020). In-vitro evaluation of botanicals and fungicides against Bipolaris sorokiniana, causing spot blotch of wheat. *Journal of Agriculture and Natural Resources*, 3(2), 296–305.

Malik, R. K., & Singh, S. (1993). Evolving strategies for herbicide use in wheat: resistance and integrated weed management. In *Proceedings of an Indian Society of Weed Science: International Symposium* (pp. 225–238), 18–20 November 1993, Hisar, India.

Malik, R. K., & Singh, S. (1995). Littleseed canarygrass (Phalaris minor) resistance to isoproturon in India. *Weed Technology*, 9(3), 419–425.

Marenya, M. O., & du Plessis, H. M. (2006). Torque requirements and forces generated by a deep tilling down-cut rotary tiller. In *2006 ASAE Annual Meeting* (p. 1). New York: American Society of Agricultural and Biological Engineers.

McMullen, M., Bergstrom, G., De Wolf, E., Dill-Macky, R., Hershman, D., Shaner, G., & Van Sanford, D. (2012). A unified effort to fight an enemy of wheat and barley: Fusarium head blight. *Plant Disease*, 96(12), 1712–1728.

Meena, R. P., Karnam, V., Tripathi, S. C., Jha, A., Sharma, R. K., & Singh, G. P. (2019). Irrigation management strategies in wheat for efficient water use in the regions of depleting water resources. *Agricultural Water Management*, 214, 38–46.

Mehta, C. R., Singh, K., & Selvan, M. M. (2011). A decision support system for selection of tractor–implement system used on Indian farms. *Journal of Terramechanics*, 48(1), 65–73.

Mehta, Y. R. (1993). Spot blotch. In *Seedborne Diseases and Seed Health Testing of Wheat* (pp 105–112). Copenhagen: Jordburgsforlaget.

Mitra, M. (1931). A new bunt on wheat in India. *Annals of Applied Biology*, 18, 178–179.

Mobolade, A. J., Bunindro, N., Sahoo, D., & Rajashekar, Y. (2019). Traditional methods of food grains preservation and storage in Nigeria and India. *Annals of Agricultural Sciences*, 64(2), 196–205.

Mohapatra, D., Kar, A., & Giri, S. K. (2015). Insect pest management in stored pulses: An overview. *Food and Bioprocess Technology*, 8(2), 239–265.

Mojid, M. A., & Hossain, A. Z. (2013). Conjunctive use of saline and fresh water for irrigating Wheat (*Triticum aestivum L.*) at different growth stages. *The Agriculturists*, 11(1), 15–23.

Mondal, S., Chakraborty, D., Das, T. K., Shrivastava, M., Mishra, A. K., Bandyopadhyay, K. K., Aggarwal, P., & Chaudhari, S. K. (2019). Conservation agriculture had a strong impact on the sub-surface soil strength and root growth in wheat after a 7-year transition period. *Soil and Tillage Research*, 195, 104385.

Morrison III, W. R., Bruce, A., Wilkins, R. V., Albin, C. E., & Arthur, F. H. (2019). Sanitation improves stored product insect pest management. *Insects*, 10(3), 77.

Moumouni, D. A., Doumma, A., & Seyni, I. S. (2014). Effect of solar drying on the biological parameters of the cowpea weevil, *Callosobruchus maculatus* Fab. (Coleoptera-Bruchinae), in Sahelian area. *Journal of Applied Biosciences*, 84, 7723–7729.

Mrema, G., Soni, P., & Rolle, R. S. (2014). A regional strategy for sustainable agricultural mechanization: sustainable mechanization across agri-food chains in Asia and the Pacific region. Food and Agriculture Organization of the United Nations Regional Office for Asia and the Pacific [RAP Publication (2014/24)].

Murumkar, R. P., Dongarwar, U. R., Borkar, P. A., Pisalkar, P. S., & Phad, D. S. (2014). Performance evaluation of self-propelled vertical conveyor reaper. *International Journal of Science, Environment and Technology*, 3(1), 1701–1705.

Nadeem, M., Iqbal, M., Farooque, A., Munir, A., Ahmad, M., & Zaman, Q. (2015). Design and evaluation of self propelled reaper for harvesting multi crops. In *2015 ASABE Annual International Meeting* (p. 1). New York: American Society of Agricultural and Biological Engineers.

Nalavade, P. P., Salokhe, V. M., Niyamapa, T., & Soni, P. (2010). Performance of free rolling and powered tillage discs. *Soil and Tillage Research*, 109(2), 87–93.

Nataraj, E., Sarkar, P., Raheman, H., & Upadhyay, G. (2021). Embedded digital display and warning system of velocity ratio and wheel slip for tractor operated active tillage implements. *Journal of Terramechanics*, 97, 35–43.

Oerke, E. C. (2006). Crop losses to pests. *The Journal of Agricultural Science*, 144(1), 31–43.

Ogendo, J. O., Deng, A. L., Belmain, S. R., Walker, D. J., & Musandu, A. A. O. (2004). Effect of insecticidal plant materials, *Lantana camara* L. and *Tephrosia vogelii* Hook, on the quality parameters of stored maize grains. *Journal of Food Technology in Africa*, 9(1), 29–35.

Pampolino, M., Majumdar, K., Jat, M. L., Satyanarayana, T., Kumar, A., Shahi, V. B., Gupta, N., & Singh, V. (2012). Development and evaluation of nutrient expert for wheat in South Asia. *Better Crops*, 96(3), 29–31.

Pandey, M. M., Majumdar, K. L., Singh, G., & Singh, G. (1997). *Farm Machinery Research Digest. Agriculture* (Technical Bulletin No: CIAE/FIM/1997/69, p. 318). Bhopal: Central Institute of Agricultural Engineering.

Parida, B. C. (2008). Evaluation, constraints and acceptability of different types of vertical conveyer reaper for harvesting rice in coastal Orissa, India. *Agricultural Mechanization in Asia, Africa & Latin America*, 39(1), 29.

Parihar, R. K., Devedee, A. K., & Verma, S. (2018). Leaf colour chart: a resource conservation technology. *Agrobios Newsletter*, XVII(1), 28.

Patel, S. K., & Varshney, B. P. (2014). Modeling of wheat crop harvesting losses. *Agricultural Engineering International: CIGR Journal*, 16(2), 97–102.

Pawar, C. S., Shirsat, N. A., & Pathak, S. V. (2008). Performance evaluation of combine harvester and combination of self-propelled vertical conveyor reaper with thresher for wheat harvesting. *Agriculture Update*, 3(1–2), 123–126.

Petrov, A., Saveliev, Y., Ishkin, P., & Petrov, M. (2020). Soil tillage energy efficiency increase. *BIO Web Conferences*, 17, 1–4.

Prasad, J. (1996). A comparison between a rotavator and conventional tillage equipment for wheat-soybean rotations on a vertisol in Central India. *Soil and Tillage Research*, 37(2–3), 191–199.

Prescott, J. M., Burnett, P. A., Saari, E. E., Ransom, J. K., Bowman, J. D., De Milliano, W., . . . & Geleta, A. B. (1986). *Wheat Diseases and Pests: A Guide for Field Identification* (p. 135). Mexico: CIMMYT.

Qiu, G. Y., Wang, L., He, X., Zhang, X., Chen, S., Chen, J., & Yang, Y. (2008). Water use efficiency and evapotranspiration of winter wheat and its response to irrigation regime in the north China plain. *Agricultural and Forest Meteorology*, 148(11), 1848–1859.

Raheman, H., & Roul, A. K. (2013). Combination tillage implement for high horse power 2WD tractors. *Agricultural Mechanization in Asia, Africa and Latin America*, 44(3), 75–79.

Rahman, M. M., Islam, W., & Ahmed, K. N. (2009). Functional response of the predator Xylocoris flavipes to three stored product insect pests. *International Journal of Agriculture and Biology*, 11(3), 316–320.

Raj, S. (2021). *Studies on Bipolaris sorokiniana, the fungal incitant of spot blotch of wheat (Triticum aestivum L.).* M.Sc. Thesis, CCS University, Meerut, p. 50.

Rajendran, S., & Sriranjini, V. (2008). Plant products as fumigants for stored-product insect control. *Journal of Stored Products Research,* 44(2), 126–135.

Reena, V. C., Dhyani, S. C., Chaturvedi, S., & Gouda, H. S. (2017). Growth, yield and nitrogen use efficiency in wheat as influenced by leaf colour chart and chlorophyll meter based nitrogen management. *International Journal of Current Microbiology and Applied Sciences,* 6(12), 1696–1704.

Safdar, M. E., Asif, M., Ali, A., Aziz, A., Yasin, M., Aziz, M., Afzal, M., & Ali, A. (2011). Comparative efficacy of different weed management strategies in wheat. *Chilean Journal of Agricultural Research,* 71(2), 195.

Saharan, M. S., Kumar, J., Sharma, A. K., & Nagarajan, S. (2004). Fusarium head blight (FHB) or head scab of wheat-A review. *Proceedings of the National Academy of Sciences, India Section B,* 3, 255–268.

Sahoo, R. K. (2005). *Development and performance evaluation of combination tillage implements for 2wd tractors.* Doctoral dissertation, IIT, Kharagpur.

Sankar, M. V., & Abideen, S. (2015). Pesticidal effect of green synthesized silver and lead nanoparticles using *Avicennia marina* against grain storage pest *Sitophilus oryzae. International Journal of Nanomaterials and Biostructures,* 5(3), 32–39.

Sarkar, D., Datta, V., & Chattopadhyay, K. S. (2013). Assessment of pre and post harvest losses in rice and wheat in West Bengal. Agro-Economic Research Centre, Visva-Bharati, Santiniketan, Study No. 172. https://www.visvabharati.ac.in/file/Final-Report-172.pdf (Accessed 25 April 2023)

Sarkar, P., Upadhyay, G., & Raheman, H. (2021). Active-passive and passive-passive configurations of combined tillage implements for improved tillage and tractive performance: A review. *Spanish Journal of Agricultural Research,* 19(4), e02R01.

Sattar, M., U-Din, M., Ali, M., Ali, L., Waqar, M. Q., Ali, M. A., & Khalid, L. (2015). Grain losses of wheat as affected by different harvesting and threshing techniques. International *Journal of Research in Agriculture and Forestry,* 2(6), 20–26.

Saurabh, K., Rao, K. K., Mishra, J. S., Kumar, R., Poonia, S. P., Samal, S. K., . . . & Malik, R. K. (2021). Influence of tillage based crop establishment and residue management practices on soil quality indices and yield sustainability in rice-wheat cropping system of eastern Indo-Gangetic Plains. *Soil and Tillage Research,* 206, 104841.

Sessiz, A., & Ülger, P. (2003). Determination of threshing losses with a raspbar type axial flow threshing unit. *Journal of Agricultural Engineering,* 40(4), 1–8.

Shahani, W. A., Kaiwen, F., & Memon, A. (2016). Impact of laser leveling technology on water use efficiency and crop productivity in the cotton-wheat cropping system in Sindh. *International Journal of Research Granthaalayah,* 4(2), 220–231.

Shankar, M. S., Ramanjaneyulu, A., Neelima, T., & Das, A. (2015). Sprinkler irrigation–an asset in water scarce and undulating areas. In *Integrated Soil and Water Resource Management for Livelihood and Environmental Security* (pp. 259–283). Umiam: ICAR Research Complex for NEH Region.

Sharma, R. K., Chhokar, R. S., Jat, M. L., Singh, S., Mishra, B., & Gupta, R. K. (2008). Direct drilling of wheat into rice residues: experiences in Haryana and western Uttar Pradesh. *Permanent Beds and Rice-Residue Management for Rice-Wheat System of the Indo-Gangetic Plain,* pp. 147–158.

Shelar, V. R. (2008). Role of mechanical damage in deterioration of soybean seed quality during storage-a review. *Agricultural Research,* 29(3), 177–184.

Shinners, K. J., Alcock, R., & Wilkes, J. M. (1990). Combining active and passive tillage elements to reduce draft requirements. *Transactions of the ASAE,* 33(2), 400–0404.

Shukla, L. N., Dhaliwal, I. S., Verma, S. R., Tandon, S. K., & Chauhan, A. M. (2008) Tractor operated strip-till drill. In *Success Stories, AICRP on Farm Implements and Machinery*. Bhopal: Central Institute of Agricultural Engineering.

Sidhu, H. S., Jat, M. L., Singh, Y., Sidhu, R. K., Gupta, N., Singh, P., Singh, P., Jat, H. S., & Gerard, B. (2019). Sub-surface drip fertigation with conservation agriculture in a rice-wheat system: A breakthrough for addressing water and nitrogen use efficiency. *Agricultural Water Management*, 216, 273–283.

Sidhu, H. S., Singh, M., Singh, Y., Blackwell, J., Lohan, S. K., Humphreys, E., Jat, M. L., Singh, V., & Singh, S. (2015). Development and evaluation of the turbo happy seeder for sowing wheat into heavy rice residues in NW India. *Field Crops Research*, 184, 201–212.

Singano, C. D., Mvumi, B. M., & Stathers, T. E. (2019). Effectiveness of grain storage facilities and protectants in controlling stored-maize insect pests in a climate-risk prone area of Shire Valley, Southern Malawi. *Journal of Stored Products Research*, 83, 130–147.

Singh, B., Humphreys, E., Gaydon, D. S., & Eberbach, P. L. (2016). Evaluation of the effects of mulch on optimum sowing date and irrigation management of zero till wheat in central Punjab, India using APSIM. *Field Crops Research,* 197, 83–96.

Singh, D. P., Sharma, A. K., Pankaj, K., Chowdhury, A. K., Singh, K. P., Mann, S. K., ... & Tewari, A. N. (2008). Management of leaf blight complex of wheat (*Triticum aestivum*) caused by Bipolaris sorokiniana and Alternaria triticina in different agroclimatic zones using an integrated approach. *Indian Journal of Agricultural Sciences*, 78(6), 513–517.

Singh, D. V., Agarwal, R., Shrestha, J. K., Thapa, B. R., & Dubin, H. J. (1989). First report of Tilletia indica on wheat in Nepal. *Plant Disease*, 73(3), 273.

Singh, D. V., Pande, S. K., Kavita, Y. J., & Kumar, S. (2018). Bioefficacy of *Trichoderma* spp. against Bipolaris sorokiniana causing spot blotch disease of wheat and barley. *International Journal of Current Microbiology and Applied Sciences*, 7(3), 2322–2327.

Singh, R., & Singh, B. (2004). Effect of irrigation time and weed management practices on weeds and wheat yield. *Indian Journal of Weed Science*, 36(1–2), 25–27.

Singh, R., Singh, A., Kumar, S., Sheoran, P., Rai, A. K., Rani, S., & Yadav, R. K. (2021). Mini-sprinkler irrigation influences water and nitrogen use efficiency and wheat yield in western indo-gangetic plains of India. *Journal of Soil Salinity and Water Quality*, 13(2), 191–197.

Singh, R. V., Singh, A. K., Ahmad, R., & Singh, S. P. (1998). Influence of agronomic practices on foliar blight, and identification of alternate hosts in rice-wheat cropping system. *Helminthosporium Blights of Wheat: Spot Blotch and Tan Spot*, pp. 346–348.

Singh, S., Kirkwood, R. C., & Marshall, G. (1999). Biology and control of Phalaris minor Retz.(littleseed canarygrass) in wheat. *Crop Protection*, 18(1), 1–16.

Singh, S., Malik, R. K., Panwar, R. S., & Balyan, R. S. (1995). Influence of sowing time on winter wild oat (*Avena ludoviciana*) control in wheat (*Triticum aestivum*) with isoproturon. *Weed Science*, 43(3), 370–374.

Singh, T. P. (2016). *Farm Machinery*. New Delhi: PHI Learning Pvt. Ltd.

Singh, V., Singh, B., Thind, H. S., Singh, Y., Gupta, R. K., Singh, S., & Balasubramanian, V. (2014). Evaluation of leaf colour chart for need-based nitrogen management in rice, maize and wheat in north-western India. *Indian Journal of Research*, 51(3), 4.

Srivastava, A. K., Mahoney, W. T., & West, N. L. (1990). The effect of crop properties on combine performance. *Transactions of the ASAE*, 33(1), 63–72.

Stack, R. W., & McMullen, M. (1988). *Root and Crown Rots of Small Grains* (p. 8). Fargo: NDSU Extension Service.

Sur, H. S., Prihar, S. S., & Jalota, S. K. (1980). Effect of rice-wheat and maize-wheat rotations on water transmission and wheat root development in a sandy loam of the Punjab, India. *Soil and Tillage Research*, 1, 361–371.

Teli, B., Chattopadhyay, A., Meena, S. C., Gangwar, G. P., & Pandey, S. K. (2016). Present status of Fusarium head blight of wheat and barley in India. *Diseases of Wheat and Their Management*, 79–92.

Teli, B., Purohit, J., Rashid, M. M., Jailani, A. A. K., & Chattopadhyay, A. (2020). Omics insight on fusarium head blight of wheat for translational research perspective. *Current Genomics*, 21(6), 411–428.

Tfwala, C. M., Mengistu, A. G., Haka, I. B. U., van Rensburg, L. D., & Du Preez, C. C. (2021). Seasonal variations of transpiration efficiency coefficient of irrigated wheat. *Heliyon*, 7(2), e06233.

Timsina, J., & Connor, D. J. (2001). Productivity and management of rice–wheat cropping systems: issues and challenges. *Field Crops Research*, 69(2), 93–132.

Timsina, J., Godwin, D., Humphreys, E., Kukal, S. S., & Smith, D. (2008). Evaluation of options for increasing yield and water productivity of wheat in Punjab, India using the DSSAT-CSM-CERES-wheat model. *Agricultural Water Management*, 95(9), 1099–1110.

Tiwari, R. K., & Chauhan, S. K. (2018). Effect of round spiked threshing cylinder geometry on the threshing performance of wheat crop. *Indian Journal of Hill Farming*, 31(1), 61–67.

Tiwari, R. K., Din, M., & Kumar, M. (2018). Power threshers for effective threshing of crops since green revolution-a review. *International Journal of Agriculture Sciences*, 10(15), 6793–6795.

Tiwari, R. K., Singh, Y., Din, M., Chauhan, S. K., & Namdev, A. (2019). Effect of threshing cylinder configuration on wheat straw quality in Indo-Gangetic plains. *International Journal of Agriculture Sciences*, 11(3), 7813–7817.

Tripathi, S. C., Mongia, A. D., Sharma, R. K., Kharub, A. S., & Chhokar, R. S. (2005). Wheat productivity at different sowing dates in various agroclimatic zones of India. *SAARC Journal of Agriculture*, 3, 191–201.

Umair, M., Hussain, T., Jiang, H., Ahmad, A., Yao, J., Qi, Y., Zhang, Y., Min, L., & Shen, Y. (2019). Water-saving potential of subsurface drip irrigation for winter wheat. *Sustainability*, 11(10), 2978.

Upadhyay, G., & Raheman, H. (2018). Performance of combined offset disc harrow (front active and rear passive set configuration) in soil bin. *Journal of Terramechanics*, 78, 27–37.

Upadhyay, G., & Raheman, H. (2019). Comparative analysis of tillage in sandy clay loam soil by free rolling and powered disc harrow. *Engineering in Agriculture, Environment and Food*, 12(1), 118–125.

Upadhyay, G., & Raheman, H. (2020a). Comparative assessment of energy requirement and tillage effectiveness of combined (active-passive) and conventional offset disc harrows. *Biosystems Engineering*, 198, 266–279.

Usaborisut, P., & Prasertkan, K. (2019). Specific energy requirements and soil pulverization of a combined tillage implement. *Heliyon*, 5(11), e02757.

Walker, G. W., Kookana, R. S., Smith, N. E., Kah, M., Doolette, C. L., Reeves, P. T., Lovell, W., Andersen, D. J., Turney, T. W., & Navarro, D. A. (2017). Ecological risk assessment of nano-enabled pesticides: A perspective on problem formulation. *Journal of Agricultural and Food Chemistry*, 66(26), 6480–6486.

Walter, G. H., Chandrasekaran, S., Collins, P. J., Jagadeesan, R., Mohankumar, S., Alagusundaram, K., Ebert, P. R., Daglish, G. J., Nayak, M. K., Mohan, S., Srivastava, Chitra, Chadda, I. C., Rajagopal, A., Reid, R., & Subramanian, S. (2016). The grand challenge of food security: general lessons from a comprehensive approach to protecting stored grain from insect pests in Australia and India. *Indian Journal of Entomology*, 78(special), 7–16.

Watts, C. W., & Patterson, D. E. (1984). The development and assessment of high speed shallow cultivation equipment for autumn cereals. *Journal of Agricultural Engineering Research*, 29(2), 115–122.

Wegulo, S. N., Baenziger, P. S., Nopsa, J. H., Bockus, W. W., & Hallen-Adams, H. (2015). Management of Fusarium head blight of wheat and barley. *Crop Protection*, 73, 100–107.

World Bank Report, (1999) Post-harvest management – fights hunger with FAO. *India Grains*, 4(3), 20–22.

Zahir, A. A., Bagavan, A., Kamaraj, C., Elango, G., & Rahuman, A. A. (2012). Efficacy of plant-mediated synthesized silver nanoparticles against *Sitophilus oryzae*. *Journal of Biopesticides*, 5, 95–102.

4

Contribution of Wheat in Global Food Security in Changing Climatic Conditions

Challenges Ahead and Coping Strategies

**Anil Kumar Dixit, Biswajit Sen, Shweta Bijla,
Sanjit Maiti, and Babita Kathayat**

CONTENTS

4.1 Introduction

4.1.1 Dimensions of Food Security

Food security is a dynamic concept, and its definition has evolved over time depending upon the various facets of socio-political development shaped by human values. The term "food security" stemmed out first time in World Food Conference in mid-1970s and connoted only the availability dimension. Food security initially meant

DOI: 10.1201/9781003307938-4

'availability at all times of adequate world food supplies of basic foodstuffs to sustain a steady expansion of food consumption and to offset fluctuations in production and prices' (United Nations, 1975). The definition was constrained by the availability of food supplies in a macro perspective of food security. So, food security was viewed from the 'production led' phenomena. Subsequently, the micro approach of food security – 'ensuring that all people at all times have both physical and economic access to the basic food that they need' – initiated a 'production-distribution led' dimension and included the concepts of accessibility and availability (FAO, 1983). Later, the World Bank's Poverty and Hunger Report included a 'temporal effect' dimension to food security, distinguishing it as chronic food insecurity (long-term) and transitory food insecurity (short-term) (World Bank, 1986).

4.1.2 The Bi-Modal Problem of Hunger and Food Security

Maslow's Need Triangle Theory considered food as basic necessity of life (Maslow and Lewis, 1987). According to studies, between 2014 and 2016, over 11% of the world population's daily dietary requirements were severely inadequate (FAO, 2018). This problem could get worse by trapping more people in the cycle of food insecurity due to rising population and stagnant food production. By 2050, it is anticipated that there will be more than ten billion people on the planet. Such a population growth scenario may knock the tipping point of food insecurity earlier contrasting the projection of several literatures (von Braun and Birner, 2017; Dillard, 2019).

Climate shocks have a significant impact on global food production system in terms of growing variability in production pattern (Ziervogel et al., 2006; Wossen et al., 2018). The current global food production system is majorly affected by climatic shocks in terms of increasing variability in production pattern (Ziervogel et al., 2006; Wossen et al., 2018). These climatic shocks lead to increased production risk, besides, depleting soil health and water quality create detrimental environmental effect (Dillard, 2019). Further, genetic improvement programmes that aim to increase production have also

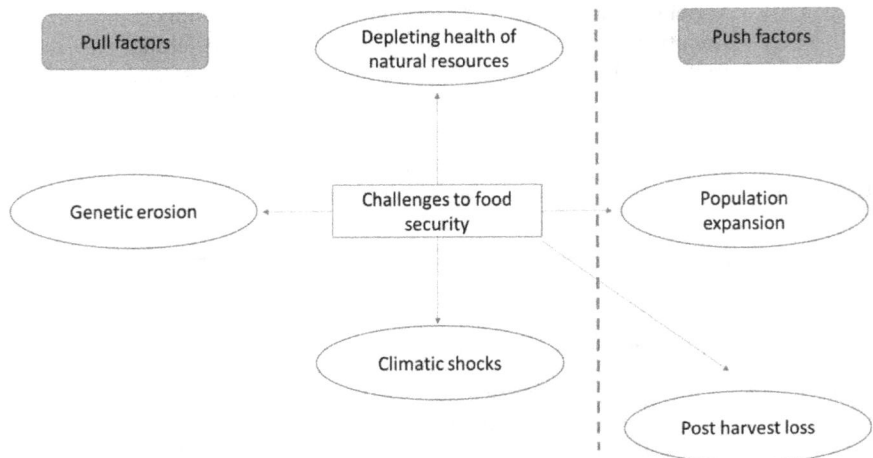

FIGURE 4.1   Threats to food security.

created genetic erosion and a reduction in the genetic diversity of the crop species. This led to an increase in susceptibility and less tolerance to various environmental shocks (Esquinas-Alcázar, 2005; Brussaard et al., 2010). These factors jointly create the 'pull factors' that pose challenges to food security. In addition to these pull variables, several directly human-led factors (also known as push factors), such as population expansion and post-harvest loss, also pose a threat to food security (Figure 4.1).

4.2 Materials and Methods

4.2.1 Data

The present study used different parameters of food and nutritional security across for major 204 countries (Appendix 4.1) for the reference year 2018–2019 (www.fao. org.in). To estimate the role of wheat crop in determining food security of a country, time series of latest 21-year data for wheat yield (reference year: 2000 to 2020) was collected and correlated with food insecurity parameters.

4.2.2 Methodological Design

The study assumed food insecurity in any particular region as a multifaceted problem; hence, it cannot be identified separately. To establish the assumption, Zellner and Theil's (1992) Three Stage Least Square (3SLS) technique was adopted in the study to identify the factors that influence the prevalence of severe food insecurity, share of dietary energy supply derived through cereals and cereal import dependency ratio. 3SLS technique uses full information characteristic so that the estimation of coefficient gains efficiency even though the structural disturbances form non-diagonal moment matrix (Greene, 2018).

The system of equation framed is represented here.

$$Prevalence\ of\ serve\ food\ insecurity_i = \alpha_{0i} + \alpha_{1i}(ADE)_i + \alpha_{2i}(Y\_wheat)_i$$
$$+ \alpha_{3i}(incm)_i + \alpha_{4i}(rtl\_loss)_i + \alpha_{5i}(cpi)_i + \epsilon_{1i} \quad (4.1)$$

$$Share\ of\ dietary\ energy\ supply\ derived\ through\ cereals_i$$
$$= \beta_{0i} + \beta_{1i}(ADE)_i + \beta_{2i}(import)_i + \beta_{3i}(pol)_i + \epsilon_{2i} \quad (4.2)$$

$$Cereal\ import\ dependency\ ratio_i = \theta_{0i} + \theta_{1i}(cpi)_i + \theta_{2i}(Y\_wheat)_i$$
$$+ \theta_{3i}(Share\ of\ dietary\ energy\ supply \quad (4.3)$$
$$derived\ through\ cereals)_i + \theta_{4i}(incm)_i + \epsilon_{3i}$$

The system of Equations (4.1–4.3) can be framed as

$$\begin{bmatrix} y_1 \\ y_2 \\ \vdots \\ y_M \end{bmatrix} = \begin{bmatrix} Z_1 & 0 & \cdots & 0 \\ 0 & Z_2 & \cdots & 0 \\ \vdots & \vdots & \ddots & \vdots \\ 0 & 0 & \cdots & Z_M \end{bmatrix} \begin{bmatrix} \beta_1 \\ \beta_2 \\ \vdots \\ \beta_M \end{bmatrix} + \begin{bmatrix} \epsilon_1 \\ \epsilon_2 \\ \vdots \\ \epsilon_M \end{bmatrix}. \quad (4.4)$$

where y elements describe endogenous left hand side dependent variable list, and Z elements represent both endogenous and exogenous variable list on right hand side. The total number of structured equations (M) are 3. The variables included in the model (Equations (4.1–4.4)) are described below in Table 4.1.

The endogenous and exogenous variables included in the empirical econometric model are given as:

Endogenous variables: Prevalence of severe food insecurity in total population, Share of dietary energy supply derived through cereals, Cereal import dependency ratio

Exogenous variables: Average dietary energy supply adequacy, Wheat yield, Per capita income, Incidence of caloric losses at retail, Consumer price index value, Political stability and absence of violence

TABLE 4.1

Variable Description

Variables	Description
Cereal import dependency ratio	*Cereal import dependency ratio* $$= \frac{(\text{Cereal import dependency ratio})}{(\text{Cereal production} + \text{Cereal imports} - \text{Cereal exports})} \times 100$$ The indicator assumes only values of 100. Negative values indicate that the country is a net exporter of cereals.
Share of dietary energy supply derived through cereals	It highlights the energy supply (in kcal/caput/day) derived from cereals, roots, and tubers as a percentage of the total Dietary Energy Supply (DES) (in kcal/caput/day). The database used the FAOSTAT food balance sheets to retrieve country-wise respective data.
Prevalence of severe food insecurity in total population	As per FAO, 'prevalence of severe food insecurity is an estimate of the percentage of people in the population who live in households classified as severely food insecure'. One-parameter logistic Item Response Theory model (the Rasch model) was used to estimate the probability to be food insecure. The FIES global reference scale was used for cross-country thresholds of comparison. The threshold to classify "severe" food insecurity indicates the item "having not eaten for an entire day" on the global FIES scale.
Average dietary energy supply adequacy	The indicator expresses the Dietary Energy Supply (DES) as a percentage of the Average Dietary Energy Requirement (ADER).
Wheat yield	It is the ratio between productions to area under particular crop.
Per capita income	GDP per capita in purchasing power parity (PPP) terms have been used as measure of per capita income. Data are in constant dollars prices with reference to 2011 level.
Incidence of caloric losses at retail	Incidence of caloric losses at retail distribution level.
Consumer price index value	Weighted average of consumer prices.
Political stability and absence of violence	It measures the 'perceptions of the likelihood that the government will be destabilized or overthrown by unconstitutional or violent means, including politically-motivated violence and terrorism'.

Source: www.fao.org.in

In this study, the prevalence of severe food insecurity along with share of dietary energy supply derived through cereals and cereal import dependency ratio is assumed as endogenous variable in the system of equation. The list of Z variable elements is described in Table 4.1. As endogeneity exists in the framed model, it is assumed that disturbances of the equations are correlated so that

$$E(\epsilon\epsilon') = \Sigma \qquad (4.5)$$

And their expected value is zero. To solve the system of equation, the instruments are derived by linearly regressing all endogenous variables on exogenous variable list so that,

$$\widehat{z}_i = X(X'X)^{-1} X'z_i \quad \forall i \qquad (4.6)$$

where $\widehat{Z}$ contains the instrumented values.

Although 3SLS framework factors in endogeneity bias, it is still subjected to different assumptive limitedness and computational difficulty. The major rigid assumption in linear regression framework is constant marginal effect of covariates across the covariate space (Ferwerda et al., 2015). In addition, the construction of correct functional form is also difficult and, thus, misspecification may result in constructed econometric model leading to biased estimate (Hainmueller and Hazlett 2014). So, following Rifkin et al. (2003), Cawley and Talbot (2002), and Ferwerda et al. (2015), non-parametric Kernel-Based Regularized Least Squares (KRLS) framework was adopted to identify the factors impacting the prevalence of severe food insecurity, share of dietary energy supply derived through cereals, and cereal import dependency ratio (Eqs. (4.1)–(4.3)), assuming the target function as

$$y = f(x) \qquad (4.7)$$

And approximating it as

$$f(x) = \sum_{i=1}^{N} c_i k(x, x_i) \qquad (4.8)$$

where $k(x, x_1)$ captures the similarity between (x), that is, point of interest and covariate vector [Insert Equation Here] out of N vectors by assigning weight c_i.

Equations (4.1)–(4.3) were estimated using the KRLS framework, and reported results were presented in Table 4.2 along with 3SLS estimates for comparison.

4.3 Results and Discussion

4.3.1 Situation Assessment of Food and Nutritional Security: Worldwide Perspective

It is evident from the results that, on average, 0 to 60% of the overall worldwide population still experiences acute food insecurity (Figure 4.2). However, the global

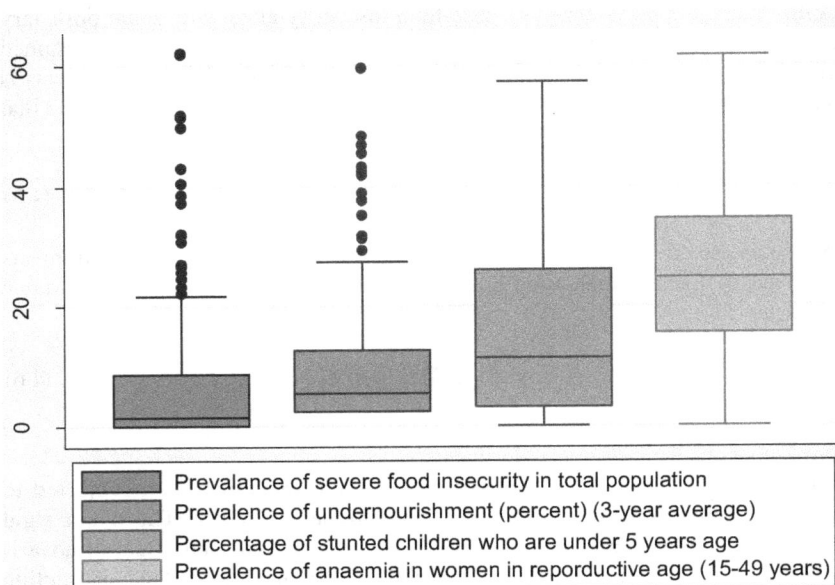

FIGURE 4.2 Worldwide food and nutritional security scenario.

average of 7.58% indicates that there are comparably fewer people who experience extreme food insecurity. However, when expressed in absolute terms, it equates to 560 million people who are severely food insecure due to a lack of access to, availability of, or affordability of food. Country-wise prevalence of undernourishment was also found to be high with country-specific 2.5 to 59.5% of its undernourished population. The global average of 10.09% undernourishment with standard deviation of 11.34 indicates quite large cross-country variations. Additionally, it was found that, on average, 15.13% children worldwide under the age of five suffered from stunting. For certain nations, the percentage is as high as 57.2%, although it is 0 for some nations. Further, it was observed that anaemia affects roughly 26% of women globally between 15 to 49 years of age and particularly during reproductive age. Surprisingly, the figure rose up to 61.5% for specific countries (Niger, Gabon, Nigeria, Senegal, Mali, etc.). Nonetheless, in developed countries like New Zealand, the United Kingdom, and France, the incidences of anaemia were very less.

4.3.2 Association Between Different Parameters of Food Security

The association between different food and nutritional security parameters was further assessed to identify rank correlation and degree of association between the identified parameters (Table 4.2). Kendal Tau statistic revealed that there exists high rank order correlation between undernourishment of population and percentage population of stunted children (tau = 0.579). Further, the prevalence of anaemia among women in reproductive age was also found to be highly associated with the percentage of undernourished population in the country (0.460) and percentage of stunted children (tau = 0.499).

TABLE 4.2

Association Between Different Food and Nutritional Security Parameters

	Prevalence of Severe Food Insecurity in Total Population	Prevalence of Undernourishment	Percentage of Stunted Children	Prevalence of Anaemia in Women in Reproductive Age
Prevalence of severe food insecurity in total population	1			
Prevalence of undernourishment	0.177	1		
	(0.004)			
Percentage of stunted children	0.291	0.579	1	
	(0.001)	(0.001)		
Prevalence of anaemia in women in reproductive age	0.138	0.460	0.499	1
	(0.019)	(0.001)	(0.001)	

4.3.3 Role of Wheat in Food Security

Wheat constitutes a significant portion in the food consumption basket of the majority of the population. It can meet the daily dietary energy requirement of the population with superior nutritive value (Shewry and Hey, 2015). Numerous studies have demonstrated how important wheat is in reducing hunger and malnutrition. Wheat constitutes a balanced diet comprising adequate quantity of quality carbohydrates, protein, fat and other micronutrients (Collins, 2007; Yonar et al., 2021).

So, to estimate the role of wheat in determining food insecurity, a system of equation was framed (Eqs. (4.1)–(4.3)) and estimated using 3SLS framework (Table 4.3). The Kernel-Based Regularized Least Square technique provided estimates for mean population, first quartile (lower quartile), median population (50 percentile), and third quartile (upper quartile) (Table 4.3). According to the estimates, average dietary energy supply adequacy rate reduces food insecurity. It was also found that per capita income has a significant negative impact on food insecurity and it enhances food security. Higher consumer price index has been found to worsen food security.

The cereal import dependency ratio was found to positively affect the share of dietary energy supply derived from cereals. The countries importing wheat have a strong dependence on cereal in terms of eating habits (Kataki et al., 2001). Political instability had a considerable negative impact on food security. Further, it was observed that wheat yield of a particular country and the share of dietary energy supply derived through cereals describe its cereal import dependency ratio.

The estimated model was tested for adequacy, and the overall model fit was found satisfactory (Table 4.4).

TABLE 4.3

Estimates of Parameters Determining Food Insecurity, Dietary Energy Supply, and Cereal Import

	Coefficient	z	Avg.	P25	P50	P75
$\hat{Y}$ = Prevalence of severe food insecurity						
Average dietary energy supply adequacy	0.025	0.920	−0.042*	0.069	−0.045	−0.014
Wheat yield	0.001	0.630	0.001	0.001	0.001	0.001
Log (per capita income)	−4.836***	−6.040	−2.712***	−4.822	−2.039	−0.612
Incidence of caloric losses at retail	−0.596	−0.670	0.297	−0.232	0.431	1.025
Consumer price index value	0.061***	4.150	0.067***	0.007	0.057	0.114
Intercept	43.119***	5.51				
$\hat{Y}$ = Share of dietary energy supply derived through cereals						
Average dietary energy supply adequacy	0.126***	4.160	−0.145***	−0.302	−0.202	−0.045
Cereal import dependency ratio	0.158***	2.920	−0.007	−0.069	−0.010	0.061
Political stability and absence of violence	−7.646***	−5.500	−7.430***	−11.925	−7.884	−3.420
Intercept	26.286***	7.460				
$\hat{Y}$ = Cereal import dependency ratio						
Consumer price index value	−0.091	−0.880	−1.010***	−1.781	−0.779	−0.071
Wheat yield	0.001*	−1.810	−0.001**	−0.001	−0.001	0.000
Share of dietary energy supply derived through cereals	1.117*	1.870	1.309***	−0.028	0.839	2.713
Log (per capita income)	−3.752	−0.480	9.201	−−2.982	8.352	17.917
Intercept	24.760	0.270				

Note: *, **, *** indicates significance at 10%, 5% and 1% level respectively.

TABLE 4.4

Model Adequacy Test

Model adequacy test			
	Equation (4.1)	Equation (4.2)	Equation (4.3)
R square (3SLS)	0.34	0.12	0.05
Chi square	80.38***	58.46***	20.30***
R square (KRLS)	0.51	0.62	0.41
Leave one out error loss	1202	1817	8602

Note: *, **, *** indicates significance at 10%, 5% and 1% level respectively.

The relation between wheat yield and food insecurity (the prevalence of severe food insecurity) was modelled using nonlinear and nonparametric regression using locally weighted scatterplot smoothing (LOWESS) estimator by computing an estimate of the location of wheat yield level (y) within a specific band of food insecurity level (x) (Cleveland, 1979).

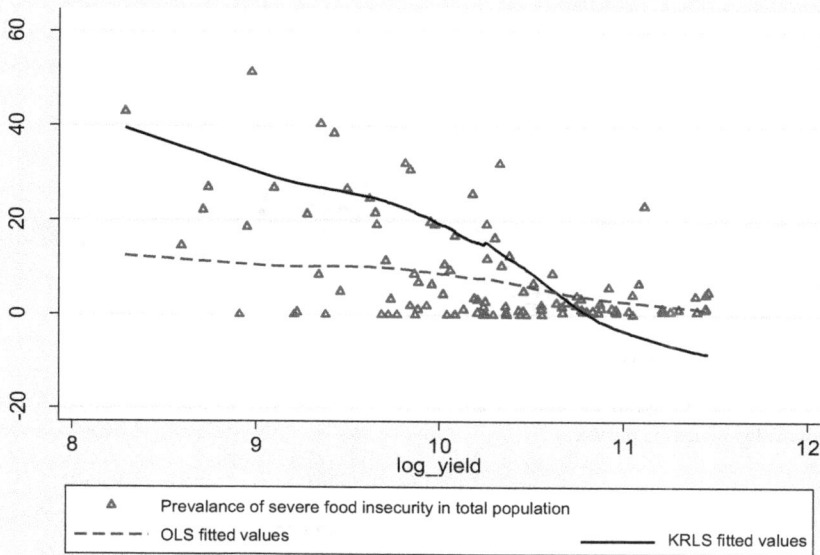

FIGURE 4.3 LOWESS estimate of prevalence of severe food insecurity in total population with respect to wheat yield.

$$(x) = (y \mid x) \tag{4.9}$$

The estimated effect of wheat yield on food insecurity was found negative as indicated by negative slope of the fitted line (Figure 4.3). In other words, we can say wheat yield has positively contributed towards food security. However, the slope of the fitted line derived through the estimates of Kernel-Based Regularized Least Square technique was found comparatively higher than the estimated fitted line derived through 3SLS. The better model fit in KRLS indicates the reason behind it (Rifkin et al., 2003; Cawley and Talbot, 2002; and Ferwerda et al., 2015).

4.3.4 Wheat Yield and Yield Gap

Wheat yield was found to be highly region specific, and the variation across countries was found very high (Figure 4.4A). The worldwide average yield was estimated to be 3 tonnes/ha. For some regions, viz. Egypt, China, Mexico etc., wheat yield level was found higher than 5 tonnes/ha for specific year. Figure 4.4B depicts the yield gap in wheat across the wheat-growing countries. The distribution indicate that the modal yield gap in wheat is approximately 3 tonnes/ha.

Further, the yield level of wheat was found positively dependent on time (t). Wheat yield is observed to be positively impacted by innovations and technology adoptions, as 't' implies technological or institutional innovation. Further, the average temperature shift was also found to have a negative impact on yield level. Wheat yield level was observed to be significantly impacted by an increase in temperature (Figure 4.5).

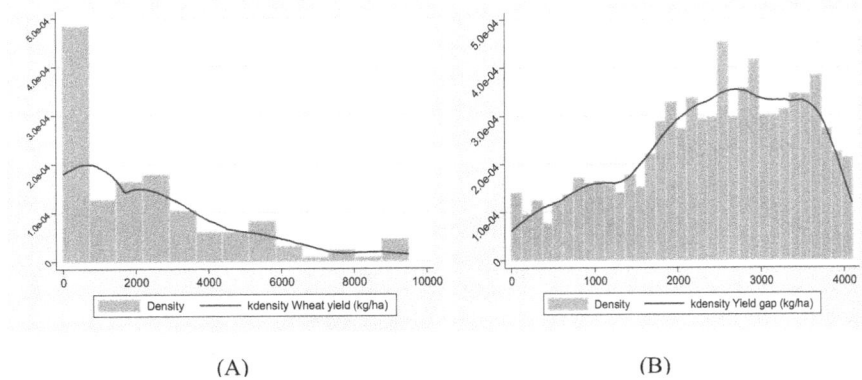

(A)   (B)

FIGURE 4.4 (A) Yield distribution of wheat across the globe; (B) yield gap in wheat across the wheat-growing regions in the world.

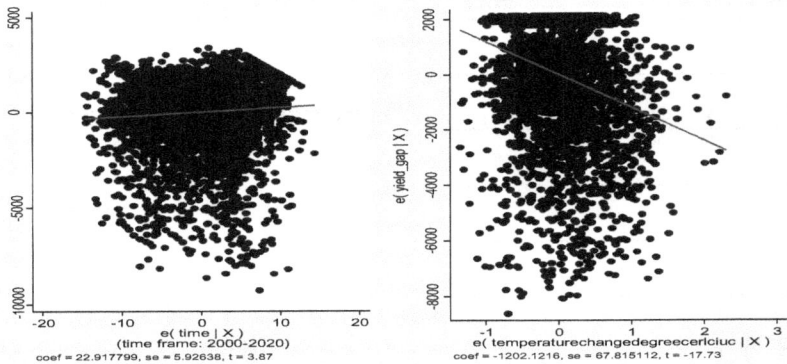

FIGURE 4.5   Effect of temporal and temperature component on the yield level of wheat.

4.4 Measures to Maintain and Enhance Wheat Yield

4.4.1 Varietal Improvement and Training – Institutional Support

The adoption of proven agronomic and management practices helps in enhancing and sustaining wheat productivity.

Government and Research and Development (R&D) organizations are promoting modern varieties of wheat – staple food crop for ensuring food security. A nationally representative of farm households' data suggest that around three-fourth wheat producers have adopted improved varieties of wheat accounting for 83% of total wheat area in the country. Large and medium farmers are typically the ones who are early adopter of improved varieties; these farmers benefit from higher pricing and better extension services than non-adopters. It is also important to know the characteristics of the households which act as push factors to adopt any new intervention. In this section, we have succinctly outlined the review of the previously completed research work. Improved wheat varieties have a considerable and favourable influence on food security in rural Ethiopia – a 1% increase in their area reduces consumption costs and increases food security by 2.9 and 45%, respectively.

In major wheat-growing regions of Mebya district in Tanzania, the adopters preferred a wheat variety on the basis of primary characteristics such as high yield, marketability, grain colour, and early maturity. Moreover, farm size, family size, and the use of hired labour significantly affected the proportion of land allocated to improved wheat varieties (Mwanga et al., 1999). In the same line, a cross-sectional study in Meket district of Amhara National Regional State in Ethiopia used 214 agricultural households which suggested that livestock number, assets ownership, access to extension services, and literacy considerably affect the adoption of improved wheat varieties (Siyum et al., 2022).

In Hula Woreda of Ethiopia, a study of 124 randomly selected households identified the factors which influence the adoption of improved wheat varieties. The factors of household income, fertilizer use, and credit access influence such decisions significantly. The study ascertained the profitability of improved wheat varieties with the help of partial budgeting technique and showed their impact with the increase in the production of improved varieties and better income of the adopter farmers (Degu et al., 2006). Matuschke et al. (2007) assessed the adoption and impact of hybrid wheat in India using contingent valuation method and instrument variable approach. They found noteworthy yield and quality advantage over traditional varieties. The input intensity was also less in hybrid wheat and hence affordable to small farmers. It provided greater economic benefits to smallholders as compared to their large counterparts. The small farmers gained more (₹2,048/acre) as compared to medium (₹1,924/acre) and large farmers (₹1,466/acre). The factors such as easy access to information and income can affect the adoption of hybrid wheat in the study area. A similar study by Tesfaye et al. (2016) in Ethiopia Arsi Zone used Propensity Score Matching method to recognize the impact of improved varieties of wheat on farm productivity and income. The overall adoption rate was 56% in the study area. The results of probit model showed that the adoption of improved wheat varieties is much more likely in farms with the male head (decision-maker) of household, frequent occurrence of diseases on farms (traditionally growing varieties), and livestock ownership. Wheat yield was reported to be higher in case of adopters, and they also earn 35 to 50% higher income than non-adopters. Using Propensity score matching (PSM), Mulugeta and Hundie (2012) examined that technology advancements in wheat have a significant impact on daily food consumption per person in Southeast Ethiopia. The estimated Average Treatment Effect on the Treated (ATT) was found to range from 377.37 to 603.16 calories per day which signified that improved wheat technology dissemination has contributed towards food security of farm households.

4.4.2 Adoption and Impact of Improved Technologies (Including Biofortified Varieties) and Natural Resource Practices

4.4.2.1 Natural Resource Practices and Zero Tillage

Conservation agriculture and zero-tillage techniques can aid to a greater extent in reducing the negative externalities – arising due to continuous rice–wheat cropping system in India. By using a displacement model, Krishna et al. (2022) revealed that the economic impact of technology is more pronounced in the state of Bihar (less developed) than that of Punjab (agriculturally developed state). A similar micro level study by Krishna and Veettil (2014) found that with the implementation of zero-tillage in wheat cultivation, costs were reduced by almost 14%, productivity increased by 5%, while technical efficiency climbed by 1%. Further, the resource use efficiency and the net returns were highest in zero tillage (ZT) in Faisalabad, Pakistan (Farooq and Nawaz, 2014).

In a study conducted by Ladha et al. (2009) on rice–wheat System across Indo-Gangetic Plains (IGP), the laser land levelling techniques increased farmers' net income by 145–300 USD/ha. This is due to the improvement in wheat productivity by 7–20% and saving the irrigation water use by 10–25%. Sarwar and Goheer (2007) analysed the impact of zero tillage (ZT) practice on wheat crop in Sheikhupura district, Punjab (Pakistan). The yield from zero tillage (3.4 tons/ha) was considerably more than conventional techniques (CT) (3.1 tons/ha). The Benefit Cost Ratio (BCR) was also high for zero tillage (1.68) than conventional methods (1.12) as zero tillage is a natural practice and significantly decreases the working cost of cultivation. A similar study done in Karnal district, Haryana (India), also reported higher yield and net returns from ZT over CT.

4.4.2.2 Biofortification

Now as the world has achieved food security, accomplishing nutritional security is the new area of concern (Stein et al., 2007). The difference between 'overt' and 'hidden' hunger is highlighted in 'United Nations Millennium Project Task Force on Hunger' (Sanchez and Swaminathan, 2005). Besides, others suffer from hidden hunger owing to micronutrient and vitamin deficiencies such as vitamin-A-deficiency (150 million), iodine-deficiency (two billion), iron-deficiency (4–5 billion), and about 2.7 billion are zinc-deficient (2.7 billion) (World Health Organization, 2002).

Biofortification – a breeding method with the potential to boost yield and improve mineral content (Welch and Graham, 2000) – could be a solution to alleviate hunger among the poor in a cost-effective manner. Since wheat is one of the world's most important staples, feeding 35% of the population and serving around 20% of dietary energy, it can act as an affordable means of providing much-needed nutrition to disadvantaged sections in developing countries. In Southern Asia, Zn-fortified wheat led to an increase in Zn intake and reduction in child morbidity (Singh et al., 2017). First biofortified wheat variety (WB 02) developed in India, which is now cultivated in 1.44 lakh hectares, producing an economic surplus of ₹50.54 million (at 2011–12 constant prices). Despite its ability to eliminate hidden hunger and having potential economic benefits, the adoption rate is slow due to many socio-economic constraints and institutional bottlenecks (Cariappa et al., 2022).

Stein et al. (2005) used disability-adjusted life years (DALYs) to assess the current burden of ZnD (zinc deficiency) and the potential impact (ex-ante) of biofortification. India loses 2.8 million DALYs annually as a result of ZnD. Zinc biofortification of the staples, i.e. wheat and rice, could reduce this burden by 20–51% and save 0.6–1.4 million DALYs annually. According to WHO standards, the cost to prevent one DALY was to the tune of $0.73, the results demonstrating the cost-effectiveness of biofortification.

Further, Stein et al. (2007) reported dual benefits (i.e. nutritional security and resource saving) of biofortified wheat. This hidden hunger, when left unchecked, results in child malnutrition. National Family Health Survey conducted in 2015–2016 across all districts of India depicts that we, as a nation, rank first in terms of lower-than-normal birth weight of infants (44%), and around 36% of under five children are stunted. With staples having low Zn content, mild-to-moderate Zn deficiency is common in Indians. In one of the largest state of India, Uttar Pradesh, 48% under five children suffer from Zn deficiency. By using DALYs method, Tewari et al. (2017) reported that the current burden of ZnD (zinc deficiency) in Uttar Pradesh is 0.91 million DALYs. Among this, 0.89 million DALYs were lost due to mortality and 0.02 million DALYs were lost due to morbidity. They also compared the economics of traditional and biofortified wheat varieties in which they found no noticeable difference among the cultivation cost and other factors such as taste, appearance, and cooking quality. They also reported that with the maximum adoption of Zn-fortified variety, that is, by 30%, the burden of ZnD will drop. They stated that the burden of ZnD will decrease by 11%, with 30% increase in the adoption of Zn-fortified varieties. The cost (₹79 to ₹177) of avoiding one DALY loss using Zn-fortified wheat is a promising means of ensuring the state's nutritional security.

Micronutrient deficits are quite common not only in India but also in China. According to a national census conducted in China, there are 250 million people who are iron-deficient, including 30% of children and 20% of women. About 100 million people, mostly in rural regions, are affected by ZnD, which causes stunting (low height for age). To combat these deficiencies, several interventions were identified such as supplementation, dietary diversification, and food biofortification. According to Ma et al. (2008), the biofortification of staples such as wheat and rice exhibited the lowest per capita cost 0.01 I$ (international dollars) in addressing these deficiencies. Ex ante economic impact of wheat fortified with zinc was examined in blast-affected districts of Bangladesh by Mottaleb et al. (2019). The study predicted that 24% area will be under new biofortified variety (BARI Gom 33) with the adoption rate of 30% by 2027. The study finds considerable monetary gains with 5.2% higher yields than existing varieties. Net production gains from this new variety would be 135.6 thousand metric tons between 2017 and 2030, which would be worth USD 20.4 million.

4.5 Conclusions and Policy Implications

An attempt has been made to understand how wheat crop plays a vital role in mitigating food and nutritional insecurity. Recent reports show that there is prevalence of food insecurity, as 11% of world's population is underfed. Nevertheless, wheat is one of the most essential foods in the world, providing 20% of dietary energy and feeding 35% of the world's population. Tracing back to the problem of hunger and food security at global level, we developed a system of equations, viz., (i) prevalence of severe food insecurity, (ii) share of dietary energy supply derived through cereals, (iii) cereal import dependency ratio, and estimated using 3SLS framework. The estimated effect of wheat yield on food insecurity was found to be negative (results of 3SLS). In comparison to 3SLS, the estimations from the Kernel-Based Regularized Least Square technique were robust – confirming that wheat is positively contributing to food security. The other variables like political instability and consumer price had a considerable negative impact on food security. However, wheat faces significant obstacles, notably lower farm productivity in some countries. The worldwide average yield was estimated to be 3 tonnes/ha, while it was more than 5 tonnes/ha for some countries, indicating scope for productivity improvement in less endowed regions through technological interventions and appropriate policy support. Further, wheat yield level was observed to be significantly impacted by climatic variables (such as temperature). Mitigating the effects of extreme climatic temperature could shore up wheat productivity, contributing to fighting food insecurity at the global level. In this direction, a number of studies noted that biofortification, conservation techniques, and varietal development might all significantly and favourably impact food security.

Finally, we can show a strong link between wheat productivity and food security at the global level. Nonetheless, climate-smart technological interventions to wheat production might significantly reduce food insecurity.

REFERENCES

Brussaard, L., Caron, P., Campbell, B., Lipper, L., Mainka, S., Rabbinge, R., & Pulleman, M. (2010). Reconciling biodiversity conservation and food security: scientific challenges for a new agriculture. Current Opinion in Environmental Sustainability, 2(1–2), 34–42.

Cariappa, A. G., Ramasundaram, P., Gupta, V., Gupta, O. P., Kumar, A., Singh, S., & Singh, G. P. (2022). *Biofortification in wheat: research progress, potentia l impact, and policy imperatives*. https://ssrn.com/abstract=4087960.

Cawley, G. C., & Talbot, N. L. (2002). Improved sparse least-squares support vector machines. Neurocomputing, 48(1–4), 1025–1031.

Cleveland, W. S. (1979). Robust locally weighted regression and smoothing scatter-plots. Journal of the American Statistical Association, 74(368), 829–836.

Collins, S. (2007). Treating severe acute malnutrition seriously. Archives of Disease in Childhood, 92(5), 453–461.

Degu, G., Dadi, L., & Negatu, W. (2006). Adoption and impact of improved wheat technology: the case of Hula Woreda, Ethiopia. Ethiopian Journal of Development Research, 28(2), 1–29.

Dillard, H. R. (2019). Global food and nutrition security: from challenges to solutions. Food Security, 11(1), 249–252.

Esquinas-Alcázar, J. (2005). Protecting crop genetic diversity for food security: political, ethical and technical challenges. Nature Reviews Genetics, 6(12), 946–953.

FAO (1983). *World Food Security: A Reappraisal of the Concepts and Approaches (Director Generals Report)*. Rome: FAO.

FAO (2018). *Save food: global initiative on food loss and waste reduction*. Retrieved from www.fao.org/save-food/resources/keyfindings/en/. Accessed 28 June 2022.

Farooq, M., & Nawaz, A. (2014). Weed dynamics and productivity of wheat in conventional and conservation rice-based cropping systems. Soil and Tillage Research, 141, 1–9.

Ferwerda, B., Yang, E., Schedl, M., & Tkalcic, M. (2015). Personality traits predict music taxonomy preferences. In *Proceedings of the 33rd Annual ACM Conference Extended Abstracts on Human Factors in Computing Systems* (pp. 2241–2246), New York: ACM.

Greene, W. H. (2018). *Econometric Analysis/Limdep Users Manual*. New York: Econometric Software.

Hainmueller, J., & Hazlett, C. (2014). Kernel regularized least squares: Reducing misspecification bias with a flexible and interpretable machine learning approach. Political Analysis, 22(2), 143–168.

Kataki, P. K., Hobbs, P., & Adhikary, B. (2001). The rice-wheat cropping system of South Asia: Trends, constraints and productivity—a prologue. Journal of Crop Production, 3(2), 1–26.

Krishna, V. V., Keil, A., Jain, M., Zhou, W., Jose, M., Padmaja, S. S., Escoto, L. B., Jat, M. L., & Erenstein, O. (2022). Conservation agriculture benefits Indian farmers, but technology targeting needed for greater impacts. Frontiers in Agronomy, 24.

Krishna, V. V., & Veettil, P. C. (2014). Productivity and efficiency impacts of conservation tillage in northwest Indo-Gangetic Plains. Agricultural Systems, 127, 126–138.

Ladha, J. K., Kumar, V., Alam, M. M., Sharma, S., Gathala, M., Chandna, P., Saharawat, Y. S., & Balasubramanian, V. (2009). Integrating crop and resource management technologies for enhanced productivity, profitability, and sustainability of the rice-wheat system in South Asia. Integrated Crop and Resource Management in the Rice–Wheat System of South Asia, 69–108.

Ma, G., Jin, Y., Li, Y., Zhai, F., Kok, F. J., Jacobsen, E., & Yang, X. (2008). Iron and zinc deficiencies in China: what is a feasible and cost-effective strategy? Public Health Nutrition, 11(6), 632–638.

Maslow, A., & Lewis, K. J. (1987). Maslow's hierarchy of needs. Salenger Incorporated, 14(17), 987–990.

Matuschke, I., Mishra, R. R., & Qaim, M. (2007). Adoption and impact of hybrid wheat in India. World Development, 35(8), 1422–1435.

Mottaleb, K. A., Govindan, V., Singh, P. K., Sonder, K., He, X., Singh, R. P., Joshi, A. K., Barma, N. C., Kruseman, G., & Erenstein, O. (2019). Economic benefits of blast-resistant biofortified wheat in Bangladesh: the case of BARI Gom 33. Crop Protection, 123, 45–58.

Mulugeta, T., & Hundie, B. (2012). Impacts of adoption of improved wheat technologies on households' food consumption in southeastern Ethiopia (No. 1007–2016–79620). International Association of Agricultural Economists (IAAE).

Mwanga, J., Mwangi, W. M., Verkuijl, H., & Mussei, A. (1999). Adoption of improved wheat technologies by small scale farmers in Mbeya district of southern highlands, Tanzania. *Regional Wheat Workshop for Eastern, Central and Southern Africa*, Stellenbosch, 14–18 September, p. 603.

Rifkin, R., Yeo, G., & Poggio, T. (2003). Regularized least-squares classification. Nato Science Series Sub Series III Computer and Systems Sciences, 190, 131–154.

Sanchez, P. A., & Swaminathan, M. S. (2005). Enhanced: cutting world hunger in half. Science, 307, 357–359.

Sarwar, M. N., & Goheer, M. A. (2007). Adoption and impact of zero tillage technology for wheat in rice-wheat system—water and cost saving technology. A case study from Pakistan (Punjab). In *International Forum on Water Environmental Governance in Asia* (pp. 14–15). Princeton, NJ: Citeseer.

Shewry, P. R., & Hey, S. J. (2015). The contribution of wheat to human diet and health. Food and Energy Security, 4(3), 178–202.

Singh, R., Govindan, V., & Andersson, M. S. (2017). *Zinc-biofortified Wheat: Harnessing Genetic Diversity for Improved Nutritional Quality (No. 2187–2019–666, Science Brief: Biofortification Series)*. Mexico: International Maize and Wheat Improvement Center.

Siyum, N., Giziew, A., & Abebe, A. (2022). Factors influencing adoption of improved bread wheat technologies in Ethiopia: empirical evidence from Meket district. Heliyon, 8(2), e08876.

Stein, A. J., Meenakshi, J. V., Qaim, M., Nestel, P., Sachdev, H. P. S., & Bhutta, Z. A. (2005). *Analysing Health Benefits of Biofortified Staple Crops by Means of the Disability-Adjusted Life Years Approach: A Handbook Focusing on Iron, Zinc and Vitamin A (Technical Monograph No. 4)*. Washington, DC: HarvestPlus.

Stein, A. J., Nestel, P., Meenakshi, J. V., Qaim, M., Sachdev, H. P. S., & Bhutta, Z. A. (2007). Plant breeding to control zinc deficiency in India: how cost-effective is bio-fortification? Public Health Nutrition, 10(5), 492–501.

Tesfaye, S., Bedada, B., & Mesay, Y. (2016). Impact of improved wheat technology adoption on productivity and income in Ethiopia. African Crop Science Journal, 24(s1), 127–135.

Tewari, H., Rani, R., Singh, H. P., Singh, R., & Singh, P. K. (2017). Comparative study of biofortified and non-biofortified wheat in Uttar Pradesh, India: combating nutritional security through biofortification. International Journal of Agricultural and Statistical Sciences, 13(1), 365–370.

United Nations. (1975). Report of the World Food Conference, Rome, 5–16 November 1974. New York: UN.

von Braun, J., & Birner, R. (2017). Designing global governance for agricultural development and food and nutrition security. Review of Development Economics, 21(2), 265–284.

Welch, R. M., & Graham, R. D. (2000). A new paradigm for world agriculture: productive, sustainable, nutritious, healthful food systems. Food and Nutrition Bulletin, 21, 361–366.

World Bank. (1986). *Poverty and Hunger: Issues and Options for Food Security in Developing Countries*. Washington, DC: World Bank.

World Health Organization (WHO). (2002). *The World Health Report*. Geneva: WHO.

Wossen, T., Berger, T., Haile, M. G., & Troost, C. (2018). Impacts of climate variability and food price volatility on household income and food security of farm households in East and West Africa. Agricultural Systems, 163, 7–15.

Yonar, A., Yonar, H., Mishra, P., Kumari, B., Abotaleb, M., & Badr, A. (2021). Modeling and forecasting of wheat of South Asian region countries and role in food security. Advances in Computational Intelligence, 1(6), 1–8.

Zellner, A., & Theil, H. (1992). Three-stage least squares: simultaneous estimation of simultaneous equations. In *Henri Theil's Contributions to Economics and Econometrics* (pp. 147–178). Dordrecht: Springer.

Ziervogel, G., Nyong, A., Osman, B., Conde, C., Cortés, S., & Downing, T. (2006). *Climate Variability and Change: Implications for Household Food Security*. Washington, DC: Assessment of Impacts and Adaptations to Climate Change (AIACC).

APPENDIX 4.1

List of countries included in the study

Afghanistan	Bolivia	Cook Islands	Finland	Iraq	Malaysia
Albania	Bosnia and Herzegovina	Costa Rica	France	Ireland	Maldives
Algeria	Botswana	Cote d'Ivoire	French Polynesia	Israel	Mali
American Samoa	Brazil	Croatia	Gabon	Italy	Malta
Andorra	Brunei Darussalam	Cuba	Gambia	Jamaica	Marshall Islands
Angola	Bulgaria	Cyprus	Georgia	Japan	Mauritania
Antigua and Barbuda	Burkina Faso	Czechia	Germany	Jordan	Mauritius
Argentina	Burundi	Democratic People's Republic of Korea	Ghana	Kazakhstan	Mexico
Armenia	Cabo Verde	Democratic Republic of the Congo	Greece	Kenya	Micronesia (Federated States of)
Australia	Cambodia	Denmark	Greenland	Kiribati	Mongolia
Austria	Cameroon	Djibouti	Grenada	Kuwait	Montenegro
Azerbaijan	Canada	Dominica	Guatemala	Kyrgyzstan	Morocco
Bahamas	Central African Republic	Dominican Republic	Guinea	Lao	Mozambique
Bahrain	Chad	Ecuador	Guinea-Bissau	Latvia	Myanmar
Bangladesh	Chile	Egypt	Guyana	Lebanon	Namibia
Barbados	China, Hong Kong SAR	El Salvador	Haiti	Lesotho	Nauru
Belarus	China, Macao SAR	Equatorial Guinea	Honduras	Liberia	Nepal
Belgium	China, mainland	Eritrea	Hungary	Libya	Netherlands
Belize	China, Taiwan Province	Estonia	Iceland	Lithuania	New Caledonia
Benin	Colombia	Eswatini	India	Luxembourg	New Zealand
Bermuda	Comoros	Ethiopia	Indonesia	Madagascar	Nicaragua
Bhutan	Congo	Fiji	Iran	Malawi	Niger

Nigeria	Saint Kitts and Nevis	Switzerland	Venezuela
Niue	Saint Lucia	Syrian Arab Republic	Viet Nam
North Macedonia	Saint Vincent and the Grenadines	Tajikistan	Yemen

(Continued)

APPENDIX 4.1 (*Continued*)

Norway	Samoa	Thailand	Zambia
Oman	Sao Tome and Principe	Timor-Leste	Zimbabwe
Pakistan	Saudi Arabia	Togo	
Palau	Senegal	Tokelau	
Palestine	Serbia	Tonga	
Panama	Seychelles	Trinidad and Tobago	
Papua New Guinea	Sierra Leone	Tunisia	
Paraguay	Singapore	Türkiye	
Peru	Slovakia	Turkmenistan	
Philippines	Slovenia	Tuvalu	
Poland	Solomon Islands	Uganda	
Portugal	Somalia	Ukraine	
Puerto Rico	South Africa	United Arab Emirates	
Qatar	South Sudan	The United Kingdom of Great Britain and Northern Ireland	
Republic of Korea	Spain	United Republic of Tanzania	
Republic of Moldova	Sri Lanka	The United States of America	
Romania	Sudan	Uruguay	
Russian Federation	Suriname	Uzbekistan	
Rwanda	Sweden	Vanuatu	

5

Nutritional Composition, Bioactive Compounds, and Phytochemicals of Wheat Grains

Akhlash P. Singh

CONTENTS

DOI: 10.1201/9781003307938-5

5.1 Introduction

Wheat (*Triticum aestivum* L.) is a main food crop that provides nearly 20% of the total calories' need to man every day. About 796 million tons of wheat grain were expected to be produced in 2023 a little bit higher of 2022's wheat production, attributed to better climatic conditions (Angelino et al., 2017; www.fao.org/worldfoodsituation/csdb/en/2023). Currently, the human population is growing very fast. Therefore, it has become essential to enhance wheat crop production by nearly 60% to fulfill the food requirements by 2050 (www.fao.org/temprefdocrep/fao/meeting/018/k6021e.pdf).

Wheat grain is a major source of carbohydrates and provides 55% of total carbohydrates obtained by food, so it offers humans most of the food calories (Liu, Yu, and Wu, 2020). The main wheat-cultivating countries are India, Russia, Ukraine, China, Turkey, Australia, the United States, and Pakistan. These countries collectively produce about 80% of the total wheat crop production world over. There are a few major world regions that consume most of the wheat grain in the form of different food items. These regions are West Asia, Western Europe, North Africa, sub-Saharan, Gulf countries, Egypt, South East Asia, China, and Mexico (I. Sharma, Tyagi, and Singh, 2015). In the last two decades, rapid urbanization and industrialization of cities have enhanced the demand for wheat as a major energy source in the world.

Although a large number of ancient and newly developed wheat cultivars are available, only a few wheat species are mainly cultivated in the wheat-growing regions, such as *Triticum aestivum*, a hexaploid species also known as "common" or "bread" wheat. Another is *T. turgidum* var. durum, a tetraploid species that grows in hot, dry climatic conditions. Often, the "durum wheat variety" is used for making pasta; therefore, this variety is also known as "pasta wheat". There are many ancient wheat varieties, namely einkorn (diploid *T. monococcum var.* monococcum), emmer (tetraploid *T. turgidum v*ar. dicoccum), and spelt (*T. aestivum* var. spelta), which are known for their better protein quality, high level of bionutrients, and mineral content, and, thus, these varieties are gaining popularity in the current era too (Gałkowska, Witczak, and Witczak, 2021). There are a few wheat varieties that are not significant for crop production, but they are cultivated for cultural reasons in very small arable regions.

Wheat grains are used by billions of people as staple diets in the form of chapati, pasta, bread, noodles, and other regional preparations around the world. It can be converted into flour to make numerous food item, including bread slices, pasta, instant noodles, and other sundry food products. Wheat grains and their based foods have become very significant since the inception of agricultural societies (Hayakawa, Tanaka, Nakamura, and Endo, 2004). Therefore, wheat grain has emerged and will remain relevant because the wheat crop offers raw materials for the food and feed industry in the near future. Additionally, wheat grain is major source for energy, raw materials, bioactive, and nutritive components that are associated with health benefits for human populations all over the world. The bioactive and nutritive compounds are not evenly distributed in the wheat grains but rather unevenly distributed in the various fractions of the grain such as bran, germ, aleurone layers, and endosperm (see Figure 5.1). Wheat bran is an important provider of phytochemicals, especially phenolic acids (phenolics, carotenoids, flavonoids, alkylresorcinols, tocopherols, phytosterols, benzoxazinoids (BXs)), and other bioactive components and nutritive components, including water and fat-soluble vitamers, for example, vitamins E and its vitamers (Saini et al., 2021). Many clinical and nutritional studies have proven that bioactive components present in whole wheat grains are effective against many lifestyle-related chronic diseases such as obesity, metabolic syndrome, cardiovascular diseases, diabetes mellitus, and cancers (Călinoiu and Vodnar, 2018). The potential effects of whole-grain-based diets against these diseases are endorsed by various human and animal-based experimental studies as well as World Health Organization (WHO) reports. Therefore, it is advised that the consumption of whole wheat grains is positively related to chronic diseases. It is well known that whole wheat grain diets contain many bioactive compounds, but their digestion and absorption and, thus, final bioavailability in the cellular systems/tissues, is the major issue (Angelino et al., 2017). Because the production of whole wheat grains and flour and different value-added food items has undergone tremendous changes via many food-processing and manufacturing stages, the content of bioactive nutrients present in the cereal grains has substantially reduced. Many factors affect the direct and indirect bioavailability of grain-derived functional nutritive compounds in the human body, such as downstream processes, germination, farming, debranning, milling, and fermentation of wheat grains (Angelino et al., 2017).

The main aims of this chapter are to describe the bioactive components, micronutrients, and macronutrients present in whole wheat grain and their potential beneficial effects on chronic diseases that are known to damage human health substantially. Furthermore, current chapter also focuses on the numerous post-harvesting and food-processing methods that influence the bioaccessibility and bioavailability of bioactive components, especially phenolic compounds, in the human digestive system. Finally, the current chapter encourages future studies on whole grains and their nutritional advantages.

5.2 Nutritional Components of Wheat Grain

Wheat is an inevitable part of a staple diet in the world because of its biochemical composition and balanced nutritive composition. Wheat grain provides most of the energy and nutrients for animals and the human population (Katileviciute et al., 2019). The nutritional composition analysis of wheat grains is essential to test the efficacy of various crop improvement programs and to ensure the suitability of wheat grains for human consumption. Therefore, nutritional analysis is a very appropriate option employed in wheat crop production, which shows that wheat grains offer a balanced amalgam of macro and micronutrients for the human diet (Ma, Wang, Feng, and Xu, 2021). A common wheat grain proximate analysis shows that it contains total carbohydrates, percentage of crude proteins, moisture, ether extract, crude fiber, and total ash (Table 5.1). It is well known that thousands of wheat varieties grow in the world, and their biochemical compositions are influenced by the types of varieties, cultivars, local growing conditions, application of fertilizers, and post-harvest downstream processing. For example, normal bread and durum wheat grains have lower levels of bioactive and micronutrients than colored or pigmented wheat grains. Quantitative and qualitative data related to the protein in common wheat showed that almost 11 to 18% higher protein levels, 7.31 to 18.13% more essential amino acids, and 8.88 to 18.91% more amino acids are measured in the pigmented wheat species (Saini et al., 2021). The proximate composition is also employed for approximating the nutritive value of foods such as wheat, rice, pulses, and other food items by using various biochemical quantitative tests. Wheat is an important agricultural commodity, and information related to the nutritive analysis of different wheat varieties is available in abundance, which shows that it offers a balanced diet in terms of proteins, fats, carbohydrates, dietary fiber, and minerals. The average values

TABLE 5.1

Comparison of Biochemical Composition (g/100 g) in the Wheat Grain

Name of Species	Total Carbohydrates	Crude Protein (N × 5.7)	Ether Extract	Dry Matter	Crude Fiber	Crude Ash
Common wheat (*Triticum aestivum*)	74.5	11.0	1.72	90.5	1.78	1.52
Einkorn wheat (*Triticum monococcum*)	71.9	12.8	2.17	90.7	1.96	1.86
Emmer wheat (*Triticum dicoccon*)	65.9	15.4	2.28	90.9	5.03	2.16
Spelt wheat (*Triticum spelta*)	62,3	18.1	2.43	90.4	5.19	2.65

Note: Common wheat (*Triticum aestivum*), einkorn wheat (*Triticum monococcum*), emmer wheat (*Triticum dicoccon*), and spelt wheat (*Triticum spelta*) source (Biel et al., 2021).

of nutritional components are given in Table 5.1 (Biel, Jaroszewska, Stankowski, Sobolewska, and Kępińska-Pacelik, 2021), which shows a typical proximate analysis of a few wheat varieties.

5.2.1 Carbohydrate

Wheat grain is the final product of wheat crop farming and harvesting. Many dietary carbohydrates are present in wheat grain, ranging from simple sugars to non-digestible complex polysaccharides, including both digestible and resistant starch, glucans, and lignin. Furthermore, wheat grain also contains smaller and lower molecular weight compounds such as monosaccharides, disaccharides, and oligosaccharides (Lineback and Rasper, 1988). Each type of carbohydrate requires different levels of digestion and metabolism in the human body, hence their bioavailability and bioaccessibility are major issues that will be discussed in the later part of this chapter. The total carbohydrates present in whole wheat grain can broadly be classified into two groups, for example, digestible and non-digestible.

5.2.1.1 Digestible Carbohydrates

Wheat grains consist of carbohydrates that can be digested in the mouth and gut called "digestible carbohydrates", for example, starch and many sugars.

5.2.1.1.1 Starch

Wheat grains are major sources of starch, a main dietary carbohydrate that provides a big chunk of cellular fuel responsible for energy production in cellular respiration. At the time of maturity, wheat grain consists of a total of 85% (w/w) carbohydrates, mainly comprising starch. Starch is mainly present in the endosperm part and is a major fraction of the whole grain (D'Appolonia and Rayas-Duarte, 1994a; Gooding and Shewry, 2022). Starch is a primary source of energy and is eaten in the form of mainly chapati, bread, and other wheat-based food items all over the globe.

The starch consists of two types of glucose-based polymers that are amylose and amylopectin in a 1:3 ratio in most wheat varieties. The amylose is a single long unbranched chain of many thousands of glucose residues linked via 1–4 glycosidic bonds. While amylopectin is an intensely branched polymer containing hundreds of glucose residues joined together by 1–6 glycosidic linkages, and overall, both polymers constitute starch, which has 100,000 glucose unit residues arranged in a highly intricate branching polymer that is further aggregated to form characteristic starch granules in wheat grain (Shevkani, Singh, and Bajaj, 2017).

5.2.1.1.2 Monosaccharides, Disaccharides, and Oligosaccharides

In addition to high molecular weight polysaccharides, wheat grain also comprises low molecular weight monosaccharides (glucose), disaccharides, and oligosaccharides, which constitute approximately 7% of the total carbohydrates of wheat grain. These sugars mainly present in the different fractions of wheat grain, for example, aleurone, endosperm, and embryonic tissues. Fructans is another important sugar present in the starchy endosperm and bran, which are two major parts of the whole grain (Shevkani et al., 2017).

5.2.1.2 Non-Digestive Carbohydrates

Non-digestible carbohydrates are another category of carbohydrates provided by the wheat grains, for example, resistant starches (RSS), chemically modified starches, complex non-starchy polysaccharides, and cell wall polysaccharides. Among them, resistance starch is very significant and constitutes the major chunk of dietary fiber.

Resistant starch is not digested by gut enzymes due to its entrapment in the food structure. This can be further divided into five different categories, that is, ranging from RS 1 to RS5 types. Generally, the resistance characteristic of starch in the digestion system is mainly inferred by the size, shape, composition, and crystal structure of starch granules. Furthermore, the characteristics of starch resistance are influenced by the composition of amylose and amylopectin, proteins, lipid-based compounds, and phosphate ions present in the grain (D'Appolonia and Rayas-Duarte, 1994b). Resistant starch is part of dietary fiber and has the same properties as the non-starchy polysaccharides (NSP), which are fermented by gut microorganisms and metabolized into short chain fatty acids (SCFAs), for instance, acetate, butyrate, and propionate. These metabolites are highly significant for human and animal metabolisms. Whole-grain bread is the major source of dietary fiber (Ferguson and Harris, 1999; Niba and Niba, 2003). Currently, dietary fiber is considered to be a major agent that protects the human population from many types of chronic diseases such as diabetes, heart diseases, and cancer of the large intestine.

In addition to resistant starch, cell-wall-based polysaccharides are also the major part of grain dietary fiber. Cell wall polysaccharides are about 12% and are mainly present in all tissues of wheat plants. These polysaccharides are conjugated with other components like proteins, ions, and other branched polymers of plant cells and make an intricate interwoven network of dietary fiber. Hence, dietary fiber is resilient to digestion process in the gut. The major cell wall-based polysaccharides are cellulose, lignin, pectin, fructans, and dimers of ferulic acid and other polysaccharides. The other cell wall polysaccharides of the wheat grain are arabinoxylan and β-D-glucan, which contain a cellulose-like structure and are also known as wheat bran β-glucan. Additionally, cell wall polysaccharide also contains small amounts of cellulose ((1→4)-β-D-glucan), glucomannan, callose ((1→3)-β-D-glucan), xyloglucan, and pectins. Lignin is also an important cell wall glucan consisting of monomeric units of aromatic alcohols. Lignin is mainly present in the pericarp/seed coat of a whole grain, which makes bran fraction a rich source of dietary fiber, but it is generally not present in white flour (Parker, Ng, and Waldron, 2005).

The total dietary fiber is classified into two main classes based on its water-binding capacities inside the human gut. First is soluble dietary fiber (SDF), which is soluble in water and mainly presents in bran which accounts for 82% of total soluble dietary fiber. Grain analysis shows that about 6.5 to 52.4 g/100 g of dry weight, dietary fiber has been quantified in the wheat bran (Liu et al., 2020). While the amount of dietary fiber in the grain part is about 11.6 to 17.0 g per 100 g of DM, which is substantially lower than the bran fraction (Liu et al., 2020).

Second, the insoluble dietary fiber, which is not soluble in water, includes fructans, cellulose, arabinoxylan, and lignin. The amounts of insoluble dietary fiber are highly variable in terms of amounts among the various wheat varieties. DFs from wheat bran can influence the growth of specific bacteria, which often promotes the growth of SAFCs forming bacteria. These microbes produce specific types of metabolites via fermentation that further promote the growth of mucus-producing bacteria. Mucus-producing bacteria improve the intestinal barrier and regulate the function of the host immune system (Yao, Gong, Li, Hu, and You, 2022). Many scientific investigations have shown that gut bacteria also reduce inflammation by reducing cytokine levels.

5.2.2 Proteins

Grain protein is classified as a macronutrient and, thus, somehow, fulfills protein requirements in wheat-consuming population. In most wheat varieties, the amount of protein varies from 10 to15% of the dry weight and is distributed unevenly in various parts of the grain. For example, protein content in the various parts of grain was 5.1%, 5.7 %, 22.8%, and 34.1% in pericarp, testa, aleurone, and germ, respectively (Shewry and Hey, 2015). It is noticeable that the protein quantity of the wheat grain composition is influenced by both genetic and environmental factors, for example, the types of cultivars and especially the application of nitrogen fertilizers. Generally, quality of wheat protein is mainly determined by the content of essential amino acids. Because a complete standard set of 20 amino acids is not produced by human cells, essential amino acids must be added to the human diet. Being a cereal, wheat often lacks lysine amino acid, but the rest of the other necessary amino acids are present in sufficient quantities (El-Naggar, de Neergaard, and El-Araby, 2009). So, wheat grain can be considered a good source of protein. Wheat grain possesses a unique amino acid profile, characterized by a high content of glutamine and proline residues and a lower lysine concentration.

Wheat grain also expressed highly specific protein profiles, including water-soluble albumins, globulins, gliadins, and glutenin. Among them, two proteins, albumin and globulin, account for 20 to 25% of the whole proteins present in grain, which are mainly present in the endosperm fraction of grain. Both water-soluble albumin and globulin are responsible for enhancing the biological value of cereal protein, which is

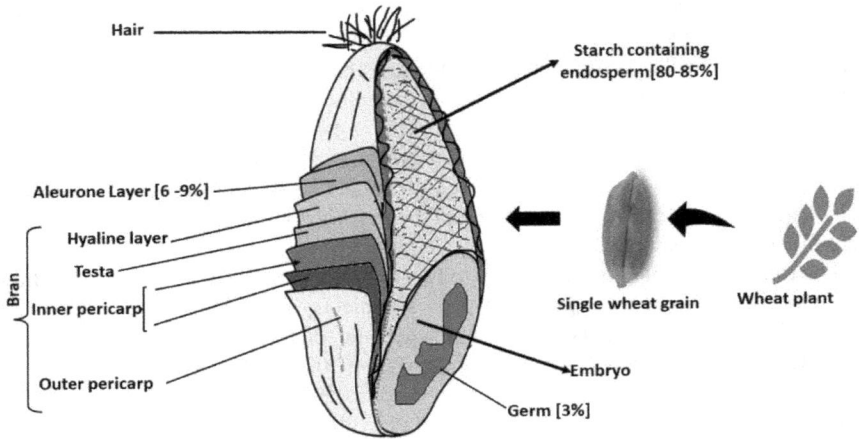

FIGURE 5.1   The histological structure of a single whole wheat grain is partitioned into three important fractions that are endosperm, germ, and bran. The bran part is further divided into many layers.

known as superior nutritional indicator (Merlino, Leroy, Chambon, and Branlard, 2009). Moreover, both albumin and globulin fractions indirectly influence the bread-producing process because both are important for the normal baking properties of bread. Albumins and globulins contain more necessary amino acids in comparison to both glutenin and gliadin, such as arginine, aspartic acid, lysine, threonine, and tryptophan amino acids.

Wheat gluten, a complex protein composed of glutenin and gliadin, also known collectively as gluten proteins, accounts for 63–90% of total grain proteins. Two proteins gliadin and glutenin are crucial for the bread production qualities of wheat-based flour because both proteins are important for water and gas retention at the time of bread and chapati preparation (Mills et al., 2020). Therefore, they determine the physio-chemical properties of wheat-based food products, including many bread qualities. Moreover, aforementioned proteins create a continuous, cohesive network that encircles the starch molecules, which is very crucial for dough formation and bread quality because it influences loaf area, breadcrumbs, and the overall texture of the bread (Brouns et al., 2022).

In some rare health conditions, the consumption of wheat seed proteins can cause a variety of immunological diseases, such as gluten-sensitive enteropathy, also known as celiac disease (CD), allergy and asthma, and wheat-dependent exercise-induced anaphylaxis (WDEIA) in genetically susceptible people (Wieser, Koehler, and Scherf, 2020). Globulin was the first storage protein discovered in wheat that was found to cause diabetes (T1D) and celiac disease simultaneously. Likewise, gliadins, a fraction of gluten, are known to have epitopes that elicit immune responses and cause celiac disease. The main reason for the development of CD is that gliadin cannot be digested in the gastrointestinal tract due to its unique sequence of amino acids that are not acted upon by enzymes (Wieser et al., 2020). Therefore, partially digested peptides can elicit an innate immune response by physically interacting with the small intestine mucosa and causing the CD.

5.2.3 Lipid

Wheat grain is not a good source of fat in the human diet and only contains about 2.2% of its dry weight. Wheat fatty acids can be divided into various groups based on their nutritional values. For example, wheat grains contain 1.4 g of saturated fatty acids/100 g of dry weight, 1.1 g of monounsaturated fatty acids/100 g of dry weight, and 4.3 g of polyunsaturated fatty acid/100 g of dry weight (Boukid, Folloni, Ranieri, and Vittadini, 2018). Although wheat lipids do not contribute nutritionally, their interactions with protein and carbohydrates impart texture and quality to bread and wheat-based food. In wheat grain, fatty acids are available in the form of triacyl glycerols (TAGs). On the other hand, bound-form fatty acids are present in the form of phosphatidylcholine, phosphatidyl ethanolamine, and phosphatidyl serine, as well as lysophosphatidyl derivatives. Many biochemical analyses show that wheat grain comprises a substantial amount of unsaturated fatty acids, for example, oleic, linoleic, and α-linoleic acids. Recently, studies on hamsters indicate that wheat bran oil (WBO) is equally significant as rice bran oil (RBO) by inhibiting both cholesterol and fatty acid synthesis inside the liver by decreasing the activities of hepatic 3-hydroxy-3-methylglutaryl-CoA reductase and fatty acid synthase enzymes. This suggests that wheat bran oil can replace rice bran oil in the human diet (Lei, Chen, Liu, Wang, and Zhao, 2018).

So far, the aforementioned discussion was based on the whole grain, but the total nutrients in wheat grain are divided into the various fractions of wheat grain, which are mentioned later in the chapter. Commercially, wheat kernels undergo many industrial and downstream processes that give rise to many food products. The complete wheat kernel can be processed into bran, germ, refined flour, and whole-grain flour-based food items that contain an uneven distribution of nutrients and bioactive components (Katileviciute et al., 2019). Table 5.2 shows the presence of the bionutrients distributed in the different

TABLE 5.2

Comparison of Nutritional Contents Extracted from Different Parts of the Whole Grain and Refined Flour in Wheat (µg/100 g)

Content	Aleurone	Bran	Germ	Whole Wheat Grain	White Wheat Flour
Carbohydrates	23	64.5	51.3	62.2	71
Proteins	23	15.5	28.1	10	12.6
Fats	23.6	4.25	9.6	2	1.1
DF	-	42.5	12.3	11	4
Thiamine 1	1.4	.523	1.45	.4	.07
Riboflavin 2	.2	.577	.61	.15	.04
Niacin	32.9	13.5		5.7	1
Pyridoxine 6	1.3	1.303	1.42	.35	.12
Folates	.2	79	-	37	22
Iron	26	10.57	5	4	.8
Zinc	14	7.27	17.8	2.9	.64
Magnesium	-	611	259	124	20
Sodium	1.2	2	5.5	5	2
β-Glucan	-	3	-	.7	.08

Source: (Balandrán-Quintana et al., 2015; Cheng et al., 2022)

parts of wheat grain. Therefore, it becomes essential to discuss various parts/fractions of the grain and their nutritive values, which are currently exploited in the protection of many chronic ailments such as cancers, diabetes, and cardiovascular diseases. Structurally, a complete wheat kernel can be divided into endosperm, bran, and germ (Figure 5.1).

1. **Endosperm**: The most of the starch and proteins exist in the endosperm which constitutes nearly 80–85% of the grain. The endosperm is wrapped in numerous outer layers, also referred to as the aleurone part. The aleurone contains living cells that are very rich in bioactive substances with proteins, arabinoxylans, and β-glucans specifically present in the cell wall.

2. **Bran**: The wheat bran contains pericarp and testa, which are about 13–17% of the whole grain. The testa is a cuticular part that is lignin-rich, water-insoluble, characterized by alkylresorcinols, lipid-rich compounds that occur in the cuticle of grain. The grain pericarp constitutes the outermost layers of the whole grain, which is further classified into inner pericarp, outer pericarp testa, hyaline layer, and endocarp layers.

3. **Wheat germ (WG)**: Wheat germ makes up 2–3% of the grain and includes the embryo and the scutellum. Both parts of the embryo and the scutellum are isolated by aleurone from the endosperm. The scutellum part of the grain is considered a storage region that lies between the pericarp and the embryo (Boukid et al., 2018; Panato, Antonini, Bortolotti, and Ninfali, 2017; Turnbull and Rahman, 2002). During the milling process of grain, the scutellum is excluded, and the germ is retained as an embryo.

5.3 Micronutrients in Wheat Grains

Micronutrients are required for growth and development of the human body and thus need to be included sufficiently in the human diet. The deficiency of micronutrients is the major nutritional problem in the world, particularly in developing nations where children and women do not get the daily recommended amount (RDA) of vitamins. Generally, vitamins are produced by human cells, while plants and microorganisms can synthesize a complete spectrum of water-soluble vitamins. Micronutrients have a high level of modulatory effects on the biochemical reactions inside the cell, especially in the human context (Xu, An, Li, and Xu, 2011a). For instance, many coenzymatic forms of vitamins and metal ions act as coenzymes and cofactors, respectively, in enzyme-mediated biochemical reactions. Micronutrients also act as antioxidant agents against oxidation of biomolecules caused by reactive oxygen species (ROS) attributed to an oxidative environment in a cellular milieu. Generally, micronutrients are mostly vitamins and minerals (Hu and Schmidhalter, 2001). Vitamins are divided into two categories: water-soluble and fat-soluble vitamins. Most vitamins are present in whole wheat grains.

5.3.1 Vitamins

Vitamins contribute significantly to maintaining a good health in animals and humans as well. They are required in minute quantities in the human diet every day. Whole grains provide most of the water and fat-soluble vitamins. Water-soluble vitamins include vitamins-B (B1 thiamine), B2 (riboflavin), B3 (niacin), B6 (pyridoxine), B9 (folate), and vitamin E (tocopherol/tocotrienol). In addition to B-complex vitamins, humans also require fat-soluble vitamins in a minute quantity for vital biochemical reactions, for example, vision, blood clotting, and reproduction. There are a few lipid-soluble vitamins like provitamin A (β-carotene), vitamin D (calciferol), and vitamin K (phylloquinone) occurring in whole wheat grains (Dhua, Kumar, Kumar, Singh, and Sharanagat, 2021; Shewry et al., 2011).

5.3.1.1 Water-Soluble Vitamin B Complex

Many vitamins which belong to the B-complex are present in wheat grain, such as thiamine, riboflavin, pyridoxine, niacin, pantothenic acid, biotin, and folates. In most cases, vitamins are generally present in a group of whole grains. These vitamin molecules are water-soluble, a common feature among them, but otherwise they play highly diverse biochemical roles, particularly in the carbohydrates' metabolism (thiamine), and proteins, and fats (riboflavin and pyridoxine), and folates in iron metabolism. Vitamins are mainly available in the bran and germ fractions of the grain, for instance, aleurone layer. Cereals and cereal-based products provide around 30% of the daily need of vitamins in the diet. Many factors affect the availability of vitamins in wheat-based foods, for example, wheat varieties, growing location, use of manure, fertilizers or herbicides, storage conditions, milling process, and bread making (Heshe, Haki, Woldegiorgis, and Gemede, 2016). This fact is endorsed by the experiment conducted on whole meal flours of 24 winter wheat varieties grown in four different locations, that is, the United Kingdom, Poland, France, and Hungary. Two spring varieties were grown on the same sites, and the total contents of thiamine, riboflavin, pyridoxine and the biological form of niacin (B3) were quantified. The values of B1 (5.53 to 13.55 g/g DW), B2 (0.77 to 1.40 g/g DW), and B6 (1.27 to 2.97 g/g DW) were in complete agreement with earlier reports. The milling process significantly reduced the number of vitamins, by about 68% of the total thiamine, 58.65% of the riboflavin, and 85% of the pyridoxine (Shewry et al., 2011). It would be noteworthy here that the use of natural manure increases the B vitamins, while the application of fertilizers or herbicides reduces the riboflavin concentration in wheat crops.

5.3.1.2 Thiamine (B1)

Wheat grains and its based food products are the main sources of vitamin B1 or thiamine. Hence, it is recommended that there should be the inclusion of whole wheat grains and bran in human and animal diets because the outer coat of the seed is very rich in thiamine. In the case of wheat grain, thiamine is mainly present in the scutellum in the free form. Thiamine pyrophosphate (TPP) is a coenzymatic active form of thiamine that mainly participates in carbohydrate metabolism. The deficiency of thiamine in the diet promotes body fat accumulation that is ascribed to the diversion of excessive carbohydrates toward fat biosynthesis due to sluggish carbohydrate metabolism, including glycolysis and TCA (Goyer, 2010). Hence, the deficiency of thiamine promotes obesity, metabolic syndrome, and CVD. Beriberi is the main disease that has been documented in human subjects who have a deficiency of thiamine. In the biochemical analysis of several wheat genotypes (among them, 46 bread wheat,

1 durum bread, and 2 spelt), the average value of thiamine was 3.82 g/g dry matter (DM), which was mainly reduced due to the fine grinding and milling process of whole grains (Batifoulier, Verny, Chanliaud, Rémésy, and Demigné, 2006).

5.3.1.3 Riboflavin (B2)

Wheat grain consists of riboflavin, a vitamin that plays a very significant role in cellular metabolism, particularly the metabolism of carbohydrates and proteins. There are two coenzymatic forms of riboflavin: (1) flavin adenine dinucleotide (FAD) and (2) flavin mononucleotide (FMN). Both of these forms act as intermediates of hydrogen donors and acceptors in the electron transport chain (ETC), where electrons travel through various proteins in cellular respiration, which is catalyzed by a plethora of mitochondrial enzymes. Although dairy products are a very rich source of riboflavin, whole wheat grain is provider of riboflavin; especially, bran fraction offers up to 20% riboflavin in the daily intake (Hrubša et al., 2022; Thakur, Tomar, Singh, Mandal, and Arora, 2017). The average riboflavin value recorded in the wheat grain was 0.73 g per 100 DM, with a range of 0.48 g per 100 DM ('Victo' cultivar) to 1.07 g per 100 DM (blue cultivar 'Meropa'). Genetic variability, down processing, and milling processes are a few factors that substantially influence the availability of riboflavin in diets (Shewry et al., 2011). The deficiency of riboflavin can affect lipid peroxidation and oxidative stress inside the cell. Hence, the proper intake of riboflavin reduces the occurrence of some cancers, and it is also implicated in some neurological disorders.

5.3.1.4 Niacin (B3)

Niacin, or nicotinic acid, is widely distributed in plants and animals. Whole cereals and wheat grains comprise a high amount of niacin. Biochemically, niacin is known as nicotinic acid (pyridine-3-carboxylic acid), and its biologically active form is nicotinamide and its derivatives. In humans, niacin is synthesized in the hepatocytes from tryptophan, an aromatic amino acid. Niacin has two most important coenzymatic forms, namely NAD(H) (nicotinamide adenine dinucleotide) and NADP(H) (nicotinamide adenine dinucleotide phosphate), which involve nearly 200 different metabolic reactions, particularly oxidation and reduction reactions in which electrons are accepted and donated, especially in the case of glycolysis and the TCA cycle. Persistent deficiency of niacin can lead to dyslipidemia and atherosclerosis in humans. Furthermore, NADPH indirectly protects the cell from oxidative stress via reducing the enzyme glutathione reductase, which requires the regeneration of GSH to act as an antioxidant against ROS and LDL oxidation (Ganji, Kamanna, and Kashyap, 2003; Garg et al., 2017). In a very significant experiment, the bioavailability of niacin was estimated in 24 wheat varieties that are grown at various locations in four countries. The average amount of niacin was measured at about 0.161.74 µg/g DW. Niacin is mainly concentrated in the bran and/or germ, but downstream processes such as milling degrade up to 68% of the total amount of this vitamin (Batifoulier et al., 2006).

5.3.1.5 Pantothenic Acid (B5)

Pantothenic acid is a major water-soluble vitamin that is related to the vitamin B-complex. This vitamin is mainly synthesized in plants and is never present in human tissues. But whole grains and whole-grain-based diets are rich sources of pantothenic acid, especially in the brain, which is the major site of its occurrence. Pantothenic acid is the main precursor of the most important biomolecules that play a central role in the synthesis of central metabolites like acetyl CoA and an acyl carrier protein (ACP). Both metabolites, acetyl CoA and an acyl carrier protein (ACP), are involved in the oxidative metabolism of carbohydrates, transfer reactions in the citric acid cycle, and fat metabolism. It is already mentioned that glutathione dehydrogenase (GDH) acts as an antioxidant agent, and its production is promoted by pantothenic acid. The deficiency of B5 causes fatigue, sleep apnea, a weak digestive process, and neurol disorders (Tahiliani and Beinlich, 1991). The average amount of B5 in whole wheat ranged from 2.128 (emmer wheat grass) to 10.294 µg per gram of dry weight. In whole grains, average B5 content was recorded ranging from 3.296 to 10.294 µg per gram of dry weight in einkorn (IZA), emmer, durum, and bread wheat varieties (Pehlivan Karakas, Keskin, Agil, and Zencirci, 2021).

5.3.1.6 Pyridoxine (B6)

Pyridoxine, or vitamin B6, is collectively used for three related compounds, namely pyridoxine, pyridoxal, and pyridoxamine. It mainly occurs in plant-based foods, especially wheat grain and wheat grain foods, which are considered excellent sources of vitamin B6. In the wheat grain, pyridoxine is present in the aleurone layer and germ. Pyridoxal phosphate (PLP) is the main coenzymatic form of pyridoxine that particularly contributes to protein metabolism in transamination, decarboxylation, and racemization reactions. PLP also plays a crucial role in the synthesis of many vital biochemicals, such as serotonin, histamine, and niacin, from amino acids (Mateo Anson, Hemery, Bast, and Haenen, 2012). The average value of pyridoxine was 2.2 g per 100 DM, with a range of 1.45 to 3.16 g per 100 DM in different wheat cultivars (Mateo Anson et al., 2012).

5.3.1.7 Biotin (B7)

Biotin is also known as vitamin H, and it is mainly present in the free or bound form in wheat grain. Biotin is a heterocyclic sulfur-containing monocarboxylic with imidazole and thiophene rings, mainly attached to protein lysyl residues (biocytin) (McMahon, 2002). Normally, 11 µg/100 and 10.7 µg/100 biotin are present in whole grains and wheat flour, respectively, which may be present in free or conjugated form (Bryden, Mollah, and Gill, 1991). The milling process substantially reduces up to 70% to 80% of the biotin content present in whole cereals such as wheat, rice, rye, and maize. Biotin is very crucial in fatty acid metabolism and acts as a coenzyme for the enzyme acetyl-CoA carboxylase that catalyzes the carboxylation reactions, an initial step in fat biosynthesis. Biotin deficiency can lead to a loss of appetite and dermatitis in humans. The deficiency of biotin generally influence the transcription of genes that further affect pro-inflammatory processes like NF-kB (McMahon, 2002).

5.3.1.8 Folate (Vitamin 9)

Folates are classified under the vitamin B complex. It is also named as vitamin B9 and includes tetrahydrofolate and its derivatives (collectively termed folates), which are

considered essential micronutrients for human health due to their not being synthesized in human cells. Often, 400 µg of folate is recommended for daily consumption,
while pregnant women need 600 µg. Mostly, folates exist in reduced forms (tetrahydrofolate) rather than as folic acid (pteroylmonoglutamic acid) in whole wheat grain.
Folates are essential micronutrients for humans and participate in many metabolic
processes. In particular, folates play an important role in the one-carbon metabolism. There are many forms of folate, but 5-methyltetrahydrofolate is a biologically
active form that often participates in metabolic pathways. Folates act as the main
donors of one-carbon units and contribute mainly to many biochemical processes
that require single-carbon transfer, including purine biosynthesis and thymidylate.
It also participates in the synthesis of methionine, serine, pantothenate, and transfer RNA. Moreover, folate also provides methyl groups to the cellular methylation
process. Furthermore, folates also participate in scavenging activities such as the
removal of free radicals, as proven by many *in vitro* models. Indirectly, they also help
in the reduction of homocysteine (Ohrvik and Witthoft, 2011; Stanger et al., 2009).
Wheat contains a small amount of folate, ranging from 30.58 to 40.11 µg/100 g in
whole grains, while in flour it was in the range of 9.29 to 10.64 µg/100 g. It is reduced
significantly under various processing methods, such as storage (26%), milling (71%),
boiling (13%), and baking (16%). These methods reduce the folate content in wheat
and wheat products, while yeast-based fermentation improves the folate content in
foods (Liang, Wang, Shariful, Ye, and Zhang, 2020).

5.3.2 Fat-Soluble Vitamins

Many vitamins are soluble in fat and organic solvents but insoluble in water. Lipid
or fat-soluble vitamins are present in various foods, including whole wheat grains.
Vitamins like A, D, E, and K and their vitamers are classified as fat-soluble vitamins.
Fat-soluble vitamins are used in diets, cosmetics, pharma, and healthcare products
(Borel and Desmarchelier, 2018).

5.3.2.1 *Vitamin E (Tocotrienols)*

Vitamin E is a lipid-soluble vitamin that acts as the most active antioxidant available
in nature. This vitamin possesses the biochemical activity of α-tocopherol. Generally,
Vitamin E includes a group of different eight vitamers: four types of tocopherols
(α, β, γ, and δ) and four different corresponding tocotrienols (α, β, γ, and δ), and
these vitamers are not chemically identical (Jensen and Lauridsen, 2007). The α- and
γ-tocopherol forms are mainly present in bound form with transporting proteins and
are also present in the hepatocytes due to their excessive involvement in metabolic
reactions. However, the precise mechanism of action is not known, but experimental
results show that vitamin E plays a protective role against ROS produced during
low-LDL and PUFA oxidation, cancer, and cardiovascular diseases (CVD). The antioxidant potential of vitamin E is attributed to its refined phenolic chromanol head
group that is associated with plasma membrane (Azzi, 2007). Because α-tocopherol
is preferentially maintained in the plasma membrane and reduces the concentration
of peroxyl radicals, hence, vitamin E is also known as a very strong chain-breaking
antioxidant and is actively involved in scavenging properties against ROS and other
RNS. Vitamin E protects the plasma membrane and ultimately protects the RBC

from hemolysis. Currently, the supercritical fluid extraction (SFE) process is combined with carbon dioxide ($SFE-CO_2$) for the extraction of vitamin E from wheat germ, which is considered a best source of vitamin E (T. Wang and Johnson, 2001) and contains about (α-tocopherol 1159 µg/g dry wt.) which is a substantially higher amount than that present in different parts of wheat grain (Capitani, Mateo, and Nolasco, 2011).

However, fat-soluble vitamins such as provitamin A, vitamin D, and vitamin K are present in a very small amount in the pigmented rice varieties but are very feeble or not detected in the bread wheat varieties. For instance, black wheat contains 11.47 mg/kg of vitamin K, seven times greater than conventional wheat. Vitamin K is implicated in the blood-clotting process and bone metabolism (Suttie, 1992).

5.3.3 Betaine and Choline

Whole wheat grain is the main source of betaine and choline. Several experimental studies show that whole-grain-based diets contain a sufficient amount of betaine and choline. Both betaine and choline play a defensive role against lifestyle-related diseases. Choline is categorized as an essential nutrient and is synthesized via *de novo* synthesis pathways inside the cells. Choline is very important for living cells because its derivatives are mainly existing in the form of lipoprotein molecules, blood, and membranous lipids, and it acts as a precursor of the neurotransmitter acetylcholine. Choline and betaine are both providers of one carbon unit in the folate-dependent metabolism (Likes, Madl, Zeisel, and Craig, 2007). In the case of wheat grain, choline is distributed in different fractions, such as bran, germ, and flour. Currently, LC-dilution mass spectroscopy is used to estimate the amount of both choline and betaine in the different fractions of hard wheat (*Triticum aestivum* L. cv. Tiger) grown in the winter season. The amount of choline in aleurone, bran, and flour was 209 mg/100 g, 102 mg/100 g, and 28 mg/100 g, respectively. Individual concentrations of betaine are also quantified in aleurone, bran, and flour and consist of 1553 mg/100 g, 867 mg/100 g, and 23 mg/100 g of sample, respectively (Likes et al., 2007). Although the exact measurement of choline is difficult to make because it is present in the forms of glycerophosphocholine, phosphocholine, and phosphatidylcholine, which are major constituents of the plasma membrane in the cell.

5.3.4 Minerals

Minerals are another class of micronutrients that are very essential for proper biochemical and physiological functions in humans as well as animals. Wheat grain provides many minerals, which are required for nutritional purposes in the human diet. Most of the mineral content is present in the bran fraction of the wheat grain (Cilla et al., 2019). Mineral content in the wheat crop is influenced by a wide range of environmental and genetic factors, as well as fertilizer and mineral content in the soil and their interaction. Moreover, the milling process substantially reduces the mineral content of wheat grains, which is reflected in white or refined wheat flour. The pearling process improves the mineral content because it releases the minerals entrapped in polysaccharides and components, such as hemicellulose, cellulose, lignin, and phytate, in the cell wall.

5.3.4.1 Phosphate (P)

Several mineral ions are always bound with phytic acid, which constitutes the phytate salts, for example, Fe, Ca, Mg, and Zn. In wheat grains, P is mainly present in the conjugated form of phytate, especially in the aleurone layer. Most of the phosphate content, up to 84%, is concentrated in the wheat bran. Phytate accounts for about 84% of the total phosphate present in the wheat bran. Moreover, phosphate is also associated with protein-rich globular structures, which are a type of granule. Thus, the bioavailability of phosphate from wheat-based food is poorly absorbed in the human digestive system. Experimental analysis shows that ancient wheat varieties are rich sources of various minerals, except for phosphate (Igrejas, Ikeda, and Guzmán, 2020). The common wheat variety contains nearly 5.16 g/kg phosphate, while three ancient wheat varieties contain phosphate in the range of 3.71 to 4.74 g/ kg. Phosphorus deficiency substantially influences bone health in humans as well as animals. Recently, a low-phytate wheat mutant (*lpa1–1*) has been developed by using the mutagenesis method, which reduces the phytate content in the grain by up to 35%, so the bioaccessibility of inorganic phosphate in wheat grain is enhanced (Venegas, Guttieri, and Jr, 2022).

5.3.4.2 Iron (Fe)

In wheat grains, Fe is mainly stored in the aleurone part, although its amount is higher in the crease region. Naturally, the crease region, which is surrounded by aleurone cells, offers minerals such as Cu, Fe, and Mn. The nucellar part abundantly contains Zn, K, and Ca. In the wheat grains, iron mainly occurs in the aleurone layer (bran) and often forms a complex with phytate in the form of myo-inositol phosphate 1, 2, 3, 4, 5, 6-hexa-kisphosphate that limits the bioavailability of iron in the diet of both humans and animals. Because mineral-phytate complexes are insoluble, they restrict the bioavailability of iron to humans and livestock. Currently, scientists are using many transgenic approaches to express indigenous phytase enzymes in the developing grain so that the mineral bioavailability can be increased (Brier et al., 2015). The interest in finding a heat-stable phytase enzyme is greater as it catalyzes the phytate complex hydrolysis at the time of food processing. Another alternative is to enhance the concentration of Fe in grains via crop-breeding programs. A comparison between common wheat and a few ancient grain varieties shows that the latter is richer and comprises iron up to 54.4 g/kg to 94.5 g/kg than common wheat, with 60.5 g/kg (Younas, Sadaqat, Kashif, Ahmed, and Farooq, 2020).

5.3.4.3 Potassium (K)

The potassium ion is very important for facilitating many biochemical reactions and participates in numerous enzymatic reactions. Potassium is an integral part of the structural components of many biomolecules; hence it supports cell integrity. Moreover, potassium is also important for nerve impulse or nerve conduction, which is extremely important in multicellular organisms. The wheat grains tested for potassium have been proved to be a good source of potassium with other elements too, which are essential for many biochemical reactions in human metabolism (Bahmanyar and Ranjbar, 2008). The common wheat (*Triticum aestivum*) consists of 4.74 g/kg of potassium, which is quite a lesser amount than ancient varieties (4.54 to 6.55 g/kg), and this amount is even greater than the RDA, that is, 4 g/kg (Zhang et al., 2010).

5.3.4.4 Zinc (Z)

Zinc is another essential micronutrient for the growth of humans and animals. The deficiency of zinc can lead to the weakness of appetite, poor growth, delayed skin repair, and a weak immune system. So, an essential daily zinc amount must be supplemented to wheat-based foods. Although the concentration of Zn in wheat is relatively low, the aleurone and bran layers comprise zinc (Kaur et al., 2021). Downstream processing of wheat is another important reason that can substantially reduce the amount of zinc in wheat-based flour, and, ultimately, the intake of zinc in the diet must be increased. Worldwide analysis of wheat grain composition indicates that the average Zn content is only 28.48 mg/kg. This amount is much less than the advised intake of zinc in the diet. The distribution of zinc varies in the three fractions, that is, bran, embryo, and endosperm, which are 14–16, 2–3, and 81–84% in the wheat grain, respectively. Some biochemical analysis also shows that the Zn content in bran is three times greater than that in the endosperm. Studies also show that the milling process reduces zinc content up to six times in refined flour (M. Wang, Kong, Liu, Fan, and Zhang, 2020). It is also observed that the Zn content of wheat grains also varies in the different genotypes grown in different locations in various countries.

5.3.4.5 Selenium (Se)

To investigate the variation in the average amount of selenium among 150 bread wheat lines, different varieties of wheat for selenium content were analyzed; einkorn (278.9 µg/kg), emmer (229.2 µg/kg), and spelt (209.0 µg/kg). In wheat, the higher quantities of Se were in grains than in bread (80.8 µg/kg) (Zhao et al., 2009). Generally, selenium occurs in the form of selenomethionine (Se-Met) in wheat grains, which is mainly incorporated into proteins in the place of methionine. Selenium is also part of many crucial proteins involved in redox reactions, for example, selenoprotein occurs in the plasma membrane where it acts as a carrier of Se. Some enzymes that are involved in the neutralization of ROS, such as thioredoxin reductase, iodothyronine deiodinases, and glutathione peroxidases, mainly depend on the selenium ion for their catalytic functions. Being an essential trace element, the Se scarcity is responsible for many metabolic diseases (Mateo Anson et al., 2012). Selenium also protects against oxidative stress and cardiovascular ailments.

5.3.4.6 Manganese (Mn)

Manganese (Mn) is another important trace element that contributes to the growth and development of humans. It supports many enzymes by acting as cofactors, which are very crucial in metabolic reactions. Like other nutritive components, Mn is present in a high quantity in the outermost layer of wheat grain, which means it is mostly present in the bran. Manganese deficiency influences mainly carbohydrate and lipid metabolism and ultimately affects the growth and development and reproduction processes in animals and man (Xu et al., 2011b). Generally, the Mn concentration in wheat is in the range of 32.2–70.6 mg/kg in 250 Chinese genotypes of wheat. However, genotype and environmental interactions are a very crucial factor that decides the Mn content in wheat grain. Currently, experimental studies have proved that the use of high-level phosphate-based fertilizer in wheat crops grown on calcareous soil enhanced the Mn amount in the wheat grain (Barman et al., 2017).

5.3.4.7 Copper (Cu)

Although copper deficiency does not occur in humans, its prolonged deficiency in the human diet can cause many adverse effects, for example, a low count of RBC and more frequent bacterial infections. Copper deficiency can be associated with defects in the connective, vascular tissue, and skeletal systems, and it helps to maintain strong bones and the immune system in humans. It also affects iron metabolism, which can lead to CVD and Alzheimer's disease. The common wheat variety consists of a substantially higher concentration of copper (2.345 mg/kg), which is almost 53 times more than ancient species of wheat (Saini et al., 2021).

Mineral deficiencies have very serious effects on health, productivity, and mental development that can last a lifetime. Moreover, the recommended amounts of minerals are not essential; rather, they must be available in a certain ratio in the human diet. Therefore, to fix these problems, the whole scientific community must work together.

5.4 Bioactive Phytochemicals

Diets containing whole grains (WG), fruit, and vegetables are responsible for reducing the incidences of many noncommunicable diseases in humans, for example, obesity, metabolic syndrome, diabetes, cardiovascular disease (CVD), and cancer. A whole-grain diet is proven to improve gut health and, simultaneously, reduce the risk of cancers, particularly colorectal cancer. The protective effects of whole grains can be attributed to the fact that whole grains contain a large number of bioactive compounds that positively impact overall metabolism and neutralize the effects of reactive oxygen species (ROS) or other harmful free radicals/ions. These ROS generally damage the genetic materials, proteins, and other vital biomolecules that are shielded by the bioactive components of whole grains. Bioactive compounds act as antioxidants inside the cellular environment and hence retain the vitality of the living system. Therefore, whole-grain-containing diets are recommended by dieticians and different agencies all over the world. Because wheat grain and its various fractions (bran, aleurone, and endosperm) consist of bioactive and nutritive compounds, they are linked to good human health and a lower risk of chronic diseases (Angelino et al., 2017; Ktenioudaki, Alvarez-Jubete, and Gallagher, 2015). However, major sources of bioactive and nutritive components are plants, animals, whole grains, fruit, vegetables, nuts, and oils. But there are certain groups of whole grains that are rich sources of bioactive agents.

Bioactive and nutritive components are chemical agents or substances that mould biological or metabolic reactions inside a cell and living system. Initially, this definition was only applied to drugs that were used to treat illness in humans. But later on, based on antioxidants protective effects against the harmful effects of free radicals, changing energy, utilized, reducing inflammatory pathways, and control metabolic disorders, this definition is also applied to nondrug compounds. Due to their protective effects, phenolic compounds, flavonoids, carotenoids, vitamins, and minerals are now known as bioactive and nutritive components. But a pertinent question arises as to why plants have a very rich repertoire of phenolic-based bioactive compounds (Saini et al., 2021; Witkamp, 2022).

Plants live in a sedentary mode of life to overcome negative effects induced by adverse environmental conditions in the surrounding ecosystem. Hence, plants have evolved various biochemical strategies by using bioactive components to neutralize the adverse effects of ROS, often produced by biotic and abiotic factors. In this section, the occurrence and distributions of bioactive and nutritive components present in the whole wheat grain and its fractions such as pericarp, germ, and endosperm will be discussed.

5.4.1 Phytochemicals

Phytochemicals are synthesized in the primary or secondary metabolism of plants. Generally, phytochemicals are mainly involved in defense mechanisms to protect plants against microbes and insects (https://en.wikipedia.org/wiki/Phytochemical). Wheat grains consist of diverse types of phytochemicals, including phenolic compounds, flavonoids, carotenoids, and other aromatic compounds. Currently, sophisticated analytical techniques such as reversed-phase UHPLC/ESI-QTOF-MS are used to discover a wide variety of phytochemicals in wheat grain (Aloo, Ofosu, and Oh, 2021). In the case of whole wheat grain, an abundance of phytochemicals is mainly distributed in the bran and germ.

5.4.1.1 *Phenolic Compounds*

Phenolic-based compounds are a very large group of aromatic compounds with a phenolic ring. Phenolic compounds act as bioactive compounds and are produced as metabolites in secondary metabolic pathways (e.g., the shikimic acid pathway, phenylpropanoid pathway, and flavonoid pathway) with high levels of antioxidant, antimicrobial, and anti-inflammatory activity that reduce the risk of noncommunicable lifestyle-related diseases and aging. Whole cereal grains are very rich sources of phenolic compounds, particularly wheat-grain-based foods. The names of few important phenolic compounds and their distributions are given in the Table 5.3. In wheat grain, an abundance of the phenolic compounds is present in the outer layers of the grains. After the industrial revolution, the milling process was extensively used to produce refined flour or white flour from the endosperm part, containing starch and mostly lacking phenolic compounds and vitamins that are lost during the downstream process of milling (Dhua et al., 2021; M. Sharma et al., 2016).

Biochemically, phenolic compounds contain phenolic groups as a common structural feature. The biosynthesis of phenolic compounds is started by aromatic amino acids like phenylalanine and tryptophan. The deamination of phenylamine is catalyzed by an enzyme called phenylalanine ammonia lyase (PAL), which is a regulatory enzyme of the aromatic acid biosynthesis pathway. Consequently, t-cinnamic acid is produced. This t-cinnamic acid is changed into p-coumaric acid by an important regulatory enzyme, cinnamate-4-hydroxylase (C4H). In the same metabolic pathway, many other phenolic compounds are produced, such as phenolic acids, coumarins, flavonoids, and lignans. Because phenolic compounds are a diverse group of aromatic compounds (Ma et al., 2022; H. Zhu, Liu, Yao, Wang, and Li, 2019), it becomes essential to first become familiar with the basic structure of phenolic structures and their

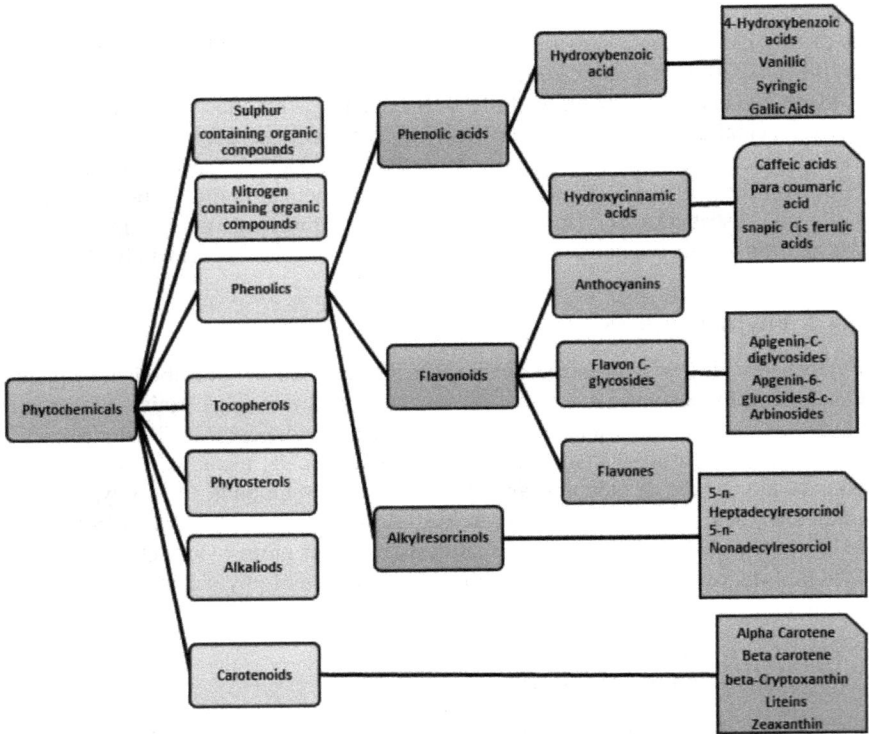

FIGURE 5.2 Detailed classification of phytochemicals in the whole wheat grain. It shows different phenolics, carotenoids, flavonoids, and alkylresorcinols.

parent compounds and, simultaneously, with the most extended classification of phytochemicals, including other phenolic compounds (see Figure 5.2).

Phenolic acids are major building blocks for intricate cell wall matrices that exist in a variety of cereal grains, legumes, and other seeds. However, there are three main types of phenolic acids found in grains in free, conjugated, and bound forms. But the total phenolic compounds distributed in the wheat grains are divided into two major groups: (i) Hydroxybenzoic acids and (ii) hydroxycinnamic acids. The nomenclature is based on positions when any specific functional groups are attached to the C1–C6 and C3–C6 phenol rings. Phenols are linked with a carboxylic acid known as phenolic acid, and if a carboxylic acid as a functional group is linked to the phenol ring, then hydroxybenzoic acid is formed. This type of phenolic compound is known as **hydroxybenzoic** acid. But when a phenol ring and a carboxylic acid functional group are separated via double carbon bonds (a C=C bond), then phenolic compounds are called **hydroxycinnamic acids**. There are a large number of derivatives of hydroxybenzoic acids, for example, p-hydroxybenzoic, vanillic, syringic, and gallic acids. Contrarily, derivatives of hydroxycinnamic acid are para-coumaric, ferulic, caffeic, and synaptic acid (Tufarelli, Casalino, D'Alessandro, and

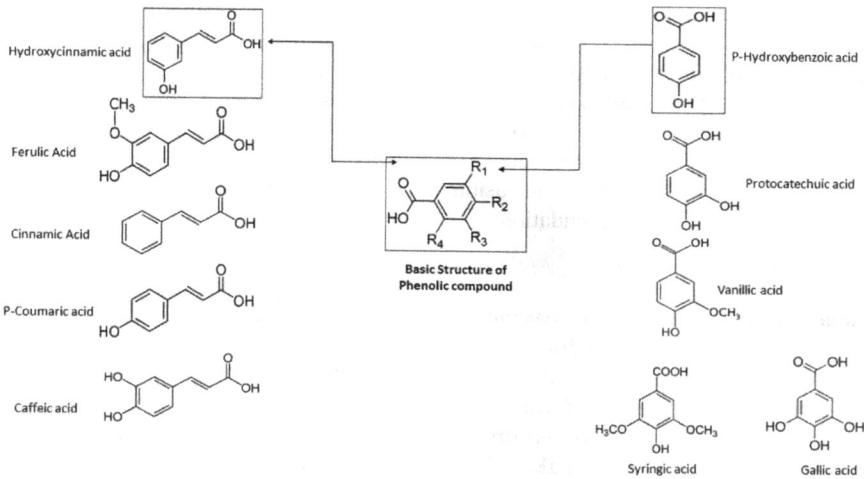

FIGURE 5.3 Two major types of phenolic compounds, which are divided into various types of phenolic compounds present in the various parts of fractions of whole wheat grains.

Laudadio, 2017), and their classification is given in Figure 5.2. The chemical structures of both types of phenolic compounds present in the whole wheat grain are shown in Figure 5.3.

5.4.1.1.1 Ferulic Acid

Among the phenolic acids, ferulic acid is more abundant than other acids like oxalic acid and p-coumaric acid, and it is present as esters and glycosides in whole wheat grain. Ferulic acid is always covalently conjugated to sugars, polysaccharides, glycoproteins, polyamines, and lignin in the plant cell wall. Ferulic acid is mainly distributed in the different parts of the whole grain and, more specifically, the embryo and pericarp, but a meager amount is also stored in the starch-rich endosperm (see Table 5.3). However, the wheat bran is considered a rich source of ferulic acid, where it is linked to hemicellulose, a cell wall polysaccharide, via ester bonds in cell walls (Dhua et al., 2021).

Wheat grains contain about 0.8–2 g of ferulic acid per kg of dry wt., which accounts for 90% of all phenolic compounds. It is already noted that cereals contain ferulic acid in three different forms: free, soluble, and conjugated. Insoluble or bound forms of ferulic acid make up roughly 90–95% of the total amount (Lu et al., 2014), so it is concluded that wheat comprises a much higher amount of bound ferulic acid than the free forms. In comparison to the hard wheat variety, soft varieties appear to have a larger concentration of ferulic acid. In plant cell walls, ferulic acid and other phenolic acids produce a sort of chemical and physical barrier by cross-linking saccharides/glycans that act against microbial infections and free radical damage. Ferulic acid also possesses astringent properties that keep animals and insects away from plants and seeds. Ferulic acid is a strong antioxidant like other phenolic compounds. Its mode of action is mediated by donating one hydrogen atom from its phenolic hydroxyl group in response to free radicals (Kumar and Pruthi, 2014). The two most potent effects of ferulic acid on human health are free radical scavenging and anti-inflammation, and both have been proven by many scientific investigations. Ferulic acid has been proven to be very effective in treating noninfectious diseases such as Alzheimer's disease, heart diseases, type-2-diabetes, cancers, and skin diseases. But there are certain challenges – for example, its proper inclusion in the diet and the quantification of the exact amount for a daily recommendation.

5.4.1.1.2 Sinapic Acid

Sinapic acid is the main hydroxycinnamic acid that has a very predominant presence in the plant kingdom, including fruits and vegetables, whole cereal grains, oil-containing seeds, spices, and medicinal plants, and, thus, it is a very common phenolic acid in the human diet. Generally, sinapic acid is always found in the form of esters of sugar (glycosides) and organic compounds, for example, sinapoyl esters and sinapoyl malate (Saini et al., 2021). After ferulic acid, sinapic is the most important phenolic acid present in wheat, which accounts for 10 to 12% of total phenolic compounds (see Table 5.3). Canadian wheat varieties comprise up to 17 to 31 µg/g of sinapic acid in the wheat flour, but vegetables from the Brassicaceae family contain a very high amount – nearly 100 µg/g. Sinapic acid and its derivatives belong to the Brassicaceae family (Tian, Chen, Tilley, and Li, 2021). It shows a high level of antioxidant activity against oxidative stress and antimicrobial, anti-inflammatory, anticancer, and anti-anxiety activity. There are two main sinapic-acid-based derivatives, such as 4-vinylsyringol and sinapine or sinapoyl choline, which are responsible for overcoming the aforementioned health problems in humans hence, currently being added to food processing, beverages, health products, and pharmaceuticals (Nićiforović and Abramovič, 2014).

5.4.1.1.3 P-Coumaric Acid

p-Coumaric acid is a hydroxyl-group-containing cinnamic acid. It is exists in low quantity at the center of the grain and has a higher amount in the outer layers of whole wheat grain. This phenolic acid is of high significance due to its chemoprotective and antioxidant properties. Most recently, the amounts of various phenolic compounds have been estimated by using the LC-ESI-MS/MS analysis. p-Coumaric acid up to 1.360 µg/g DW was present in the highest amounts in the grain and grass of traditional and modern wheat varieties (Aslam et al., 2021).

5.4.1.1.4 Caffeic Acid

Caffeic acid has a great potential to scavenge harmful reactive oxygen species (ROS). Caffeic acid is a naturally occurring cinnamic acid that is largely involved in lignin production. In several studies, caffeine has been linked to the regulation of cell development, turgor balance, photoperiodism, and plant–water relationships. Caffeic acid and its derivatives have a higher antioxidant capacity to protect plants from biotic and abiotic stresses – for example, the adverse effects of high temperature, water scarcity, heavy metal ions, and saline stress by regrowing roots and by upregulating antioxidant enzymes that modify the transcription of salt-tolerant genes. It also improves plant and water interactions and effectively scavenges ROS in plants. But the mechanism responsible for inferred tolerance against many forms of adverse environmental conditions is still poorly understood (Ma et al., 2021).

Phenolic acids are quantified in the two wheat varieties. The results show that ferulic acid has a high concentration, followed by other phenolic acids such as sinapic, syringic, vanillic, and p-coumaric acids see Table 5.3. Although the concentration of these phenolic compounds is very much influenced by the environmental conditions, the content of ferulic and sinapic acids was higher in the winter variety than that in the spring variety (Ahmad and Al-Shabib, 2020). The concentrations of caffeic acid, ferulic acid, p-coumaric acid, syringic acid, vanillic acid, and sinapic acid were 107, 268, 18.5, 24.4, 29.1, and 121 µg/g, respectively. Moreover, it was also observed that ferulic acid accounts for 48% and 60% of the bran of two spring and winter wheat grains, respectively (Katileviciute et al., 2019).

5.4.2 Carotenoid

Carotenoids are a major group of lipophilic bioactive compounds with many conjugated double bonds known as carotenoids. The number of carotenoids determines the pigment colors, which are responsible for the color of wheat grain. The major source of carotenoids is bran because of its excessive presence in the outer seed coat. Foods prepared from whole wheat grain have a bright yellow color due to the presence of carotenoid pigment, which is also an important parameter for food quality. The average carotenoid value is about 6.2 mg per kg of dry weight in the case (Borel and Desmarchelier, 2018; Mateo Anson et al., 2012) of durum wheat.

Chemically, carotenoids are 40 carbon-containing tetraterpenoid compounds containing a very reactive system of conjugated double bonds. This conjugation of double bonds makes carotenoids very interactive with ROS, and, thus, carotenoids are responsible for color and antioxidant activities. Carotenoids are synthesized from the very basic compound isoprenoids, which have a specific linear and symmetric arrangement of eight isoprenes. The basic cyclic structure of carotenoids can undergo many chemical changes, for example, dehydrogenation, hydrogenation, cyclization, and oxidation reactions. Therefore, it efficiently neutralizes the harmful RNS and ROS synthesized in response to biotic and abiotic stress inside the living systems (Ficco et al., 2014). In wheat grain, carotenoids are estimated, such as lutein, β-carotene, zeaxanthin, β-cryptoxanthin, β-apocarotenal, and antheraxanthin (see Figure 5.4). Food-processing methods, both milling and cooking, substantially damage the carotenoid pigments, but, still, some pigments are stable even at a higher temperature. During the storage of wheat grains and flour, carotenoids are reduced due to enzymatic and nonenzymatic oxidation reactions. Due to the bond conjugation process, double bonds absorb light. Hence, both temperature and light affect the biochemical activity of various types of carotenoids (Abdel-Aal, Young, and Rabalski, 2007).

5.4.2.1 Lutein

Among the major carotenoids, lutein is very significant, and it accounts for 70–80% of the total carotenoids present in the wheat grain, while the rest are zeaxanthin and β-carotene. For example, einkorn, Khorasan, and durum wheat varieties and

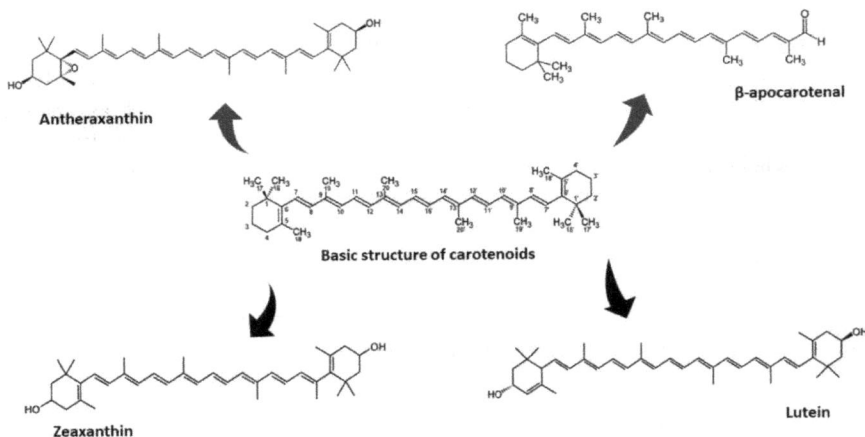

FIGURE 5.4 Structure of various types of carotenoids occurring in the whole wheat grain and its various fractions.

corn are rich sources of lutein and zeaxanthin. Bread made from modern wheat varieties contains small amounts of lutein and zeaxanthin, whereas bread made from ancient wheat varieties comprises large amounts (Melini, Melini, Luziatelli, and Ruzzi, 2020). The highest value of zeaxanthin was noticed in einkorn wheat, followed by the other four durum, emmer, spelt, and bread wheat. Ancient wheat varieties are also rich sources of carotenoids. Einkorn (*T. monococcum*), ancient diploid wheat, provides lutein up to 8.41 mg/kg on a dry weight basis. The amount of lutein in various fractions of wheat grain, especially the aleurone layer, starchy endosperm, and germ, comprises about 0.425, 0.557, and 2.157 mg per kg, respectively (Oduro-Obeng, Apea-Bah, Wang, Fu, and Beta, 2022). Many experiments show that high amounts of lutein in wheat flour induce the yellowish color of wheat-flour-based noodles.

5.4.2.2 Zeaxanthin

Zeaxanthin is a great ROS quencher and protects against the harmful effects of oxidative stress, hence it is categorized under potential bioactive compounds. Other benefits of zeaxanthin are also documented, such as the reduction of age-related ailments, antiretroviral, malignant tissue, and ocular disorders. Therefore, the supplementation of lutein and zeaxanthin into foods, drugs, nutraceuticals, and cosmetics is gaining ground at the current time. The einkorn and durum wheat varieties contain high amounts of carotenoids, but zeaxanthin is dominant, followed by α and β-carotene, β-cryptoxanthin, and lutein (Garcia Molina et al., 2021; Paznocht, Kotíková, Orsák, Lachman, and Martinek, 2019). Biochemically, zeaxanthin can change into moon-epoxidized xanthophyll antheraxanthin and another xanthophyll-based violaxanthin, but it could not be estimated in wheat grain. Different parts of wheat grains contain various amounts of zeaxanthin, for instance, the fractions, that is, endosperm, aleurone layer, and germ contain 0.557, 0.425, and 2.157 mg/kg of grain, respectively. In other experiments, zeaxanthin was estimated in aleurone and germ fractions with its amounts recorded as 0.776 and 3.094 mg/kg, but it is lacking in the endosperm completely (Lachman, Martinek, Kotíková, Orsák, and Šulc, 2017).

Moreover, colored wheat varieties, for example, purple wheat, consist of a higher quantity of lutein, particularly in the aleurone part (0.534 mg/kg) and lowest in the germinal part (1.714 mg/kg). The content of carotenoids is more in colored wheat cultivars as compared to noncolored wheat varieties. These results can be ascribed to color, which indicates high amounts of carotenoids, especially in the colored wheat variety (Lachman et al., 2017).

5.4.2.3 β-Cryptoxanthin

β-Cryptoxanthin is the most important oxygenated carotenoid and bioactive agent containing provitamin A activity. It is mainly present in fresh fruits, vegetables, and whole grains. The high concentration of β-cryptoxanthin in the serum is linked with a lower mortality rate in humans and is a great protectant against free radicals. β-Cryptoxanthin is utilized as a food supplement and additive and is also used as a natural food colorant. Biochemical analysis of 11 wheat varieties shows differences in the quantities of lutein, zeaxanthin, and β-cryptoxanthin by 5-, 3-, and 12-fold, respectively (Titcomb et al., 2018). The β-cryptoxanthin amount in wheat varieties ranged from 1.12 (0.13 g of β-cryptoxanthin/100 g of grain for W7985) to 13.28 (0.430 μg of β-cryptoxanthin/100 g of grain for Stoa variety with a 12-fold difference (p < 0.01)). Free β-cryptoxanthin and β-cryptoxanthin esters are distributed in the grain (Titcomb et al., 2018), but their distribution depends upon the cultivars, environmental conditions, growing sessions, extraction techniques used, and storage temperature of grains.

5.4.3 Flavonoids

Flavonoids are another major group of phenolic compounds that occur in wheat grains. Currently, about 5,000 types of flavonoids and related compounds have been identified and stored in databases. There are major flavonoids occurring in the whole wheat grain, and its various fractions are shown in Figure 5.5. Chemically, flavonoids contain a 15-carbon skeleton or parent compound consisting of two phenyl rings (referred to as "A and B rings") and a heterocyclic ring (the C ring), which is given in Figure 5.5. Flavonoids are divided into six different classes on the basis of their molecular structure, namely anthocyanidins, flavan-3-ols, flavanols, flavanones, flavones, and isoflavones, which are part of human diet. Flavonoids occur in the form of C-glycoside, and various types of flavone-C-glycosyl derivatives are present in the whole grain and its different fractions, such as germ, endosperm, and bran, so that this flavone-C-glycoside can act as biomarkers to identify the products produced from whole-wheat flour or refined flour (Dhua et al., 2021). These flavonoids are also responsible for the yellow pigment, which gives a specific color to wheat grain because anthocyanins are mainly present in the outer layers. It is well known that wheat grains are also colorful or pigmented – for example, black-grained wheat cultivars comprise a high content of anthocyanins, like blue- and purple-colored wheat grain with a purple pericarp. Whole wheat grains with yellow pigments comprise mainly four groups of flavonoids – for example, flavones, flavonols, flavanones, and flavans (Singla et al., 2019).

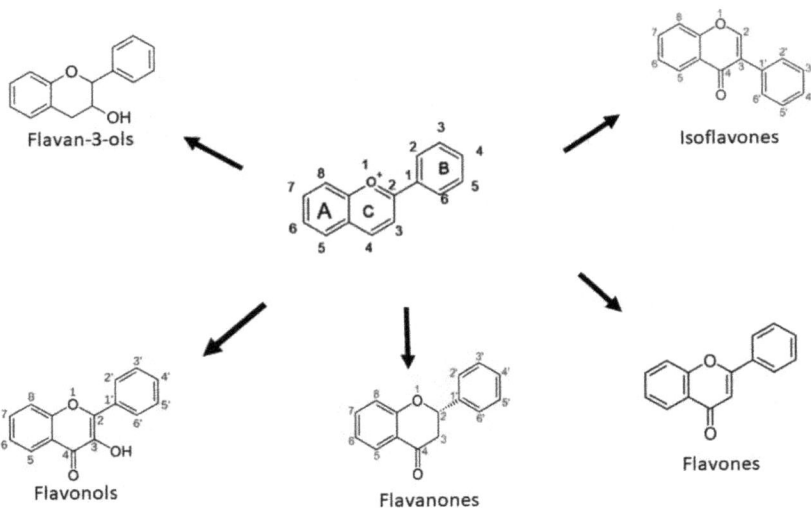

FIGURE 5.5 There are many types of flavonoids, such as anthocyanidins, flavan-3-ols, flavonols, flavanones, flavones, and isoflavones, which are mainly present in wheat grain. Nowadays, flavonoids are also included in the human diet because of their health benefits.

5.4.3.1 Flavan-3-ols

Generally, three polyphenols-based compounds, flavan-3-ols, flavan-4-ols, and flavan-3,4-diols, are synthesized in a flavonoid metabolic pathway. Flavan-3-ols with proanthocyanidins are responsible for giving rise to brown-color compounds by oxidation reaction. Some wheat varieties, particularly red-grained wheat coats, contain reddish brown-colored hydrophobic phlobaphenes (deoxy flavonoids) produced by oxidation and polymerization reactions (Leváková and Lacko-Bartošová, 2017).

5.4.3.2 Flavonols

One of the major flavonols is the apigenin C-glycosides that are mainly present in the durum and bread wheat varieties. For example, tricin, the most important flavonoid, is often used as a nutraceutical and is mainly present in the pericarp and aleurone layers of wheat grain (Lachman et al., 2017). Recently, different 32 flavones have been identified in wheat grain samples by using the UPLC-QTOF-MS which mainly includes 21 C-glycosidic forms, 7 O-glycosidic forms, and 4 aglycone compounds (Liyana-Pathirana and Shahidi, 2006). Several experiments show that red wheat contains a higher content of flavonoids, particularly flavonols, than white wheat. For example, it contains flavones in the range of 41.2 to 1126.0 µg/kg mg/kg flavones (apigenin, luteolin) (Lachman et al., 2017).

5.4.3.3 Flavanones

The bran fraction of wheat grain comprises flavanones, for example, naringenin. Additionally, wheat bran also contains minute amounts of catechin and its di-, tri-, and oligomeric proanthocyanidins. Two major techniques, such as multistage high-resolution mass spectroscopy and mass defect filtering, are used to extract flavanones from wheat germ (Anunciação et al., 2017). Flavanones are also used as biomarkers, which are generally used to discriminate between whole grains and refined wheat flour.

5.4.3.4 Flavones

Flavones are very important flavonoids in plants, which mainly act as natural pesticides against insects and fungi. Blue-colored flavones protect plants from UV-based damage because they consist of a double bond between C2 and C3 in the flavonoid skeleton. However, there is no functional group with C3 position. The main flavone, apigenin C-glycosides, is mainly present in the wheat variety of *Triticum durum* (Geng et al., 2016).

5.4.3.5 Isoflavones

Isoflavones participate in many antimicrobial mechanisms against many types of fungal pathogens. Isoflavonoid phytoalexins isoflavones are polyphenols mainly available in legumes, fruit and vegetables, and nuts. Although soybeans and soya-based foods are the main sources of isoflavones which are conjugated with glycosides, for example, daidzin, genistein, biochanin A, and formononetin. Generally, isoflavones are hardly detectable but if supplemented exogenously for example, soybeans and tofu-based okara that substantially enhanced the nutrition value of wheat-based food (Rinaldi et al., 2000).

5.4.3.6 Anthocyanin

Anthocyanins are major water-soluble aromatic compounds that act as natural pigments and mainly impart colors to grains, fruits, and vegetables. Anthocyanins are flavonoids represented by the aglycones and their glycosylated and acylated forms. Aglycone is a parent compound of anthocyanins that are attached to various types of sugars via hydroxyl groups at different positions. Individual anthocyanins differ from each other in respect of the number of hydroxy- or methoxy-groups of aglycones and particularly linked sugars by using O-glycosidic linkage at various positions (Figure 5.6). Furthermore, aliphatic or aromatic acids are linked with these sugars (He and Giusti, 2010).

5.4.3.7 Proanthocyanidin

Proanthocyanidin is a major class of flavonoids that are oligomers of catechin and epicatechin and esters of gallic acid. Phlobaphenes are mainly tannins, which are associated flavonoids synthesized by the nonenzymatic oxidation reactions of colorless polymers. The content and composition of various anthocyanins like delphinidin-3-glucoside, delphinidin-3-rutinoside, and malvidin-3-glucoside are varied in various colored wheat varieties. Ancient wheat varieties are a very rich source of proanthocyanidins, especially in bran (Ma et al., 2021). Wheat bran contains a very small amount

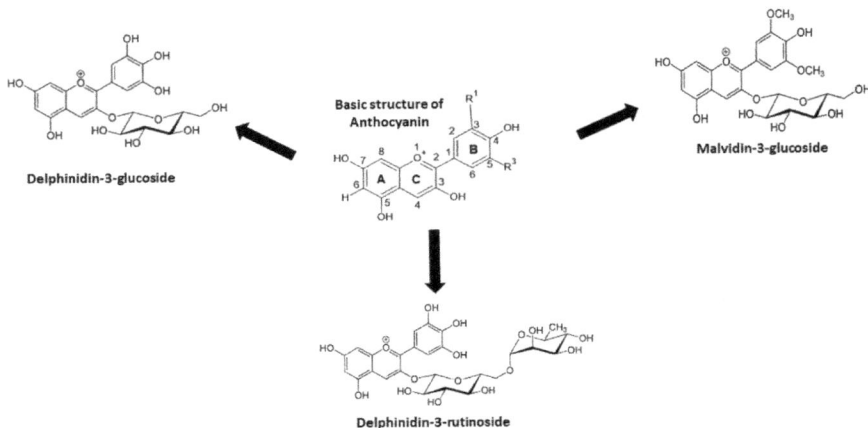

FIGURE 5.6 Structure of anthocyanins present in wheat grains.

of catechin procyanidin units. Although dimeric proanthocyanidins also include di-, tri-, and oligomeric proanthocyanidins, their detection is interfered with by methoxy hydroquinone glycosides (Dinelli et al., 2011). Hence, the detection of these compounds is very complicated.

The composition of anthocyanins is mainly determined by environmental and genetic factors, such as the combination of genes and shade, which gives rise to a very dark or black color for grains. Several factors, such as the spike, shading, magnesium-based fertilizers, and harvesting time of wheat crops, are also responsible for the amount of anthocyanin in plants (Matus-Cádiz et al., 2008). Recently, genetic and breeding programs have been launched to enhance the anthocyanin content in wheat varieties.

5.5 Benzoxazinoids (BXs)

Benzoxazinoids are other important phytochemicals present in whole wheat grain and rye, but the bran fraction is the major site where they are present in abundance. Currently, very sophisticated techniques like UPLC-qTOF-MS and LCMS are used to identify a range of benzoxazinoid metabolites, especially the hydroxamic acids (2,4-dihydroxy-1,4-benzoxazin-3-one based compounds such as DIBOA; 2,4-dihydroxy-7-methoxy-1,4-benzoxazin-3-one, DIMBOA); lactams (2-hydroxy-1,4-benzoxazin-3-one, HBOA); and benzoxazolinone (1,3-benzoxazol-2-one, BOA) (Shavit, Batyrshina, Dotan, and Tzin, 2018). Initially, the benzoxazinoid metabolites were found to be very effective against plant pathogens and acted as allelochemicals. The allelochemicals are natural products that help in communication among plants, with other plants, insects, or microbes, and are very important in plant microbes interactions. Several investigations have proved that benzoxazinoids are effective in anti-cancer, anti-allergy, and anti-inflammation. More recently, they found them to be quite effective against several cancerous cell lines – for example, the prostate cancer cell line. Furthermore, the most abundant water-soluble compound, 2,4-dihydroxy-1,4-benzoxazin-3-one (DIBOA),

TABLE 5.3

The Amount of Phenolic Acid in Various Fractions of Wheat Grain (µg/g)

Phenolic Compounds	Whole Grain	Wheat Bran	Wheat Aleurone
Ferulic acid	399–870	4,610–5,670	6,440–7,980
Ferulic acid dimers	19–150	780–1,550	360–950
Sinapic acid	13–18	115–276	269–353
p-Coumaric acid	15–28	130–162	160–288
Syringic acid	13–18	57.2	90.3
Vanillic acid	4.9–21	16.5	20.0

Source: Data from Barron et al. (2007); Hemery et al. (2009, 2010, 2011); Li et al. (2008); Noort et al. (2010); and Zhou et al. (2004)

caused the cell death of prostate cancer cells and ultimately reduced the growth of cancer tumors (Hanhineva et al., 2011; Landberg et al., 2019). The information related to their bioavailability, stability, absorbability, chemical modifications such as glucurono and sulfoconjugation, and the influence of gut microbial processing of benzoxazinoid metabolites is still not known.

5.6 Lignans

It is already mentioned before that lignin is the polymer of p-coumaryl, coniferyl, and sinapyl alcohols, which get polymerized through enzyme-based processes. Both lignans and lignin phenolic-based compounds occur in the bran part of wheat grain, which accounts for 3 to 7% of total dietary fiber. Owing to their polyphenolic structure, lignin and lignans are responsible for a high level of antioxidant activity, for example, protecting DNA against oxidative damage in cells (Eriksen et al., 2020). Lignin is the most important component of insoluble dietary fiber, which turns into the lignans and has been proven effective against colon cancer because of its effectiveness in disposing of excretory materials from the gut.

Plant lignans belong to the phytosterols, which are chemically plant-based diphenolic compounds. They resemble the 17-estradiol types of hormones. Plant lignans are responsible for reduced incidence of chronic diseases if included in human diets. The high concentrations of up to 6,700 µg/g and 2,270 µg/g of lignans are noticed in rye and wheat, respectively (Eriksen et al., 2020).

5.7 Alkylresorcinols

Alkylresorcinols are another class of phenolic compounds that occur in the external layer of whole wheat grains. They are also known as resorcinol lipids and are made up of resorcinol-type phenolic rings with odd-numbered aliphatic chains. The saturated aliphatic side chain in wheat-based ARs often contains 17, 19, 21, 23, or 25 carbons (Figure 5.7). Additionally, ARs also have unsaturated side chains, which have also been discovered in many other kinds of cereal (Mateo Anson et al., 2012). Alkylresorcinols are sort of biomarkers to identify whole-grain-based food products.

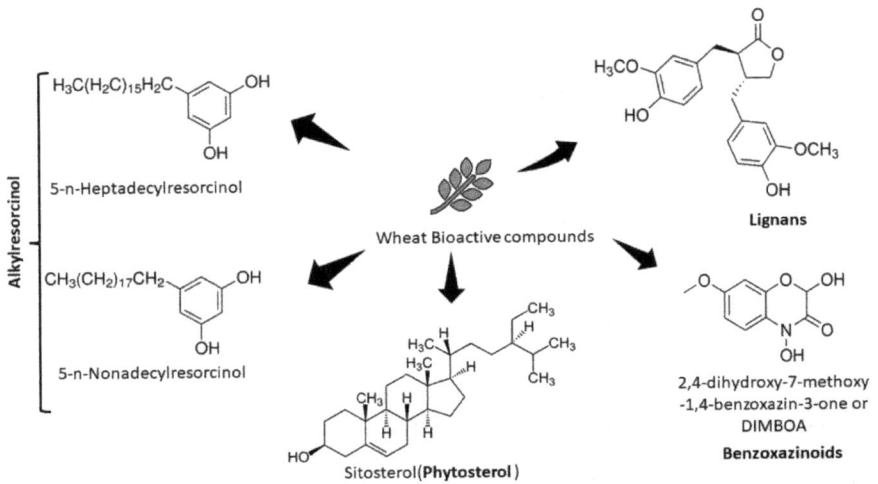

FIGURE 5.7 Alkylresorcinols, lignans, and benzoxazinoids are very significant bioactive compounds that are currently used as biomarkers and nutraceuticals. Phytosterol is another significant compound that shows a great resemblance to cholesterol, behaves like the hormones in the plant, and has several biological properties that protect against cardiovascular diseases.

Because alkylresorcinols (ARs) are present in abundance in the outermost layer of grains, they are also found in whole-grain barley, wheat, and rye. So far, 100 different types of ARs have been identified in plants and microbes (Landberg et al., 2019). To track the consumption of whole-grain products in the diet, ARs and their derivatives act as very promising biomarkers (McKeown et al., 2016). ARs are extracted in organic solvents – for example, acetone, ethyl acetate, and alcohol. The amount of ARs in various 175 genotypes of wheat ranged between 191 and 741 μg per gram of flour. Given the current scenario, future research related to ARs must be focused on how many food-making processes such as baking affect AR content, particularly those with unsaturated side chains. The concentration of ARs also depends upon genetics and environmental conditions. Therefore, creating new wheat genotypes with increased total AR content is an interesting new area of food research.

5.8 Phytosterol

Structurally, phytosterol shows a great resemblance to cholesterol, which is 27 carbon compounds associated with cardiovascular diseases although phytosterols contain an additional methyl or ethyl group linked to the sterol ring (Yoshida and Niki, 2003). In the case of wheat, almost 50% of all sterols come from phytosterols and sitosterol, but other sterols like campesterol, sitostanol, and campestanol are also present. GC-MS analysis (Prinsen, Gutierrez, Faulds, and Del Rio, 2014) clearly shows that phytosterols are present in either free or bound form with glycoside and ester conjugates. Biochemical analysis of 175 different wheat genotypes revealed total phytosterol concentrations in whole-wheat flour ranging from 670 to 1,187 μg/g. Total phytosterol content is comparable to phenolic acid concentration, which is commonly accepted as the main phytochemical in whole wheat grains. Phytosterols have a reducing effect on cholesterol and cancerous growth (Ramprasath and Awad, 2015; Tian et al., 2022). However, compared to phenolic acids, their general health advantages have received very little attention.

5.9 The Significance of Bioactive Compounds in Human Health

5.9.1 Obesity and Weight Management

In the above discussion, it is mentioned that wheat grains have numerous bioactive complexes which are responsible for reducing the possibility of the onset of obesity. Certain noncommunicable diseases such as metabolic syndrome, insulin resistance, diabetes mellitus, high blood pressure, CVDs, lack of sleep, and cancers are associated with excess body weight or obesity. Generally, triacylglycerols (TAGs) are long-term fuel in the human body, which provides more energy than carbohydrates and protein after their complete oxidation in the cell. But the excessive deposition of TAGs in the tissues under the subcutaneous layer of skin is harmful, which is also known as obesity or being overweight (Chooi, Ding, and Magkos, 2019; Kopelman, 2000). Many international agencies have tried to present a unitary definition of obesity based on bodily symptoms.

According to the National Institutes of Health (NIH) and the World Health Organization (WHO), obesity can be defined as having a body mass index (BMI) of over 25, and a man having it is considered overweight, and if with BMI over 30, he is being considered obese [*www.who.int/westernpacific/health-topics/obesity*]. Currently, obesity is one of the most serious public health challenges of this century, which is present in most countries worldwide. Earlier, obesity was more prevalent in developed countries after the great industrial revolution, but now it equally affects all countries. The occurrence of obesity has increased threefold in the last four decades, considerably spoiling public health (Ogden, Yanovski, Carroll, and Flegal, 2007). Extreme obesity can be attributed to the eating of refined flour in the form of fast foods, which certainly lack bioactive and traditional nutritive compounds that are generally present in whole grains.

Many short- [3 years] and long-term [12 years] experimental studies indicate that the consumption of a wheat-grain-based diet is inversely proportional to the incidence of obesity and being overweight. In similar studies, characteristics such as weight change, waist perimeter, body adiposity index, and consumption of dietary phytochemical index were measured for 3 years, reflecting the less frequent occurrence of obesity in adults who consumed whole grains (Liu et al., 2020; Witkamp, 2022). Several experimental proofs are put forward in support of the consumption of whole wheat grain-based diets, which create certain physiological effects that help in body weight management. Whole wheat grain is provider of traditional nutritive agents such as vitamins and their coenzymatic forms that enhance carbohydrate and lipid metabolic rates. Moreover, wheat grain diets provide more dietary fiber, nutrients, and phytochemicals with lower energy density; increase satiety; improve the glycemic response; and reduce insulin resistance. Furthermore, it was also observed that wheat consumption also reduces concentrations of biochemical factors such as insulin, C-peptide, and leptin.

In a highly significant study, whole-wheat bread was fed to two groups of 50 Japanese human subjects with BMI more than 23 kg/m². One group fed on 100 g of refined flour while the other group fed on whole-wheat bread and refined wheat bread for 12 weeks (Liu et al., 2020). After that, it was observed that whole-grain consumption promotes less weight gain, particularly visceral fat. Experimental studies also show that a whole-grain diet also improves several signal molecules such as adiponectin, which are produced in the adipose tissues and lead to weight loss and have a positive effect on both fat and carbohydrate metabolism, energy balance, and biomolecule homeostasis, which consequently reduces the deposition of fat in the body.

Whole-grain diets have a low glycemic index, which is further characterized by high satiety and low digestion and absorption that can be attributed to dietary fiber. Dietary fiber mainly binds with water and makes a thick cluster-type structure that delays the absorption of glucose, thereby reducing the requirement for insulin and maintaining glucose homeostasis. It is also observed that two main digestive hormones secreted from the small intestine help in the maintenance of postprandial glucose and, ultimately, glucose and fat metabolism (Witkamp, 2022).

Recently, many studies have linked gut microbiota and obesity, which shows that whole cereal grain consumption positively promotes the good gut microflora composition that feeds on short-chain fatty acids (SCFAs) and hormones that regulate appetite management, given that the wheat bran is provider of dietary fibers (DFs) and its consumption modulates gut microbiota. Moreover, microflora produces specific types of metabolites that promote the growth of mucus-producing microbes to improve the intestinal barrier, which is very important for the regulation of body weight and composition by acting as metabolism-based energy sources. A few experiments show that the whole grain contains prebiotics that positively affect the growth of gut bacteria that lower fat accumulation through the production of a variety of lower molecular weight unique metabolites (Esposito et al., 2005; Khan et al., 2022).

Additionally, SCFAs play very important roles in the regulation of liver and blood glucose and lipid catabolism and thus promote the secretion of two of the most important gut hormones, peptide-YY and glucagon-like peptide 1 (GLP-1), hence they regulate fat synthesis in the body (Mansour, Hosseini, Larijani, Pajouhi, and Mohajeri-Tehrani, 2013). The consumption of wheat bran dietary-based fiber promotes the growth of *Akkermansia muciniphila* in the gut of rats. One of the most common gut bacteria, for example, *Prevotella*, is linked with the improvement of glucose metabolism, which depends upon the amount and type of dietary fiber. The ratio of the two most important gut strains, Prevotella and *Bacteroides*, increases on the dietary fiber of the wheat bran via fermentation inside the gut (Prasoodanan P K et al., 2021). Moreover, Prevotella and *Bacteroides* secrete antibacterial agents that reduce harmful bacterial growth and also prove helpful in the reduction of body fat attributed to low appetite or satiety, slow gastrointestinal transit, and a change in glucose metabolism.

5.9.2 Metabolic Syndrome

Metabolic syndrome is a major medical problem in middle-aged people and affects almost 25% of the population worldwide, especially in developed countries. Metabolic syndrome is a prior medical condition characterized by glucose intolerance, pre-diabetic, hypertension, CVD, fatty liver, excessive central belly fat, insulin resistance, and hyperlipidemia. Additionally, metabolic syndrome is also linked with inflammation, hyper-cholesterolemia, nonalcoholic fatty liver, gallstone disease, and impotency in men (Samson and Garber, 2014). These conditions can lead to many lifestyle-related problems such as heart disease, T2D, and stroke, which are responsible for most deaths in middle age. Metabolic syndrome is characterized by high plasma triacylglycerol (TAG) ($\geq$150 mg/dL), fasting glucose ($\geq$100 mg/dL), low level of high-density lipoproteins (HDL) (<40 men; <50 mg/dL for women), high blood pressure ($\geq$130/$\geq$85 mm Hg), and large waist circumference (>102 cm for men; >88 cm for women) (Eckel, Grundy, and Zimmet, 2005). To overcome the problem of metabolic syndrome, several measures are recommended, which include drugs, lifestyle changes, and a plant-based diet, particularly the dietary fiber-rich diet provided by whole cereals. Plant-based diets have proved to be highly effective against metabolic diseases. In 13-week long experimental study carried out on 29 individuals, the results showed a lower body weight and improved levels of HDL cholesterol, thus once again underlining the significance of a whole-grain-rich diet (McGrath and Fernandez, 2022).

Given the above discussion, whole wheat grain comprises many bioactive components as mentioned in the table which is proven effective in various human diseases related to lifestyle, but they are equally effective in treating metabolic syndrome. In a total of 14 studies on humans, it showed that eating a whole-grain-based diet was inversely linked with a less number of metabolic syndromes, while refined grain eating is positively linked with metabolic syndrome. In a significant study carried out on the Taiwan population, three types of diets were supplied to 26,016 people (Guo, Ding, Liang, and Zhang, 2021). These diets contained meat-rich and rapid food dietary systems, vegetable-based seafood dietary patterns, and cereal and dairy dietary patterns. Among them, the cereal–dairy dietary pattern involved six food items such as flour products, whole-grain diets (including whole-wheat bread, brown rice, mixed cereal grains, and oat-based meal), root crops, and dairy products. Their consumption decreases the occurrence of metabolic syndrome in the subjects. This can be linked to the cereal dairy dietary pattern, which comprises complex glucans, dietary fiber, phytochemicals, antioxidants, prebiotics, and probiotics (Giacco et al., 2014; Syauqy, Hsu, Rau, and Chao, 2018).

Although similar types of studies are also conducted on the Iranian and French populations, metabolic syndrome affects them irrespective of all geographic locations, food types, and food habits. In both studies, fiber-rich diets were accountable for the lower incidence of metabolic syndrome. The presence of 3-(3,5-dihydroxyphenyl)-1-propanoic acid (DHPPA) in blood plasma, a well-recognized biomarker that reflects the consumption of a whole-grain wheat diet. Higher DHPPA amounts in the plasma are negatively related to metabolic syndrome when compared to subjects with lower DHPPA amounts in their blood plasma. Authors have ascribed these findings to whole-grain consumption because whole grains offer vitamins, minerals, dietary fiber, phytochemicals, and antioxidants that protect against metabolic syndrome (L. Zhou et al., 2022). Moreover, alkylresorcinols include a large number of phenolic lipid compounds present in the bran fraction of wheat grains. However, alkylresorcinols contain fewer hydrogen donation and peroxyl-scavenging capabilities, but they are still effective against lipid peroxidation in plasma membranes due to the long alkyl chain in alkylresorcinols (L. Zhou et al., 2022).

Inflammation is another sign or symptom that appears in metabolic syndromes, including CVD and insulin resistance. Inflammation is also responsible for creating prediabetes and cardiovascular diseases. But bioactive components and nutritive compounds available in the wheat grain, particularly in the bran, the outer layers, play a very important role in reducing inflammation and offering antioxidant power.

5.9.3 Cardiovascular Diseases (CVD)

CVD is a group of diseases that mainly include coronary artery diseases (CAD), for example, angina, heart attack, stroke, weak heart, cardiomyopathy, genetic heart disease, and valvular heart disease. Higher consumption of whole wheat grains and their derived products was associated with a reduced risk of CVD. The incidence of CVD can be reduced by almost 20 to 30% by the eating a whole-grain diet (Chapagai and Fink, 2022).

5.9.3.1 Hypertension

Hypertension is a medical condition that is mainly responsible for the damage of vital body organs for example, the heart, brain, and kidneys and is a major reason of premature deaths worldwide. According to an estimate, 1.28 billion people aged between 30 and 79 years will have hypertension in developing countries. Notably, 46% of affected people do not know they are suffering from high blood pressure, so it is also known as "a silent killer". According to WHO, "hypertension is diagnosed if, when it is measured on two different occasions, the systolic blood pressure readings on both days is ≥140 mmHg and/or the diastolic blood pressure readings on both days is ≥90 mmHg" (www.who.int/news-room/fact-sheets/detail/hypertension).

A few epidemiological and human-based studies in the United States were conducted on 29,000 women who consumed a minimum of four servings of whole grains. Parallelly studies were also carried out on women who were eating a 23% whole-grain-based diet (Poutanen et al., 2022). The experimental data show that the first group was less prone to hypertension. In an important study with 88 subjects, whole-grain-based diet was tested, whereby twice as many subjects eating whole oat cereal promoted good health and reduced their hypertension medication when compared with the control group that did not consume whole oats. In a nutritional approach called the Dietary Approaches to Stop Hypertension (DASH) trial, whole grains were included in the various dietary formulations to lower blood pressure (Jones and Engleson, 2010).

In the case of hypertension, many experiments were carried out to assess the effects of a whole-grain-based diet on hypertension in comparison to other chronic diseases. But both intervention and epidemiological studies indicate the same type of results for whole-grain consumption. But still, there are many unresolved issues related to whole wheat grain consumption and the effectiveness of the holistic effect of all the bioactive and nutritive components present in the whole grain against diseases. Furthermore, a pertinent question arises about how much contribution of individual components for instance, dietary fiber, phenolic compounds, vitamins, and mineral is needed to resist the CVD. Many researchers concluded that a higher intake of whole grains promotes a lower reading of blood pressure, which can be attributed to the fact that whole grains contain a higher level of dietary fiber, which is responsible for lowering hypertension (Saleh, Wang, Wang, Yang, and Xiao, 2019) although a few investigations are carried out on individual components of wheat grain, such as phytosterols and alkylresorcinols, as a dietary formulation. There is also a need to study the combined effects of whole grains and lifestyle-related changes on CVDs. The exact biochemical mechanisms that are responsible for the relationship between WGs and hypertension remain unknown (Liu et al., 2020). An important study was conducted on whole grains, which significantly influenced blood pressure.

In addition to fiber, whole-grain diets also comprise magnesium, potassium, and some protein, which can reduce systolic and diastolic blood pressures. Furthermore, whole grains increase blood concentrations of hormones such as adiponectin, which are anti-inflammatory and help in lowering blood pressure (Gupta, Meghwal, and Prabhakar, 2021).

The epidemiological studies suggested that whole wheat grain eating reduces the incidence of cardiovascular diseases. In a study conducted on 822 diabetic women in the United States, results indicate that the consumption of a whole-grain-based diet is helpful in the reduction of cardiovascular disease-related mortality among diabetic women. A group of 82 subjects, who also eat three or more servings of whole-grain-based diets per day, lower down up to 20 to 30% possibilities of heart-related diseases as compared to those who do not consume whole grains. The bran, a fraction of the whole grain, has been shown to play a significant role in reducing the incidence of cardiovascular disease (Khan et al., 2022). In a 21-day animal experiment, rats were fed either whole-wheat bread or refined bread *ad libitum*. The plasma cholesterol that was measured was dramatically reduced in the whole-grain-fed group while the total steroid excretion was substantially increased, indicating a positive benefit of the whole wheat diet in the prevention of cardiovascular disease.

Recently, a cardioprotective mechanism has been put forward in which three important components of a whole-grain diet are identified that are phytosterol, phytochemical, and lipidomic compounds alkylresorcinols. The combined effect of dietary fiber and phytosterol affects fat metabolism, reduces atherosclerosis activity, and improves lipid profiles in cardiac patients. At the same time, phytosterol influences many intermediates of lipid metabolism and exerts high levels of antioxidant effects that regulate the vascular endothelium and finally reduce the endothelial dysfunction that promotes atherosclerosis, which is considered a major risk factor for CVD (Rosa-Sibakov, Poutanen, and Micard, 2015).

Moreover, other biochemical parameters such as TAGs and LDL cholesterol were reduced, while HDL cholesterol (HDL-C) improved in whole-grain-fed rabbits. The apparent effect is ascribed to the biochemical mechanism that favors healthy effects attributed to whole wheat. The consumed dietary fibers undergo bacterial-mediated fermentation in the human large intestine and enhance the generation of SCFAs. In addition, a recent study indicated that a transient 0.4% (w/w) AR-based diet significantly enhanced the disposal of fecal cholesterol, and this decreased the blood cholesterol level in an HFHSD-induced mouse model. Additionally, it also decreased the rate of cholesterol absorption in the digestive tract. In addition, epidemiological studies have linked increased flavonoid consumption to a decreased risk of cardiovascular diseases and death due to cardiovascular diseases (Oishi et al., 2015). Wheat flavonoids and intermediate metabolites may mitigate oxidative stress and inflammation by lowering down nitric oxide synthase (Enos) and reactive oxygen species (ROS) levels in endothelial, while enhanced extract of whole wheat reduced Dox-induced redox stress in rat cardiac cells (Yamagata and Yamori, 2020). It is also observed that the consumption of wheat-based polyphenols and fiber grains reduce the risk of cardiovascular diseases.

Recently, dysbiosis in the gut was held responsible for bowel inflammation that enhances the permeability of the gut membrane and further increases the risk of CVD, but the consumption of whole grains restored the natural structure of gut barriers and gut microbiota because of presence of prebiotics in wheat bran. The detailed microflora-based mechanisms are still not known completely.

5.9.3.2 Stroke

Stroke has caused a large number of mortality and morbidity in the developing world. The survivors of strokes often live with deformities and other serious complications. Therefore, improved diet and lifestyle are very important to reduce the incidence of stroke in the human population. A link between whole-grain consumption in the diet and a lower risk of stroke has been established. It is also observed that there is a strong connection between high blood pressure and stroke. In an epidemiological study, a diet rich in cereal fiber, vitamin K, and antioxidants and a healthy lifestyle were linked to a lower risk of stroke. An 18-year study conducted on a population of 39,053 healthy women over 45 years old, using female subjects, endorses this claim (Oh et al., 2005; Kurth et al., 2006). The risk of hemorrhagic and total stroke was found to be inversely related to DF consumption. It was noticed that the risk of stroke was lowered overall by 36%, and the risk of hemorrhagic stroke was reduced by almost 50% for those who consumed the most cereal fiber. Numerous reports concluded that both WGs and cereal fiber were probably the two dietary components that reduced the risk of stroke. But still, substantial evidence is not available and data is not credible, rather associational only. More information is required to fully comprehend the consequences of all WG-based diets, and their effects on cellular metabolism. To unravel the connection between WGs and incidence of stroke, other lifestyle factors and WG biochemical pathways must be considered, just like in the case of other diseases. Furthermore, it remains to be seen whether the reduction in diseases is due to the synergetic actions of functional components or any individual component of whole grains.

5.9.4 Type II Diabetes

Type II diabetes (T2D) is a serious global health problem that has a negative socio-economic impact, particularly in developing countries. According to estimates, by 2030, 439 million people will be affected by T2D globally. In the last 20 years, the incidence of T2D in adults has been raised to 69%, especially in poor nations, and by 20% in developed countries. Recently, many scientific investigations have shown that eating whole-grain cereals-based diets is linked with a decreasing risk of T2D (Lachman et al., 2017). Long-term studies are carried out on 90,000 women and about 45,000 men. Both studies show a nearly 30% lower possibility of the development of T2D in subjects who consume more wheat-based dietary fiber compared to others (Salmerón et al., 1997). In a different study, it was also found that those who eat mostly refined grains and little number of whole grains were 57% more prone to develop T2D than people who consumed mostly whole grains. In addition, a 42,898-man follow-up study by researchers found a 37% lower risk of the occurrence of type II diabetes in subjects who consumed three meals of wheat grains per day (Khan et al., 2022). These findings showed that whole-grain cereal consumption lowers the relative risk of T2D by 30% when combined with prospective cohort studies (Fung et al., 2002; Y. Wang, Duan, Zhu, Fang, and He, 2019). According to randomized, controlled dietary trials conducted on human subjects and other experimental studies, whole-grain eating and the occurrence of T2D are very much associated. A diet rich in whole-grain foods also reduces postprandial (PP) concentration of insulin and plasma triglyceride levels in the blood of persons suffering from metabolic syndrome by 29 and 43%, respectively; and the findings of the study suggest that the relationship between cereal consumption and a lower risk of type-2-diabetes and cardiovascular diseases has been explained by the amount of insulin hormone and plasma TAGs after 2 hours of taking a diet (PP) (Khan et al., 2022).

However, studies in this area have produced widely divergent results, which sufficiently demonstrate the need for greater investigation. While the dietary fibers and phenolic acids present in whole-grain bran offer great health benefits of eating whole grains, more research is necessary to fully know the effects. More recently, the combined effect of phytic acid, phenolic acids, and Ars has been studied that reveals their contribution in regulation of blood glucose reading, insulin resistance, and hyperinsulinemia. The prescription of a 0.05% ferulic acid (FA) diet affects the activities of three main enzymes involved in the regulation of glucose metabolism, that is, glucose-6-phosphatase (G-6-pase), phosphoenolpyruvate carboxykinase (PEPCK), and glucokinase (GK) activity and insulin secretion. Among them, the first two enzymes were stimulated while the last one was inhibited by an FA-based diet (Tian et al., 2022).

5.9.5 Cancers

According to World Health Organization (WHO), cancer is a cause of prolonged illness and premature death. Every year, nearly 14 million new cancer incidences and eight million deaths are reported, and the number is constantly increasing (Parascandola and Pearlman, 2022). According to a study by the World Cancer Research Foundation, eating 90 grams of whole grains per day is strongly associated with a lower incidence of cancer, particularly colorectal cancer. Consequently, it is generally recommended that people eat more whole grains and fruits and vegetables. The logic behind this advice is that eating whole grains may stimulate the growth of healthy intestinal microbiota that produce short-chain fatty acids, reducing the transit time of food through the gut or preventing insulin resistance (Khan et al., 2022). Dietary fiber is also significant to protect against cancers by its rapid excretion from the body via binding with carcinogens and also by regulating glycemic response. Currently, whole grains and their based products are becoming more popular among the population due to their protective roles in cancer prevention. Three major wheat components, such as phytochemicals, antioxidants, and vitamins, which fight against oxidative stress in the cellular environment, give them an anti-cancer effect.

Moreover, phytochemicals in whole grains regulate cellular signal transduction processes and the behavior of cancer cells, such as their growth and death or apoptosis. Antioxidants in wheat bran protect normal body cells and stop the damage from oxidation at the cellular level. Therefore, it can be inferred that a variety of phenolic acids can stop cancerous growth because of their potential to reduce oxidative damage to cells and organelles. The whole grain has proved effective, particularly in the case of colorectal cancer. In an experiment conducted on 25 subjects and control studies, a relationship has been established between whole-grain consumption and the occurrence of rectum cancer (Haas, Machado, Anton, and Silva, 2009). The results showed that eating a whole-grain-rich diet and cereal fiber was linked to a lower risk of colorectal cancer. In a population-based study of about 60,000 women, the susceptibility to colorectal cancer and the consumption of whole-grain diets were related (Ringland, Arkenau, O'Connell, and Ward, 2010). The results showed that women who consume whole grains abundantly (4.5 servings/day) were less susceptible to getting colon cancer in comparison to those who consumed a small number of whole grains (1.5 servings/day).

It is already known that phytosterol lowers the chance of developing cancer. The potential role of phytosterol has been noticed in methylnitrosourea-fed rats. It is well known that methylnitrosourea is a cancer-causing chemical agent (Ramprasath and Awad, 2015). Furthermore, the results also revealed that eating 0.2% sitosterol in the diet for 28 weeks substantially lowered the risk of rectum cancer by up to 24% and also reduced the severity of tumor formation in methylnitrosourea-affected animals. Wheat bran is a major component of the whole grain that consists of phenolics and dietary fiber that promote the synthesis of short-chain fatty acids (SCFAs) via gut microbe fermentation (Adebo and Gabriela Medina-Meza, 2020). Generally, cancer patients have lower growth of *lactobacillus, Clostridium, and Roseburia* spp., but dietary fiber and phytochemicals promote their growth that plays a significant role in the repair of colon tissues and maintaining homeostasis in the digestive system (Liu et al., 2020). More recently, the individual component alkylresorcinol has been tested against colon cancer. The experiments were conducted on the cancerous cell lines, HCT-116 and HT-29, which show that bioactive components in wheat bran reduce the growth of colon cancer cells which is attributed to cellular apoptosis, autophagy, and metabolic pathways inside the endoplasmic reticulum (ER) (Y. Zhu, Conklin, Chen, Wang, and Sang, 2011).

Based on many research reports, it can be inferred that eating sufficient amounts of whole grains makes a person less susceptible to cancer. But we still require more studies to find a stronger link between whole grains and lower mortality with cancers that can be attributed to their intricate molecular structure, and bound phenolic acids offer health benefits by reaching the colon mostly undigested, where they act as antioxidants and anti-inflammatory agents.

5.9.6 Effect on Overall Mortality

Lifestyle or noncommunicable diseases (NCDs) are responsible for the premature deaths of 41 million people (71% of total deaths) between the ages of 30 to 69 years every year in low-income countries, which mainly include cardiovascular diseases, cancers, respiratory diseases, and diabetes ("Noncommunicable diseases: Mortality", n.d.; www.who.int/data/gho/data/themes/topics/topic-details/GHO/ncd-mortality). Epidemiological studies reveal that individuals who consumed mostly whole grains, fruit, vegetables, and regular exercise had a 20% lower death risk than those who consumed the least bioactive-rich diets. Consequently, fewer deaths from CHD, CVD, and cancer resulted in substantially lower mortality rates. This can be endorsed by the studies conducted in Iran and the study carried out by the Atherosclerosis Risk in Communities (ARIC) Study on mixed populations (Weng et al., 2013).

5.10 Bioavailability and Bioaccessibility of Bioactive Compounds in Wheat Grains

Although the bioavailability concept was first used in the pharmaceutical industry, it is mainly applied to pharmaceutical agents that reach blood circulation after an absorption process in the human digestive system. In nutritional science, bioavailability is referred to as the efficient absorption of a nutrient in the blood circulation and then its entry into the cellular systems. Many factors often decide the bioavailability of nutrients, such as the compound proportion that is released of bionutrients from the food matrix and then absorbed in intestine (Angelino et al., 2017) – for example, wheat bran, which is a rich source of bionutrients but entrapped in a very intricate interwoven network of many cell wall polysaccharides and polyphenols that are released and available after wheat bran is fine ground, hence nutrients are efficiently bioavailable in the gut. It is already mentioned before that whole wheat grains and their related diets contain a large number of bioactive components which are associated with many health benefits in the human population (Katileviciute et al., 2019). Hence, the bioavailability of bioactive compounds inside the tissues is of the utmost priority. But many factors affect the bioavailability of these bioactive compounds, directly or indirectly in the human digestive system and, finally, their absorption in the blood. There are many downstream processes for grains, including germination, farming, debranning, milling, fermentation, etc., which will be discussed next.

5.10.1 Wheat Farming

The aforementioned discussion and numerous scientific investigations clearly show that genotypes influence the composition of bioactive components (phenolics and micro-nutrients) in the wheat grain, such as phenolic compounds, concentration of thiamine, riboflavin, and minerals. Additionally, farming conditions are also very crucial parameters that affect the composition of wheat content. Simultaneously, it also affects the antioxidant potential and scavenging properties of phenolic compounds and nutrients. Environmental and farming conditions like production year and crop pattern, irrigation conditions, abiotic and biotic stresses, and genotypes all affect the bioactive components in a grain. In a highly significant experiment, it has been shown that the chelating potential and radical scavenging properties of 2,2-diphenyl-1-picrylhydrazyl (DPPH) are affected by the environmental conditions that indirectly indicate the change in the composition of bioactive compounds (Yu, Nanguet, and Beta, 2013). Furthermore, strong action of 2,2'-azinobis (3-ethylbenzothiazoline-6-sulfonic acid) (ABTS) against free radicals was recorded in a crop that was grown under stressful growing conditions. In winters, wheat varieties and three durum wheat varieties contain almost the same amount of phenolics. But it was also noticed that the phenolic amount of the wheat genotypes grows under organic farming, which consist of higher amounts of phenolic compounds and total antioxidant properties than when grown in conventional ways by using a large number of fertilizers and pesticides. Moreover, the composition of free and conjugated phenolic components (tocols and alkylresorcinols) is also influenced by environmental conditions (K. Zhou, Yin, and Yu, 2005). The substantial effect of farming conditions on the content of carotenoids, an important phenolic compound, is affected by environmental and genotype interactions.

5.10.2 Germination

Germination is used to enhance the nutritional composition of whole-grain products. This is attributed to the action of some endogenous enzymes which break down or metabolize large molecules into small ones, for example, monosaccharides and essential amino acids, gamma-aminobutyric acid (GABA), fat-soluble vitamins, micronutrients, phytochemicals, and soluble dietary fiber. Scientific investigations have shown that germinated wheat grains consist of high levels of bioactive components such as phenolic compounds, micronutrients, and antioxidant properties that provide better health benefits. To confirm these findings, wheat grains were germinated for various periods of 24, 48, 72, and 96 hours, and then phenolic compounds, antioxidants, and related proteins and enzymes were estimated by using 2D gel electrophoresis and peptide mass fingerprinting (PMF) (Kim, Kwak, and Kim, 2018). Results show that 15 proteins/enzymes (such as glutathione-S-transferase, starch synthase, and β-glucosidase) out of 85 proteins are linked with a high potential to neutralize reactive oxygen species. That capability was highest in the 96 hours germinated sample. Moreover, high amounts of phenol-based acids such as gallic acid, 4-hydroxybenzoic acid, vanillic acid, caffeic acid, syringic acid, ferulic acid, p-coumaric acid, and GABA (gamma-aminobutyric acid) were recorded in the germinated samples than in the control. The improvements in the antioxidant potential of germinated wheat grains were also observed in the Canadian Western Red Spring (CWRS) and Chinese wheat varieties (Chen et al., 2017).

5.10.3 Debranning

After the industrial revolution, the debranning of wheat grains was started before the milling process. Debranning is a downstreaming process in which external layers of the whole wheat grain are sequentially removed by using abrasion and friction techniques. The main objectives of the debranning process are to improve the refinement of hard grains and decrease contamination from the outer layers of the grain because the outer layers of wheat grains comprise pesticides and toxic traces, elements, heavy metals, and mycotoxins that can be deleterious to health (Ficco et al., 2020). Moreover, debranning is also carried out to isolate the bioactive components from the grain to prepare nutritious food and to meet the quality parameters of refined flour. But the debranning process leaves behind a very degradative effect in terms of reduced bioactive components in debranning grains and their based foods (Barroso Lopes, Salman Posner, Alberti, and Mottin Demiate, 2022). For example, the debranning of durum wheat varieties reduced the phenolic compounds anthocyanin and polyphenols by 45 and 15%, respectively. On the other hand, nutritive and essential trace elements such as minerals were retained up to 90%, and toxic elements were reduced up to 11%. Hence, debranning is a good technology for the preparation of quality refined flour, and it is considered a better process than milling to provide bioactive compounds, low contamination, and safety of flour. The level of debranning affects the mixograph properties such as mixing time, dough strength and pasting parameters (final, breakdown, and peak viscosity) (Ladhari et al., 2022). While cooking properties (lighter crumb and crust color) are improved at 4% of the debranning level. Recently, Chinese steamed bread (CSB) was analyzed for its sensory qualities. The results showed that bread texture, taste, and color improved up to 7%, but the antioxidant potential was reduced very much in the 16% degree of debranning wheat flour sample. Although the excessive debranning process is deleterious to the bioactive components in grain, the moderate level of debranning 0–5% proved highly effective to reduce microbial load up to 80.1%, and 15.82% of deoxynivalenol and aleurone layer was intact; hence, bran-associated micronutrients could be retained in the flour (Sun, Liu, Zhang, Liu, and Hou, 2022). Therefore, moderate peeling is better for economic and nutritional reasons, simultaneously reaping the benefits of milling wheat grains.

5.10.4 The Effect of Grain Milling and Grinding

The milling process is a prehistoric process that was practiced in all old civilizations of the world and is used to grind starch and plant materials. Since the invention, milling technology has changed substantially and is carried out by very elaborate machinery. The process of wheat milling is affected by grain hardness and biochemical factors (such as protein content), which are the most important parameters for wheat quality. Milling is also influenced by the moisture content (typically 16%) and composition of the cell wall of wheat grain before the pre-milling process, that is, tempering. Given that whole grains and whole-grains-based products offer more nutritive and health benefits than refined flour (RF), whole-wheat flour poses many quality problems that are in high demand by consumers the world over, such as functionality and quality. Moreover, whole-grain bread is low in quality, for instance, bread consisting of a lower specific volume and a denser crumb texture (Jones, Adams, Harriman, Miller, and Van der Kamp, 2015).

Another factor is that the particle size of whole-wheat flour also affects food product quality and the functionality of the wheat flour. Many studies have focused to evaluate the effects of milling on whole-wheat flour and its particle size and on various thermo-mechanical, and physicochemical properties by using Mixolab, Farinograph, Mixograph, and Consistograph techniques. It has been observed that flour particle size influences the damaged starch and solvent retention capacity (SRC) profiles of dough (Bressiani, Oro, and Da Silva, 2019).

The effect of milling and ultrafine grinding on the nutritive values of wheat-based products such as bread, pasta, and bulgur has been discussed in various scientific investigations. This provides deep insight and offers a mixed bag of results, for example, some reports indicate that the whole grain provides bioactive components, but at the same time, a few reports also show that the fine milling process decreases the particle size of bran and starch that improve bioaccessibility in the human digestive system and thus ensure more nutritive value (Igrejas et al., 2020). Therefore, the milling process improves the digestibility and the bioavailability of amino acids, minerals, and vitamins. Because ultrafine grinding of wheat bran improves the surface area up to three times the ultrafine bran fraction with a particle size of 30 micrometers, it releases the free and bound phenolics from the cell wall matrix, thus improving antioxidant activities from 30 to 45 mmol TEAC/kg (Rosa, Barron, Gaiani, Dufour, and Micard, 2013). Another study substantiates the fact that endogenous enzymes in the flour are inactivated, hence reducing the rancidity, breakdown of lipids, and antioxidants so that refined flour is more stable as compared to whole-grain flour. Large sizes of bran particles influenced the concentration, for example, of phytic acid, the main anti-nutrient conjugated with most minerals due to being a strong metal chelator, Therefore, the bioavailability of iron, zinc, and calcium can be impeded in the digestive system (Santos, Saraiva, Vicente, and Moldão-Martins, 2019).

The milling process alone does not affect the nutritional status, but the type of milling technology itself used for milling wheat flour is also highly significant. In an important study, different types of milling methods were used for the preparation of whole-wheat flour. Then various physicochemical properties and concentrations of bioactive components were also evaluated. When comparing nutrient concentration influenced by different milling techniques, some studies show that stone milling is better than other methods such as over roller, plate, or hammer milling in respect of nutrients. In a comparison study between stone- (SWF) and roller-milling (RWF) techniques in the bread-making process, results indicate that no palpable differences were noticed in the proximate composition of the nutrients of four Bolero cv CWRS BonaVita cv Skorpion cv wheat varieties, but the RWF is proved most suitable for rheological, nutritional technological properties and gluten aggregation kinetics in terms of bread-making process (M A Pagani, Giordano, Cardone, and Pasqualone, 2020). Moreover, the extraction level is also responsible for the bioavailability of bioactive components, which means the percentage of extraction is also important for the availability of nutrients.

5.10.5 Bread Making

The bread-making industry was established in ancient civilizations. However, wheat grain and flour are used to produce a large number of food items, including bread, chapati, pasta, various types of noodles, and other food products. But there are millions of food lovers who mainly consume bread as a staple food, particularly in South Asia, Africa, and various regions of the world. Recently, consumers from all over the world are demanding bread made from whole-grain wheat flour which contributes to the health benefits of people because a major fraction of whole wheat grain, that is, bran and germ, consists of a large number of bioactive components. But now a question has emerged about bread making and its effects on the available bioactive components in whole-grain flour and their bioavailability in the human digestive system (Pagani et al., 2020). A plethora of studies has been carried out to address this particular question, ranging from flour to storage of bread and consumption. A few studies have suggested that the use of whole grains improves the nutritional quality of bread by enhancing the quantity of proteins, micronutrients, fatty acids, and both soluble and insoluble dietary fiber compared to bread produced from refined wheat flour. Moreover, both the formulation and downstream processing of bread flour and the bread-making process itself affect the composition and bioavailability of micronutrients and phenolic acids in the bread and their absorption in the digestive system (Pagani et al., 2020).

Bread making is a multistep process that includes dough mixing, yeast-based fermentation and proofing, and baking that affect phenolic profiles and their antioxidant potential, proteins, and dietary fibers. To reap the maximum benefits of whole-grain bread and whole wheat products, it is essential to understand the mechanisms of bread-making methods that influence bioactive components (Pagani et al., 2020).

Initially, flour is mixed to make the distribution of ingredients even. However, flour composition is affected by the types of genotypes and other preprocessing methods. Inside the bread, the interaction between water and other biochemical molecules is very crucial – for example, proteins – which leads to gluten formation which constitutes a network structure containing more cohesiveness, non-stickiness, homogeneity, and good rheological characteristics. These physicochemical properties are very important for dough formation. However, it depends upon the type of dough-making procedure and the equipment. The gluten's viscoelasticity influences the available composition of phenolic compounds in bread because phenolic compounds are conjugated with proteins using many hydrogen bonds between the functional groups (OH) of phenolic acids and the carbonyl group of the peptide (Brouns et al., 2022).

In an experiment, breads were made from four different hard wheat varieties. The quality parameters such as complete phenolics, total flavonoid content (TFC), antioxidant properties, and amounts of different types of phenolic acids were measured and changed substantially. For example, the phenolic profile and antioxidant activities of whole wheat were improved. Among the three important phenolic acids, syringic acid concentration decreased but enhanced the concentration of vanillic acid and ferulic acid during the fermentation. The quantity of ferulic acid and sinapic acid ranged from 278.4 to 488.9 µg/g and from 8.6 to 159.0 µg/g, respectively (Tian et al., 2021).

Another step of bread making is dough mixing, which causes the loss of up to 50% of phenolic acids, including bound ferulic acid, sinapic acid, and caffeic acid, but overall improves the bioaccessibility of bioactive components in the digestive system. Although the effects of the hydrolytic enzymes, for example, oxygenase and peroxidase, that are mixed in flour, become active after the mixing and dough making, which are also responsible for lowering the concentration of total phenolic acid. The next important step is the leaving and fermentation, which increases the size of the bread and improves the phenolic acid content due to the fermentation process. It can be due to the activation of enzymes present in the wheat flour and fermentation causing microorganisms which act on the cell wall and thus release the bound form of phenolic acids (Dahiya, Bajaj, Kumar, Tiwari, and Singh, 2020). However, the obtained results do not show complete agreement and consistency across the literature regarding the above discussion.

5.10.6 Effect of Fermentation on the Whole Grain

Fermentation is the centuries-old food-processing method that has been used for the transformation or modification of raw food materials or substrates into editable or value-added forms by using a variety of microbes in different civilizations. Generally, the fermentation process positively influences the functionality, nutritional composition, appearance, flavor, color, and textures of fermented foods that can be attributed to microbial actions. As a comparatively less intricate process, fermentation turns whole grains into editable food and increases the bioavailability of bioactive components, especially the antioxidant properties, which are health-promoting in nature. It was also proved that the fermentation process improved the phenolic compounds because of enzymatic activities that are responsible for conjugating phenolic compounds into media (Adebo and Gabriela Medina-Meza, 2020).

There are two types of fermentation processes: (1) natural (also called spontaneous) and (2) controlled fermentation is used in whole wheat grain fermentation, but the latter proved highly beneficial for improving detoxification, organoleptic properties, aroma, texture, flavor, and thus improving bioaccessibility. There are fermented foods prepared from whole wheat grains, including bread, boza, sourdough bread, and tempe. In the case of whole-grain fermentation, the outer coat acts as a strong barrier for amino acids and small-size sugars, which are important for the fermentation process. In an important study, TPC, TFC, and TNC were decreased because of the degradation and hydrolysis of the phenolics (Montemurro, Pontonio, Gobbetti, and Rizzello, 2019) The increase in the concentration of catechin, gallic acid, and quercetin was due to the breakdown of complex phenolic compounds in *Lactobacillus*-based fermentation. The production of tempe from whole grains brings about many changes, for example, enhanced antioxidant activities, water-soluble bioactive compounds, small-sized peptides, and many oligosaccharides that attributed to the fermentation process. Similar types of results were also reported in many experiments, where ABTS, OH-scavenging, and FCRS-RP assays were used for the estimation of antioxidants during the conversion of wheat flour into tempe (Adebo and Gabriela Medina-Meza, 2020).

In whole wheat grains, most of the bioactive ingredients are centered in the bran. To improve the health benefits, exogenous bran is mixed with wheat flour in a specific ratio to make whole wheat grains, as recommended by various food agencies. But the addition of bran produced a few undesirable properties such as negative sensory quality, dough rheology, and technological properties. Furthermore, whole-wheat flour has a low shelf-life and increases in anti-nutrient and total phenolic compounds. The short shelf life is because of the breakdown of lipid-based compounds by the few hydrolytic enzymes that also lead to the rancidity of flour. To overcome these problems, a few biotechnological interventions have been suggested, such as combining germination and sourdough fermentation, which have been used in many experimental studies. Sourdough fermentation is carried out by using *Lactobacillus rossiae* LB5, *Lactobacillus plantarum* 1A7, and *Lactobacillus sanfranciscensis* DE9, which increase the concentration of phytic acid condensed tannins, raffinose, and trypsin inhibitory activity. While bread made from fermented sprouted flour has an enhanced protein digestibility and low starch availability, and provide high concentration of peptides, free amino acids, and γ-aminobutyric acid (Montemurro et al., 2019).

5.10.7 Enzyme Technology

Given the above discussion, it is very palatable that most of the bioactive components of whole grains are trapped in the bran or outer layers of cereals. The consumption of whole grains offers bioavailability of bioactive components in the digestive system that is affected by the entrapment of polyphenolic compounds in the cell wall-based dietary fiber, and these bioactive compounds must be released from a highly intricate network of cell wall structures. However, various processing methodologies have been applied to improve the bioaccessibility via thermal treatment, ultra-milling, and enzymatic treatment of hydrolytic enzymes, which are provided by microbial sources in fermentation, indigenously present in the flour, and by exogenous additions. In addition to polyphenolic compounds, most of the minerals are bound in the form of phytates, so it becomes essential to use food-processing techniques that improve the stability, bioavailability, and biological activity of polyphenols in the biological systems and, simultaneously, the availability of minerals (Dahiya, Bajaj, Kumar, Tiwari, and Singh, 2020).

Currently, researchers have suggested adding a variety of enzymes, such as alpha-amylase, at the mixing level to increase the properties of bread in terms of fermentable sugars. Many other enzymes, especially cell wall degrading enzymes, such as glucose oxidase, hemicellulose, phospholipase plus hemicellulose, and xylanases, are used to improve the quality of dough and bread prepared from WWF. Bran contains more arabinoxylan content than endosperm. Moreover, enzymes are more important in the whole-wheat bread-making process. Another very important enzyme is phytases used to free the bound form of minerals from the phytase – for example, the release of calcium in the presence of thermostable α-amylase. However, the bioavailability of minerals and bioactive compounds depends upon the types of fermentation and used microbes like lactic acid bacteria that provide the phytase enzymes (Gómez, Gutkoski, and Bravo-Núñez, 2020).

Few studies have tried to unravel the mode of action of hydrolytic enzymes on the bran and the release of entrapped polyphenolic compounds from the cell wall. Generally, most polyphenolic compounds are bound with proteins and cell-wall polysaccharides (arabinoxylans), which mainly comprise xylose residues linked together by β-(1,4)-linkage and thus form a long backbone that is further joined with arabinose residues. Most of the ferulic acid is conjugated to the C(O)-5 position of arabinose. Few studies show that endo-b-(1,4)-D-xylanase application releases the xylan residues. Another enzyme, β-D-xylosidase, acts on the non-reducing end of the polymeric chain of xylose. Two other important enzymes, B-L-arabinofuranosidase and ferulic acid esterases, have been used to free arabinose and ferulic acid, respectively (Angelino et al., 2017; Tian et al., 2021).

The main aim of enzymatic treatments is to reorganize wheat bran structure and release trapped compounds, for example, feruloylated oligosaccharides and feruloyl oligosaccharides, which are implicated in the very important antioxidant-protecting biomolecules. The effects of some commercially prepared enzyme preparations, including β-glucanase, cellulase, polygalacturonase, and aminopeptidase, on wheat bran are studied. Results show that up to 50% of phenolic compounds, both free and bound forms, are released from the bran. In the same experiment, the antioxidant potential of the enzyme-treated and fermented bran is measured by using the DPPH scavenging, ABTS, DPPH, FRAB, and ROS-reducing potential assays (Tian et al., 2021). Moreover, the bioaccessibility assay was also conducted to test the TNO Intestinal Model of the digestive system. Although bioaccessibility is improved in the gastrointestinal model, we still need to further test the bioaccessibility by using better model systems in animals and humans (Carbonell-Capella, Buniowska, Barba, Esteve, and Frígola, 2014).

5.11 Conclusions and Prospects

Whole wheat grains are a rich source of many important nutrients and bioactive components. Many research reports have proved that wheat bioactive components are effective to prevent lifestyle-related diseases in the human population because they all act as strong antioxidants and sources of co-enzymes and cofactors that support metabolic reactions hence improving overall human metabolism. But it remains to be seen whether all bioactive components are effective synergically or act individually to reduce the possibility of chronic diseases. In addition, there are many factors such as farming conditions, cultivars, application of fertilizers or growing organically, various types of biotic and abiotic stress, and downstream processing substantially impacting the quantity of bioactive compounds in the human digestive system. The food-processing method such as the milling and debranning process also influenced the bioaccessibility and bioavailability of nutrients. For example, the milling process substantially reduced the number of vitamins and phenolic compounds present in the bran of the wheat grain. Bran is a major source of most of the vitamin- and phenol-based nutraceuticals hence generally recommended for the consumption of whole-grain-based diets. The two most important food-manufacturing processes, for example, fermentation and bread-making, also affect the absorption and digestion of wheat diets although fermentation improved the nutrient content of whole-wheat flour during the bread-making process. Few experiments palpably show that fine grinding makes available the amount of minerals and phenolic compounds in the refined flour which is a poor source of dietary fiber, which regulates the growth of beneficial microflora in the colon. The healthy microflora produces the very unique metabolites which positively regulate the host metabolism and are helpful in the prevention of modern lifestyle diseases such as obesity, metabolic syndrome, cardiovascular diseases, and cancers.

Enzymes have been an unavoidable part of food sciences, so they equally were applied in the improvement of wheat-based foods – for instance, the addition of enzymes further increased the availability of dietary fibers across the large intestine hence leading to more growth of beneficial bacteria. The addition of phytase is an effective way to free more minerals from bond forms and enhanced the bioavailability of minerals without impacting the quality of bran.

But in the near future, researchers need to develop effective and suitable assays to test the bioaccessibility of nutrients in humans and rats. At the same time, there is also a need to develop food-processing technologies that will help to retain the properties of grains as a whole, and, simultaneously, food must be tastier and more palatable and enjoyable in comparison to refined flour which is often used in processed foods.

BIBLIOGRAPHY

Abdel-Aal, E. S. M., Young, J. C., & Rabalski, I. (2007). Identification and quantification of seed carotenoids in selected wheat species. *Journal of Agricultural and Food Chemistry, 55*(3), 787–794.

Adebo, O. A., & Gabriela Medina-Meza, I. (2020). Impact of fermentation on the phenolic compounds and antioxidant activity of whole cereal grains: A mini review. *Molecules (Basel, Switzerland), 25*(4). https://doi.org/10.3390/molecules25040927.

Ahmad, S., & Al-Shabib, N. A. (2020). *Functional Food Products and Sustainable Health.* Springer.

Aloo, S.-O., Ofosu, F.-K., & Oh, D.-H. (2021). Effect of Germination on Alfalfa and Buckwheat: Phytochemical Profiling by UHPLC-ESI-QTOF-MS/MS, Bioactive compounds, and in-vitro studies of their diabetes and obesity-related functions. *Antioxidants (Basel, Switzerland), 10*(10). https://doi.org/10.3390/antiox10101613.

Angelino, D., Cossu, M., Marti, A., Zanoletti, M., Chiavaroli, L., Brighenti, F., . . . Martini, D. (2017). Bioaccessibility and bioavailability of phenolic compounds in bread: A review. *Food & Function, 8*(7), 2368–2393. https://doi.org/10.1039/c7fo00574a.

Anunciação, P. C., Cardoso, L. de M., Gomes, J. V. P., Della Lucia, C. M., Carvalho, C. W. P., Galdeano, M. C., . . . Pinheiro-Sant'Ana, H. M. (2017). Comparing sorghum and wheat whole grain breakfast cereals: Sensorial acceptance and bioactive compound content. *Food chemistry, 221*, 984–989. https://doi.org/10.1016/j.foodchem.2016.11.065.

Aslam, A., Zhao, S., Lu, X., He, N., Zhu, H., Malik, A. U., . . . Liu, W. (2021). High-throughput LC-ESI-MS/MS metabolomics approach reveals regulation of metabolites related to diverse functions in mature fruit of grafted watermelon. *Biomolecules, 11*(5). https://doi.org/10.3390/biom11050628.

Azzi, A. (2007). Molecular mechanism of alpha-tocopherol action. *Free Radical Biology & Medicine, 43*(1), 16–21. https://doi.org/10.1016/j.freeradbiomed.2007.03.013.

Bahmanyar, M. A., & Ranjbar, G. A. (2008). The role of potassium in improving growth indices and increasing amount of grain nutrient elements of wheat cultivars. *Journal of Applied Sciences, 8*(7), 1280–1285.

Balandrán-Quintana, R. R., Mercado-Ruiz, J. N., & Mendoza-Wilson, A. M. (2015). Wheat bran proteins: A review of their uses and potential. *Food Reviews International, 31*(3), 279–293. https://doi.org/10.1080/87559129.2015.1015137.

Barman, A., Pandey, R. N., Singh, B., & Das, B. (2017). Manganese deficiency in wheat genotypes: Physiological responses and manganese deficiency tolerance index. *Journal of Plant Nutrition, 40*(19), 2691–2708. https://doi.org/10.1080/01904167.2017.1381717.

Barron, C., Surget, A., & Rouau, X. (2007). Relative amounts of tissues in mature wheat (*Triticum aestivum* L.) grain and their carbohydrate and phenolic acid composition. *Journal of Cereal Science, 45*(1), 88–96.

Barroso Lopes, R., Salman Posner, E., Alberti, A., & Mottin Demiate, I. (2022). Pre milling debranning of wheat with a commercial system to improve flour quality. *Journal of Food Science and Technology, 59*(10), 3881–3887 https://doi.org/10.1007/s13197-022-05411-6.

Batifoulier, F., Verny, M. A., Chanliaud, E., Rémésy, C., & Demigné, C. (2006). Variability of B vitamin concentrations in wheat grain, milling fractions and bread products. *European Journal of Agronomy, 25*(2), 163–169. https://doi.org/10.1016/j.eja.2006.04.009.

Biel, W., Jaroszewska, A., Stankowski, S., Sobolewska, M., & Kępińska-Pacelik, J. (2021). Comparison of yield, chemical composition and farinograph properties of common and ancient wheat grains. *European food research and technology = Zeitschrift fur Lebensmittel-Untersuchung und -Forschung. A, 247*(6), 1525–1538. https://doi.org/10.1007/s00217-021-03729-7.

Borel, P., & Desmarchelier, C. (2018). Bioavailability of fat-soluble vitamins and phytochemicals in humans: Effects of genetic variation. *Annual Review of Nutrition, 38,* 69–96. https://doi.org/10.1146/annurev-nutr-082117-051628.

Boukid, F., Folloni, S., Ranieri, R., & Vittadini, E. (2018). A compendium of wheat germ: Separation, stabilization and food applications. *Trends in Food Science & Technology, 78,* 120–133. https://doi.org/10.1016/j.tifs.2018.06.001.

Bressiani, J., Oro, T., & Da Silva, P. (2019). Influence of milling whole wheat grains and particle size on thermo-mechanical properties of flour using Mixolab. *Czech Journal of Food Science, 37*(4), 276–284.

Brier, N. D., Gomand, S. V., Donner, E., Paterson, D., Delcour, J. A., Lombi, E., & Smolders, E. (2015). Distribution of minerals in wheat grains (*Triticum aestivum* L.) and in roller milling fractions affected by pearling. *Journal of Agricultural and Food Chemistry, 63*(4), 1276–1285. https://doi.org/10.1021/jf5055485.

Brouns, F., Geisslitz, S., Guzman, C., Ikeda, T. M., Arzani, A., Latella, G., . . . Shewry, P. R. (2022). Do ancient wheats contain less gluten than modern bread wheat, in favour of better health? *Nutrition Bulletin/BNF, 47*(2), 157–167. https://doi.org/10.1111/nbu.12551.

Bryden, W. L., Mollah, Y., & Gill, R. J. (1991). Bioavailability of biotin in wheat. *Journal of the Science of Food and Agriculture, 55*(2), 269–275. https://doi.org/10.1002/jsfa.2740550212.

Călinoiu, L. F., & Vodnar, D. C. (2018). Whole grains and phenolic acids: A review on bioactivity, functionality, health benefits and bioavailability. *Nutrients, 10*(11). https://doi.org/10.3390/nu10111615.

Capitani, M., Mateo, C. M., & Nolasco, S. M. (2011). Effect of temperature and storage time of wheat germ on the oil tocopherol concentration. *Brazilian Journal of Chemical Engineering, 28*(2), 243–250. https://doi.org/10.1590/S0104-66322011000200008.

Carbonell-Capella, J. M., Buniowska, M., Barba, F. J., Esteve, M. J., & Frígola, A. (2014). Analytical methods for determining bioavailability and bioaccessibility of bioactive compounds from fruits and vegetables: A review. *Comprehensive Reviews in Food Science and Food Safety, 13*(2), 155–171. https://doi.org/10.1111/1541-4337.12049.

Chapagai, S., & Fink, A. M. (2022). Cardiovascular diseases and sleep disorders in South Asians: A scoping review. *Sleep Medicine, 100*, 139–149. https://doi.org/10.1016/j.sleep.2022.08.008.

Chen, Z., Wang, P., Weng, Y., Ma, Y., Gu, Z., & Yang, R. (2017). Comparison of phenolic profiles, antioxidant capacity and relevant enzyme activity of different Chinese wheat varieties during germination. *Food Bioscience, 20*, 159–167.

Cheng, W., Sun, Y., Fan, M., Li, Y., Wang, L., & Qian, H. (2022). Wheat bran, as the resource of dietary fiber: A review. *Critical Reviews in Food Science and Nutrition, 62*(26), 7269–7281. https://doi.org/10.1080/10408398.2021.1913399.

Chooi, Y. C., Ding, C., & Magkos, F. (2019). The epidemiology of obesity. *Metabolism: Clinical and Experimental, 92*, 6–10. https://doi.org/10.1016/j.metabol.2018.09.005.

Cilla, A., Barberá, R., López-García, G., Blanco-Morales, V., Alegría, A., & Garcia-Llatas, G. (2019). Impact of processing on mineral bioaccessibility/bioavailability. In *Innovative Thermal and Non-Thermal Processing, Bioaccessibility and Bioavailability of Nutrients and Bioactive Compounds* (pp. 209–239). Elsevier. https://doi.org/10.1016/B978-0-12-814174-8.00007-X.

D'Appolonia, B. L., & Rayas-Duarte, P. (1994a). Wheat carbohydrates: Structure and functionality. In W. Bushuk & V. F. Rasper (Eds.), *Wheat* (pp. 107–127). Springer. https://doi.org/10.1007/978-1-4615-2672-8_8.

Dahiya, S., Bajaj, B. K., Kumar, A., Tiwari, S. K., & Singh, B. (2020). A review on biotechnological potential of multifarious enzymes in bread making. *Process Biochemistry, 99*, 290–306.

De Bruijn, W. J. C., Vincken, J.-P., Duran, K., & Gruppen, H. (2016). Mass spectrometric characterization of benzoxazinoid glycosides from rhizopus-elicited wheat (*Triticum aestivum*) seedlings. *Journal of Agricultural and Food Chemistry, 64*(32), 6267–6276. https://doi.org/10.1021/acs.jafc.6b02889.

Dhua, S., Kumar, K., Kumar, Y., Singh, L., & Sharanagat, V. S. (2021). Composition, characteristics and health promising prospects of black wheat: A review. *Trends in Food Science & Technology, 112*, 780–794. https://doi.org/10.1016/j.tifs.2021.04.037.

Dinelli, G., Segura-Carretero, A., Di Silvestro, R., Marotti, I., Arráez-Román, D., Benedettelli, S., . . . Fernadez-Gutierrez, A. (2011). Profiles of phenolic compounds in modern and old common wheat varieties determined by liquid chromatography coupled with time-of-flight mass spectrometry. *Journal of Chromatography. A, 1218*(42), 7670–7681. https://doi.org/10.1016/j.chroma.2011.05.065.

Eckel, R. H., Grundy, S. M., & Zimmet, P. Z. (2005). The metabolic syndrome. *The Lancet, 365*(9468), 1415–1428. https://doi.org/10.1016/S0140-6736(05)66378-7.

El-Naggar, A., de Neergaard, A., & El-Araby, A. (2009). Simultaneous uptake of multiple amino acids by wheat. *Journal of Plant Nutrition, 32*(5), 725–740.

Eriksen, A. K., Brunius, C., Mazidi, M., Hellström, P. M., Risérus, U., Iversen, K. N., . . . Landberg, R. (2020). Effects of whole-grain wheat, rye, and lignan supplementation on cardiometabolic risk factors in men with metabolic syndrome: a randomized crossover trial. *The American Journal of Clinical Nutrition, 111*(4), 864–876. https://doi.org/10.1093/ajcn/nqaa026.

Esposito, F., Arlotti, G., Maria Bonifati, A., Napolitano, A., Vitale, D., & Fogliano, V. (2005). Antioxidant activity and dietary fibre in durum wheat bran by-products. *Food Research International, 38*(10), 1167–1173. https://doi.org/10.1016/j.foodres.2005.05.002.

FAO Cereal Supply and Demand Brief | World Food Situation | Food and Agriculture Organization of the United Nations. (n.d.). Retrieved April 14, 2023, from www.fao.org/worldfoodsituation/csdb/en/

Ferguson, L. R., & Harris, P. J. (1999). Protection against cancer by wheat bran: Role of dietary fibre and phytochemicals. *European Journal of Cancer Prevention, 8*(1), 17–25.

Ficco, Donatella B. M., Borrelli, G. M., Miedico, O., Giovanniello, V., Tarallo, M., Pompa, C., . . . Chiaravalle, A. E. (2020). Effects of grain debranning on bioactive compounds, antioxidant capacity and essential and toxic trace elements in purple durum wheats. *LWT, 118*, 108734. https://doi.org/10.1016/j.lwt.2019.108734.

Ficco, Donatella B. M., Mastrangelo, A. M., Trono, D., Borrelli, G. M., De Vita, P., Fares, C., . . . Papa, R. (2014). The colours of durum wheat: A review. *Crop and Pasture Science, 65*(1), 1. https://doi.org/10.1071/CP13293.

Fung, T. T., Hu, F. B., Pereira, M. A., Liu, S., Stampfer, M. J., Colditz, G. A., & Willett, W. C. (2002). Whole-grain intake and the risk of type 2 diabetes: A prospective study in men. *The American Journal of Clinical Nutrition, 76*(3), 535–540. https://doi.org/10.1093/ajcn/76.3.535.

Gałkowska, D., Witczak, T., & Witczak, M. (2021). Ancient wheat and quinoa flours as ingredients for pasta dough-evaluation of thermal and rheological properties. *Molecules (Basel, Switzerland), 26*(22). https://doi.org/10.3390/molecules26227033.

Ganji, S. H., Kamanna, V. S., & Kashyap, M. L. (2003). Niacin and cholesterol: role in cardiovascular disease (review). *The Journal of Nutritional Biochemistry, 14*(6), 298–305. https://doi.org/10.1016/s0955-2863(02)00284-x.

Garcia Molina, M. D., Botticella, E., Beleggia, R., Palombieri, S., De Vita, P., Masci, S., . . . Sestili, F. (2021). Enrichment of provitamin A content in durum wheat grain by suppressing β-carotene hydroxylase 1 genes with a TILLING approach. *TAG. Theoretical and Applied Genetics. Theoretische und Angewandte Genetik, 134*(12), 4013–4024. https://doi.org/10.1007/s00122-021-03944-6.

Garg, A., Sharma, A., Krishnamoorthy, P., Garg, J., Virmani, D., Sharma, T., . . . Sikorskaya, E. (2017). Role of niacin in current clinical practice: A systematic review. *The American Journal of Medicine, 130*(2), 173–187. https://doi.org/10.1016/j.amjmed.2016.07.038.

Geng, P., Sun, J., Zhang, M., Li, X., Harnly, J. M., & Chen, P. (2016). Comprehensive characterization of C-glycosyl flavones in wheat (*Triticum aestivum* L.) germ using UPLC-PDA-ESI/HRMS(n) and mass defect filtering. *Journal of Mass Spectrometry, 51*(10), 914–930. https://doi.org/10.1002/jms.3803.

Giacco, R., Costabile, G., Della Pepa, G., Anniballi, G., Griffo, E., Mangione, A., . . . Riccardi, G. (2014). A whole-grain cereal-based diet lowers postprandial plasma insulin and triglyceride levels in individuals with metabolic syndrome. *Nutrition, Metabolism, and Cardiovascular Diseases, 24*(8), 837–844. https://doi.org/10.1016/j.numecd.2014.01.007.

Gómez, M., Gutkoski, L. C., & Bravo-Núñez, Á. (2020). Understanding whole-wheat flour and its effect in breads: A review. *Comprehensive Reviews in Food Science and Food Safety, 19*(6), 3241–3265. https://doi.org/10.1111/1541-4337.12625.

Gooding, M. J., & Shewry, P. R. (2022). *Wheat: Environment, Food and Health*. Books. google.com.

Goyer, A. (2010). Thiamine in plants: Aspects of its metabolism and functions. *Phytochemistry, 71*(14–15), 1615–1624. https://doi.org/10.1016/j.phytochem.2010.06.022.

Guo, H., Ding, J., Liang, J., & Zhang, Y. (2021). Associations of whole grain and refined grain consumption with metabolic syndrome. A meta-analysis of observational studies. *Frontiers in Nutrition, 8*, 695620. https://doi.org/10.3389/fnut.2021.695620.

Gupta, R., Meghwal, M., & Prabhakar, P. K. (2021). Bioactive compounds of pigmented wheat (Triticum aestivum): Potential benefits in human health. *Trends in Food Science & Technology, 110*, 240–252. https://doi.org/10.1016/j.tifs.2021.02.003.

Haas, P., Machado, M. J., Anton, A. A., & Silva, A. S. S. (2009). Effectiveness of whole grain consumption in the prevention of colorectal cancer: Meta-analysis of cohort studies. *International Journal of Food Sciences and Nutrition, 60*(6), 1–13.

Hanhineva, K., Rogachev, I., Aura, A.-M., Aharoni, A., Poutanen, K., & Mykkänen, H. (2011). Qualitative characterization of benzoxazinoid derivatives in whole grain rye and wheat by LC-MS metabolite profiling. *Journal of Agricultural and Food Chemistry, 59*(3), 921–927. https://doi.org/10.1021/jf103612u.

Hayakawa, K., Tanaka, K., Nakamura, T., & Endo, S. (2004). End use quality of waxy wheat flour in various grain-based foods. *Cereal Chemistry, 81*(5), 666–672.

He, J., & Giusti, M. M. (2010). Anthocyanins: Natural colorants with health-promoting properties. *Annual Review of Food Science and Technology, 1*, 163–187. https://doi.org/10.1146/annurev.food.080708.100754.

Hemery, Y., Chaurand, M., Holopainen, U., Lampi, A. M., Lehtinen, P., Piironen, V., Sadoudi, A., & Rouau, X. (2011). Potential of dry fractionation of wheat bran for the development of food ingredients, part I: Influence of ultra-fine grinding. *Journal of Cereal Science, 53*(1), 1–8.

Hemery, Y., Lullien-Pellerin, V., Rouau, X., Abecassis, J., Samson, M. F., Åman, P., von Reding, W., Spoerndli, C., & Barron, C. (2009). Biochemical markers: efficient tools for the assessment of wheat grain tissue proportions in milling fractions. *Journal of Cereal Science, 49*(1), 55–64.

Hemery, Y. M., Anson, N. M., Havenaar, R., Haenen, G. R., Noort, M. W., & Rouau, X. (2010). Dry-fractionation of wheat bran increases the bioaccessibility of phenolic acids in breads made from processed bran fractions. *Food Research International, 43*(5), 1429–1438.

Heshe, G. G., Haki, G. D., Woldegiorgis, A. Z., & Gemede, H. F. (2016). Effect of conventional milling on the nutritional value and antioxidant capacity of wheat types common in Ethiopia and a recovery attempt with bran supplementation in bread. *Food Science & Nutrition, 4*(4), 534–543. https://doi.org/10.1002/fsn3.315.

Hrubša, M., Siatka, T., Nejmanová, I., Vopršalová, M., Kujovská Krčmová, L., Matoušová, K., . . . On Behalf of the Oemonom. (2022). Biological properties of vitamins of the B-Complex, Part 1: Vitamins B1, B2, B3, and B5. *Nutrients, 14*(3). https://doi.org/10.3390/nu14030484; www.fao.org/temprefdocrep/fao/meeting/018/k6021e.pdf. (n.d.). Retrieved September 2, 2022, from www.fao.org/temprefdocrep/fao/meeting/018/k6021e.pdf.

Hu, Y., & Schmidhalter, U. (2001). Effects of salinity and macronutrient levels on micronutrients in wheat. *Journal of Plant Nutrition, 24*(2), 273–281. https://doi.org/10.1081/PLN-100001387.

Igrejas, G., Ikeda, T. M., & Guzmán, C. (Eds.). (2020). *Wheat Quality for Improving Processing and Human Health.* Cham: Springer International Publishing. https://doi.org/10.1007/978-3-030-34163-3.

Jensen, S. K., & Lauridsen, C. (2007). A-Tocopherol stereoisomers. In *Vitamin E* (vol. 76, pp. 281–308). Elsevier. https://doi.org/10.1016/S0083-6729(07)76010-7.

Jones, J. M., Adams, J., Harriman, C., Miller, C., & Van der Kamp, J. W. (2015). Nutritional impacts of different whole grain milling techniques: A review of milling practices and existing data. *Cereal Foods World, 60*(3), 130–139. https://doi.org/10.1094/CFW-60-3-0130.

Jones, J. M., & Engleson, J. (2010). Whole grains: Benefits and challenges. *Annual Review of Food Science and Technology, 1*, 19–40. https://doi.org/10.1146/annurev.food.112408.132746.

Katileviciute, A., Plakys, G., Budreviciute, A., Onder, K., Damiati, S., & Kodzius, R. (2019). A sight to wheat bran: High value-added products. *Biomolecules, 9*(12). https://doi.org/10.3390/biom9120887.

Kaur, N., Singh, B., Kaur, A., Yadav, M. P., Singh, N., Ahlawat, A. K., & Singh, A. M. (2021). Effect of growing conditions on proximate, mineral, amino acid, phenolic composition and antioxidant properties of wheatgrass from different wheat (*Triticum aestivum* L.) varieties. *Food Chemistry, 341*(Pt. 1), 128201. https://doi.org/10.1016/j.foodchem.2020.128201.

Khan, J., Khan, M. Z., Ma, Y., Meng, Y., Mushtaq, A., Shen, Q., & Xue, Y. (2022). Overview of the composition of whole grains' phenolic acids and dietary fibre and their effect on chronic non-communicable diseases. *International Journal of Environmental Research and Public Health, 19*(5). https://doi.org/10.3390/ijerph19053042.

Kim, M. J., Kwak, H. S., & Kim, S. S. (2018). Effects of germination on protein, γ-aminobutyric acid, phenolic acids, and antioxidant capacity in wheat. *Molecules (Basel, Switzerland), 23*(9). https://doi.org/10.3390/molecules23092244.

Kopelman, P. G. (2000). Obesity as a medical problem. *Nature, 404*(6778), 635–643. https://doi.org/10.1038/35007508.

Ktenioudaki, A., Alvarez-Jubete, L., & Gallagher, E. (2015). A review of the process-induced changes in the phytochemical content of cereal grains: The breadmaking process. *Critical Reviews in Food Science and Nutrition, 55*(5), 611–619. https://doi.org/10.1080/10408398.2012.667848.

Kumar, N., & Pruthi, V. (2014). Potential applications of ferulic acid from natural sources. *Biotechnology Reports (Amsterdam, Netherlands), 4*, 86–93. https://doi.org/10.1016/j.btre.2014.09.002.

Kurth, T., Moore, S. C., Gaziano, J. M., Kase, C. S., Stampfer, M. J., Berger, K., & Buring, J. E. (2006). Healthy lifestyle and the risk of stroke in women. *Archives of Internal Medicine, 166*(13), 1403–1409.

Lachman, J., Martinek, P., Kotíková, Z., Orsák, M., & Šulc, M. (2017). Genetics and chemistry of pigments in wheat grain – A review. *Journal of Cereal Science, 74*, 145–154. https://doi.org/10.1016/j.jcs.2017.02.007.

Ladhari, A., Corrado, G., Rouphael, Y., Carella, F., Nappo, G. R., Di Marino, C., . . . Palatucci, D. (2022). Chemical, functional, and technological features of grains, brans, and semolina from purple and red durum wheat landraces. *Foods, 11*(11). https://doi.org/10.3390/foods11111545.

Landberg, R., Hanhineva, K., Tuohy, K., Garcia-Aloy, M., Biskup, I., Llorach, R., . . . Kolehmainen, M. (2019). Biomarkers of cereal food intake. *Genes & Nutrition, 14*, 28. https://doi.org/10.1186/s12263-019-0651-9.

Lei, L., Chen, J., Liu, Y., Wang, L., & Zhao, G. (2018). Dietary wheat bran oil is equally as effective as rice bran oil in reducing plasma cholesterol. *Journal of Agricultural and Food Chemistry, 66*(11), 2765–2774.

Leváková, Ľ., & Lacko-Bartošová, M. (2017). Phenolic acids and antioxidant activity of wheat species: A review. *Agriculture (Poľnohospodárstvo), 63*(3), 92–101. https://doi.org/10.1515/agri-2017-0009.

Li, L., Shewry, P. R., & Ward, J. L. (2008). Phenolic acids in wheat varieties in the HEALTHGRAIN diversity screen. *Journal of Agricultural and Food Chemistry, 56*(21), 9732–9739.

Liang, Q., Wang, K., Shariful, I., Ye, X., & Zhang, C. (2020). Folate content and retention in wheat grains and wheat-based foods: Effects of storage, processing, and cooking methods. *Food Chemistry, 333*, 127459. https://doi.org/10.1016/j.foodchem.2020.127459.

Likes, R., Madl, R. L., Zeisel, S. H., & Craig, S. A. S. (2007). The betaine and choline content of a whole wheat flour compared to other mill streams. *Journal of Cereal Science, 46*(1), 93–95. https://doi.org/10.1016/j.jcs.2006.11.002.

Lineback, D. R., & Rasper, V. F. (1988). Wheat carbohydrates. In Y. Pomeranz *Wheat: Chemistry and Technology* (vol. I., 3rd ed., pp. 277–372). American Association of cereals chemists, St Paul Minnesota, USA.

Liu, J., Yu, L. L., & Wu, Y. (2020). Bioactive components and health beneficial properties of whole wheat foods. *Journal of Agricultural and Food Chemistry, 68*(46), 12904–12915. https://doi.org/10.1021/acs.jafc.0c00705.

Liyana-Pathirana, C. M., & Shahidi, F. (2006). Importance of insoluble-bound phenolics to antioxidant properties of wheat. *Journal of Agricultural and Food Chemistry, 54*(4), 1256–1264. https://doi.org/10.1021/jf052556h.

Lu, Y., Lv, J., Yu, L., Fletcher, A., Costa, J., Yu, L. and Luthria, D. (2014). Phytochemical composition and antiproliferative activities of bran fraction of ten Maryland-grown soft winter wheat cultivars: Comparison of different radical scavenging assays. *Journal of Food Composition and Analysis, 36*(1–2), 51–58.

Ma, D., Wang, C., Feng, J., & Xu, B. (2021). Wheat grain phenolics: A review on composition, bioactivity, and influencing factors. *Journal of the Science of Food and Agriculture, 101*(15), 6167–6185. https://doi.org/10.1002/jsfa.11428.

Ma, D., Xu, B., Feng, J., Hu, H., Tang, J., Yin, G., . . . Wang, C. (2022). Dynamic metabolomics and transcriptomics analyses for characterization of phenolic compounds and their biosynthetic characteristics in wheat grain. *Frontiers in Nutrition, 9*, 844337. https://doi.org/10.3389/fnut.2022.844337.

Mansour, A., Hosseini, S., Larijani, B., Pajouhi, M., & Mohajeri-Tehrani, M. R. (2013). Nutrients related to GLP1 secretory responses. *Nutrition, 29*(6), 813–820. https://doi.org/10.1016/j.nut.2012.11.015.

Mateo Anson, N., Hemery, Y. M., Bast, A., & Haenen, G. R. M. M. (2012). Optimizing the bioactive potential of wheat bran by processing. *Food & Function, 3*(4), 362–375. https://doi.org/10.1039/c2fo10241b.

Matus-Cádiz, M. A., Daskalchuk, T. E., Verma, B., Puttick, D., Chibbar, R. N., Gray, G. R., . . . Hucl, P. (2008). Phenolic compounds contribute to dark bran pigmentation in hard white wheat. *Journal of Agricultural and Food Chemistry, 56*(5), 1644–1653. https://doi.org/10.1021/jf072970c.

McGrath, L., & Fernandez, M.-L. (2022). Plant-based diets and metabolic syndrome: Evaluating the influence of diet quality. *Journal of Agriculture and Food Research, 9*, 100322. https://doi.org/10.1016/j.jafr.2022.100322.

McKeown, N. M., Marklund, M., Ma, J., Ross, A. B., Lichtenstein, A. H., Livingston, K. A., Jacques, P. F., Rasmussen, H. M., Blumberg, J. B. & Chen, C. Y. O. (2016). Comparison of plasma alkylresorcinols (AR) and urinary AR metabolites as biomarkers of compliance in a short-term, whole-grain intervention study. *European Journal of Nutrition, 55*, 1235–1244.

McMahon, R. J. (2002). Biotin in metabolism and molecular biology. *Annual Review of Nutrition, 22*, 221–239. https://doi.org/10.1146/annurev.nutr.22.121101.112819.

Melini, V., Melini, F., Luziatelli, F., & Ruzzi, M. (2020). Functional ingredients from agrifood waste: Effect of inclusion thereof on phenolic compound content and bioaccessibility in bakery products. *Antioxidants (Basel, Switzerland), 9*(12). https://doi.org/10.3390/antiox9121216.

Merlino, M., Leroy, P., Chambon, C., & Branlard, G. (2009). Mapping and proteomic analysis of albumin and globulin proteins in hexaploid wheat kernels (*Triticum aestivum* L.). *TAG. Theoretical and Applied Genetics. Theoretische und Angewandte Genetik, 118*(7), 1321–1337. https://doi.org/10.1007/s00122-009-0983-8.

Mills, E. N. C., Wellner, N., Salt, L. A., Robertson, J., & Jenkins, J. A. (2020). Wheat proteins and bread quality. In Stanley P. Cauvain, *Breadmaking* (pp. 109–135). Woodhead Publishing.

Montemurro, M., Pontonio, E., Gobbetti, M., & Rizzello, C. G. (2019). Investigation of the nutritional, functional and technological effects of the sourdough fermentation of sprouted flours. *International Journal of Food Microbiology, 302*, 47–58. https://doi.org/10.1016/j.ijfoodmicro.2018.08.005.

Niba, L. L., & Niba, S. N. (2003). Role of non-digestible carbohydrates in colon cancer protection. *Nutrition & Food Science, 33*(1), 28–33. https://doi.org/10.1108/00346650310459545.

Nićiforović, N., & Abramovič, H. (2014). Sinapic acid and its derivatives: Natural sources and bioactivity. *Comprehensive Reviews in Food Science and Food Safety, 13*(1), 34–51. https://doi.org/10.1111/1541-4337.12041.

Noncommunicable Diseases: Mortality. (n.d.). Retrieved September 5, 2022, from www.who.int/data/gho/data/themes/topics/topic-details/GHO/ncd-mortality

Noort, M. W., van Haaster, D., Hemery, Y., Schols, H. A., & Hamer, R. J. (2010). The effect of particle size of wheat bran fractions on bread quality–Evidence for fibre–protein interactions. *Journal of Cereal Science, 52*(1), 59–64.

Oduro-Obeng, H., Apea-Bah, F. B., Wang, K., Fu, B. X., & Beta, T. (2022). Effect of cooking duration on carotenoid content, digestion and potential absorption efficiencies among refined semolina and whole wheat pasta products. *Food & Function, 13*(11), 5953–5970. https://doi.org/10.1039/d2fo00611a.

Ogden, C. L., Yanovski, S. Z., Carroll, M. D., & Flegal, K. M. (2007). The epidemiology of obesity. *Gastroenterology, 132*(6), 2087–2102. https://doi.org/10.1053/j.gastro.2007.03.052.

Oh, K., Hu, F. B., Cho, E., Rexrode, K. M., Stampfer, M. J., Manson, J. E., Liu, S., & Willett, W. C. (2005). Carbohydrate intake, glycemic index, glycemic load, and dietary fiber in relation to risk of stroke in women. *American Journal of Epidemiology, 161*(2), 161–169.

Ohrvik, V. E., & Witthoft, C. M. (2011). Human folate bioavailability. *Nutrients, 3*(4), 475–490. https://doi.org/10.3390/nu3040475.

Oishi, K., Yamamoto, S., Itoh, N., Nakao, R., Yasumoto, Y., Tanaka, K., . . . Takano-Ishikawa, Y. (2015). Wheat alkylresorcinols suppress high-fat, high-sucrose diet-induced obesity and glucose intolerance by increasing insulin sensitivity and cholesterol excretion in male mice. *The Journal of Nutrition, 145*(2), 199–206. https://doi.org/10.3945/jn.114.202754.

Pagani, M. A., Giordano, D., Cardone, G., Pasqualone, A., Casiraghi, M. C., Erba, D., ... Marti, A. (2020). Nutritional features and bread-making performance of whole-wheat: Does the milling system matter? *Foods, 9*(8), 1035–1043.

Panato, A., Antonini, E., Bortolotti, F., & Ninfali, P. (2017). The histology of grain caryopses for nutrient location: a comparative study of six cereals. *International Journal of Food Science & Technology, 52*(5), 1238–1245. https://doi.org/10.1111/ijfs.13390.

Parascandola, M., & Pearlman, P. C. (2022). The development of global cancer research at the United States National Cancer Institute. *Journal of the National Cancer Institute, 114*(9), 1228–1237.

Parker, M. L., Ng, A., & Waldron, K. W. (2005). The phenolic acid and polysaccharide composition of cell walls of bran layers of mature wheat (*Triticum aestivum* L. cv. *Avalon*) grains. *Journal of the Science of Food and Agriculture, 85*(15), 2539–2547.

Paznocht, L., Kotíková, Z., Orsák, M., Lachman, J., & Martinek, P. (2019). Carotenoid changes of colored-grain wheat flours during bun-making. *Food chemistry, 277*, 725–734. https://doi.org/10.1016/j.foodchem.2018.11.019.

Pehlivan Karakas, F., Keskin, C. N., Agil, F., & Zencirci, N. (2021). Profiles of vitamin B and E in wheat grass and grain of einkorn (Triticum monococcum spp. Monococcum), emmer (*Triticum dicoccum* ssp. Dicoccum Schrank.), durum (*Triticum durum* Desf.), and bread wheat (*Triticum aestivum* L.) cultivars by LC-ESI-MS/MS analysis. *Journal of Cereal Science, 98*, 103177. https://doi.org/10.1016/j.jcs.2021.103177.

Poutanen, K. S., Kårlund, A. O., Gómez-Gallego, C., Johansson, D. P., Scheers, N. M., Marklinder, I. M., . . . Landberg, R. (2022). Grains—a major source of sustainable protein for health. *Nutrition Reviews, 80*(6), 1648–1663. https://doi.org/10.1093/nutrit/nuab084.

Prasoodanan P. K., V., Sharma, A. K., Mahajan, S., Dhakan, D. B., Maji, A., Scaria, J., & Sharma, V. K. (2021). Western and non-western gut microbiomes reveal new roles of Prevotella in carbohydrate metabolism and mouth-gut axis. *Npj Biofilms and Microbiomes, 7*(1), 77. https://doi.org/10.1038/s41522-021-00248-x.

Prinsen, P., Gutierrez, A., Faulds, C., & Del Rio, J. C. (2014). A comprehensive study of valuable lipophilic phytochemicals in wheat bran. *Journal of Agricultural and Food Chemistry, 62*(7), 1664–1673. https://doi.org/10.1021/jf404772b.

Ramprasath, V. R., & Awad, A. B. (2015). Role of Phytosterols in Cancer Prevention and Treatment. *Journal of AOAC International, 98*(3), 735–738. https://doi.org/10.5740/jaoacint.SGERamprasath.

Rinaldi, V. E. A., Ng, P. K. W., & Bennink, M. R. (2000). Effects of extrusion on dietary fiber and isoflavone contents of wheat extrudates enriched with wet okara. Cereal Chemistry, 77(2), 237–240.

Ringland, C. L., Arkenau, H. T., O'Connell, D. L., & Ward, R. L. (2010). Second primary colorectal cancers (SPCRCs): experiences from a large Australian Cancer Registry. *Annals of Oncology, 21*(1), 92–97. https://doi.org/10.1093/annonc/mdp288.

Rosa, N. N., Barron, C., Gaiani, C., Dufour, C., & Micard, V. (2013). Ultra-fine grinding increases the antioxidant capacity of wheat bran. *Journal of cereal science, 57*(1), 84–90. https://doi.org/10.1016/j.jcs.2012.10.002.

Rosa-Sibakov, N., Poutanen, K., & Micard, V. (2015). How does wheat grain, bran and aleurone structure impact their nutritional and technological properties? *Trends in food science & technology, 41*(2), 118–134. https://doi.org/10.1016/j.tifs.2014.10.003.

Saini, P., Kumar, N., Kumar, S., Mwaurah, P. W., Panghal, A., Attkan, A. K., . . . Singh, V. (2021). Bioactive compounds, nutritional benefits and food applications of colored wheat: a comprehensive review. *Critical reviews in food science and nutrition, 61*(19), 3197–3210. https://doi.org/10.1080/10408398.2020.1793727.

Saleh, A. S. M., Wang, P., Wang, N., Yang, S., & Xiao, Z. (2019). Technologies for enhancement of bioactive components and potential health benefits of cereal and cereal-based foods: Research advances and application challenges. *Critical reviews in food science and nutrition, 59*(2), 207–227. https://doi.org/10.1080/10408398.2017.1363711.

Salmerón, J., Manson, J. E., Stampfer, M. J., Colditz, G. A., Wing, A. L., & Willett, W. C. (1997). Dietary fiber, glycemic load, and risk of non-insulin-dependent diabetes mellitus in women. *The Journal of the American Medical Association, 277*(6), 472–477. https://doi.org/10.1001/jama.1997.03540300040031.

Samson, S. L., & Garber, A. J. (2014). Metabolic syndrome. *Endocrinology and Metabolism Clinics of North America, 43*(1), 1–23. https://doi.org/10.1016/j.ecl.2013.09.009.

Sankar, R. (n.d.). The Comprehensive Review on Fat Soluble Vitamins.

Santos, D. I., Saraiva, J. M. A., Vicente, A. A., & Moldão-Martins, M. (2019). Methods for determining bioavailability and bioaccessibility of bioactive compounds and nutrients. In *Innovative Thermal and Non-Thermal Processing, Bioaccessibility and Bioavailability of Nutrients and Bioactive Compounds* (pp. 23–54). Elsevier. https://doi.org/10.1016/B978-0-12-814174-8.00002-0.

Sharma, I., Tyagi, B. S., & Singh, G. (2015). Enhancing wheat production-A global perspective. *Indian Journal of.*

Sharma, M., Sandhir, R., Singh, A., Kumar, P., Mishra, A., Jachak, S., . . . Roy, J. (2016). Comparative Analysis of Phenolic Compound Characterization and Their Biosynthesis Genes between Two Diverse Bread Wheat (Triticum aestivum) Varieties Differing for Chapatti (Unleavened Flat Bread) Quality. *Frontiers in plant science, 7*, 1870. https://doi.org/10.3389/fpls.2016.01870.

Shavit, R., Batyrshina, Z. S., Dotan, N., & Tzin, V. (2018). Cereal aphids differently affect benzoxazinoid levels in durum wheat. *Plos One, 13*(12), e0208103. https://doi.org/10.1371/journal.pone.0208103.

Shevkani, K., Singh, N., & Bajaj, R. (2017). Wheat starch production, structure, functionality and applications—a review. *International Journal of . . . .*

Shewry, P. R., & Hey, S. J. (2015). The contribution of wheat to human diet and health. *Food and Energy Security, 4*(3), 178–202. https://doi.org/10.1002/fes3.64.

Shewry, P. R., Van Schaik, F., Ravel, C., Charmet, G., Rakszegi, M., Bedo, Z., & Ward, J. L. (2011). Genotype and environment effects on the contents of vitamins B1, B2, B3, and B6 in wheat grain. *Journal of Agricultural and Food Chemistry, 59*(19), 10564–10571. https://doi.org/10.1021/jf202762b.

Singla, R. K., Dubey, A. K., Garg, A., Sharma, R. K., Fiorino, M., Ameen, S. M., . . . Al-Hiary, M. (2019). Natural polyphenols: chemical classification, definition of classes, subcategories, and structures. *Journal of AOAC International, 102*(5), 1397–1400. https://doi.org/10.5740/jaoacint.19-0133.

Stanger, O., Fowler, B., Piertzik, K., Huemer, M., Haschke-Becher, E., Semmler, A., . . . Linnebank, M. (2009). Homocysteine, folate and vitamin B12 in neuropsychiatric diseases: review and treatment recommendations. *Expert Review of Neurotherapeutics, 9*(9), 1393–1412. https://doi.org/10.1586/ern.09.75.

Sun, Y., Liu, Y., Zhang, J., Liu, J., & Hou, H. (2022). Effects of wheat debranning on the sensory quality and antioxidant activity of Chinese steamed bread. *Cereal Chemistry.*

Suttie, J. W. (1992). Vitamin K and human nutrition. *Journal of the American Dietetic Association, 92*(5), 585–590.

Syauqy, A., Hsu, C.-Y., Rau, H.-H., & Chao, J. C.-J. (2018). Association of Dietary Patterns with Components of Metabolic Syndrome and Inflammation among Middle-Aged and Older Adults with Metabolic Syndrome in Taiwan. *Nutrients, 10*(2). https://doi.org/10.3390/nu10020143.

Tahiliani, A. G., & Beinlich, C. J. (1991). Pantothenic acid in health and disease. *Vitamins and Hormones, 46,* 165–228.

Thakur, K., Tomar, S. K., Singh, A. K., Mandal, S., & Arora, S. (2017). Riboflavin and health: A review of recent human research. *Critical reviews in food science and nutrition, 57*(17), 3650–3660. https://doi.org/10.1080/10408398.2016.1145104.

Tian, W., Chen, G., Tilley, M., & Li, Y. (2021). Changes in phenolic profiles and antioxidant activities during the whole wheat bread-making process. *Food chemistry, 345,* 128851. https://doi.org/10.1016/j.foodchem.2020.128851.

Tian, W., Zheng, Y., Wang, W., Wang, D., Tilley, M., Zhang, G., . . . Li, Y. (2022). A comprehensive review of wheat phytochemicals: From farm to fork and beyond. *Comprehensive Reviews in Food Science and Food Safety, 21*(3), 2274–2308. https://doi.org/10.1111/1541-4337.12960.

Titcomb, T. J., Sheftel, J., Sowa, M., Gannon, B. M., Davis, C. R., Palacios-Rojas, N., & Tanumihardjo, S. A. (2018). B-Cryptoxanthin and zeaxanthin are highly bioavailable from whole-grain and refined biofortified orange maize in humans with optimal vitamin A status: a randomized, crossover, placebo-controlled trial. *The American Journal of Clinical Nutrition, 108*(4), 793–802. https://doi.org/10.1093/ajcn/nqy134.

Tufarelli, V., Casalino, E., D'Alessandro, A. G., & Laudadio, V. (2017). Dietary phenolic compounds: biochemistry, metabolism and significance in animal and human health. *Current Drug Metabolism, 18*(10), 905–913. https://doi.org/10.2174/138920 0218666170925124004.

Turnbull, K. M., & Rahman, S. (2002). Endosperm texture in wheat. *Journal of cereal science, 36*(3), 327–337. https://doi.org/10.1006/jcrs.2002.0468.

Venegas, J., Guttieri, M. J., & Jr, J. B. (2022). Genetic architecture of the high-inorganic phosphate phenotype derived from a low-phytate mutant in winter wheat. *Crop . . . .*

Wang, M., Kong, F., Liu, R., Fan, Q., & Zhang, X. (2020). Zinc in wheat grain, processing, and food. *Frontiers in nutrition, 7,* 124. https://doi.org/10.3389/fnut.2020.00124.

Wang, T., & Johnson, L. A. (2001). Refining high-free fatty acid wheat germ oil. *Journal of the American Oil Chemists' Society, 78*(1), 71–76. https://doi.org/10.1007/s11746-001-0222-2.

Wang, Y., Duan, Y., Zhu, L., Fang, Z., & He, L. (2019). Whole grain and cereal fiber intake and the risk of type 2 diabetes: a meta-analysis. *International Journal of . . . .*

Weng, L.-C., Steffen, L. M., Szklo, M., Nettleton, J., Chambless, L., & Folsom, A. R. (2013). A diet pattern with more dairy and nuts, but less meat is related to lower risk of developing hypertension in middle-aged adults: The Atherosclerosis Risk in Communities (ARIC) study. *Nutrients, 5*(5), 1719–1733. https://doi.org/10.3390/nu5051719.

Wieser, H., Koehler, P., & Scherf, K. A. (2020). The two faces of wheat. *Frontiers in Nutrition, 7,* 517313. https://doi.org/10.3389/fnut.2020.517313.

Wikipedia Contributors. (2022, May 31). Phytochemical. In *Wikipedia, The Free Encyclopedia.* Retrieved November 18, 2022, from https://en.wikipedia.org/w/index.php?title=Phytochemical&oldid=1090806218

Witkamp, R. F. (2022). Bioactive components in traditional foods aimed at health promotion: A route to novel mechanistic insights and lead molecules? *Annual Review of Food Science and Technology, 13,* 315–336. https://doi.org/10.1146/annurev-food-052720-092845.

Xu, Y., An, D., Li, H., & Xu, H. (2011). Review: Breeding wheat for enhanced micronutrients. *Canadian Journal of Plant Science, 91*(2), 231–237. https://doi.org/10.4141/CJPS10117.

Yamagata, K., & Yamori, Y. (2020). Inhibition of endothelial dysfunction by dietary flavonoids and preventive effects against cardiovascular disease. *Journal of Cardiovascular Pharmacology, 75*(1), 1–9. https://doi.org/10.1097/FJC.0000000000000757.

Yao, W., Gong, Y., Li, L., Hu, X., & You, L. (2022). The effects of dietary fibers from rice bran and wheat bran on gut microbiota: An overview. *Food Chemistry: X, 13,* 100252. https://doi.org/10.1016/j.fochx.2022.100252.

Yoshida, Y., & Niki, E. (2003). Antioxidant effects of phytosterol and its components. *Journal of Nutritional Science and Vitaminology, 49*(4), 277–280.

Younas, A., Sadaqat, H. A., Kashif, M., Ahmed, N., & Farooq, M. (2020). Combining ability and heterosis for grain iron biofortification in bread wheat. *Journal of the Science of Food and Agriculture, 100*(4), 1570–1576. https://doi.org/10.1002/jsfa.10165.

Yu, L., Nanguet, A. L., & Beta, T. (2013). Comparison of antioxidant properties of refined and whole wheat flour and bread. *Antioxidants, 2*(4), 370–383.

Zhang, Y., Song, Q., Yan, J., Tang, J., Zhao, R., Zhang, Y., . . . Ortiz-Monasterio, I. (2010). Mineral element concentrations in grains of Chinese wheat cultivars. *Euphytica, 174*(3), 303–313. https://doi.org/10.1007/s10681-009-0082-6.

Zhao, F. J., Su, Y. H., Dunham, S. J., Rakszegi, M., Bedo, Z., McGrath, S. P., & Shewry, P. R. (2009). Variation in mineral micronutrient concentrations in grain of wheat lines of diverse origin. *Journal of Cereal Science, 49*(2), 290–295. https://doi.org/10.1016/j.jcs.2008.11.007.

Zhou, K., Laux, J. J., & Yu, L. (2004). Comparison of Swiss red wheat grain and fractions for their antioxidant properties. *Journal of Agricultural and Food Chemistry, 52*(5), 1118–1123.

Zhou, K., Yin, J. J., & Yu, L. (2005). Phenolic acid, tocopherol and carotenoid compositions, and antioxidant functions of hard red winter wheat bran. *Journal of Agricultural and Food Chemistry, 53*(10), 3916–3922.

Zhou, L., Hu, S., Rong, S., Mo, X., Wang, Q., Yin, J., . . . Liu, L. (2022). DHPPA, a major plasma alkylresorcinol metabolite reflecting whole-grain wheat and rye intake, and risk of metabolic syndrome: a case-control study. *European Journal of Nutrition, 61*(6), 3247–3254. https://doi.org/10.1007/s00394-022-02880-5.

Zhu, H., Liu, S., Yao, L., Wang, L., & Li, C. (2019). Free and bound phenolics of buckwheat varieties: HPLC characterization, antioxidant activity, and inhibitory potency towards α-glucosidase with molecular docking analysis. *Antioxidants (Basel, Switzerland), 8*(12). https://doi.org/10.3390/antiox8120606.

Zhu, Y., Conklin, D. R., Chen, H., Wang, L., & Sang, S. (2011). 5-alk(en)ylresorcinols as the major active components in wheat bran inhibit human colon cancer cell growth. *Bioorganic & Medicinal Chemistry, 19*(13), 3973–3982. https://doi.org/10.1016/j.bmc.2011.05.025.

6

Wheat-Based Anti-Nutritional Factors and Their Reduction Strategies

An Overview

**Vanita Pandey, Ajeet Singh, Neha Patwa, Ankush,
Om Prakash Gupta, Gopalareddy K., Sunil Kumar,
Anuj Kumar, Sewa Ram, and Gyanendra Pratap Singh**

CONTENTS

DOI: 10.1201/9781003307938-6

6.1 Introduction

Cereals such as rice, wheat, and maize are staple food crops providing food and nutrition to human population across the globe. Cereal grains are vital for human nutrition providing adequate amounts of carbohydrates, protein, vitamins, minerals, and very essential dietary fiber required for the maintenance of growth and development (Nadeem et al., 2010). Consumption of whole-grain foods is associated with the prevention of several lifestyle disorders such as cardiovascular diseases (Jacobs and Gallaher, 2004), type II diabetes (Murtaugh et al., 2003) as well as few types of cancers (Larsson et al., 2005).

Wheat (*Triticum aestivum* L.), a major cereal grain, is part of daily diet of approximately one-third of global populace. A very versatile crop, wheat is grown in nearly every region of the six continents with roughly 777 million-tones production (FAOSTAT, 2022). Wheat is a nutritive cereal crop with a good composition of essential macronutrients like carbohydrates, proteins, and fats and, additionally, micronutrients like vitamins and minerals (FAO 2018a, b). In the case of cereal-based diets, wheat is a major provider of dietary energy and protein to its consumers globally. Wheat is a vital crop in terms of production and consumption across the world (FAO, 2009). In recent times, wheat-based products have received widespread interest in the global consumer market as well as from the food companies due to their superior nutritional profile. Cereals in general contain lower ratio of protein in comparison to carbohydrates. Wheat flour is made up of 87% carbohydrates and nearly 11% protein, in comparison to legumes and millets which have a higher protein content (Kavitha and Parimalavalli, 2014; Malik 2015). Wheat like other cereals is limiting in essential amino acid lysine, which is present in sufficient quantities in legumes. A possible solution can be supplementing wheat flour with legumes to enhance lysine as well as other proteins (Katina et al., 2005; Awolu et al., 2017).

6.1.1 Anti-Nutritional Factors

Antinutrients or anti-nutritional factors (ANFs) are chiefly substances or compounds, which impede the absorption or assimilation of nutrients, thus reducing nutrient intake, digestion, and utilization, additionally causing adverse health effects. Antinutrients are naturally synthesized by several plants and hence are commonly linked with raw, vegan, or plant-based diets (Gemede and Ratta, 2014). Consumption of antinutrients may sometimes cause adverse effects like nutritional deficiencies, bloating, nausea, rashes, and headaches. (Essack et al., 2017), while sensitivity to ANFs may differ in people and several times use of sufficient food-processing techniques can initially reduce its content (Soetan and Oyewole, 2009). ANFs once introduced or consumed cannot be eliminated from the body. In order to study the effects or symptoms produced by a particular ANF, removing and reintroducing the ANF through the specific food can be an effective strategy. Several studies are being carried out in this regard to study the biochemical effects of the ANFs on human body (Aletor, 1993; Petroski and Minich, 2020). Several secondary metabolites, behaving as ANFs, may elicit dangerous biological response; however, numerous of them are commonly utilized as pharmaceutical drug or have role in nutrition (Soetan, 2008).

However, these ANFs are generally vital to plant and may have a role in plant growth, development, defense, etc. ANFs in general are found in high concentration in beans, grains, nuts, and legumes in comparison to roots, leaves, and fruits in the plants they occur. Several ANFs are found in plants having toxic potential to humans as well as animals. ANFs are divided into two categories as heat labile or heat stable on the basis of heat sensitivity. Major ANFs of plant origin are phytic acid, oxalates, tannins, saponins, gossypol, lectins, antivitamin factors, protease inhibitors, amylase inhibitors, metal binding ingredients, goitrogens, etc. (Figure 6.1) .

Diets based solely on uncooked or raw plant-based foods like whole grains, legumes, vegetables, and nuts are susceptible to toxic effects of ANFs. Oxalates, for example, are known to bind with calcium and inhibit its absorption in the body. Raw kale, spinach, broccoli, and soybean generally contain oxalates (Savage and

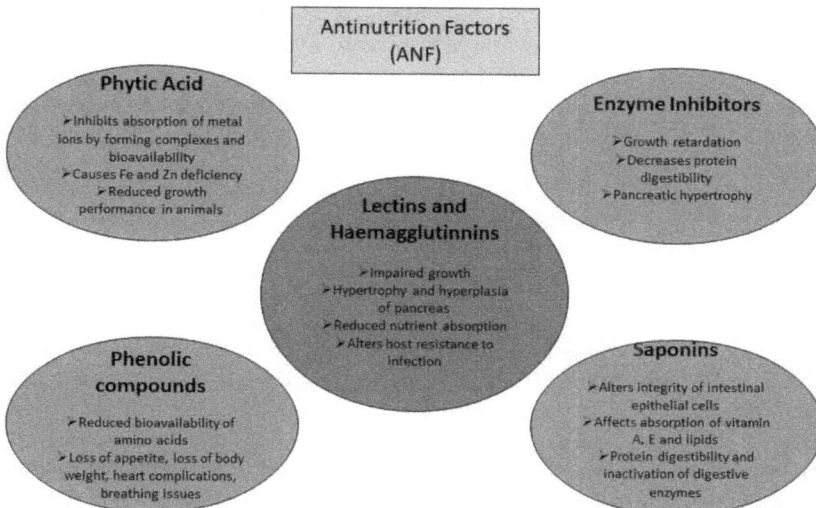

FIGURE 6.1 A brief outline of the adverse effects of major ANFs on humans and animals present in plants (Gemede & Ratta, 2014; Samtiya et al., 2020).

Klunklin, 2018). The consumption of excessive tannins found in certain fruits is known to inactivate enzymes associated with protein absorption. Phytic acid or phytates are a very important class of ANFs found in cereal grains, legumes, and nuts which are associated with chelation of divalent cations and thus impeding their absorption. Lectins found in tomatoes, brinjals, and peppers are known to cause severe reactions to the body (Pandey et al., 2016). The assimilation of saponins is known to cause enzyme inhibition, red blood cells damage, and intervention in thyroid function (Fan et al., 2013).

6.1.2 Antinutrients and Human Health

ANFs are generally associated with harmful effects on humans as well as animals; however; documented research also shows certain beneficial effects on humans. The effect of ANF is in general concentration-dependent. ANFs are important active ingredients in several drinks and food. When used at low levels, phytic acid, lectins and phenolic compounds as well as enzyme inhibitors and saponins have reduced blood triglycerides, plasma cholesterols, and blood glucose. Additionally, saponins have role in preventing osteoporosis, maintaining liver function, and preventing platelet agglutination (Kao et al., 2008). ANFs like phytic acid, saponins, phenolic compounds, protease inhibitors, phytoestrogens, and lignans of plant origin have roles in reducing the risk of cancers. Tannins are another such compounds found to have a role in probable antibacterial, antiparasitic, and antiviral effects (Lu et al., 2004; Akiyama et al., 2001; Kolodziej and Kiderlen, 2005). Certain ANFs like lignans and phytoestrogens have been associated with the onset of sterility in humans. Hence, a cautious approach in ascertaining all facets of ANFs, their beneficial effects, concentrations for toxic effects, and quantification methods needs to be examined stringently (Popova and Mihaylova, 2019).

The aforementioned implies that antinutrients could be valuable tools for managing various diseases. They might not always be harmful even though they lack nutritional value. What is most important is focusing on dosage intake in order to find the balance between beneficial and hazardous effects of plant bioactives and antinutrients, in addition to the chemical structure, time of exposure, and interaction with other dietary components. Many factors influence their activity. They can both be considered as ANFs with negative effects or nonnutritive compounds with positive health effects. Consumers' awareness is crucial, especially when abnormal health conditions are established.

There are several approaches to negate harmful effects of ANFs. Modern biotechnology's techniques could reduce the level of certain allergens and anti-nutrients in food. Genome editing biotechnology can create mutations and substitutions in plant and other eukaryotic cells on the basis of nuclease-based forms of engineering such as the TALENS (Transcription Activator-Like Effector Nucleases) or the CRISPR (Clustered Regularly Interspaced Short Palindromic Repeats)/CRISPR-Associated Systems (CAS) (Gaj et al., 2013; Jankele and Svoboda, 2014). Providing an enhanced level of prebiotic in the body can positively influence the effects of anti-nutrients. A classic approach to remove anti-nutrients is to treat the product thermally and use methods such as extrusion, autoclaving, hydro techniques, and enzymatic and harvest treatments (Gibson et al., 2006).

The nutritional value of foods strongly depends on their nutrients as well as anti-nutrients composition. This review was designed to focus on the occurrence of anti-nutrients in wheat, clear their effect on the human body, and commemorate possible paths to disable them. A major drawback by the consumption of ANFs, especially phytic acid, oxalates etc., present in cereals like wheat and rice is inhibiting the absorption and assimilation of divalent cations like iron, zinc, magnesium, and calcium. With iron and zinc, malnutrition is of grave concern due to numerous health hazards associated with it, the main being the effects on children and pregnant women.

6.1.3 Anti-Nutritional Factors in Wheat

Several ANFs of dietary elements are found in wheat, which might significantly affect the digestibility of wheat items and, hence, influence human health (Table 6.1). Hence, the fundamental focal point of this chapter is to discuss about different anti-nutrient factors present in wheat and furthermore evaluate traditional or advanced handling strategies that can be utilized to diminish the concentration of anti-nutritional factors such as phytate, saponins, polyphenols and protease inhibitors.

6.1.3.1 Phytic Acid (PA)

Phytic acid (PA) or phytate is a plant-based secondary metabolite present in seeds of cereals, legumes, peanuts, oilseeds, etc. It is also known as *myo* inositol-1,2,3,4,5,6-hexakis dihydrogen phosphate and generally occurs in the range of 0.1 to 6.0% in plants (Lolas, 1976; García-Estepa et al., 1999). PA is unstable in its acid form and is typically found as its calcium or calcium magnesium salt (phytate) inside the kernel. As phytate, this compound is thought to represent up to 80% of the phosphorus present inside the cereal tissue. Its appropriation varies relying upon the types of plant considered (Gupta et al., 2015). Subsequently, while it is mainly concentrated in the germ of maize, in wheat it occurs mainly in the aleurone layers, and in pearl millet it is distributed throughout the whole kernel. PA acts as an anti-nutrient and shows resistance against the endogenous compounds of the mammalian gastrointestinal-digestive system. The nutritional significance of phytate relies particularly upon both the population and the item they eat. Phytate-sequestration of iron is regarded to be an issue because of complementary food varieties planned for the alleviation of malnourished people (Hurrell et al., 2003). For the production of white bread from wheat, bran and germ layer are removed from wheat kernel and it effectively removes the

TABLE 6.1

Content of Few Anti-Nutritional Factors in Wheat

S. No.	Anti-Nutrient	Content	Reference
1	Tannin	1.43–1.83 mg/g	Singh et al. (2012)
2	Phytic acid	7.95–8.00 mg/g	Singh et al. (2012)
3	Total polyphenols	379 (mg GAE/100 g)	Gunashree et al. (2014)
4	Flavonoids	36.3 (mg CAE/100 g)	Gunashree et al. (2014)
5	Trypsin inhibitors	226.3 mg/g	Gunashree et al. (2014)

Source: Popova and Mihaylova (2019); Samtiya et al. (2020)

phytate present in wheat kernel. In a few cases, phytates contain around 50 to 80% of the total phosphorous in seeds (Lott et al., 2000; Raboy, 2000). The plant-based food sources have more concentration of PA than animal-based foods; and vegetarian diet culture in emerging nations contribute to high ingestion levels (Kwun and Kwon, 2000; Amirabdollahian and Ash, 2010). Studies show that in stomach and small intestine, PA hinders the activity of protein-digesting enzymes (Kies et al., 2006). PA mainly affects the absorption of minerals and significantly affects children, lactating as well as pregnant women when enormous amount of cereal-based foods are consumed (Al Hasan et al., 2016). During the germination of seeds, a few native enzymes are activated, which degrades PA (Kaukovirta-Norja et al., 2004).

In monocotyledons crops like wheat and rice, phytates are present in the bran and aleurone layer and can be removed during processing. In dicotyledons like legumes, oilseeds, and nuts, phytates are found in close relationship with proteins, which decreases the simplicity of division by a basic handling strategy like processing (Sinha and Khare, 2017). PA is a negatively charged structure, which for the most part binds with positively charged divalent ions, for example, Zn, Fe, Mg, and Ca to make complexes and diminishes the absorption of these minerals causing low absorption rates. Mainly because of this chelating property, PA is considered as an effective anti-nutrient in food varieties and a reason for the deficiency of mineral particles in animals and human nourishment (Grases et al., 2017). Wheat flour contains relatively high levels of PA (600–1,000 mg/kg), while in refined flour, the concentration reduces to 200–400 mg/kg. The concentration of PA is lower in commercially milled flours than in flours produced in a domestic environment (Febles et al., 2002). The role of PA is beneficial as well as shows anti-nutritional effects. Phytate works in a broad pH range as a highly negatively charged ion, and therefore its presence in the diet has a negative impact on the bioavailability of divalent and trivalent mineral ions (such as Zn^{2+}, $Fe^{2+/3+}$, Ca^{2+}, Mg^{2+}, Mn^{2+}, and Cu^{2+}) (Lönnerdal, 2002; Fredlund et al., 2006).

6.1.3.1.1 Methods for Estimation of Phytic Acid (PA)

Several methodologies have been developed to quantify PA content in plants. Precipitation procedures and potentiometric titration are the old-style strategies for PA assessment. With the appearance of innovation, spectroscopy-based (UV-vis, NMR, ICP-OES, ICP-MS, and so on) and sensor-based (electro biosensor and nanobiosensor) instruments have been utilized to estimate the different types of phytates. Ionic chromatography-based strategy was also developed for the recognition of different groups of phosphates including PA and higher inositol phosphates (InsP5-InsP3) in legumes by utilizing an extra purification step bringing about the expulsion of lower inositol phosphates (InsP1 and InsP2) (Burbano et al., 1995). The initial methods depended on relationship between phytic acid and inorganic phosphorus, as well as that between PA and iron. These techniques were very relentless and needed precision. Later on, profoundly exact techniques like HPLC, SAX-HPLC, capillary electrophoresis, and NMR spectroscopy were developed (Pandey et al., 2016; Perera et al., 2018). High-performance ion-exchange chromatography (HPIC) has been the best strategy concerning efficiency as well as precision, as it permits the separation of most isomers of inositol phosphates, excluding the stereoisomers. Moreover, a colorimetric strategy (called Wade's reagent) based on the reaction between iron, sulfosalicylic acid, and PA was developed to estimate the PA in legumes and other food. This test depends on the principle that there is an inverse relationship between the phosphorus content and interaction between iron and sulfosalicylic acid bringing about a reduction of color intensity with an increase in phosphorus content.

Wade's reagent was applied likewise for the estimation of phytates in different kinds of food samples, like corn, soybeans, wheat, sunflower, oats, and rye (Agostinho et al., 2016). Recently, one more colorimetric test based upon the inhibition of color formation by the interaction between glyoxal-bis (2-hydroxianiline) (GBHA) and calcium ions in the presence of PA has been developed. A comparison of different techniques for PA quantification determined that the modified Wade's measure was simpler, precise, and economical (Gao et al., 2007). In this way, Wade's measure could be utilized to screen the germplasm and mutant sources of various crops. Methods based upon electrophoresis have also been developed – e.g., capillary isotachophoresis (cITP) was utilized to decide the phytate content in various cereal crops including rice, wheat, barley (Krishnan et al., 2015), and legumes. In these strategies, the sodium salt forms of the plant extracts were utilized for electrophoretic partitioning of various groups of phosphates including inositol phosphates and orthophosphate. Capillary zone electrophoresis (CZE) was utilized for phytic acid assessment in soybeans (Perera et al., 2018). Novel techniques with high awareness have been developed that can recognize low phytic acid content. These strategies depend on the concurrent utilization of square wave voltammetry and quick Fourier change (FFT-SWV). Additionally, nano-sensor-based strategies for the recognition of low degrees of PA were also developed.

6.1.3.1.2 Measures for Reducing Phytic Acid

Removal of phytic acid increases bioavailability of many cations and thus the nutritional value of meal. There are several methods which have been developed for the removal of phytic acid from grains. The physical methods of PA reduction are discussed later in the chapter. Here, we discuss the genetic and molecular basis of PA reduction.

6.1.3.1.2.1 Application of Phytase Phytase is an enzyme which catalyzes the hydrolysis of phytate to orthophosphate and lower substituted inositol phosphates and further release of chelated cations. It is one of the most effective methods for reducing PA content in seeds without altering the seed mineral content. Phytase enzyme serves many roles like the removal of phytate in food industries and feed, against environmental phosphorus pollution, promotion of plant growth, and the preparation of special myo-inositol phosphates as tools for biochemical investigation (Greiner and Carlsson 2006; Singh, 2011).

Segueilha et al. (1993) used phytase from yeast (*S. castelii*) to remove PA in wheat bran. Commercially, phytase is added to fish, swine, and poultry feed to enhance the bioavailability of minerals, phosphate, amino acids, and energy. Enzymatic action of phytase on PA releases the bound nutrients, which are otherwise not available for absorption in the digestive tract. Mineral bioavailability is affected by splitting of PA during the process of bread making (Mollgaard, 1946). As a result, bread-making processes are designed with the aim to reduce PA content in the final product. Phytase was isolated commercially from wheat and added to whole-meal wheat flour during bread making, and naturally occurring phytase was activated by soaking and grain malting (Knorr et al., 1981).

6.1.3.1.2.2 Screening of Micronutrient Dense Germplasm   Staple foods of plant origin can be made vitamin- or mineral-rich by using techniques like conventional plant breeding methods or through the use of transgenic techniques. For these techniques, a good amount of genetic variation is required in the germplasm, which can be achieved by utilizing TILLING (Till et al., 2007). The technique is efficient in creating a genetically variant population. The transgenic techniques can complement the current breeding efforts and expedite the immediately required biofortified crops to fulfill the nutrition demands of the developing world population.

Wheat generally contains a low amount of Fe and Zn (21–32 mg kg^{-1} and 15–22 mg kg^{-1}), respectively (Rawat et al., 2009), which is further reduced during the various processing techniques, and hence the bioavailable amount is very low in the finished product with further presence of ANFs. Genetic biofortification techniques are cost-effective, affordable, and generally easily applicable for the intended population. However, the initial requirement of these techniques is the presence of a wide range of genetic variation in the germplasm for the concerned trait like Zn or Fe concentration. Wild and primitive wheat cultivars serve as a much better genetic source of Zn content in comparison to cultivated varieties. A good variation ranging from around 14 to 190 mg of zinc kg^{-1} was found in a collection of wild emmer wheat, *Triticum turgidum* ssp. Dicoccoides. Wild emmer wheat accessions were identified having a very high Zn content up to 139 mg kg^{-1}, Fe up to 88 mg kg^{-1}, and protein up to 380 g kg^{-1} in grains. They were also reported to have high tolerance to Zn deficiency in soil and drought stress (Cantrell and Joppa, 1991; Cakmak et al., 2010).

6.1.3.1.2.3 Molecular Technologies for PA Reduction

6.1.3.1.2.3.1 Production of Mutants with Low Phytic Acid (LPA) Content   The LPA varieties can be developed by using advance molecular approaches or silencing of genes involved in the biosynthesis of PA. The method of activity of LPA crops is apparently adjusting the partitioning of phosphorous into inorganic phosphorus (Pi), PA, and inositol phosphates having five or lesser P esters (Raboy, 2002). The LPA mutants overall are not acknowledged straightforwardly to be successful because of the decrease in yield and other undesirable changes. In any case, the hindering changes can be limited through backcross breeding programs. This necessities better understanding of hereditary changes in mutants and molecular analysis of LPA traits. Recently, mutants population by compound mutagenesis in the background of high-yielding assortments was also developed – for example, PBW502 and many of the mutants showed low PA level. This is a good source for molecular studies and breeding for decreasing PA content. However, different mutants have been isolated in various cereals; and the hereditary idea of transformations has not been resolved up until this point. In maize, Lpa1 trait was created by altering one of the seven maize myo-inositol phosphate synthase genes that encode MIPS, the enzyme implicated in the underlying step of inositol phosphate biosynthesis. However, sequencing of MIPS gene, that is found adjacent to Lpa1 mutation, showed no lesions inside the coding sequence (Shukla et al., 2004). Nonlethal recessive mutations that reduce seed PA concentration have been isolated in barley (Larson et al., 1998), maize (Raboy et al., 2000), soybean (Hitz et al., 2002), rice (Larson et al., 2000), and wheat (Guttieri et al., 2004). In rice, LPA mutant was shown to be because of a single recessive mutation on chromosome 2L (Larson et al., 2000). Accordingly, additional markers were developed including the region of 47 kb containing eight putative open-reading frames. In wheat, LPA mutants have not been characterized at molecular level up to this point. Henceforth, the development and characterization of LPA mutants in wheat employing Indian wheat cultivars will be a new development. The identification of desirable mutants with LPA and their molecular characterization needs advance study (Ram et al., 2020).

6.1.3.1.2.3.2 Targeting Phytic Acid Pathway Genes   The most effective strategy for reduction of PA involves targeting of genes involved in the last steps of PA biosynthesis or transport to its organ of storage, that is, vacuoles (Shi et al., 2003; Aggarwal et al., 2018). For this purpose, two major genes IPK1 and ABC C type transporter were identified for targeting. MRP5 gene (ABC transporter) was effectively silenced in soybean with a substantial reduction of PA in the seeds. The effective reduction of PA was found associated with a concomitant increase of Zn and Fe content in seeds. Same strategy was employed and found to be effective in reducing PA content in seeds of maize, beans, and rice (Shi et al., 2007; Panzeri et al., 2011). The same IPK1 gene was effectively silenced in wheat. RNAi-mediated targeting of TaIPK1 gene in wheat resulted in the reduction of PA to the tune of approximately 56% with a concomitant increase in Zn:PA and Fe:PA molar ratios. Therefore, for reduction of PA in wheat, targeting of IPK1 gene for genome editing is a sensible choice (Aggarwal et al., 2018). List of genes targeted in wheat for PA reduction is mentioned in Table 6.2. Owing to allohexaploid nature of wheat genome, transcripts with conserved regions from A, B, and D subgenomes should be considered.

6.1.3.2 Oxalates

Oxalic acid forms salts or esters with different minerals such as Ca, K, Fe, Na, and Mg. The K and Na form soluble complexes, while the other minerals form insoluble complexes with oxalic acid. The insoluble salts like calcium oxalate have been mainly responsible for kidney stones because they cannot be excreted out if they reach the urinary tract (Popova and Mihaylova, 2019). By forming these calcium oxalate complexes, oxalate also reduces the uptake of calcium ions. In general, these salts of oxalate possess poor solubility at the intestinal pH (Campos-Vega, Loarca-Piña and Oomah, 2010). Oxalate is not a problem for many people, but, if combined with conditions like primary and enteric hyperoxaluria, the intake of oxalate needs to be reduced. Fermentation or cooking helps in decreasing the concentration of oxalate complexes (Popova and Mihaylova, 2019). Oxalate concentration of the whole grain is found higher than refined flour in case of cereals, especially wheat indicating its

TABLE 6.2

List of Genes Targeted for Lowering Phytic Acid (PA) Content in Different Crop Plants

Plant species	Targeted gene	Candidate gene	Method used	References
Triticum aestivum	*TaABCC13*	MRP	RNAi	Bhati et al. (2016)
Triticum aestivum	*TaIPK1*	*Inositol pentakisphosphate kinase-1 (IPK-1)*	RNAi	Aggarwal et al. (2018)

primary occurrence in outer layers of grain. Wheat bran has the highest soluble and total oxalate content. In hexaploid wheat *T. aestivum*, the highest total oxalate content was found in whole meal flour or whole-grain flakes in comparison to the whole grain (Siener et al., 2006). Studies showed that a total oxalate content of whole-wheat flour ranged from 67 to 70 mg/100 g dry matter (DM) (Siener et al., 2001). In contrast, the total oxalate content of white flour ranged from 16.8 to 45.0 mg/100 g DM while brown rice flour contained 37 mg/100 g DM (Boontaganon et al., 2009). Although, the oxalate content of plants is mainly a species characteristic, considerable variations can occur within the same species depending on the age of the plant, maturity, season, and soil conditions during growth. Moreover, Siener et al. (2006) compared the soluble and total oxalate contents among cereals and cereal products such as wheat, rye, oat, barley, maize, and rice. The result showed that wheat bran had the highest concentrations (131.2 and 457.4 mg/100 g DM) for soluble and total oxalate contents, respectively, whereas whole-grain rice flakes contained the lowest amount compared to the other whole-grain cereals (soluble oxalate 4.2 mg/100 g DM and 12.2 mg/100 g DM for total oxalates). Therefore, Siener et al. (2006) studied that a high concentration of oxalate was present in the outer layer of cereal grains. Judprasong et al. (2006) observed that a very low concentration of oxalate was present in whole-grain rice both before and after boiling (<3 mg/100 g DM).

At present, it is not clear to what extent insoluble calcium oxalate dissociates in the gut prior to absorption and whether calcium oxalate could be absorbed as intact salt in the human intestine (Hanes et al., 1999). Moreover, the percentage of oxalate may vary depending on the intake of oxalate and the composition of the diet. A low intake of calcium increases intestinal absorption and urinary excretion of oxalate (Borghi et al., 2002; von Unruh et al., 2004). To limit intestinal absorption of oxalate, a normal calcium diet (1,000 mg/day) should be recommended for calcium oxalate stone patients with hyperoxaluria. An adequate supply of calcium can be obtained with lean dairy products. Dietary calcium should be ingested with oxalate-containing meals to maximize the oxalate-binding effect of calcium in the intestine. Cereals and cereal products contribute to the daily oxalate intake to a considerable extent. Vegetarian diets may contain high amounts of oxalate when whole-grain wheat and wheat products are ingested. Recommendations for the prevention of recurrence of calcium oxalate stone disease have to take into account the oxalate content of these foodstuffs. Further research should determine the oxalate bioavailability of cereals and cereal products.

6.1.3.3 Enzyme Inhibitors

Proteinase enzymes are crucial for the nourishing and useful properties of many protein molecules (Salas et al., 2018). Protease inhibitors carry out a variety of proteolytic functions, including signal initiation, transmission, and cellular death, inflammatory response, blood coagulation, and some role in hormone processing (Gomes et al., 2011). Plant serpins, one of the largest families of protease inhibitors, are mostly found in cereal seeds. These compounds, often known as 'self destruction inhibitors', have a subatomic mass of 39 to 43 kDa and are present in a wide range of plant species (Haq et al., 2004). Serpins are the effective inhibitors that primarily prevent trypsin and chymotrypsin activities by pursuing their overlapping of enzyme reactive sites (Dahl et al., 1996). Because of their potent method of limiting enzyme activity by creating protein–protein interactions, protease inhibitors are naturally occurring plant inhibitors that have developed into an important research area. By inhibiting the active site of chemical, they reduce enzyme activity in a synergistic manner. Protease inhibitors' N- or C-end and exposed loop are frequently regarded as the key features for the inhibition of enzyme activity (Otlewski et al., 2005). Ragi has a bifunctional inhibitor that, through a trimetric complex interaction with trypsin and α-amylase, respectively, inactivates substances that break down proteins and starches (Shivaraj and Pattabiraman, 1981). High levels of protease inhibitors, α-amylase inhibitors, and lectins found in legumes may result in decreased mineral bioavailability as well as reduced nutritional absorption and digestibility (Bajpai et al., 2005; Yasmin et al., 2008).

Cereals contain substantially low levels of such digestive inhibitors than legumes, mainly those that work against the amylases as well as proteases (Nikmaram et al., 2017). The 'Kunitz trypsin inhibitor' and the 'Bowman–Birk inhibitor' typically originate in soybeans nonetheless are not immediately rendered inactive by the heat treatment due to the 'disulfide bonds' (Liu, 1997; Van Der Ven et al., 2005). There are many different enzyme inhibitors found in agro food sources, but the two main types are those that affect the trypsin and α-amylase activity and are present in nearly all legume-based foods and cereals. The main function of α-amylase is to control the conversion of polysaccharides into oligosaccharides, a type of carbohydrate. Therefore, by delaying the digestion of carbohydrates, enzyme inhibitors that selectively decrease α-amylase activity will lengthen the time needed for the carbohydrates' absorption. The normal postprandial plasma glucose level is impacted by the decreased glucose retention rate caused by the prolonged time needed to digest carbohydrates (Bhutkar and Bhise, 2012). Several earlier researches have demonstrated that trypsin inhibitor-rich seeds may boost the satietogenic hormone cholecystokinin (CCK), which would reduce food intake and body weight (bw) (Chen et al., 2012; Ribeiro et al., 2015; Serquiz et al., 2016). The trypsin inhibitor can reduce protein digestion and amino acid bioaccessibility in humans, as well as cause pancreatic hyperplasia, which can slow down growth (Adeyemo and Onilude, 2013). Previous research has shown that enzyme inhibition, such as that caused by alpha-amylase, alpha-glucosidase, and lipase, may also have positive effects on type II diabetes prevention and overall health (Li and Tsao, 2019).

6.1.3.3.1 Trypsin Inhibitors

Amylase/trypsin inhibitors (ATIs) are widely consumed in cereal-based foods and have been implicated in adverse reactions to wheat exposure, such as respiratory and food allergy, and intestinal responses associated with celiac disease and non-celiac wheat sensitivity. ATIs occur in multiple isoforms which differ in the amounts present in different types of wheat (including ancient and modern ones). The ATIs were first isolated from wheat in 1973 (Geisslitz et al., 2020). Wheat amylase/trypsin-inhibitors (ATIs) constitute approximately 2–4% of the total seed protein and may have role in plants' natural defense against pathogens and pests. ATIs in wheat are currently under extensive investigations owing to their suspected role in causing adverse reactions in certain people upon the consumption of wheat and its products. Although earlier wheat ATIs were known to have role in IgE-mediated allergy (specifically Bakers' asthma), some research have also involved them in triggering innate immune system, and also their possible role in causing celiac disease in certain individuals. Celiac disease is known to affect around 1% of the total population, and, very recently, ATIs are also known to cause non-celiac wheat sensitivity impacting almost 10% of the total population. These effects of ATIs present in wheat have led to extensive studies of their composition, structure, inhibitory roles, genetics, regulation of expression, processing reactions, role in adverse reactions of wheat, and, in recent times, strategy to alter or regulate their expressions in plants using gene-editing techniques.

Cereal grains are attractive to pests and pathogens because they have high contents of storage reserves (starch and protein). They have therefore evolved to contain a range of proteins which inhibit the hydrolytic enzymes of these organisms, including ATIs which are able to inhibit α-amylases from lepidopteran and coleopteran insects (Geisslitz et al., 2020). ATIs occur in multiple isoforms which differ in the amounts present in different types of wheat (including ancient and modern ones). Measuring ATIs and their isoforms is an analytical challenge as is their isolation for use in studies addressing their potential effects on the human body. ATI isoforms differ in their spectrum of bioactive effects in the human gastrointestinal (GI) tract, which may include enzyme inhibition, inflammation, and immune responses of which much is not known (Guillamon et al., 2008). It has been reported that members of the non-gluten-amylase/trypsin inhibitor (ATI) family present in wheat and related cereals are strong inducers of innate immune responses in human and murine macrophages, monocytes, and DCs. ATI family members activate the TLR4–MD2–CD14 complex and elicit strong innate immune effects not only *in vitro* but also *in vivo* after oral or systemic challenge. Nowadays, wheat has become the world's major staple crop, and its products are widely used ingredients in processed foods. Wheat consumption correlates with certain disorders like wheat allergies and especially celiac disease (Reig-Otero et al., 2018). Recent studies question the proinflammatory triggering activity of α-gliadin fraction contained in wheat, since it has been demonstrated that the amylase–trypsin inhibitors (ATIs) exert a strong activating effect on the innate immune response. The role of ATIs in the activation of innate immunity and in the development of the symptoms characteristic of non-celiac gluten sensitivity (NCGS) – 'wheat', 'gluten', and 'celiac'– is under ongoing investigation. Many studies are available on the structure, inhibition mechanism, and immune system effects of ATIs, mainly focused on IgE-mediated reactions. Recently, the increase of interest in NCGS has in turn increased the literature on the capacity of ATIs contained in wheat to activate the innate immune system. Literature published to date questions the relationship between the activation of the innate immune system and gluten in NCGS. ATIs may have acted as the interfering contaminant of gluten and appear as the potential activator of innate immunity in NCGS patients (Call et al., 2021).

Amylase–trypsin inhibitors (ATIs) are water-soluble cereal proteins, which also demonstrate good solubility in organic solvents such as chloroform–methanol (CM) mixtures (Zevallos et al., 2017). ATIs were identified as causative proteins for non-celiac wheat sensitivity (NCWS) and other wheat-related disorders (Junker et al., 2012). The diseases are characterized by both intestinal and extraintestinal symptoms that are tightly linked to the consumption of wheat and other gluten-containing foods. As there are no clear diagnostic biomarkers for NCWS, a precise number for prevalence is difficult to obtain, but based on the few data available, prevalence is estimated to range from 0.6 to 10% (Czaja-Bulsa, G., 2015). ATIs were found to initiate innate immune responses by directly activating specific pro-inflammatory receptors (i.e., TLR4) in human body cells, leading to severe inflammations and immune reactions.

Additionally, ATIs have the potential to inhibit the activity of two important digestive enzymes in the gastrointestinal system, amylase and trypsin, causing a significant impairment of digestion (Barmeyer et al., 2017). Due to this inhibitory feature, ATIs and other metabolic proteins are considered to be crucial for the natural defense mechanism of the plant itself. They are located in the endosperm of cereal seeds, where they defend starch and protein reserves by blocking amylase and trypsin activities of pathogenic fungi or invading pests. ATIs can be classified, according to their degree of aggregation, into monomeric, dimeric, and tetrameric forms. These aggregates are stabilized by non-covalent intramolecular interactions, including disulfide bonds and hydrophobic interactions, resulting in a compact 3D structure responsible for their high resistance against thermal processing and proteolysis (Geisslitz et al., 2020). Amylase–trypsin inhibitors (ATI) can be found in all gluten-containing cereals and are, therefore, the ingredient of basic foods like bread or pasta. In the gut, ATI can mediate innate immunity via the activation of the toll-like receptor 4 (TLR4) on immune cells residing in the lamina propria, promoting intestinal, as well as extraintestinal inflammation (Ziegler et ai., 2019).

6.1.3.4 Phenolic Compounds

Although phenolic and polyphenolic compounds constitute an important class of secondary metabolites that act as inhibitors of LDL, free radical scavengers have a role in DNA breakage and also cholesterol oxidation. They also form complex with essential minerals thus reducing their bioavailability upon consumption by humans. Unfortunately, iron and zinc in cereal-based foods are poorly bioavailable due to several factors that reduce the intestinal absorption which results in high rates of Fe and Zn deficiency (especially in infant, children, and pregnant women) (Sandstead, 2000; Shahidi, 2004).

In wheat, the TPC would vary according to the growth location. Processing techniques such as soaking and germination may reduce the extractable phenolics (Parker and Waldron, 2005). Mallick et al. (2013) studied few Indian wheat cultivars for TPC and found the total phenol in the range of 0.32 mg g^{-1} to 5.14 mg g^{-1}. The total phenolic compounds were found in these wheat varieties as RSP-566 (3.93 mg g^{-1}), RSP-561 (3.88 mg g^{-1}), PBW-396 (4.94 mg g^{-1}), HD-2687 (5.14 mg g^{-1}), C-306 (4.68 mg g^{-1}), PBW-175 (4.9 mg g^{-1}), RSP-81 (3.89 mg g^{-1}), PBW-550 (5.07 mg g^{-1}), DBW-17 (4.24 mg g^{-1}), WH-542 (4.77 mg g^{-1}), and CD (0.05) (0.32 mg g^{-1}).

The total phenol content in wheat varies according to its location of cultivation. Traditional processing methods like soaking and germination diminish the concentration of total extractable phenols in wheat (Parker and Waldron, 2005). Nadeem et al., (2010) observed that wheat flours produced after germination and fermentation had higher mineral extractability and lower ANFs like phytic acid and phenols.

6.1.3.5 Tannins

Tannins are secondary metabolites of plants, which are found in fruits, leaves, and bark having a molecular weight of more than 500 Da and can precipitate proteins (Timotheo and Lauer, 2018). In cereals, tannins are concentrated chiefly in the seed coat. The percentages of tannins in wheat and barley are 0.4% and 0.7%, respectively (Juliano, 1985). These compounds generally affect the protein digestibility and lead to the reduction of essential amino acids by developing reversible and irreversible tannin–protein complexes between the OH–group (hydroxyl group) of tannins and the C=O–group (carbonyl group) of proteins (Raes et al., 2014). Tannins generally form complexes with proteins that are comparatively large and hydrophobic in nature, in addition flexible, and of open structure with a high concentration of proline (Frutos et al., 2004). Naturally, there are two types of tannin groups: hydrolyzable (e.g., ellagitannin and gallotannin) and condensed (e.g., proanthocyanidin). In ruminants, the hydrolyzable type of tannins are readily broken down during the digestion, and breakdown products establish a large quantity of compounds, which can be lethal (Kumar, 1992).

6.1.3.6 Haemagglutinins and Lectins

Hemagglutinins and lectins are a form of sugar-binding proteins, which easily attach to the RBCs and cause agglutination. These compounds are mainly found in foods, which are consumed in the raw form (Hamid et al., 2013). Legumes and cereals mostly contain lectins. In addition, the transport and hydrolytic roles of the enterocyte would be decreased by the consumption of foods that contain lectins (Krupa, 2008). Phytohemagglutinin (with a molecular mass of 120 kDa) is a tetrameric glycoprotein which is found in the kidney beans and also consists of two diverse subunits (Lajolo and Genovese, 2002).

   In rats, kidney bean phytohemagglutinin appears to upregulate the function and metabolism of the whole gastrointestinal tract (GIT), which includes the growth of the small intestine and the increased length of the tissue and a number of intestinal crypt cells (Bardocz et al., 1995). Phytohemagglutinins enhance the growth of rat pancreas by increasing CCK hormone release. However, an independent mechanism is responsible for the intestinal growth (Herzig et al., 1997). The 'wheat germ agglutinin' (WGA) is a specific type of lectin that has been labeled extensively for their side effects on health. Wheat germ agglutinin binds to the gut epithelium and damages the cell leading to leaky gut epithelium, which is incompetent in nutrient-uptake (Biesiekierski et al., 2011). Moreover, lectin tempts the barrier of glucose and insulin receptors, which contributes to the celiac disease (CD) and the budding of pathogenic bacteria (Cordain et al., 2000). Wheat germ constitutes merely 3% of the total whole-grain kernel weight; thus, 50 g wheat germ would display a sincere utilization of 1,666 g wheat. It is further speculated that food processing considerably affects lectin content as well as its biological activities. For instance, lectin level is significantly determined by used flour in breads and pastas, where most only contain refined flour with traces of germ, even those enriched in bran fractions. Whole-wheat flours (WWF), which are thought to be high in wheat germ agglutinin, are often reconstituted from different milling streams which impact on wheat germ agglutinin concentrations (Matucci et al., 2004).

6.1.3.7 Alkaloids

Alkaloids are the secondary metabolites, which were originally described as 'pharmacologically' natural bioactive compounds (NBCs), and are mainly nitrogenous and heterocyclic in nature (Bribi, 2018). These NBCs are most commonly synthesized from few common amino acids, that is, tryptophan, tyrosine, and lysine. In plants, more than 12,000 different alkaloids were recognized. Plant-based alkaloids are usually found as salts of organic acids, such as malic, citric, tannic, acetic, lactic, oxalic, tartaric, and numerous other organic acids. Although a limited weak basic alkaloid, that is, nicotine, arises freely in system (Ramawat et al., 2009).

6.1.3.7.1 Anti-Nutritional Effects of Alkaloids

The *Claviceps* genus (fungal pathogens) is known to synthesize a set of mycotoxins (ergot alkaloids), whose effects have been recognized since the ancient times (Richard et al., 2003). *Claviceps purpurea* from the species *Claviceps* infects more than 400 species of plants from monocotyledons, grasses, and numerous economically important cereals (i.e., millet, triticale, wheat, barley, oat, and rye) (Haarmann et al., 2009). The 'ascospores' from this pathogen are carried by the wind; they fall, attach, and germinate on the pistil surface of the crop at anthesis, starting host–pathogen contact that results in sclerotia generation. Fruiting bodies of this pathogen are 'sclerotia', which attain maturation succeeding to five weeks of fungal infection and are the reservoir of alkaloids. The production of the sclerotia causes double damage to the crops: first they have a damaging effect hindering the value of grains due to the incidence of diverse classes of alkaloids, considered disadvantageous compounds for human and animal health because of their pharmacological effects, and second they affect the crop yield. The commonly produced Eas are ergotamine, ergosine, ergometrine, ergocornine, ergocristine, ergocryptine, and their analogous – inine (S)-epimers (White et al., 2003; Beuerle et al., 2012).

Around 60 samples of wheat and barley from Algeria, when investigated for the natural occurrence of six major ergot alkaloids, ergosine, ergometrine, ergotamine, ergocryptine, ergocornine, and ergocristine, along with their epimers, revealed infection to the range of 20% in the samples. Wheat samples were highly contaminated (26.7% incidence of ergot alkaloids) in comparison to barley. Total ergot alkaloid concentration ranged from 3.66 to 76.0 µg/kg for wheat and from 17.8 to 53.9 µg/kg for barley samples. Ergosine, ergocryptine, and ergocristine were more commonly found in wheat grains (Carbonell-Rozas et al., 2021).

6.1.3.8 Saponins

Saponins are widely distributed plant natural products with a vast structural and functional diversity. They are typically composed of a hydrophobic aglycone, which is extensively decorated with functional groups prior to the addition of hydrophilic sugar moieties to result in surface-active amphipathic compounds. The saponins are broadly classified as triterpenoids, steroids, or steroidal glycoalkaloids, based on the aglycone structure from which they are derived. The saponins and their biosynthetic intermediates display a variety of biological activities of interest to the pharmaceutical, cosmetic, and food sectors. Although their relevance in industrial applications has long been recognized, their role in plants is under explored. Recent research on modulating native pathway flux in saponin biosynthesis has demonstrated the roles of saponins and their biosynthetic intermediates in plant growth and development. Saponins are naturally occurring structurally and functionally diverse phytochemicals that are widely distributed in plants. They are a complex and chemically varied group of compounds consisting of triterpenoid or steroidal aglycones linked to oligosaccharide moieties. The combination of a hydrophobic aglycone backbone and hydrophilic sugar molecules makes saponins highly amphipathic and confers foaming and emulsifying properties. These surface-active molecules have important roles in plant ecology and are also exploited for a wide range of commercial applications in the food. Saponins are also the major constituents of traditional folk medicines such as the ginsenosides produced by Panax species (Qi et al., 2011). A number of therapeutic properties have been ascribed to saponins. These molecules are potent membrane-permeabilizing agents. They are also immune stimulatory, hypocholesterolemic, anti-carcinogenic, anti-inflammatory, anti-microbial, anti-protozoan, and molluscicidal and have anti-oxidant properties. Furthermore, they can impair gut protein digestion and uptake of vitamins and minerals. In general, the saponins and their biosynthetic intermediates display an array of properties that can have positive or negative effects in different animal hosts. They are generally considered to have important roles in defense of plants against pathogens, pests, and herbivores due to their antimicrobial, antifungal, antiparasitic, insecticidal, and anti-feedant properties (Moses et al., 2014).

The biological activity of saponins is normally attributed to the amphipathic properties of these molecules, which consist of a hydrophobic triterpene or sterol backbone and a hydrophilic carbohydrate chain, although some saponins are known to have potent biological activities that are dependent on other aspects of their structure. Saponins are glycosides of triterpenes and steroids. Steroidal glycoalkaloids are sometimes also referred to as saponins. The triterpene and steroid backbones are both derived from the mevalonic acid pathway, the common precursor being 2,3-oxidosqualene. The name 'saponin' derives from the soap-like cinalis, which was historically used as a source of detergent. Saponins represent a sizable proportion of the number of known plant natural products. While plant natural products used to be regarded as waste products of a "luxurious metabolism", they are now accepted as the products of natural selection with diverse biological activities and important ecological roles. The structural diversity of saponins is reflected in the array of different biological activities associated with these compounds, and these diverse compounds provide a significant resource for drug and agro-chemical discovery. Indeed, many plant-derived saponins are currently used as important pharmaceuticals in the treatment of a range of diseases in conventional and traditional medicine. Research into the functions and synthesis of saponins has provided a wealth of information on the properties of this important group of compounds, both for human use and in plants. Saponins are exploited as important pharmaceuticals and for a variety of other industrial uses. The triterpenoid-ginsenoside-saponins, for example, ginsenoside Rb1 are the major bioactive components of ginseng, the roots of which are widely used in traditional Chinese medicine. Ginsenosides have multiple pharmacological properties, including anti-tumor, immune modulatory, and neurological activity (Mugford et al., 2013). Saponins have found wide applications in beverages and confectionery, as well as in cosmetics and pharmaceutical products. Due to the presence of a lipid-soluble aglycone and water soluble sugar chain(s) in their structure (amphiphilic nature), saponins are surface active compounds with detergent, wetting, emulsifying, and foaming properties.

Saponins were treated as toxic because they seemed to be extremely toxic to fish and cold-blooded animals, and many of them possessed strong hemolytic activity. Saponins, in high concentrations, impart a bitter taste and astringency in dietary plants. The bitter taste of saponin is the major factor that limits its use. In the past, saponins were recognized as anti-nutrient constituents due to their adverse effects such as growth impairment and reduction in food intake due to the bitterness and throat-irritating activity of saponins. In addition, saponins were found to reduce the bioavailability of nutrients and decrease enzyme activity, and it affects protein digestibility by inhibiting various digestive enzymes such as trypsin and chymotrypsin. Saponins also attract considerable interest as a result of their beneficial effects in humans. Recent evidence suggests that saponins possess hypocholesterolemic, immune stimulatory, and anti-carcinogenic properties. In addition, they reduce the risk of heart diseases in humans consuming a diet rich in food legumes containing saponins. Saponin-rich foods are important in human diets to control plasma cholesterol, preventing peptic ulcer and osteoporosis and to reduce the risk of heart disease. Saponins are used as adjuvants in viral (e.g., Quillaja saponaria-21) and bacterial vaccine (e.g., Quillaja saponins) applications. A high saponin diet can be used in the inhibition of dental caries and platelet aggregation in the treatment of hypercalciuria in humans and as an antidote against acute lead poisoning. In epidemiological studies, saponins have been shown to have an inverse relationship with the incidence of renal stones (Gemede and Fekadu, 2014). In wheat, saponins are generally concentrated in the roots and leaves of the plant; hence, low content of the same is reported from wheat bran and barley husk (López-Perea et al., 2019).

6.2 Strategies for the Reduction of Anti-Nutrients

6.2.1 Physical Methods

6.2.1.1 Autoclaving and Heat treatment

Autoclave is an instrument typically used for heat treatment. This instrument, when used for cereals and other plant-based dietary sources, initiates the phytase enzyme and enhances its activity (Ertop and Bektaş, 2018). When consumed after autoclaving, the majority of food variety exhibited health benefits. For instance, boiling cereal grains reduced the amount of antinutrients they contained, improving their nutritional value (Rehman and Shah, 2005). Additionally, heating and soaking significantly reduced the amount of PA in legume grain (Vadivel and Biesalski, 2012). Legumes used as food are typically pressure cooked or boiled before consumption. Additionally, previous studies showed that cooking or boiling food has a significant impact on its dietary benefits by decreasing their anti-nutritional (e.g., tannins and trypsin inhibitors) content (Patterson et al., 2017). In yet another study, Vadivel and Biesalski (2012) found that heating and soaking legume grains lowered their PA content. When lentil seeds were treated with soaking followed by quick cooking, Vidal-Valverde et al. (1994) found a significantly lower concentration of PA in the lentils. When whole-wheat bread was autoclaved and microwaved, as demonstrated by Mustafa and Adem (2014), the amount of PA reduced while the amount of total free (unbound) mineral content increased. This is because PA has the ability to chelate minerals; as a result, a decrease in phytates reduces the amount of minerals that are bound while increasing the amount of free minerals. Another research showed that the nutritional quality of legumes was significantly improved after cooking because of the decrease in the content of lectins and saponins (Maphosa and Jideani, 2017).

It was studied that antinutrients of black grams and mung beans were diminished by pressure-cooking when compared with ordinary cooking treatment (Kataria et al., 1989). Shah (2001) studied that pressure-cooking diminished the tannins' content, which prompted better black gram protein digestibility. Savage and Mårtensson (2010) found that oxalate content of taro leaves was decreased by 47% when boiled in water for 40 min, despite the fact that there was no critical decrease seen in oxalate content subsequent to baking for 40 min at 180 °C. Roasting method additionally diminished the trypsin inhibitor action fundamentally in soybean meal (Vagadia et al., 2017). Another review saw that few antinutrients of wholesome elements were decreased subsequent to autoclaving, soaking, and cooking of legumes (Torres et al., 2016). Previous researches also showed that autoclaving is the best technique to decrease levels of a few anti-nutritional compounds when contrasted with other processing strategies (Shimelis and Rakshit, 2007; Vadivel et al., 2008; Doss et al., 2011).

The saturated heat treatment (home preparing or industrial food processing) is broadly used to prepare various plant-based food items for human utilization. Moist heat is a successful method for decreasing protease inhibitors' activity and refining the nutritional quality of the plant-based protein (Rackis and McGhee, 1975). Loss of movement is connected with temperature, moisture conditions, span of intensity therapy, and molecule size. Up to 95% activity can be obliterated by heating at 100°C for 15 min, while autoclaving for 15–30 min has displayed to annihilate most trypsin inhibitor's action in a scope of grain legume species.

6.2.1.2 Milling

Milling is most commonly used to isolate the wheat bran layer from the grains. This method is used for grinding of grains into flour. The processing strategy eliminates antinutrients (e.g., PA, lectins, and tannins), which are present in the aluerone layer of grains; however, this procedure has a disadvantage that it likewise eliminates significant minerals (Gupta et al., 2015). Studies conducted on millet processing showed that the chemical composition of pearl millets was changed because of processing process. Then, again, not much change was seen in pearl millet flour when processed through baking. The processing and heating methods during the making of chapati diminished the PA and polyphenol content and improved starch and protein absorption (Chowdhury and Punia, 1997). In another examination, two varieties of pearl millets were utilized for assessing their nutrients, anti-nutrients, and mineral bioavailability after milling them into entire flour, grain-rich portion, and semi-refined flour. The results of nutrient composition showed that no distinction was found in semi-refined flour and entire flour aside from fat substance, which was 1.3%. The content of phytate and oxalate was viewed to be low in semi-refined flour when compared with whole flour, because of the removal of the wheat bran portion (Suma and Urooj, 2014).

6.2.1.3 Soaking

Soaking is used for eliminating anti-nutrient content of food sources since it likewise diminishes the cooking time. Soaking enhances the release of enzymes (e.g., endogenous phytases), which are present in plant food sources like almonds and different nuts and grains. Soaking generally provides essential moist conditions in nuts, grains, and other edible seeds, which are required for their germination and related decreases in degree of compound inhibitors as well as other antinutrients to enhance absorbability and dietary benefit (Kumari, 2018). Soaking is likewise required for fermentation, which can likewise be utilized to diminish the degree of different antinutrients in food sources (Gupta et al., 2015). A considerable lot of the antinutrients are water soluble in nature, which enhance their expulsion from food varieties through leaching. Soaking increases the hydration level of legumes and cereals, which makes them delicate and furthermore activates endogenous enzymes like phytase to improve the ease of further processing such as cooking or heating. A previous study stated that 6 hours of soaking reduced 27.9% and 24 hours of soaking reduced 36.0% of phytic acid at room temperature in *Mucuna flagellipes* (Udensi et al., 2008). Due to soaking, the activity of phytase increased, which reduced the phytate component present in the grains. Because of soaking and fermentation, phytochemicals are reduced due to leaching of water-soluble vitamins and minerals in grains and legumes (Ogbonna et al., 2012; Kruger et al., 2014). However, soaking commonly reduces the content of antinutrient phytochemicals like phytate and tannins. Therefore, due to these benefits, it was recommended that wheat and barley should be consumed after soaking for a period of time (Gupta et al., 2015), especially 12 to 24 hours (Ertaş and Türker, 2014; Mahgoub and Elhag, 1998; Onwuka, 2006). For example, during soaking, endogenous or exogenous phytase enzymes could enhance the *in vitro* solubility of minerals such as zinc and iron by 2 to 23% (Vashishth et al., 2017). A previous research carried out by Greiner and Konietzny (2006) showed that soaking reduced phytate content significantly at 45°C and 65°C. Soaking of grains and beans was found much effective to enhance the minerals concentration and protein availability, accompanied by reductions in phytic acid level (Coulibaly et al., 2011). Another study reported that phytic acid concentration in chickpea was decreased by 47.45 to 55.71% when the soaking time was increased from 2 to 12 hours (Ertaş and Türker, 2014). Pressure-cooking, fermentation, and germination potentially increase the major nutrients. There was a significant reduction of PA by 61.5% and 51.8% in roasted finger millet and wheat, respectively. Total polyphenols significantly reduced by 65.8 and 87.1% in blanched finger millet and roasted wheat, respectively. A significant reduction of tannins by 73.6 and 85.1% was found in soaking followed by pressure-cooked finger millet and fermentation followed by germinated wheat, respectively. Flavonoids reduced by 51% in germinated and kilned finger millet and wheat, while 61.5 and 52.3% reduction in trypsin inhibitors was noticed in roasted and pressure-cooked finger millet and pressure-cooked wheat, respectively (Gunashree et al., 2014).

6.2.2 Chemical Methods

6.2.2.1 Gamma Radiation

Gamma radiation appeared to be a good procedure to decrease the level of trypsin inhibitor, phytic acid, and oligosaccharides of broad bean between 5 and 10% (Al-Kaisey et al., 2003). However, Hassan et al. documented that a 2 kGy dose had no significant change in the tannin content of two maize cultivars. El-Niely (2007) and Fombang et al. (2005), reported similar observations. Low doses of gamma irradiation (0.5 and 1.0 kGy) on faba bean seeds significantly reduced ANFs such as tannin and phytic acid (Osman et al., 2014). Gamma radiation can be applied as a safe postharvest method to minimize anti-nutrients of millet grains (Mahmoud et al., 2015).

6.2.2.2 Fermentation

Fermentation may be a useful strategy for reducing bacterial contamination of foods. For treating diarrhea in young children, fermented millet products are recommended as probiotics (Manisseri and Gudipati 2012; Nduti et al., 2016). Fermentation is a metabolic process in which sugars are oxidized to produce energy; it also improves the absorption of minerals from the plant-based foods. Fermentation is one of the processing methods, which is used in Africa to make cereals crops edible and also increase the nutritional quality as well as safety aspects of these foods, because cereals are not easily consumed in natural/raw forms (Galati et al., 2014).

In cereals, phytic acid normally forms complexes with the metal cations including iron, zinc, calcium, and proteins. These complexes are generally degraded by enzymes, which require an optimum pH maintained by fermentation. Thus, this kind of degradation decreases the phytic acid content and liberates soluble iron, zinc, and calcium, which enhances the nutritional level of food grains (Gibson et al., 2006). Fermentation of cereals by lactic acid bacteria (LAB) has been reported to increase free amino acids and their derivatives by proteolysis and by metabolic synthesis. Fermentation has been shown to improve the nutritional value of grains by increasing the content of essential amino acids such as lysine, methionine, and tryptophan (Mohapatra et al., 2019).

Previous reports observed that several anti-nutrients including protease inhibitors, phytic acids, and tannins were reduced due to millet grain fermentation for 12 and 24 hours (Coulibaly et al., 2011). Fermentation is such an important process, which significantly lowers the content of anti-nutrients such as phytic acid, tannins, and polyphenols of cereals (Simwaka et al., 2017). Fermentation also provides optimum pH conditions for enzymatic degradation of phytate, which is present in cereals in the form of complexes with polyvalent cations such as iron, zinc, calcium, magnesium, and proteins. Such a reduction in phytate may increase the amount of soluble iron, zinc, and calcium by several folds (Gupta et al., 2015). Tannin levels may be reduced as a result of lactic acid fermentation, leading to an increased absorption of iron, except in some high tannin cereals, where little or no improvement in iron availability has been reported (Ray and Didier, 2014). Reduction in polyphenols may be happening because of the presence of phenolic oxidase during germination (Tajoddin et al., 2014; Tian et al., 2019). In most of the cereals, *Lactobacillus* spp. plays a major role in fermentation (Bhatia, 2016). *Lactobacillus* spp. and *Streptococcus* spp. are not very suitable bacteria for rice fermentation because they lack amylase, which is necessary for starch saccharification (Ray et al., 2016). In cases when cereal grains are used as natural medium for lactic acid fermentation, amylase needs to be added before or during fermentation or amylolytic bifido bacteria need to be used because these bacteria contain enough amylase, which is necessary for saccharification of the grain starch (Kim et al., 2003). A study reported that when germinated millets sprouts were fermented at 30°C with mixtures of probiotics culture consisting of *Saccharomyces diastaticus, S. cerevisiae, Lactobacillus brevis*, and *L. fermentum* for 72 hours, approximately 88.3% reduction of phytic acid content was observed (Khetarpaul and Chauhan, 1990). Ragon et al. (2008) concluded that phytic acid (IP6) was reduced into lower forms, such as IP5, IP4, IP3, IP2, IP1, and myo-inositol by the action of microbial enzymes when rice flour was subjected to natural fermentation. In a latest study, maize flour was fermented with a consortium of lactic acid bacteria by standard method in 12-hour intervals to check the effect of fermentation on anti-nutritional factors (Ogodo et al., 2019). The results showed that with increasing fermentation period, significant (p < 0.05) reductions in anti-nutrients, including tannin, polyphenol, phytate, and trypsin inhibitor activity, were observed in the fermented maize. Results concluded that the anti-nutritional contents were reduced more in LAB-consortium fermentation compared to spontaneous fermentation (Ogodo et al., 2019).

Etsuyankpa et al. (2015) evaluated the effect of microbial fermentation on antinutritional composition of local cassava products. Results of the study emphasized that fermentation by microorganisms significantly decreased (P < 0.05) the level of cyanide, tannins, phytate, oxalate, and saponins by 86, 73, 72, 61, and 92%, respectively in the cassava products. A study by Samia et al. (2005) reported that fermentation and germination could enhance the nutritional level of cereals and legumes by altering the chemical composition and reduce the level of anti-nutritional factors.

6.2.3 Biotechnological Interventions

Genetic engineering techniques have been used to eliminate the genes involved in the metabolic pathways for reducing the production and/or inactivation of antinutrients. Major biotechnological tools have been utilized for reduction of phytic acid, the major ANF of wheat, which has already been discussed in the PA section.

With the advent of advance technologies like genome-editing, new opportunities have opened for manipulating the expression of gene families. Recent CRISPR-Cas9-based genome-editing technology has turned into the principal novel technique for the development of crops, being a safer method for the development of novel genome-edited genotypes with no fear of transgenic genes (Jouanin et al., 2019). Additionally, CRISPR technique boasts of the development of newer crops by including breeding with the added advantage of high accuracy and efficiency of gene editing along with multiplexing. With minimized chance of off-target mutations and the prospect to attain transgene-free mutant plants in the first generation (Zhang et al., 2016) that make them non-GMO, this is still a matter of debate at legislative level.

It has been ascertained that structural and metabolic proteins, like α-amylase/trypsin inhibitors (ATIs), are involved in the onset of wheat allergies (bakers' asthma) and probably Non-Celiac Wheat Sensitivity (NCWS). The ATIs are a group of exogenous protease inhibitors, which are encoded by a multigene family dispersed over several chromosomes in durum and bread wheat. WTAI-CM3 and WTAI-CM16 subunits are considered among the main proteins involved in the onset of bakers' asthma and probably NCWS. A CRISPR-Cas9 multiplexing strategy was used to edit the ATI subunits WTAI-CM3 and WTAI-CM16 in the grain of the Italian durum wheat cultivar Svevo with the aim to produce wheat lines with reduced amount of potential allergens involved in adverse reactions. Using a marker-gene-free approach, whereby plants are regenerated without selection agents, homozygous mutant plants without the presence of CRISPR vectors were obtained directly from T_0 generation. This study demonstrates the capability of CRISPR technology to knock out immunogenic proteins in a reduced time compared to conventional breeding programs. The editing of the two target genes was confirmed either at molecular (sequencing and gene expression study) or biochemical (immunologic test) level. Noteworthy, as a pleiotropic effect, is the activation of the ATI *0.28* pseudogene in the edited lines (Camerlengo et al., 2020).

6.3 Role of Enhancers/Promoters in Combating Anti-Nutritional Factors

Some organic compounds stimulate the absorption of essential mineral elements by humans (Table 6.3). These include ascorbate (vitamin C), β-carotene (provitamin A), protein cysteine, and various organic and amino acids. There is considerable intraspecific variation in both ascorbate and β-carotene concentrations in fruit and vegetables (Frossard et al., 2000). For example, ascorbate concentration in cassava varied by 250-fold in leaves and by 40-fold in roots among the 530 accessions of the CIAT core collection, whereas β-carotene concentration varied by 3.7-fold in leaves and by 10-fold in roots. Similarly, ascorbate concentration varied almost by 20-fold among *Dioscorea alata* accessions. There is also appreciable intraspecific variation in amino acid concentrations in edible tissues (Guzmán-Maldonado et al., 2000). However, the complement of amino acids present in different foodstuffs is constrained by evolutionary heritage such that cereal and vegetable crops contribute complimentary amino acids to the diet (White and Broadley, 2005; Singh et al., 2016)

Several factors influence the bioavailability of iron. Certain dietary factors and acids prevent the precipitation of ferric iron by reducing it to the ferrous state and by forming suitable soluble ligands that are available for absorption. Ascorbic acid is the most clearly documented enhancer of non-heme iron bioavailability. Studies showed that non-heme iron absorption was increased by a factor of 1.7 when 100 mg of vitamin C was added to the standard diet. The dose-dependent effect became clear when the absorption of iron from a semi-synthetic test meal was increased by factors of 1.9, 3.2, and 4.7 with simultaneous increasing amounts of vitamin C (Monsen, 1988). Fruit juices are known to have an enhancing effect on iron absorption. A study tested the ability of various fruit juices and fruits to modify iron absorption from a 200-g rice meal with low iron availability (Ballot et al., 1987). The ability of various fruit juices to enhance iron absorption from the rice meal was correlated with their ascorbic acid contents. The second objective of the study was to find out whether other organic acids present in fruit affected iron absorption. A direct comparison of a solution containing ascorbic acid (33 mg/100 g) with one containing the same amount of ascorbic acid and 750 mg/100 g of citric acid indicated that the citric acid had an enhancing effect on the corrected geometric mean iron absorption from the rice meal (0.114 and 0.170, respectively). Orange juice was more effective than water containing the same amount

TABLE 6.3

Nutritional Enhancers and their Antinutrients

Nutritional Enhancers	Antinutrients
β-Carotene	Oxalic acid
Inulin	Phytic acid
Long-chain fatty acids	Polyphenols
Certain amino acids (cysteine, lysine etc.)	Tannins
Certain organic acids (ascorbic acid, citrate etc.)	Others
Vitamin D	Others

Source: (White and Broadley, 2005; Welch and Graham, 2005)

of ascorbic acid (0.139 and 0.098, respectively) (Govindaraj et al., 2007). Breeding to increase bioavailability promoters is also being explored (Welch and Graham, 2004; Bouis and Welch, 2010). Small changes in the concentrations of these compounds may result in a comparatively large improvement in bioavailability. Few promoters have been identified for iron and zinc. Phytoferritin may be the most promising promoter of iron absorption, depending on the extent to which it effectively protects iron from binding to phytates. However, the knowledge of its genetic variability is limited, and thus it is still unknown whether phytoferritin concentration can be increased in foods. Another promising promoter is non-digestible carbohydrates or prebiotics, such as fructans, which promote the growth of beneficial probiotic bacteria in the hindgut that in turn may decrease inflammation and limit the growth of pathogenic bacteria. The growth of some types of such bacteria is still under investigation for their role in improvement of iron and zinc absorption. The activation of endogenous phytases can increase both iron and zinc bioaccessibility. The endogenous phytases are capable of hydrolyzing hexa- and pentaphosphates to decrease inhibitory inositol phosphates that bind to minerals (Shilpa and Lakshmi, 2012).

6.3.1 Role of Ascorbic Acid (AA) as a Bio-Enhancer

The addition of ascorbic acid to foods and beverages during processing or prior to packaging protects color, aroma, and nutrient content and improves Fe absorption. The addition of ascorbic acid to fresh flour improves its baking qualities, thus saving the four to eight weeks of maturation that flour would normally have to undergo after milling (Teucherl and Cori, 2004). The dominant form of iron in foods is ferric iron, which is much less bioavailable than ferrous iron. One of ascorbic acid's main attributes is its ability to reduce ferric to ferrous iron. Ascorbic acid undergoes a reversible two-stage redox method with a free radical intermediate. The latter reacts preferably with it, thus preventing the propagation of free radical reactions (Herbert et al., 1996). At the same time, ascorbic acid maintains a transition metal, such as Fe(III), in its reduced form Fe(II) and can promote the reaction of these ions with hydrogen peroxide to form highly reactive hydroxyl radicals in the Fenton reaction (Nappi and Vass, 2002). Such pro-oxidant activities have been demonstrated in various food matrices (Jacobsen et al., 2001) and may adversely affect shelf life. *In vivo* observations showing an enhancement of iron absorption in the presence of ascorbic acid have been attributed to AA's chelating and reducing properties (Conrad and Schade, 1968).

6.3.2 Effect of Ascorbic Acid on Non-Heme Iron Absorption in Humans

A great variety of single meal studies have shown a distinct relative increase in Fe absorption which depends on the occurrence and the type of inhibitors and other enhancers of Fe absorption. The two most important inhibitors of Fe absorption are phytate and polyphenols which are known to inhibit absorption of Fe at very low levels (like 2 mg phytate reduced absorption by about 18% and 12 mg tannic acid reduced absorption by 30%). The amount of ascorbic acid required to completely counteract the effect of 25 mg phytates was estimated to be around 80 mg. Several hundred milligrams of ascorbic acid would therefore be required to counteract the inhibition of high-phytate diets (250 mg). A minimum of 50 mg ascorbic acid was found to compensate for the negative effect of a bread meal containing in excess of 100 mg polyphenols. The phytate content of typical Western meals ranges from 10 to 100 mg, and phytate levels of more than 250 mg are present in many meals typically consumed in developing countries (Siegenberg et al., 1991; Teucherl and Cori, 2004).

6.3.3 Limitations of Food Fortification with Ascorbic Acid

The effectiveness of ascorbic acid depends on the amount added, the stability of ascorbic acid, the type and concentration of iron fortificant, and the presence of inhibitors in the diet. Ascorbic acid is not easily formulated into various completed food products due to its sensitivity to oxygen, water, and heat. The naturally occurring high ascorbic acid content of some fruits and vegetables readily reduces when exposed to air through storage and some post-harvest managements (Clydesdale et al., 1991). In the dry state, ascorbic acid is reasonably stable in air, but in solution it quickly oxidizes. On exposure to moisture, light, and heat, it may also darken. Cooking normally abolishes ascorbic acid by quickening the oxidation reaction. The instability of ascorbic acid during storage, heat processing, and cooking; the possibility of unwanted sensory changes in liquid products; and the cost of ascorbic acid itself or the cost of effective packaging are the major reasons for it not being used to enhance fortificant-iron added to food staples such as wheat flour and maize flour. The encapsulation/microencapsulation is another option, but current encapsulation techniques are unlikely to withstand the baking process. The encapsulation and microencapsulation techniques are currently under development to assist the fortification of foods with ascorbic acid and to provide new forms of consumable food supplements without causing any opposing organoleptic properties (Teucherl and Cori, 2004).

6.4 Conclusion

Anti-nutrients are inherently present in cereals like wheat and are generally associated with causing nutrient deficiencies and other adverse effects when consumed in very high quantities. Their breakdown products also cause adverse effects. Consequently, the incidence of phytic acid, tannins, lectins, alkaloids, saponins, goitrogens, inhibitors, *etc.*, in food products may induce adverse reactions in their consumers, when awareness about them is scarce. Traditionally, several measures like soaking, heating, and germinating are being used which are effective measures of ANF reduction at a small scale. However, the use of chemical methods have also been carried out. Classic approaches and modern agricultural biotechnological programs can serve as strong anti-nutritional removal tools. Recent approaches like CRIPR-based genome editing may serve as an effective strategy in lowering ANF content in wheat as well as other cereals.

6.5 Future Prospects

Rapid progress has been made in developing the tools for reverse genetics in hexaploid wheat such as genome-editing-mediated CRISPR/Cas9 and base editing. It was assessed that editing tools, coupled with the applications of nano-biotechnology, could escape the tedious and time-consuming wheat transformation procedures. The enlisted candidate genes for achieving the lowering of PA are an excellent resource for generating new germplasm that could subsequently be introgressed into the elite wheat cultivars with high productivity and rust resistance. This will help in addressing the larger goals of micro-nutrient biofortification by enhancing its bioavailability in humans.

REFERENCES

Adeyemo, S. M. & Onilude, A. A. (2013). Enzymatic reduction of anti-nutritional factors in fermenting soybeans by Lactobacillus plantarum isolates from fermenting cereals. *Nigerian Food Journal, 31*(2), 84–90.

Aggarwal, S., Kumar, A., Bhati, K. K., Kaur, G., Shukla, V., Tiwari, S. & Pandey, A. K. (2018). RNAi-mediated downregulation of inositol pentakisphosphate kinase (IPK1) in wheat grains decreases phytic acid levels and increases Fe and Zn accumulation. *Frontiers in Plant Science, 9*, 259.

Agostinho, A. J., de Souza Oliveira, W., Anunciação, D. S. & Santos, J. C. C. (2016). Simple and sensitive spectrophotometric method for phytic acid determination in grains. *Food Analytical Methods, 9*(7), 2087–2096.

Akiyama, H., Fujii, K., Yamasaki, O., Oono, T. & Iwatsuki, K. (2001). Antibacterial action of several tannins against *Staphylococcus aureus. Journal of Antimicrobial Chemotherapy, 48*(4), 487–491.

Al Hasan, S. M., Hassan, M., Saha, S., Islam, M., Billah, M. & Islam, S. (2016). Dietary phytate intake inhibits the bioavailability of iron and calcium in the diets of pregnant women in rural Bangladesh: A cross-sectional study. *BMC Nutrition, 2*(1), 1–10.

Aletor, V. A. (1993). Allelochemicals in plant foods and feeding stuffs: 1. Nutritional, biochemical and physiopathological aspects in animal production. *Veterinary and Human Toxicology, 35*(1), 57–67.

Al-Kaisey, M. T., Alwan, A. K. H., Mohammad, M. H. & Saeed, A. H. (2003). Effect of gamma irradiation on antinutritional factors in broad bean. *Radiation Physics and Chemistry, 67*(3): 493–6.

Amirabdollahian, F. & Ash, R. (2010). An estimate of phytate intake and molar ratio of phytate to zinc in the diet of the people in the United Kingdom. *Public Health Nutrition, 13*(9), 1380–1388.

Awolu, O. O., Omoba, O. S., Olawoye, O. & Dairo, M. (2017). Optimization of production and quality evaluation of maize-based snack supplemented with soybean and tigernut (*Cyperus esculenta*) flour. *Food Science & Nutrition, 5*(1), 3–13.

Bajpai, M., Pande, A., Tewari, S. K. & Prakash, D. (2005). Phenolic contents and antioxidant activity of some food and medicinal plants. *International Journal of Food Sciences and Nutrition, 56*(4), 287–291.

Ballot, D., Baynes, R. D., Bothwell, T. H., Gillooly, M., Macfarlane, J., MacPhail, A. P. & Bothwell, J. E. (1987). The effects of fruit juices and fruits on the absorption of iron from a rice meal. *British Journal of Nutrition, 57*(3), 331–343.

Bardocz, S., Grant, G., Ewen, S. W., Duguid, T. J., Brown, D. S., Englyst, K. & Pusztai, A. (1995). Reversible effect of phytohaemagglutinin on the growth and metabolism of rat gastrointestinal tract. *Gut, 37*(3), 353–360.

Barmeyer, C., Schumann, M., Meyer, T., Zielinski, C., Zuberbier, T., Siegmund, B. & Ullrich, R. (2017). Long-term response to gluten-free diet as evidence for non-celiac wheat sensitivity in one third of patients with diarrhea-dominant and mixed-type irritable bowel syndrome. *International Journal of Colorectal Disease, 32*(1), 29–39.

Beuerle, T., Benford, D., Brimer, L., Cottrill, B., Doerge, D., Dusemund, B. & Mulder, P. P. J. (2012). Scientific Opinion on Ergot alkaloids in food and feed. *EFSA Journal, 10*(7), 2798.

Bhati, K. K., Alok, A., Kumar, A., Kaur, J., Tiwari, S. & Pandey, A. K. (2016). Silencing of ABCC13 transporter in wheat reveals its involvement in grain development, phytic acid accumulation and lateral root formation. *Journal of Experimental Botany, 67*(14), 4379–4389.

Bhatia, S. C. (2016). *Food Biotechnology*. Wpi Publishing. http://www.woodheadpublishingindia.com/BookDetails.aspx?BookID=108

Bhutkar, M. A. & Bhise, S. B. (2012). *In vitro* assay of alpha amylase inhibitory activity of some indigenous plants. *International Journal of Chemical Sciences, 10*(1), 457–462.

Biesiekierski, J. R., Newnham, E. D., Irving, P. M., Barrett, J. S., Haines, M., Doecke, J. D. & Gibson, P. R. (2011). Gluten causes gastrointestinal symptoms in subjects without celiac disease: a double-blind randomized placebo-controlled trial. *Official Journal of the American College of Gastroenterology, ACG, 106*(3), 508–514.

Boontaganon, P., Jéhanno, E. & Savage, G. P. (2009). Total, soluble and insoluble oxalate content of bran and bran products. *Journal of Food, Agriculture and Environment, 7,* 204–206.

Borghi, L., Schianchi, T., Meschi, T., Guerra, A., Allegri, F., Maggiore, U. & Novarini, A. (2002). Comparison of two diets for the prevention of recurrent stones in idiopathic hypercalciuria. *The New England Journal of Medicine, 346,* 77–84.

Bouis, H. E. & Welch, R. M. (2010). Biofortification—A sustainable agricultural strategy for reducing micronutrient malnutrition in the global south. *Crop Science, 50,* S-20.

Bribi, N. (2018). Pharmacological activity of alkaloids: a review. *Asian Journal of Botany, 1*(1), 1–6.

Burbano, C., Muzquiz, M., Osagie, A., Ayet, G. & Cuadrado, C. (1995). Determination of phytate and lower inositol phosphates in Spanish legumes by HPLC methodology. *Food Chemistry, 52*(3), 321–325.

Cakmak, I., Wolfgang, H. P. & Bonnie, M. (2010). Biofortification of durum wheat with zinc and iron. *Cereal Chemistry, 87,* 10–20.

Call, L., Haider, E., D'Amico, S., Reiter, E. & Grausgruber, H. (2021). Synthesis and accumulation of amylase-trypsin inhibitors and changes in carbohydrate profile during grain development of bread wheat (*Triticum aestivum* L.). *BMC Plant Biology, 21*(1), 1–12.

Camerlengo, F., Frittelli, A., Sparks, C., Doherty, A., Martignago, D., Larré, C., . . . & Masci, S. (2020). CRISPR-Cas9 multiplex editing of the α-amylase/trypsin inhibitor genes to reduce allergen proteins in durum wheat. *Frontiers in Sustainable Food Systems, 4,* 104.

Campos-Vega, R., Loarca-Piña, G. & Oomah, B. D. (2010). Minor components of pulses and their potential impact on human health. *Food Research International, 43*(2), 461–482.

Cantrell, R. G. & Joppa, L. R. (1991). Genetic analysis of quantitative traits in wild emmer (*Triticum turgidum* L. *var. dicoccoides*). *Crop Science, 31,* 645–649.

Carbonell-Rozas, L., Mahdjoubi, C. K., Arroyo-Manzanares, N., García-Campaña, A. M. & Gámiz-Gracia, L. (2021). Occurrence of Ergot Alkaloids in Barley and wheat from Algeria. *Toxins, 13*(5), 316.

Chen, W., Hira, T., Nakajima, S., Tomozawa, H., Tsubata, M., Yamaguchi, K. & Hara, H. (2012). Suppressive effect on food intake of a potato extract (Potein®) involving cholecystokinin release in rats. *Bioscience, Biotechnology, and Biochemistry, 76*(6), 1104–1109.

Chowdhury, S. & Punia, D. (1997). Nutrient and antinutrient composition of pearl millet grains as affected by milling and baking. *Food/nahrung, 41*(2), 105–107.

Clydesdale, F. M., Ho, C. T., Lee, C. Y., Mondy, N. I., Shewfelt, R. L. & Lee, K. (1991). The effects of postharvest treatment and chemical interactions on the bioavailability of ascorbic acid, thiamin, vitamin A, carotenoids, and minerals. *Critical Reviews in Food Science & Nutrition, 30*(6), 599–638.

Conrad, M. E. & Schade, S. G. (1968). Ascorbic acid chelates in iron absorption: a role for hydrochloric acid and bile. *Gastroenterology, 55*, 35–45.

Cordain, L., Toohey, L., Smith, M. J. & Hickey, M. S. (2000). Modulation of immune function by dietary lectins in rheumatoid arthritis. *British Journal of Nutrition, 83*(3), 207–217.

Coulibaly, A., Kouakou, B. & Chen, J. (2011). Phytic acid in cereal grains: Structure, healthy or harmful ways to reduce phytic acid in cereal grains and their effects on nutritional quality. *American Journal of Plant Nutrition and Fertilization Technology, 1*(1), 1–22.

Czaja-Bulsa, G. (2015). Non coeliac gluten sensitivity–A new disease with gluten intolerance. *Clinical Nutrition, 34*(2), 189–194.

Dahl, S. W., Rasmussen, S. K., Petersen, L. C. & Hejgaard, J. (1996). Inhibition of coagulation factors by recombinant barley serpin BSZx. *FEBS Letters, 394*(2), 165–168.

Doss, A., Pugalenthi, M., Vadivel, V. G., Subhashini, G. & Subash, A. R. (2011). Effects of processing technique on the nutritional composition and anti-nutrients content of underutilized food legume *Canavalia ensiformis* L. DC. *International Food Research Journal, 18*(3).

El-Niely, H. F. (2007). Effect of radiation processing on antinutrients, in-vitro protein digestibility and protein efficiency ratio bioassay of legume seeds. *Radiation Physics and Chemistry, 76*(6), 1050–1057.

Ertaş, N. & Türker, S. (2014). Bulgur processes increase nutrition value: possible role in in-vitro protein digestibility, phytic acid, trypsin inhibitor activity and mineral bioavailability. *Journal of Food Science and Technology, 51*(7), 1401–1405.

Ertop, M. H. & Bektaş, M. (2018). Enhancement of bioavailable micronutrients and reduction of anti-nutrients in foods with some processes. *Food and Health, 4*(3), 159–165.

Essack, H., Odhav, B. & Mellem, J. J. (2017). Screening of traditional South African leafy vegetables for specific anti-nutritional factors before and after processing. *Food Science and Technology, 37*, 462–471.

Etsuyankpa, M. B., Gimba, C. E., Agbaji, E. B., Omoniyi, K. I., Ndamitso, M. M. & Mathew, J. T. (2015). Assessment of the effects of microbial fermentation on selected anti-nutrients in the products of four local cassava varieties from Niger state, Nigeria. *American Journal of Food Science and Technology, 3*(3), 89–96.

Fan, Y., Guo, D. Y., Song, Q. & Li, T. (2013). Effect of total saponin of aralia taibaiensis on proliferation of leukemia cells. *Journal of Chinese Medicinal Materials, 36*(4), 604–607.

FAO (2009). FAOSTAT. Food and Agriculture Organisation of the United Nations. *FAOSTAT.* http://faostat.fao.org/site/339/default.aspx.

FAO (2018a). www.fao.org/gsfaonline/docs/CXS_192e.pdf

FAO (2018b). *Anti-nutritional Factors within Feed Ingredients.* Rome: Aquaculture Feed and Fertilizer Resources Information System, Food and Agriculture Organizations of the United Nations www.fao.org/fishery/affris/feedresources-database.anti-nutritional-factors-within-feed ingredients/en/. Accessed 28 November 2018.

FAOSTAT (2022). Lower cereal production outlook underpins downward revisions for utilization and stocks in 2022/23, www.fao.org/worldfoodsituation/csdb/en/. Accessed 2 September 2022.

Febles, C. I., Arias, A., Hardisson, A., Rodrıguez-Alvarez, C. & Sierra, A. (2002). Phytic acid level in wheat flours. *Journal of Cereal Science, 36*(1), 19–23.

Fombang, E. N., Taylor, J. R. N., Mbofung, C. M. F. & Minnaar, A. (2005). Use of γ-irradiation to alleviate the poor protein digestibility of sorghum porridge. *Food Chemistry, 91*(4), 695–703.

Fredlund, K., Isaksson, M., Rossander-Hulthén, L., Almgren, A. & Sandberg, A. S. (2006). Absorption of zinc and retention of calcium: dose-dependent inhibition by phytate. *Journal of Trace Elements in Medicine and Biology, 20*(1), 49–57.

Frossard, E., Bucher, M., Mächler, F., Mozafar, A. & Hurrell, R. (2000). Potential for increasing the content and bioavailability of Fe, Zn and Ca in plants for human nutrition. *Journal of the Science of Food and Agriculture, 80*(7), 861–879.

Frutos, P., Hervas, G., Giráldez, F. J. & Mantecón, A. R. (2004). Tannins and ruminant nutrition. *Spanish Journal of Agricultural Research, 2*(2), 191–202.

Gaj, T., Gersbach, C. A., Barbas, C. F. III & Barbas, C. F. (2013). ZFN, TALEN, and CRISPR/Cas-based methods for genome engineering. *Trends in Biotechnology, 31*(7), 397–405.

Galati, A., Oguntoyinbo, F. A., Moschetti, G., Crescimanno, M. & Settanni, L. (2014). The cereal market and the role of fermentation in cereal-based food production in Africa. *Food Reviews International, 30*(4), 317–337.

Gao, Y., Shang, C., Maroof, M. S., Biyashev, R. M., Grabau, E. A., Kwanyuen, P., . . . & Buss, G. R. (2007). A modified colorimetric method for phytic acid analysis in soybean. *Crop Science, 47*(5), 1797–1803.

García-Estepa, R. M., Guerra-Hernández, E. & García-Villanova, B. (1999). Phytic acid content in milled cereal products and breads. *Food Research International, 32*(3), 217–221.

Geisslitz, S., Longin, C. F. H., Koehler, P. & Scherf, K. A. (2020). Comparative quantitative LC–MS/MS analysis of 13 amylase/trypsin inhibitors in ancient and modern Triticum species. *Scientific Reports, 10*(1), 1–13.

Gemede, H. F. & Fekadu, H. (2014). Nutritional composition, antinutritional factors and effect of boiling on nutritional composition of Anchote (*Coccinia abyssinica*) tubers. *Journal of Scientific and Innovative Research, 3*(2), 177–188.

Gemede, H. F. & Ratta, N. (2014). Antinutritional factors in plant foods: Potential health benefits and adverse effects. *International Journal of Nutrition and Food Sciences, 3*(4), 284–289.

Gibson, R. S., Perlas, L. & Hotz, C. (2006). Improving the bioavailability of nutrients in plant foods at the household level. *Proceedings of the Nutrition Society, 65*(2), 160–168.

Gomes, M., Oliva, M., Lopes, M. & Salas, C. (2011). Plant proteinases and inhibitors: an overview of biological function and pharmacological activity. *Current Protein and Peptide Science, 12*(5), 417–436.

Govindaraj, T., KrishnaRau, L. & Prakash, J. (2007). In vitro bioavailability of iron and sensory qualities of iron-fortified wheat biscuits. *Food and Nutrition Bulletin, 28*(3), 299–306.

Grases, F., Prieto, R. M. & Costa-Bauza, A. (2017). Dietary phytate and interactions with mineral nutrients. *Clinical Aspects of Natural and Added Phosphorus in Foods* (pp. 175–183). New York: Springer.

Greiner, R. & Carlsson, N. G. (2006). *Myo*-Inositol phosphate isomers generated by the action of a phytate-degrading enzyme from *Klebsiella terrigena* on phytate. *Canadian Journal of Microbiology, 52*, 759–768.

Greiner, R. & Konietzny, U. (2006). Phytase for food application. *Food Technology & Biotechnology, 44*(2).

Guillamon, E., Pedrosa, M. M., Burbano, C., Cuadrado, C., de Cortes Sánchez, M. & Muzquiz, M. (2008). The trypsin inhibitors present in seed of different grain legume species and cultivar. *Food Chemistry, 107*(1), 68–74.

Gunashree, B. S., Selva Kumar, R., Roobini, R. & Venkateswaran, G. (2014). Nutrients and anti-nutrients of ragi and wheat as influenced by traditional processes. *International Journal of Current Microbiology and Applied Sciences, 3*(7), 720–736.

Gupta, R. K., Gangoliya, S. S. & Singh, N. K. (2015). Reduction of phytic acid and enhancement of bioavailable micronutrients in food grains. *Journal of Food Science and Technology, 52*(2), 676–684.

Guttieri, M., Bowen, D., Dorsch, J. A., Raboy, V. & Souza, E. (2004). Identification and characterization of a low phytic acid wheat. *Crop Science, 44*(2), 418–424.

Guzmán-Maldonado, S. H., Acosta-Gallegos, J. & Paredes-López, O. (2000). Protein and mineral content of a novel collection of wild and weedy common bean (*Phaseolus vulgaris* L). *Journal of the Science of Food and Agriculture, 80*(13), 1874–1881.

Haarmann, T., Rolke, Y., Giesbert, S. & Tudzynski, P. (2009). Ergot: from witchcraft to biotechnology. *Molecular Plant Pathology, 10*(4), 563–577.

Hamid, R., Masood, A., Wani, I. H. & Rafiq, S. (2013). Lectins: proteins with diverse applications. *Journal of Applied Pharmaceutical Science, 3*(4), S93–S103.

Hanes, D. A., Weaver, C. M., Heaney, R. P. & Wastney, M. (1999). Absorption of calcium oxalate does not require dissociation in rats. *Journal of Nutrition, 129*, 170–173.

Haq, S. K., Atif, S. M. & Khan, R. H. (2004). Protein proteinase inhibitor genes in combat against insects, pests, and pathogens: natural and engineered phytoprotection. *Archives of Biochemistry and Biophysics, 431*(1), 145–159.

Herbert, V., Shaw, S. & Jayatilleke, E. (1996). Vitamin C-driven free radical generation from iron. *The Journal of Nutrition, 126*(suppl_4), 1213S-1220S.

Herzig, K. H., Bardocz, S., Grant, G., Nustede, R., Fölsch, U. R. & Pusztai, A. (1997). Red kidney bean lectin is a potent cholecystokinin releasing stimulus in the rat inducing pancreatic growth. *Gut, 41*(3), 333–338.

Hitz, W. D., Carlson, T. J., Kerr, P. S. & Sebastian, S. A. (2002). Biochemical and molecular characterization of a mutation that confers a decreased raffinosaccharide and phytic acid phenotype on soybean seeds. *Plant Physiology, 128*(2), 650–660.

Hurrell, R. F., Reddy, M. B., Juillerat, M. A. & Cook, J. D. (2003). Degradation of phytic acid in cereal porridges improves iron absorption by human subjects. *The American Journal of Clinical Nutrition, 77*(5), 1213–1219.

Jacobs, D. R. & Gallaher, D. D. (2004). Whole grain intake and cardiovascular disease: A review. *Current Atherosclerosis Reports, 6*(6), 415–423.

Jacobsen, C., Timm, M. & Meyer, A. S. (2001). Oxidation in fish oil enriched mayonnaise: ascorbic acid and low pH increase oxidative deterioration. *Journal of Agricultural and Food Chemistry, 49*(8), 3947–3956.

Jankele, R. & Svoboda, P. (2014). TAL effectors: tools for DNA targeting. *Briefings in Functional Genomics, 13*(5), 409–19.

Jouanin, A., Schaart, J. G., Boyd, L. A., Cockram, J., Leigh, F. J., Bates, R., et al. (2019). Outlook for coeliac disease patients: towards bread wheat with hypoimmunogenic gluten by gene editing of α- and γ-gliadin gene families. *BMC Plant Biology, 19*, 333. https://doi.org/10.1186/s12870-019-1889-5.

Judprasong, K., Charoenkiatkul, S., Sungpuag, P., Vasanachitt, K. & Nakjamanong, Y. (2006). Total and soluble oxalate contents in Thai vegetables, cereal grains and legume seeds and their changes after cooking. *Journal of Food Composition and Analysis, 19*(4), 340–347.

Juliano, B. O., (1985). Polysaccharides, proteins, and lipids of rice. *American Association of Cereal Chemists*, 59–174.

Junker, Y., Zeissig, S., Kim, S. J., Barisani, D., Wieser, H., Leffler, D. A. & Schuppan, D. (2012). Wheat amylase trypsin inhibitors drive intestinal inflammation via activation of toll-like receptor 4. *Journal of Experimental Medicine, 209*(13), 2395–2408.

Kao, T. H., Huang, S. C., Inbaraj, B. S. & Chen, B. H. (2008). Determination of flavonoids and saponins in *Gynostemma pentaphyllum* (Thunb.) Makino by liquid chromatography–mass spectrometry. *Analytica Chimica Acta, 626*(2), 200–211.

Kataria, A., Chauhan, B. M. & Punia, D. (1989). Anti-nutrients and protein digestibility (in vitro) of mungbean as affected by domestic processing and cooking. *Food Chemistry, 32*(1), 9–17.

Katina, K., Arendt, E., Liukkonen, K. H., Autio, K., Flander, L. & Poutanen, K. (2005). Potential of sourdough for healthier cereal products. *Trends in Food Science & Technology, 16*(1–3), 104–112.

Kaukovirta-Norja, A., Wilhelmson, A. & Poutanen, K. (2004). Germination: A means to improve the functionality of oat. *Agricultural and Food Science, 13*, 100–112.

Kavitha, S. & Parimalavalli, R. (2014). Effect of processing methods on proximate composition of cereal and legume flours. *Journal of Human Nutrition and Food Science, 2*(4), 1051.

Khetarpaul, N. & Chauhan, B. M. (1990). Effect of germination and fermentation on in vitro starch and protein digestibility of pearl millet. *Journal of Food Science, 55*(3), 883–884.

Kies, A. K., De Jonge, L. H., Kemme, P. A. & Jongbloed, A. W. (2006). Interaction between protein, phytate, and microbial phytase. *In vitro* studies. *Journal of Agricultural and Food Chemistry, 54*(5), 1753–1758.

Kim, K. I., Kim, W. K., Seo, D. K., Yoo, I. S., Kim, E. K. & Yoon, H. H. (2003). Production of lactic acid from food wastes. In *Biotechnology for Fuels and Chemicals* (pp. 637–647). Totowa, NJ: Humana Press.

Knorr, D., Watkins, T. R. & Carlson, B. L. (1981). Enzymatic reduction of phytate in whole wheat breads. *Journal of Food Science, 46*, 1866–1869.

Kolodziej, H. & Kiderlen, A. F. (2005). Antileishmanial activity and immune modulatory effects of tannins and related compounds on Leishmania parasitised RAW 264.7 cells. *Phytochemistry, 66*(17), 2056–2071.

Krishnan, V., Jain, P., Vinutha, T., Hada, A., Manickavasagam, M., Ganapathi, A. & Sachdev, A. (2015). Molecular modeling and 'in-silico' characterization of 'Glycine max' Inositol (1, 3, 4) tris 5/6 kinase-1 (Gmitpk1)-a potential candidate gene for developing low phytate transgenics. *Plant Omics, 8*(5), 381–391.

Kruger, J., Oelofse, A. & Taylor, J. R. (2014). Effects of aqueous soaking on the phytate and mineral contents and phytate: mineral ratios of wholegrain normal sorghum and maize and low phytate sorghum. *International Journal of Food Sciences and Nutrition, 65*(5), 539–546.

Krupa, U. (2008). Main nutritional and antinutritional compounds of bean seeds-a review. *Polish Journal of Food and Nutrition Sciences, 58*(2), 149–155.

Kumar, R. (1992). Anti-nutritional factors, the potential risks of toxicity and methods to alleviate them. Legume trees and other fodder trees as protein source for livestock. *FAO Animal Production and Health Paper, 102*, 145–160.

Kumari, S. (2018). *The effect of soaking almonds and hazelnuts on Phytate and mineral concentrations* (Doctoral dissertation, University of Otago).

Kwun, I. S. & Kwon, C. S. (2000). Dietary molar ratios of phytate: zinc and millimolar ratios of phytate× calcium: zinc in South Koreans. *Biological Trace Element Research, 75*(1), 29–41.

Lajolo, F. M. & Genovese, M. I. (2002). Nutritional significance of lectins and enzyme inhibitors from legumes. *Journal of Agricultural and Food Chemistry, 50*(22), 6592–6598.

Larson, S. R., Rutger, J. N., Young, K. A. & Raboy, V. (2000). Isolation and genetic mapping of a non-lethal rice (*Oryza sativa* L.) low phytic acid 1 mutation. *Crop Science, 40*(5), 1397–1405.

Larson, S. R., Young, K. A., Cook, A., Blake, T. K. & Raboy, V. (1998). Linkage mapping of two mutations that reduce phytic acid content of barley grain. *Theoretical and Applied Genetics, 97*(1), 141–146.

Larsson, S. C., Giovannucci, E., Bergkvist, L. & Wolk, A. (2005). Whole grain consumption and risk of colorectal cancer: a population-based cohort of 60 000 women. *British Journal of Cancer, 92*(9), 1803–1807.

Li, L. & Tsao, R. (2019). UF-LC-DAD-MSn for discovering enzyme inhibitors for nutraceuticals and functional foods. *Journal of Food Bioactives, 7.*

Liu, K. (1997). Chemistry and nutritional value of soybean components. In *Soybeans* (pp. 25–113). Boston, MA: Springer.

Lolas, G. M. (1976). Palamidis. N. and Markakis. *The phytic acid total P relationship in barley. oats. soybeans and wheat. Cereal Chemistry, 53*, 867.

Lönnerdal, B. (2002). Phytic acid–trace element (Zn, Cu, Mn) interactions. *International Journal of Food Science & Technology, 37*(7), 749–758.

López-Perea, P., Guzmán-Ortiz, F. A., Román-Gutiérrez, A. D., Castro-Rosas, J., Gómez-Aldapa, C. A., Rodríguez-Marín, M. L., . . . & Torruco-Uco, J. G. (2019). Bioactive compounds and antioxidant activity of wheat bran and barley husk in the extracts with different polarity. *International Journal of Food Properties, 22*(1), 646–658.

Lott, J. N., Ockenden, I., Raboy, V. & Batten, G. D. (2000). Phytic acid and phosphorus in crop seeds and fruits: a global estimate. *Seed Science Research, 10*(1), 11–33.

Lu, L., Liu, S. W., Jiang, S. B. & Wu, S. G. (2004). Tannin inhibits HIV-1 entry by targeting gp41. *Acta Pharmacologica Sinica, 25*(2), 213–218.

Mahgoub, S. E. & Elhag, S. A. (1998). Effect of milling, soaking, malting, heat-treatment and fermentation on phytate level of four Sudanese sorghum cultivars. *Food Chemistry, 61*(1–2), 77–80.

Mahmoud, N. S., Awad, S. H., Madani, R. M., Osman, F. A., Elmamoun, K. & Hassan, A. B. (2015). Effect of γ radiation processing on fungal growth and quality characteristics of millet grains. *Journal of Food Science and Nutrition, 4*(3), 342–7.

Malik, S. (2015). Pearl millet-nutritional value and medicinal uses. *International Journal of Advance Research and Innovative Ideas in Education, 1*(3), 414–418.

Mallick, S. A., Azaz, K., Gupta, M., Sharma, V. & Sinha, B. K. (2013). Characterization of grain nutritional quality in wheat. *Indian Journal of Plant Physiology, 18*(2), 183–186.

Manisseri, C. & Gudipati, M. (2012). Prebiotic activity of purified xylobiose obtained from Ragi (*Eleusine coracana*, Indaf-15) Bran. *Indian Journal of Microbiology, 52*(2), 251–257.

Maphosa, Y. & Jideani, V. A. (2017). The role of legumes in human nutrition. *Functional Food-Improve Health through Adequate Food, 1*, 13.

Matucci, A., Veneri, G., Dalla Pellegrina, C., Zoccatelli, G., Vincenzi, S., Chignola, R., Peruffo, A. D. & Rizzi, C. (2004). Temperature-dependent decay of wheat germ agglutinin activity and its implications for food processing and analysis. *Food Control, 15*(5), 391–395.

Mohapatra, D., Patel, A. S., Kar, A., Deshpande, S. S. & Tripathi, M. K. (2019). Effect of different processing conditions on proximate composition, anti-oxidants, antinutrients and amino acid profile of grain sorghum. *Food Chemistry, 271*, 129–135.

Mollgaard, H. (1946). On phytic acid, its importance in metabolism and its enzymic cleavage in bread supplemented with calcium. *Biochem Journal, 40*, 589–603.

Monsen, E. R. (1988). Iron nutrition and absorption: dietary factors which impact iron bioavailability. *Journal of the American Dietetic Association, 88*(7), 786–790.

Moses, T., Papadopoulou, K. K. & Osbourn, A. (2014). Metabolic and functional diversity of saponins, biosynthetic intermediates and semi-synthetic derivatives. *Critical Reviews in Biochemistry and Molecular Biology, 49*(6), 439–462.

Mugford, S. T., Louveau, T., Melton, R., Qi, X., Bakht, S., Hill, L., . . . & Osbourn, A. (2013). Modularity of plant metabolic gene clusters: A trio of linked genes that are collectively required for acylation of triterpenes in oat. *The Plant Cell, 25*(3), 1078–1092.

Murtaugh, M. A., Jacobs, D. R., Jacob, B., Steffen, L. M. & Marquart, L. (2003). Epidemiological support for the protection of whole grains against diabetes. *Proceedings of the Nutrition Society, 62*(1), 143–149.

Mustafa, K. D. & Adem, E. (2014). Comparison of autoclave, microwave, IR and UV-stabilization of whole wheat flour branny fractions upon the nutritional properties of whole wheat bread. *Journal of Food Science and Technology, 51*(1), 59–66.

Nadeem, M., Anjum, F. M., Amir, R. M., Khan, M. R., Hussain, S. & Javed, M. S. (2010). An overview of anti-nutritional factors in cereal grains with special reference to wheat-A review. *Pakistan Journal of Food Sciences, 20*(1–4), 54–61.

Nappi, A. J. & Vass, E. (2002). Interactions of iron with reactive intermediates of oxygen and nitrogen. *Developmental Neuroscience, 24*(2–3), 134–142.

Nduti, N., McMillan, A., Seney, S., Sumarah, M., Njeru, P., Mwaniki, M. & Reid, G. (2016). Investigating probiotic yoghurt to reduce an aflatoxin B1 biomarker among school children in eastern Kenya: Preliminary study. *International Dairy Journal, 63*, 124–129.

Nikmaram, N., Leong, S. Y., Koubaa, M., Zhu, Z., Barba, F. J., Greiner, R. & Roohinejad, S. (2017). Effect of extrusion on the anti-nutritional factors of food products: An overview. *Food Control, 79*, 62–73.

Ogbonna, A. C., Abuajah, C. I., Ide, E. O. & Udofia, U. S. (2012). Effect of malting conditions on the nutritional and anti-nutritional factors of sorghum grist. *The Annals of the University Dunarea de Jos of Galati. Fascicle VI-Food Technology, 36*(2), 64–72.

Ogodo, A. C., Agwaranze, D. I., Aliba, N. V., Kalu, A. C. & Nwaneri, C. B. (2019). Fermentation by lactic acid bacteria consortium and its effect on anti-nutritional factors in maize flour. *Journal of Biological Sciences, 19*(1), 17–23.

Onwuka, G. I. (2006). Soaking, boiling and antinutritional factors in pigeon peas (*Cajanus cajan*) and cowpeas (*Vigna unguiculata*). *Journal of Food Processing and Preservation, 30*(5), 616–630.

Osman, A. M., Hassan, A. B., Osman, G. A., et al. (2014) Effects of gamma irradiation and/or cooking on nutritional quality of faba bean (*Vicia faba* L.) cultivars seeds. *Journal of Food Science and Technology, 51*(8), 1554–60.

Otlewski, J., Jelen, F., Zakrzewska, M. & Oleksy, A. (2005). The many faces of protease–protein inhibitor interaction. *The EMBO Journal, 24*(7), 1303–1310.

Pandey, V., Krishnan, V., Basak, N., Hada, A., Punjabi, M., Jolly, M. & Sachdev, A. (2016). Phytic acid dynamics during seed development and it's composition in yellow and black Indian soybean (*Glycine max* L.) genotypes through a modified extraction and HPLC method. *Journal of Plant Biochemistry and Biotechnology, 25*(4), 367–374.

Panzeri, D., Cassani, E., Doria, E. et al. (2011). A defective ABC transporter of the MRP family, responsible for the bean lpa1 mutation, affects the regulation of the phytic acid pathway, reduces seed myo-inositol and alters ABA sensitivity. *New Phytologist, 191*(1), 70–83.

Parker, M. L., Ng, A. & Waldron, K. W. (2005). The phenolic acid and polysaccharide composition of cell walls of bran layers of mature wheat (*Triticum aestivum* L. cv. Avalon) grains. *Journal of the Science of Food and Agriculture, 85*(15), 2539–2547.

Patterson, C. A., Curran, J. & Der, T. (2017). Effect of processing on antinutrient compounds in pulses. *Cereal Chemistry, 94*(1), 2–10.

Perera, I., Seneweera, S. & Hirotsu, N. (2018). Manipulating the phytic acid content of rice grain toward improving micronutrient bioavailability. *Rice, 11*(1), 1–13.

Petroski, W. & Minich, D. M. (2020). Is there such a thing as "anti-nutrients"? A narrative review of perceived problematic plant compounds. *Nutrients, 12*(10), 2929.

Popova, A. & Mihaylova, D. (2019). Anti-nutrients in plant-based foods: a review. *The Open Biotechnology Journal, 13*(1).

Qi, L. W., Wang, C. Z. & Yuan, C. S. (2011). Ginsenosides from American ginseng: chemical and pharmacological diversity. *Phytochemistry, 72*, 689–99.

Raboy, V. (2002). Progress in breeding low phytate crops. *The Journal of Nutrition, 132*(3), 503S–505S.

Raboy, V., Gerbasi, P. F., Young, K. A., Stoneberg, S. D., Pickett, S. G., Bauman, A. T., . . . & Ertl, D. S. (2000). Origin and seed phenotype of maize low phytic acid 1–1 and low phytic acid 2–1. *Plant Physiology, 124*(1), 355–368.

Rackis, J. & McGhee, J. (1975). Biological threshold levels of soybean trypsin inhibitors by rat bioassay. *Cereal Chemistry, 52*, 85–92.

Raes, K., Knockaert, D., Struijs, K. & Van Camp, J. (2014). Role of processing on bioaccessibility of minerals: influence of localization of minerals and anti-nutritional factors in the plant. *Trends in Food Science & Technology, 37*(1), 32–41.

Ragon, M., Aumelas, A., Chemardin, P., Galvez, S., Moulin, G. & Boze, H. (2008). Complete hydrolysis of myo-inositol hexakisphosphate by a novel phytase from *Debaryomyces castellii* CBS 2923. *Applied Microbiology and Biotechnology, 78*(1), 47–53.

Ram, S., Narwal, S., Gupta, O. P., Pandey, V. & Singh, G. P. (2020). Anti-nutritional factors and bioavailability: Approaches, challenges, and opportunities. *Wheat and Barley Grain Biofortification*, 101–128.

Ramawat, K. G., Dass, S., and Mathur, M. (2009). The chemical diversity of bioactive molecules and therapeutic potential of medicinal plants. In: *Herbal Drugs: Ethnomedicine to Modern Medicine* (pp. 7–32). Berlin and Heidelberg: Springer.

Rawat, N., Tiwari, V. K., Singh, N., Randhawa, G. S., Singh, K., Chhuneja, P. & Dhaliwal, H. S. (2009) Evaluation and utilization of *Aegilops* and wild *Triticum* species for enhancing iron and zinc content in wheat. *Genetic Resources and Crop Evolution, 56*, 53–64.

Ray, M., Ghosh, K., Singh, S. & Mondal, K. C. (2016). Folk to functional: An explorative overview of rice-based fermented foods and beverages in India. *Journal of Ethnic Foods, 3*(1), 5–18.

Ray, R. C. & Didier, M. (Eds.). (2014). *Microorganisms and Fermentation of Traditional Foods*. London: CRC Press.

Rehman, Z. U. & Shah, W. H. (2005). Thermal heat processing effects on anti-nutrients, protein and starch digestibility of food legumes. *Food Chemistry, 91*(2), 327–331.

Reig-Otero, Y., Mañes, J. & Manyes, L. (2018). Amylase–trypsin inhibitors in wheat and other cereals as potential activators of the effects of nonceliac gluten sensitivity. *Journal of Medicinal Food, 21*(3), 207–214.

Ribeiro, J. A. D. N. C., Serquiz, A. C., Silva, P. F. D. S., Barbosa, P. B. B. M., Sampaio, T. B. M., Araújo Junior, R. F. D. & Morais, A. H. D. A. (2015). Trypsin inhibitor from *Tamarindus indica* L. seeds reduces weight gain and food consumption and increases plasmatic cholecystokinin levels. *Clinics, 70*, 136–143.

Richard, J. L., Payne, G. A., Desjardins, A. E., Maragos, C., Norred, W. P. & Pestka, J. J. (2003). Mycotoxins: risks in plant, animal and human systems. *CAST Task Force Report*, 139, 101–103.

Salas, C. E., Dittz, D. & Torres, M. J. (2018). Plant proteolytic enzymes: Their role as natural pharmacophores. In *Biotechnological Applications of Plant Proteolytic Enzymes* (pp. 107–127). Cham: Springer.

Samia, M., AbdelRahaman, B. & Elfadil, E. (2005). Effect of malt pretreatment followed by fermentation on antinutritional factors and HCl extractability of minerals of pearl millet cultivars. *Journal of Food Technology, 3*(4), 529–534.

Samtiya, M., Aluko, R. E. & Dhewa, T. (2020). Plant food anti-nutritional factors and their reduction strategies: an overview. *Food Production, Processing and Nutrition*, 2, 6.

Sandstead, H. H. (2000). Causes of iron and zinc deficiencies and their effects on brain. *The Journal of Nutrition, 130*(2), 347S–349S.

Savage, G. & Klunklin, W. (2018). Oxalates are found in many different European and Asian foods-effects of cooking and processing. *Journal of Food Research, 7*(3), 76–81.

Savage, G. P. & Mårtensson, L. (2010). Comparison of the estimates of the oxalate content of taro leaves and corms and a selection of Indian vegetables following hot water, hot acid and in vitro extraction methods. *Journal of Food Composition and Analysis, 23*(1), 113–117.

Segueilha, L., Moulin, G. & Galzy, P. (1993) Reduction of phytate content in wheat bran and glandless cotton flour by *Schwan niomyces castelii*. *Journal of Agricultural and Food Chemistry, 41*, 2451–2454.

Serquiz, A. C., Machado, R. J., Serquiz, R. P., Lima, V. C., de Carvalho, F. M. C., Carneiro, M. A. & Morais, A. H. (2016). Supplementation with a new trypsin inhibitor from peanut is associated with reduced fasting glucose, weight control, and increased plasma CCK secretion in an animal model. *Journal of Enzyme Inhibition and Medicinal Chemistry, 31*(6), 1261–1269.

Shah, W. H. (2001). Tannin contents and protein digestibility of black grams (Vigna mungo) after soaking and cooking. *Plant Foods for Human Nutrition, 56*(3), 265–273.

Shahidi, F. (2004). Functional foods: Their role in health promotion and disease prevention. *Journal of Food Science, 69*(5), R146–R149.

Shi, J., Wang, H., Schellin, K. et al. (2007). Embryo-specific silencing of a transporter reduces phytic acid content of maize and soybean seeds. *Nature Biotechnology, 25*(8), 930–937.

Shi, J., Wang, H., Wu, Y. et al. (2003). The maize low-Phytic acid mutant lpa2 is caused by mutation in an inositol phosphate kinase gene. Plant Physiol. 131 (2): 507–515.

Shilpa, K. S. & Jyothi Lakshmi, A. (2012). Comparison of enhancement in bioaccessible iron and zinc in native and fortified high-phytate oilseed and cereal composites by activating endogenous phytase. *International Journal of Food Science & Technology, 47*(8), 1613–1619.

Shimelis, E. A. & Rakshit, S. K. (2007). Effect of processing on anti-nutrients and in vitro protein digestibility of kidney bean (*Phaseolus vulgaris* L.) varieties grown in East Africa. *Food Chemistry, 103*(1), 161–172.

Shivaraj, B. & Pattabiraman, T. N. (1981). Natural plant enzyme inhibitors. Characterization of an unusual α-amylase/trypsin inhibitor from ragi (Eleusine coracana Geartn.). *Biochemical Journal, 193*(1), 29–36.

Shukla, S., VanToai, T. T. & Pratt, R. C. (2004). Expression and nucleotide sequence of an INS (3) P1 synthase gene associated with low-phytate kernels in maize (*Zea mays* L.). *Journal of Agricultural and Food Chemistry, 52*(14), 4565–4570.

Siegenberg, D., Baynes, R. D., Bothwell, T. H., Macfarlane, B. J., Lamparelli, R. D., Car, N. G. & Mayet, F. (1991). Ascorbic acid prevents the dose-dependent inhibitory effects of polyphenols and phytates on nonheme-iron absorption. *The American Journal of Clinical Nutrition, 53*(2), 537–541.

Siener, R., Heynck, H. & Hesse, A. (2001). Calcium-binding capacities of different brans under simulated gastrointestinal pH conditions. *In vitro* study with 45Ca. *Journal of Agricultural and Food Chemistry, 49*(9), 4397–4401.

Siener, R., Hönow, R., Voss, S., Seidler, A. & Hesse, A. (2006). Oxalate content of cereals and cereal products. *Journal of Agricultural and Food Chemistry, 54*(8), 3008–3011.

Simwaka, J. E., Chamba, M. V. M., Huiming, Z., Masamba, K. G. & Luo, Y. (2017). Effect of fermentation on physicochemical and antinutritional factors of complementary foods from millet, sorghum, pumpkin and amaranth seed flours. *International Food Research Journal, 24*(5), 1869–1879.

Singh, B., Kunze, G. & Satyanarayana, T. (2011) Developments in biochemical aspects and biotechnological applications of microbial phytases. *Biotechnology and Molecular Biology Reviews, 6*, 69–87.

Singh, D. R., Singh, S., Salim, K. M. & Srivastava, R. C. (2012). Estimation of phyto-chemicals and antioxidant activity of underutilized fruits of Andaman Islands (India). *International Journal of Food Sciences and Nutrition, 63*(4), 446–452.

Singh, U., Praharaj, C. S., Singh, S. S. & Singh, N. P. (Eds.). (2016). *Biofortification of Food Crops*. New Delhi: Springer.

Sinha, K. & Khare, V. (2017). Review on: Antinutritional factors in vegetable crops. *The Pharma Innovation Journal, 6*(12), 353–358.

Soetan, K. O. (2008). Pharmacological and other beneficial effects of antinutritional factors in plants-A review. *African journal of Biotechnology, 7*(25), 4713–4721.

Soetan, K. O. & Oyewole, O. E. (2009). The need for adequate processing to reduce the anti-nutritional factors in plants used as human foods and animal feeds: A review. *African Journal of Food Science, 3*(9), 223–232.

Suma, P. & Urooj, A. (2014). Nutrients, anti-nutrients & bioaccessible mineral content (in vitro) of pearl millet as influenced by milling. *Journal of Food Science and Technology, 51*(4), 756–761.

Tajoddin, M., Manohar, S. & Lalitha, J. (2014). Effect of soaking and germination on poly-phenol content and polyphenol oxidase activity of mung bean (*Phaseolus aureus* L.) cultivars differing in seed color. *International Journal of Food Properties, 17*(4), 782–790.

Teucherl, B. & Cori, M. O. H. (2004). Enhancers of Iron Absorption. *International Journal for Vitamin and Nutrition Research, 74*(6), 403–419.

Tian, S., Sun, Y., Chen, Z., Yang, Y. & Wang, Y. (2019). Functional properties of polyphe-nols in grains and effects of physicochemical processing on polyphenols. *Journal of Food Quality, 2019*.

Till, B. J., Cooper, J., Tai, T. H., Colowit, P., Greene, E. A., Henikoff, S. & Comai, L. (2007). Discovery of chemically induced mutation in rice by TILLING. *BMC Plant Biology, 7*, 19.

Timotheo, C. A. & Lauer, C. M. (2018). Toxicity of vegetable tannin extract from Acacia mearnsii in *Saccharomyces cerevisiae*. *International Journal of Environmental Science and Technology, 15*(3), 659–664.

Torres, J., Rutherfurd, S. M., Muñoz, L. S., Peters, M. & Montoya, C. A. (2016). The impact of heating and soaking on the in vitro enzymatic hydrolysis of protein varies in different species of tropical legumes. *Food Chemistry, 194*, 377–382.

Udensi, E. A., Arisa, N. U. & Maduka, M. (2008). Effects of processing methods on the levels of some antinutritional factors in *Mucuna flagellipes*. *Nigerian Food Journal, 26*(2).

Vadivel, V. & Biesalski, H. K. (2012). Effect of certain indigenous processing methods on the bioactive compounds of ten different wild type legume grains. *Journal of Food science and Technology, 49*(6), 673–684.

Vadivel, V. & Pugalenthi, M. (2008). Removal of antinutritional/toxic substances and improvement in the protein digestibility of velvet bean (*Mucuna pruriens*) seeds during processing. *Journal of Food Science and Technology-Mysore, 45*(3), 242–246.

Vagadia, B. H., Vanga, S. K. & Raghavan, V. (2017). Inactivation methods of soybean trypsin inhibitor–A review. *Trends in Food Science & Technology, 64*, 115–125.

Van Der Ven, C., Matser, A. M. & Van den Berg, R. W. (2005). Inactivation of soybean trypsin inhibitors and lipoxygenase by high-pressure processing. *Journal of Agricultural and Food Chemistry, 53*(4), 1087–1092.

Vashishth, A., Ram, S. & Beniwal, V. (2017). Cereal phytases and their importance in improvement of micronutrients bioavailability. *3 Biotech, 7*(1), 1–7.

Vidal-Valverde, C., Frias, J., Estrella, I., Gorospe, M. J., Ruiz, R. & Bacon, J. (1994). Effect of processing on some antinutritional factors of lentils. *Journal of Agricultural and Food Chemistry, 42*(10), 2291–2295.

von Unruh, G. E., Voss, S., Sauerbruch, T. & Hesse, A. (2004). Dependence of oxalate absorption on the daily calcium intake. *Journal of the American Society of Nephrology, 15*, 1567–1573.

Welch, R. M. & Graham, R. D. (2004). Breeding for micronutrients in staple food crops from a human nutrition perspective. *Journal of Experimental Botany, 55*(396), 353–364.

Welch, R. M. & Graham, R. D. (2005). Agriculture: the real nexus for enhancing bioavailable micronutrients in food crops. *Journal of Trace Elements in Medicine and Biology, 18*(4), 299–307.

White Jr, J. F., Bacon, C. W., Hywel-Jones, N. L. & Spatafora, J. W. (Eds.). (2003). *Clavicipitalean fungi: Evolutionary Biology, Chemistry, Biocontrol and Cultural Impacts* (Vol. 19). London: CRC Press.

White, P. J. & Broadley, M. R. (2005). Biofortifying crops with essential mineral elements. *Trends in Plant Science, 10*(12), 586–593.

Yasmin, A., Zeb, A., Khalil, A. W., Paracha, G. M. U. D. & Khattak, A. B. (2008). Effect of processing on anti-nutritional factors of red kidney bean (*Phaseolus vulgaris*) grains. *Food and Bioprocess Technology, 1*(4), 415–419.

Zevallos, V. F., Raker, V., Tenzer, S., Jimenez-Calvente, C., Ashfaq-Khan, M., Rüssel, N. & Schuppan, D. (2017). Nutritional wheat amylase-trypsin inhibitors promote intestinal inflammation via activation of myeloid cells. *Gastroenterology, 152*(5), 1100–1113.

Zhang, Y., Liang, Z., Zong, Y., Wang, Y., Liu, J., Chen, K., et al. (2016). Efficient and transgene-free genome editing in wheat through transient expression of CRISPR/Cas9 DNA or RNA. *Nature Communications, 7*, 12617.

Ziegler, K., Neumann, J., Liu, F., Fröhlich-Nowoisky, J., Cremer, C., Saloga, J. . . . & Lucas, K. (2019). Nitration of wheat amylase trypsin inhibitors increases their innate and adaptive immunostimulatory potential in vitro. *Frontiers in Immunology, 9*, 3174.

7

Wheat Milling and Recent Processing Technologies

Effect on Nutritional Properties, Challenges, and Strategies

Manju Bala, Surya Tushir, Monika Garg, Maninder Meenu, Satveer Kaur, Saloni Sharma, and Sandeep Mann

CONTENTS

DOI: 10.1201/9781003307938-7

7.1 Introduction

Wheat (*Triticum aestivum* L.) is a major primary cereal crop and a global staple food. It offers the calories and nutrients required for human growth and development (Bisht, 2021). Its milling is required to make products like bread, chapati (Indian flatbread), cookies, spaghetti, biscuits, and so forth. Majority of the wheat in India is used for making food products like grits, suji (semolina), *maida* (refined white flour), and *atta* (whole-wheat flour) (Kaukab et al., 2022). The wheat grain, also recognised as caryopsis, is of oval shape and is made up of 80–85% endosperm, 2–3% wheat germ, and 13–17% bran (Belderok et al., 2000).

All the three components have distinct physical characteristics and biochemical compositions. The bran contains a significant amount of fibre and several cell layers. The bran, which is the wheat grain's outermost structural layer and is made up of the transparent aleurone layer, the outer pericarp, and the inner pericarp, is depicted in Figure 7.1.

Furthermore, Table 7.1 represents the composition of whole wheat grain and wheat bran. Wheat grains consist of protein, carbohydrates, ash, and fat content of 12%, 70%, 1.5%, and 2%, respectively (Kumari et al., 2020). Wheat bran contains a protein content of 14%, a carbohydrate content of 27%, an ash level of 5%, and a fat content of 6% (Haque et al., 2002). The majority of the weight of wheat bran is made up of the aleurone layer, which divides the endosperm from the bran layers (48–52%). It is rich source of vitamins, proteins, and phytonutrients *viz.* ferulic acid. The tube cells, cross cells, testa, and nucellar epidermis together constitute the intermediate layer (Dornez

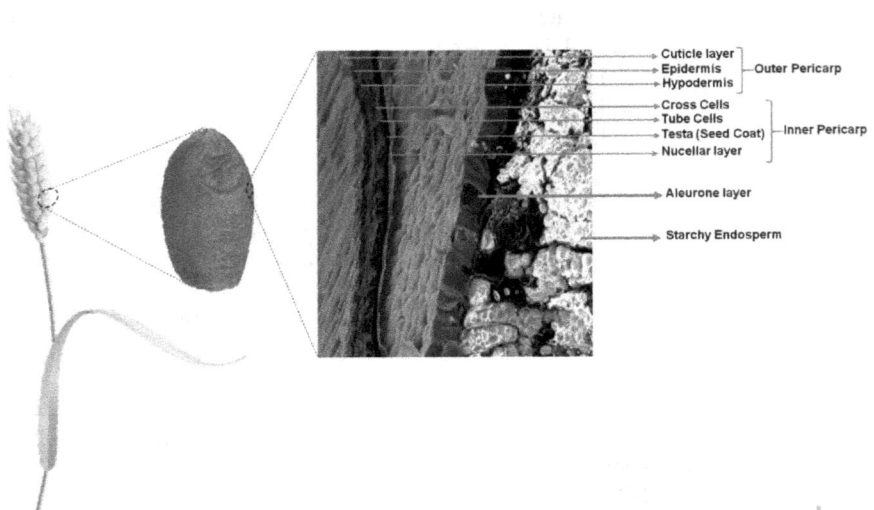

FIGURE 7.1   Structure of wheat grain and SEM showing different parts of wheat grain.

TABLE 7.1

Composition of Whole Wheat Grain, Wheat Germ, and Wheat Bran

Components	Composition				
	Whole Wheat Grain	Wheat Germ	Wheat Bran		
Moisture	12%	10.4%	8.1–12.7	-	14.6
Protein	9–10%	25%	9.9–18.6	16.5	16.6
Fat	2%	10.4%	5–6.3	4.95	4.48
Fibre	9%	10–14%	36.0–63.0	43.2	40.0
Starch	60–70%	14%	21.1–38.9	11.8	18.7
Ash	2%	4.5%	5.7–6.5	5.50	4.87
Carbohydrates	60–70%	50–51%	–	–	–
Reference	Kumari et al., 2020	Ghafoor et al., 2017	Curti et al., 2013	Yan et al., 2015	Oghbaei and Prakash, 2013

et al., 2011). The outer layer (exocarp, mesocarp) makes up around 30% of the weight of bran, and the middle layer makes up about 20% (Bao et al., 2014).

The germ part of wheat is rich source of lipids, vitamins, minerals, fibre, and phytonutrients like flavonoids. The endosperm part comprises low levels of vitamins, fibre, minerals, lipids, pigments, and phytonutrients, as well as a starch composition of roughly 75%, a protein content of 10%–14%, and other nutrients (Borneo and León, 2012).

Wheat grain generally is milled to create flour, which is then used in the preparation of different food products. Knowing the physical properties of wheat may help in understanding food processing as well as foretelling the characteristic and performance of produce during post-harvest operations (Nesvadba et al., 2004). Wheat milling is influenced by a variety of physical qualities of wheat grains, including thousand seed weight, kernel vitreousness, size, shape, colour, and hardness. The milling processes have an effect on the kernel components in the flour, which have an effect on the nutritional content of the flour. Commercial milling generally uses the endosperm of the wheat kernel to produce refined flour, along with a small quantity of germ and bran. The technological and nutritional qualities of the finished end products are greatly influenced by the flour type and particle size. The flour characteristics, in turn, depend on the part of wheat kernel utilised and determine the finished products' coarseness, texture, and colour. The majority of customers like the taste, texture, and the appearance of snacks and bakery items prepared with refined wheat flour (Boz and Murat Karaoğlu, 2013) in comparison to items prepared with whole-grain wheat flour. However, refined wheat products don't have the same nutritional advantages as whole-grain wheat products. Whole-wheat flour is nowadays gaining importance because of associated health benefits. To manufacture whole-wheat flour, whole wheat grain must be ground. Using a traditional stone mill (*chakki*), whole-wheat flour can be produced without separating the bran and germ layers. Other milling methods are being employed to obtain high-quality whole-grain flour. Whole grains contain phenolic compounds, minerals, dietary fibre, and vitamins, as well as the resultant flour constituents (Weaver, 2001). Along with health benefits, there are some challenges with storage and product development from whole-wheat flour. Different researchers have suggested strategies to cope with these challenges. Therefore, this chapter will provide an insight into physico-chemical characteristics of wheat kernel, various technologies related with wheat milling, how they impact the nutritional value of flour, the usage of whole-wheat flour, and challenges associated and strategies to cope up with those challenges.

7.2 Physico-Chemical Characteristics of Wheat Kernel

7.2.1 Physical Characteristics of Wheat Kernel

A variety of physical and compositional characteristics that define grain quality are considered by threshold standards established in accordance with end-use requirements. Understanding food processing and anticipating the quality and behaviour of produce during post-harvest operations are both aided by the knowledge of the physical qualities of wheat (Nesvadba et al., 2004). Shape, size, volume, density, surface area, and coefficient of friction of grain (wheat) are crucial in designing of equipment and structures, in analysing and determining efficacy of an operation or a machine, and for determining and maintaining the quality of the finished product (Gupta et al., 2021; Mohsenin, 1980). The physical features of grains strongly influence during design of machineries for handling, aeration, drying, storing, dehulling, and processing. Physical properties include the length, width, and thickness of each grain variant as well as their bulk density, actual density, projected area, terminal velocity, and drag coefficient. Geometric aspects, like size and form, are among the most significant physical characteristics taken into consideration for separation and cleaning of wheat kernels and are also connected to its quality. Due of their irregular shapes, grains are compared to spheres or ellipses (Gürsoy and Güzel, 2010). Bulk and true density are suitable for storage facilities as they influence how rapidly mass and heat are moved away from moisture during the aeration and drying processes (Al-Mahasneh and Rababah, 2007). In addition to the aforementioned parameters, porosity can be used to size grain hoppers and storage facilities (Bhise et al., 2014).

The grain moisture content, which also has a substantial impact on the physical properties linked to volumetric grain weight and bulk density, alters the surface characteristics of seed coat and kernel endosperm. Following their investigation on how moisture content affected several physical traits of wheat kernels, researchers came to the following conclusions: with increasing moisture content, thousand kernel weight, axial dimensions, kernel volume, porosity, and sphericity all increase, but bulk density decreases (Tabatabaeefar, 2003; Ponce-García et al., 2008; Karimi et al., 2009; Bhise et al., 2014). Since higher grain moisture content makes grains more prone to deformation, physical characteristics of cereal change with change in grain moisture content (Molenda and Horabik, 2005). Babić et al. (2011) evaluated different physical traits of one soft and two hard wheat varieties in relation to moisture content. The average kernel lengths for the Dragana, Simonida, and NS 40S cultivars were found to be 2.56, 5.46, and 2.12, respectively. Simonida's bulk density was 731.77 kg/m^3, Dragana's was 788.51 kg/m^3, and NS 40S' was 791.34 kg/m^3. The bulk densities of the investigated kernels were equally modest, in relation to their length, breadth, and thickness.

The mechanical and physical properties of roasted Zerun wheat with moisture contents varying from 8.80 to 23.40% on a wet basis revealed that angle of repose, length, width, and thickness all rose nonlinearly from 33.02 to 37.90°, 2.66 to 2.78 mm, 6.09 to 6.36 mm, 4.17 to 4.18 mm, respectively, as moisture content increased (Işıklı et al., 2014). Researchers found that green wheat kernels with moisture contents ranging from 9.3 to 41.5% had similar physical characteristics (Al-Mahasneh and Rababah, 2007). They discovered that the axial dimensions, kernel volume, static friction coefficient, surface area, and mass of 1,000 seed kernels rose when the moisture content increased from 46.65 to 45.59 (Al-Mahasneh and Rababah, 2007).

Test weight and semolina yield are related to the 1,000 kernel weight of durum wheat. The permitted weight of 1,000 kernels of durum wheat is between 35 and 40 g (Abaye et al., 1997). El-Khayat et al., 2006 assessed the physical traits of nine durum wheat genotypes to examine the relationships between kernel constituents and their quality and reported that the studied genotypes had 1,000-kernel weights (42.5–55.5 g) and high test weights (83.1–85.9 kg/hl), suggesting their good milling potential.

Vitreousness is natural kernel translucence and generally employed to describe the appearance of wheat kernel. All grains density below 1,400 kg m^{-3} were entirely mealy (non-vitreous), whereas all grains beyond this density were entirely vitreous (Dobraszczyk et al., 2002). It has been found that during tempering, denser kernels are substantially more water-resistant. Michniewicz et al. (2000) demonstrated the impact of vitreousness on break flour and size flour (r = 0.79 and 0.98, respectively).

7.2.2 Effect of Physical Characteristics on Wheat Milling and Final Product

Numerous studies on this subject have been done since the beginning of the grain-processing business. A variety of physical factors influence wheat-processing quality (Dziki and Laskowski, 2005). Physical properties of wheat grains which affect wheat milling include thousand seed weight, kernel vitreousness, size, shape, colour, and hardness. The amount of flour that may be produced from a grain is referred to as milling yield, and it depends on the mature kernel's endosperm content (75–83%) in relation to components such the seed coat (bran), aleurone, embryo (germ), and endosperm (Hammermeister, 2008). The milling yield is affected by kernel morphology, grain size, grain texture or hardness (Hrušková and Švec, 2009), test weight, shape of germ, crease, and brush (Dholakia et al., 2003).

Wheat milling and baking quality is greatly influenced by the kernel's texture. Based upon kernel hardness, wheat is classified into soft, mixed, and hard varieties. Wheat kernels that are soft are easily broken and produce fine flour with lower degradation of starch granules, while hard wheat produces rough textured flour with high damage to starch granules (Emanuelson et al., 2003; Pasha et al., 2010).

The texture, hardness, and quality of wheat grain are controlled by a special starch granule protein called friabilin. Friabilin contains two lipid-binding puroindoline proteins, PINa and PINb, which modulate the grain softness. The functionality of PIN proteins significantly affects the processing quality of wheat (Day et al., 2006; Feiz et al., 2008). The durum wheat which lacks PIN proteins is categorised into hard wheat (Morris, 2002). Wheat hardness is a factor that will significantly affect milling, so it is crucial to understand before milling. The kernels of hard wheat cultivars require more effort to crush than soft wheat cultivars (Dziki and Laskowski, 2000).

The 'test weight' is the weight of a known volume of grain represented in kilograms per hectolitre, which reveals grain bulk density. Test weight or density of wheat kernels is a physical attribute that is typically taken into consideration by flour and semolina millers. The wheat kernel mass is measured by the thousand kernel weight (TKW), a significant consideration in selecting cultivars with the best seed in terms of its physiological and physical quality. The higher TKW readings often have a positive correlation with potential yield or extraction of flour (Gutiérrez-García et al., 2006) as this trait is closely linked to endosperm to germ ratio, grain size, and the pericarp tissues (Saldivar, 2010). This method is applied by the wheat breeders and flour millers in conjunction with test weight to determine the makeup of wheat kernels and potential flour extraction (U.S. Wheat Associates, 2007). Higher test weights suggest bigger, better grain quality, while lower test weights are related with either one undersized and/or weather-damaged grain or pinched kernels with high level of moisture content (Nuttall et al., 2017).

It has been mentioned that the shallow creased kernels outperformed small, shrivelled grains with a higher ratio of bran to endosperm in terms of milling yield (Mabille and Abecassis, 2003; Marshall et al., 1984). On the other hand, Gaines et al. (1997) claimed that, aside from the fact that small kernels tended to be softer, the size of the kernel had little bearing on the characteristics of the kernel, milling performance, or the end-use attributes of soft wheat. It was shown that the milling and baking quality of small kernels were comparable to those of their larger counterparts. However, according to Morgan et al. (2000), a reduction in wheat kernel size had a detrimental effect on yield and refinement of flour (ash and colour).

Colour is a significant component that influences grain quality. Wheat caryopsis can range in colour from yellow to red-brown or light buff depending on whether or not this layer has red pigmentation (Bechtel et al., 2009). The unique grey staining mildew growth on wheat kernels causes to lower grain quality (Shahin et al., 2013) and has a negative impact on the colour of refined flours. According to El-Naggar and Mikhaiel (2011), one of the most crucial variables is the assessment of insect damage because it affects wheat milling output and colour and raises the quantity of insect pieces discovered in flours. It is generally well-known about the quality of milling and baking that it is negatively impacted by the effective enzymes generated by pests and with the respiration system of grain (Brabec et al., 2015; Singh et al., 2010). Furthermore, sprouting can happen both within the field and during storage where kernels are exposed to the appropriate temperature and moisture absorption conditions. Sprouted kernels often have darker colours because they contain substantial amounts of damaged proteins and reducing sugars that, when cooked, produce more Maillard reaction products (Saldivar, 2010). The extent of sprouted damage is frequently measured by determining the amount of amylase by means of Falling Number (FN) (Pagani et al., 2014) or else by calculating peak viscosity using Rapid Visco Analyzer (RVA) (Xing et al., 2010).

7.2.3 Chemical Properties of Wheat Kernel

Mature wheat grains contain a fairly small range of chemical compositions, with water content less than 13%. Carbohydrates, specifically starch (58%) and non-starch polysaccharides (13%), are the two largest constituent types, followed by proteins (11%). Lipids (less than 2%) and minerals (less than 2%) are essentially auxiliary components. Because of their biological significance, vitamins and phytochemicals are significant even though they are present in extremely little amounts (less than 0.1%) (Kumari et al.,

2020). Amylose and amylopectin are present, which are the two main components found in starch. Among the non-starch polysaccharides are the cell wall polysaccharides such as arabinoxylans, cellulose, glucans, fructans, and arabinogalactan peptides (dietary fibre) (Török et al., 2019). Grain proteins can be categorised into four classes based on their diverse roles: storage (gluten) proteins, metabolic proteins, protective proteins, and other proteins with unique specific functions (Wieser et al., 2020). The lipid fraction is primarily composed of three chemical classes: nonpolar lipids, phospholipids, and glycolipids. The principal minerals are potassium, phosphorus, magnesium, and calcium, with zinc, manganese, and iron being minor minerals. Wheat grains are rich in vitamin E (α-tocotrienol, α-tocopherol, β-tocopherol, and β-tocotrienol) and the B vitamins (nicotinamide, pantothenic acid, thiamine, pyridoxine, and riboflavin) (Pedersen et al., 1989). Carotenoids, flavonoids, terpenoids, and phenolic compounds are the four main families of phytochemicals (Garg et al., 2016; Garg et al., 2022; Kapoor et al., 2022).

7.2.4 Effect of Chemical Characteristics on Wheat Milling and Final Product

Wheat is milled differently depending on its starch composition, gluten content, protein content, SDS-sedimentation value, and other chemical properties. Wheat protein is a significant component that has an influence on end-use wheat quality. During milling, soft wheat ruptures into considerably smaller particles in comparison to hard wheat, and thus it gives greater 'break flour yield' upon milling. Endosperm percentage obtained during the milling process without crushing or reduction is known as break flour. Weak flour's gluten proteins are fewer and of lower quality, which prevents them from forming a continuous matrix when coupled with water.

Dough that is extensible and crumbly but lacking in strength and elasticity is encouraged when conditions for manufacturing soft wheat goods' dough are met, such as adding more sugar and less water. Dough with these characteristics expands more when baked at a higher temperature, producing more biscuits from given dough (Cauvain and Young, 2006; Edmund and Perry, 2008; Kim and Walker, 1992). Higher content of gluten proteins is undesirable for making cake because they prevent dough from spreading and make it more difficult to mould dough into exact dimensions and shapes.

Soft wheat with low-protein and low gluten and starch damage is considered superior for making cookies, biscuits, pastries, and cakes (8–10% protein content) (Pasha et al., 2010). Products developed from soft wheat produce enhanced volume and soft texture than those prepared from hard wheat.

Soft-to-medium hard wheat with medium protein content (9–11.5%) are preferred for puddings, crackers, and thickeners. Medium hard wheat with medium protein content (10.5–12.5%) is preferred for Chinese and Japanese noodles. Medium-to-hard wheat flours with medium protein content (8–12%) are preferred to produce whole wheat unfermented flat bread (chapatti).

Hard wheat with a greater protein and gluten level, as shown in Figure 7.2, is suitable for making high-quality white bread (>12% protein content). The dough is more elastic and stronger after manufacturing of sheet due to higher concentrations of gluten proteins. The glutenin proteins are responsible for the firmness and pliability of gluten or dough. Such proteins promote retention of gas, which raises the volume of baked goods. Hard wheat flours with more than 14% protein are used to make whole-wheat bread and high protein flour (Dziki and Laskowski, 2005; Pena, 2002).

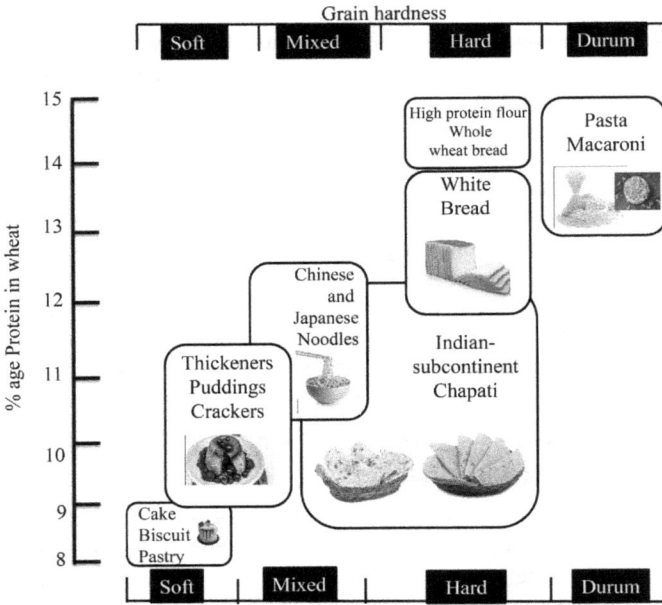

FIGURE 7.2   The suitability of different type of wheat for product development.

Durum is the hardest variety of wheat, and semolina made from it is used to make pasta. Due to its hardness and high protein content, durum wheat stands out. Pasta made from durum wheat makes final products with outstanding qualities for cooking and resistance for overcooking (Sissons, 2008). Compared to hard wheat, soft wheat is utilised to create a wide range of products. For the purpose of creating biscuits, it is generally recommended to use soft wheat flours with low damaged starch levels (1.9–3.4%), moderate SDSS volumes (20.0–32.0 mL), and intermediate protein contents (7.9–9.7%) (Dziki and Laskowski, 2005).

Dough that is extensible and crumbly but lacking in strength and elasticity is encouraged when conditions for manufacturing soft wheat goods' dough are met, such as adding more sugar and less water. Dough with these characteristics expands more when baked at a higher temperature, producing more biscuits from given dough (Cauvain and Young, 2006; Edmund and Perry, 2008; Kim and Walker, 1992). Higher content of gluten proteins is undesirable for making cake because they prevent dough from spreading and make it more difficult to mould dough into exact dimensions and shapes.

7.3 Existing Milling Technologies and Impact on Nutrients

7.3.1 Milling Technologies

The milling process is particularly crucial for the processing of wheat (Bushuk, 1998). Wheat milling is used to produce flour and other products from wheat grains. Before milling, wheat grain is analysed for the gluten content and amylase activity. In order to get flour of similar consistency, the miller may employ blending of different wheat types together, and this process is known as gristing. Once the grain is analysed and found good for milling, it is further processed using following three steps:

1. **Cleaning** – Wheat is cleaned to get rid of all the impurities.
2. **Conditioning or Tempering** – Conditioning or tempering of wheat grains is done by using appropriate moisture and temperature and time.
3. **Grinding** – It is done to achieve flour of various categories through different milling approaches.

7.3.1.1 Cleaning of Wheat

Wheat received at the mill may contain weeds, seeds from other plants, chaff, stones, and other foreign things. Before milling, it must be cleaned. Contaminants and a number of wheat physical features (weight, size, and density) affect how wheat is cleaned. Cleaning wheat helps remove numerous pollutants. Figure 7.3 represents a flow chart depicting the names of various equipment required for cleaning process.

The equipment and processes employed for cleaning are shown in the next sections.

7.3.1.1.1 Separator

The vibrating screens of separator remove bits of wood and straw and other items which are too larger or too smaller than wheat.

7.3.1.1.2 Aspirator

It makes use of air currents. The foreign matter or impurities which are lighter than wheat are sucked up and removed by strong drafts of air from the aspirator. On an oscillating

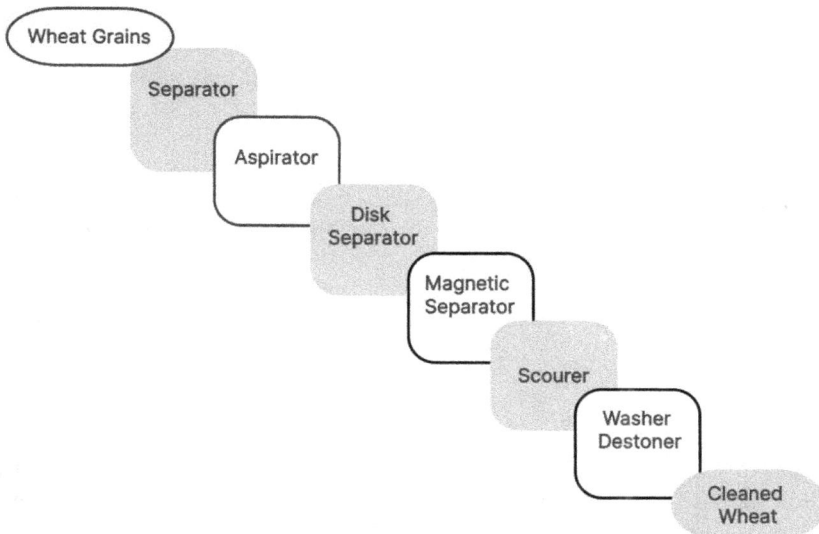

FIGURE 7.3 The whole wheat cleaning process.

board enclosed in woven wire fabric, the air is pumped through a bed of wheat. A separation is achieved using the difference between surface friction and specific gravity.

7.3.1.1.3 Disc Separator

It moves the wheat over a number of discs with groves that trap wheat grain-sized particles of wheat. As objects go across the discs, smaller or larger items are eliminated. Oats, barley, and other foreign ingredients are eliminated.

7.3.1.1.4 Magnetic Separator

Metal impurities (iron or steel) are removed from wheat with the help of magnets.

7.3.1.1.5 Scourer

It removes dirt, hair, and other debris by rubbing ferociously with the beaters. Wheat is vigorously pushed against perforated steel casings by metal beaters used in scourers. As a result, the majority of the dirt lodged in the wrinkle of the wheat grain is removed and transported by a strong air blast.

7.3.1.1.6 Washer Destoner

High-speed spinning of the wheat occurs in the water bath. A surplus of water is thrown away due to centrifugal force. When stones reach the bottom, they are removed. Only the pure wheat remains after lighter components have drifted away.

7.3.1.2 Conditioning/Tempering of Wheat

After cleaning, wheat is conditioned/tempered before milling. Conditioning makes the separation of kernel easy. For this purpose, water is added to kernel to attain desired moisture content. Conditioning toughens the bran layers and causes softening of the endosperm and, therefore, facilitates wheat milling. Wheat is tempered by placing in bins for 18–72 hours under optimised moisture and temperature conditions. (Liu et al., 2015) reported tempering for 24 hours for hard, medium-hard, and soft wheat at 16%, 15% and 14% moisture contents, respectively. It aids in separating the endosperm from the bran layer and enables complete moisture absorption into the wheat kernel. Moisture can be added at room temperature or at high temperatures (up to 47°C). Preheating, moistening, and cooling are the three stages of the conditioning process in modern mills. Another approach for conditioning is direct steaming. Heating and moistening are done simultaneously in one operation. The temperature of the grain reaches 47°C in around 20 to 30 seconds. The connection between various grain components is broken and made easier to separate by rapid heating. Furthermore, proteins and enzymes show their activities more effectively.

7.3.1.3 Grinding the Wheat

For grinding wheat, different milling technologies can be used. As per Liu et al. (2015), stone milling (SM), roller milling (RM), ultrafine milling (UM), and hammer milling (HM) are the four major milling methods. However, new technologies are also evolving, and jet milling is an example.

7.3.1.3.1 Stone Milling

In this milling process, the original stone abrasion mills for grinding wheat into flour are employed. This is thought of as a standard milling method. These mills employ a range of physical forces, like abrasion, compression, and shear. Stone mills employ chiselled emery stone. A theoretical extraction rate of 100% is suggested for these mills (Kihlberg et al., 2004). Stone milling is very simple, fast, and the easiest method to make flour from whole wheat (Zhang et al., 2018). It also known as single stream milling. During stone milling, wheat is crumpled between the two stones (Figure 7.4),

FIGURE 7.4   Traditional stone mill.

and as a result whole-wheat flour with bran, endosperm, middlings, and germ fractions is obtained.

In modern systems, composite millstones, connected with metal plates, are being used (Doblado-Maldonado et al., 2012). One of the plates is fixed, while the other plate rotates. Prabhasankar and Rao (2001) used plate mill, in which plates have about 120 corrugations of 3–4 mm in depth. Electric motors (e.g. 7.5 bhp motor for wheat grinding capacity of 100 kg/h) are used to power these mills. The clean grain is fed into the gap between the two plates using a hopper, where it is sheared and ground into flour by frictional forces. The flour enters through the exit at the bottom of the plate. The stone or plate mill is also known as *chakki*.

Due to the stone mill's ability to generate whole-wheat flour and consumers' growing knowledge of the health advantages of whole wheat, flour making by stone milling has recently undergone a renewal. Further, small entrepreneurs, householders, and food artisans prefer its use due to its simple and cheap process, no grain conditioning requirements (Posner and Hibbs, 2005), and relatively low capital inputs. The majority of the flour produced by stone milling produced particle size less than 85 μm, and about 35% particles were in the size range of >85 μm to 363 μm (Carcea et al., 2020). Irrespective of the wheat type used, stone-milling resulted in a high number of particles of smaller size range (<85 μm).

To generate refined white flour, the resulting whole-wheat flour is next sifted or sieved. Centrifugal sifters or plansifters are used for this in tiny/medium-sized, or larger facilities, respectively. The cylindrical sieve of the centrifugal sifter revolves, and an internal paddle turns counterclockwise. When the product is dropped against sieve by paddle, the fine fraction is forced all through, even though the coarse fraction tails off (Posner, 2003). Three distinct fractions are accumulated in end of the process: refined white flour, bran, and middlings. Due to friction, stone mills generate a lot of heat. It has been asserted that flour ground in plate and stone mills develops at a greater temperature (90–98°C) than flour ground in other types of mills. This can significantly impair protein, unsaturated fatty acids, and starch content compared to other grinding methods (Prabhasankar and Rao, 2001).

7.3.1.3.2 Roller Milling (Modern Flour Milling)

The milling process leads to a slow reduction of the wheat kernels to make flour. Germ is rich in fat content, and if it is not removed, whole-wheat flour is more prone to rancidity; hence, it is removed during milling. Bran is usually dark in colour and, if crushed, imparts dark colour to resulting flour and, hence, is removed during milling. In order to get the most of white flour from wheat endosperm while excluding the bran and germ, roller milling is commonly utilised (Campbell, 2007). Roller milling creates granules of different sizes that represent the bran and endosperms, which are then, further, separated using sifters and purifiers. Using a series of reduction rollers, the endosperm is ground into flour. Roller milling turns kernels into refined flour, which makes up the majority of the product (72%) and the coarse fraction (28%), which is composed of bran (12%), germ (3%), and residual endosperm (15%). Sifters are positioned in between the rolls, used to prepare the wheat (Doblado-Maldonado et al., 2012). Flat belts are used to power rolls, and their speed can be adjusted. Depending on the stage of the procedure, rolls can be either smooth or fluted. Wheat grains are fed by two feed wheels into machine (Meuser, 2003). Wheat is ground between two rolls that are positioned horizontally and rotate in opposition to one another (Figure 7.5).

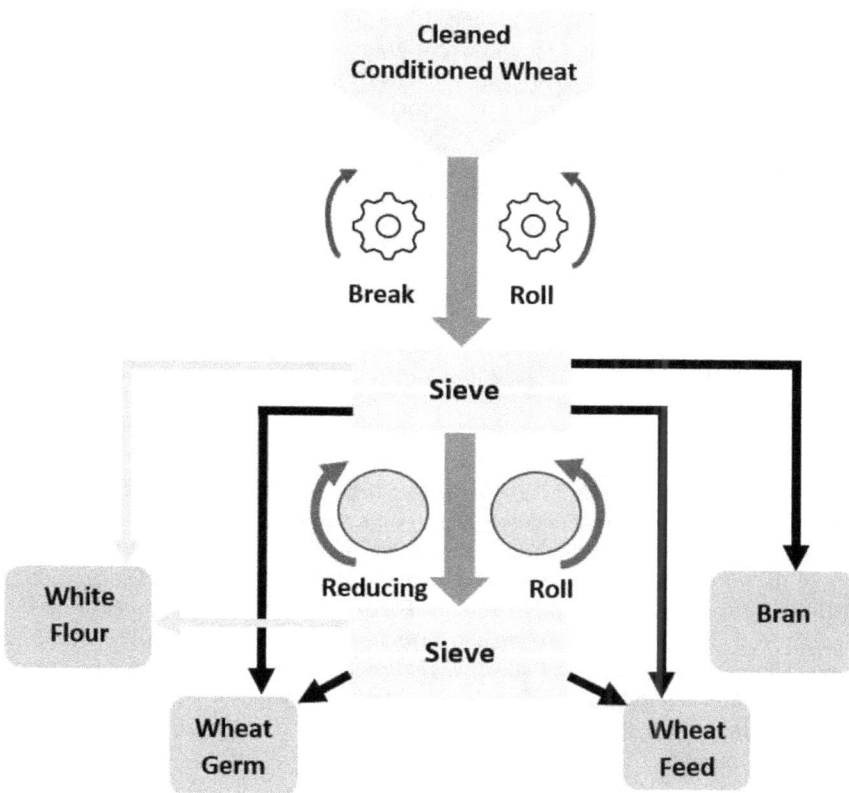

FIGURE 7.5 Modern Flour milling process.

Commercial rollers typically range from 180 to 350 mm in diameter and up to 1,500 mm in length. Smaller-diameter rollers are intended for shear processes, and tear apart the endosperm and bran, while, larger-diameter rolls work through compression. However, the majority of modern roller mills utilise rollers of same length and diameter so as to attain maximum uniformity and to do away with the need to maintain an inventory of several types of replacement rollers (Posner, 2003). High pressure during milling results in a temperature increase. Running cold water through hollow centre of rollers can fix this problem. Additionally, the processed material may adhere to the surface of the roller, blocking the flutes of corrugated rolls and causing an uneven expansion of effective diameter of the smooth rollers. Corrugated rollers cannot be used to remove the stuck-on flour; only the smooth rollers need to be scraped (Rosentrater and Evers, 2017). According to Rosentrater and Evers (2017), roller milling can be described in four phases as is described next.

The endosperm is separated from the germ and bran during the first phase, also identified as the break stage (Rosentrater and Evers, 2017). The basic concept is separation of wheat kernel and scraping of endosperm to separate the bran part (Campbell et al., 2012; Posner, 2003). The system comprises four break rolls and sieves. If the number of break rolls used is eight, then sieves are not employed. Flutes (3.5–1 cm) of the first break rolls shear open the wheat grain. Overtails are fed into the second break of the roller mill (Posner, 2003), which has less coarsely fluted rolls (5.5–1 cm). As grains migrate from the first to the second break and so on, the space between rolls gets narrower. The next sieving stage transfers the large particles to the third break, where they are grounded by smaller gap, finer-fluted rollers before moving on to the fourth break. Three groups are formed from the output of the breaker rollers after they have passed through metal sieves and an air current. The purest material resembles coarse flour and is referred to as middlings or farina. The larger interior portions are referred as semolina. The third category includes interior components that are still connected to the bran. During this process, most of the endosperm is separated from the germ and bran (Campbell et al., 2012). Endosperm that has been acquired in this manner is dispersed into large bits and introduced into the purification and reduction mechanism.

During second phase, in some milling systems, sizing rollers, which are small, fluted rollers, can be used to treat the widest endosperm particles. It is intended for scraping of bran from endosperm to easily separate the two components before continuing with the reduction of endosperm (Rosentrater and Evers, 2017).

The third phase refers to the use of reduction rollers. Reduction rollers reduce the size of the particles in coarse flour to produce fine and smooth flour (Posner, 2003). Milling is done with smooth metal rollers. These rollers differ from break rollers in that they roll slowly at 500–550 rpm and have a smooth surface. The degree to which the two forces of crushing and shearing the grain are balanced in this mechanism depends on how flat the rolls surface is (Rosentrater and Evers, 2017).

The fourth and last tailing system comes into force. During this, the residual heterogeneous mixture from every stage is transported to plansifter, a sifting apparatus (Campbell et al., 2012; Posner, 2003). The container is a sealed system and can accommodate up to ten stacks of 32 sieves (compartments). Depending on the size of the raw material to be sieved and the industry requirement, the number of frames may vary. Metal wire, nylon, and silk sieves are used to screen coarse and fine flour. Sifting, separating, and regrinding the flour can produce numerous distinct grades of flour at once. They combine 6 to 12 milling cycles with sifters to remove the flour generated by the previous grind (Campbell et al., 2012; Rosentrater and Evers, 2017).

The selection of suitable corrugations and roller speed in roller mill leads to the reduction of shear and compressive forces during grinding, causing minimal damage to nutritive value of the resulting flour (Prabhasankar and Rao, 2001). The roller mills are considered more economical, more productive, and more efficient and give persistent results with time (Posner and Hibbs, 2005; Ross and Kongraksawech, 2018). Because the endosperm fraction is separated from germ fraction and coarse bran by the roller mill, the milling process may be successfully handled to produce flour that satisfies the demands of different consumers. As a result, wheat flour that has been refined and stripped of its bran and germ is produced (Campbell et al., 1991). Most of the commercial flour is currently made in roller mills. Typically, to make whole-wheat flour, coarse bran is micronised and re-mixed with endosperm (Tian et al., 2020). To avoid the rancidity problem in wheat flour, germ part being rich in fat is usually not included (Campbell et al., 1991). Flour particle size distribution in roller milled flour revealed the particles, which are more than 363 µm and in range of 129 to 85 µm, together represent 55% of the flour (Carcea et al., 2020). The roller-milled flours, however, displayed bimodal curve with bigger particle size distribution (>1,000 μm) and fine particles (<250 μm) (Pagani et al., 2020).

7.3.1.3.3 Hammer Milling

Another type of milling is hammer milling, which is among the oldest and widely used mills. It is composed of many rectangular hardened steel pieces, or hammers, each of which is housed in a sturdy metal container or drum and connected to a central shaft. Figure 7.6 represents schematic diagram of hammer mill. There are sometimes four or more hammers in a drum. It works on the principle of impact; when wheat grains are hit with a hammer against a wall, the particle size is decreased (Posner, 2009).

   The hammers spin rapidly inside the chamber. These hammers make a dynamic contact with the wheat grains while turning quickly. The grain is then in touch with a moveable screen that has a "tight tolerance" gap between it and the rotating hammers.

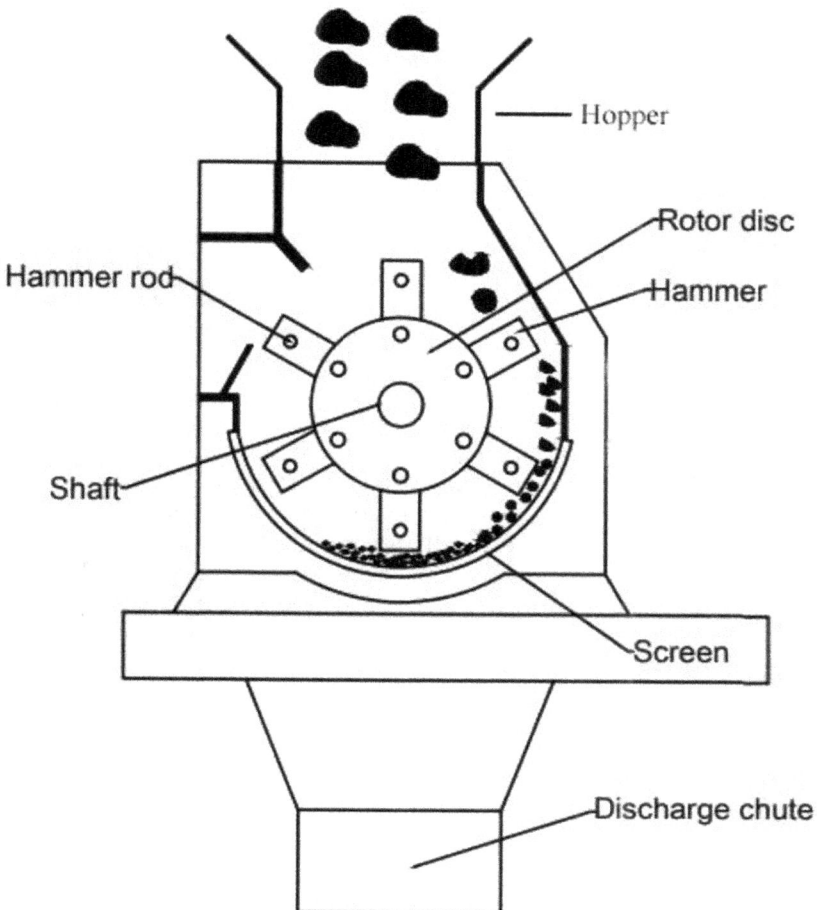

FIGURE 7.6   Schematic diagram of hammer mill.

Sizing by abrasion and shear occurs in the second zone, where there is minimal room between the hammer and the screen bar, producing particles with a range of 15–50 μm. The type of hammers utilised, the rotor speed, and the size and shape of the screen opening are three essential variables that all affect the grinding process. Prabhasankar and Rao (2001) ground two kinds of commercially available wheat in a hammer mill with an 800 μm sieve size. Hammer mills have higher capital cost than other mills; however, they require less horsepower per ton. Whole-grain flours might theoretically be made using hammer mill or impact, although not often done (Doblado-Maldonado et al., 2012). But nowadays, with increasing demand for whole-grain flours, the use of hammer mill for producing wheat flour is also in demand. Grinding of dry grains including wheat, rye, oats, and barley can be achieved using such mills. During wheat milling through hammer mill, the flour gets heated up and loses moisture content (Posner and Hibbs, 2005).

7.3.1.3.4 Novel Technologies

Superfine and ultrafine grinding procedures are being employed for size reduction of fibre in wheat bran. Superfine grinding can generate micron-sized powder, whereas ultrafine grinding can decrease particles to the submicron range (100–1,000 nm). The resulting powders are known to improve the nutritional value and cooking abilities of products (Du et al., 2020; Xu et al., 2020). Superfine grinding, as opposed to regular grinding, is a sort of cellular level pulverising that uses mechanical or hydrodynamic processes to decrease material from 0.5 to 5 mm to less than 40 μm (Chamayou and Dodds, 2007; Ou and Wang, 2010). Commercially, various machines such as SPAZIES SY-1200 Commercial Superfine Grain Cereal Mill, customised GMP superfine grinding universal crusher machine, and superfine rice or wheat flour grinding machine are available for superfine grinding of wheat. During this, wheat bran's aleurone layer is broken down, assisting in the breakdown of cell walls and a more efficient release of nutrients. Additionally, the outer pericarp layer and intermediate layer of wheat bran including fibrous structures can be broken up to boost the bran's value and improve its palatability (Hemery et al., 2011; Rosa et al., 2013; Zhang et al., 2006). The macro crushing aspects of the crushing device, such as the crushing load, crushing form, grinding temperature, grinding medium, and so on, limit the crushing performance of wheat bran cell tissue. Finished product quality is impacted by the physicochemical properties of the fine powder and the reduction in particle size of the grain flour (Drakos et al., 2017; Muttakin et al., 2015; Protonotariou et al., 2016). Findings recommend that grinding with the MLU-202 Automatic Laboratory Mill might improve the whole-wheat flour quality significantly by reducing its particle size (Niu et al., 2014). Protonotariou et al. (2014) claimed that the superfine grinding process produces particles with altered dough rheological properties and thermodynamic of starch that may significantly influence the final product, as well as improve water-holding capacity. The seeds were first ground in disc mill, and coarse samples were then re-milled for varying lengths of time (5–20 min) in an impact superfine grinder to make four different varieties of whole-wheat flour (Niu et al., 2014). During the superfine grinding process, the steel bars collided with each other or with the inside wall of the grinding bowl, reducing the size of the flour particles. They employed a circulating water bath to lower the abnormally high temperature of the grinding bowl.

7.3.1.3.4.1 Ball Milling Ball milling makes use of mechanical and thermal effects to change the physical and chemical properties of raw materials (Shi et al., 2015). It works on the principle of intensive impact, attrition grinding force, and the pressure and friction of the grinding medium (balls) to achieve the purpose of size reduction. Ball milling is an efficient, cost-effective, and powerful non-equilibrium grinding method, which can mechanically reduce the size of solids to an ultrafine scale (Lyu et al., 2018). It has limited capacity of grinding cavity, but is considered as successful technique tool because of effective particle size reduction (Burmeister et al., 2018). A ball mill is a heavy mechanical equipment that depends on its own rotation to drive the steel balls inside to impact and grind materials with high consistency. The disadvantages include low working efficiency and high energy consumption. It cannot be employed for routine flour preparation purpose but can be used for physical alterations in food for specific food applications. Thanatuksorn et al. (2009) studied the effect of the ball-milling process on glass transition temperature (Tg) of wheat flour constituents. They reported that the decrease in Tg, as a function of milling time, could be attributed to conformational changes in gluten protein subunits as well as depolymerisation of the non-gluten protein fraction. They concluded that ball-mill can be employed to physically modify wheat flour ingredients to find different food applications.

7.3.1.3.4.2 Jet Milling As the demand for whole-wheat flour is increasing, thereby, greater attention is being given to unconventional milling processes and micronising machineries for producing whole-wheat flours with enhanced functional properties (Sakhare et al., 2013, Protonotariou et al., 2014; de La Hera et al., 2013). Direct pulverisation of whole wheat grains by employing jet mills is one such technology for the production of whole-wheat flour.

Jet milling is a type of impact milling that uses fluid energy. Such mills use high air pressure to produce flour with particle sizes of less than 10 μm or 100 nm (Midoux et al., 1999; Palaniandy et al., 2008, Guan et al., 2020; Nakach et al., 2019). The flour thus produced is called ultrafine flour. Usually ultrafine powders are employed in chemical, pharmaceutical, and mineral industries. Nowadays, jet milling through the production of whole-wheat flour is gaining practicality in food applications also (Gao et al., 2020). To produce ultrafine powders, particles are accelerated by air stream of high velocity (Figure 7.7); particle size reduces due to collisions or contacts with solid surfaces (Létang et al., 2002). The nozzle size, feedstock size, feed rate (FR), grinding air pressure (AP), (Angelidis et al., 2015), feeder vibration rate, and feedback – all affect the size of the flour particle (Protonotariou et al., 2014).

Jet milling in combination with air classification has been employed for starch separation from protein for producing fine flours rich in starch (Graveland and Henderson, 1991). Protonotariou et al. (2021) milled soft wheat flour in a jet mill at various feed rates (2–18 kg/h) and air pressures (4 and 8 bar, respectively). Lowest value of 10.62 μm for finer flour was achieved as a result of a reduction in feed rate and an increase in grinding air pressure. Soft wheat was ground with minimal energy use at a feed rate of 70%. As the damaged starch content was reduced, a brief grinding at a high feed rate and grinding air pressure is also preferred. About 10% of the starch in ultrafine flour has reportedly been destroyed.

The insoluble fibre components of wheat bran fibre were changed into soluble fractions by grinding them to a submicron scale (Zhu et al., 2010). The formation of more

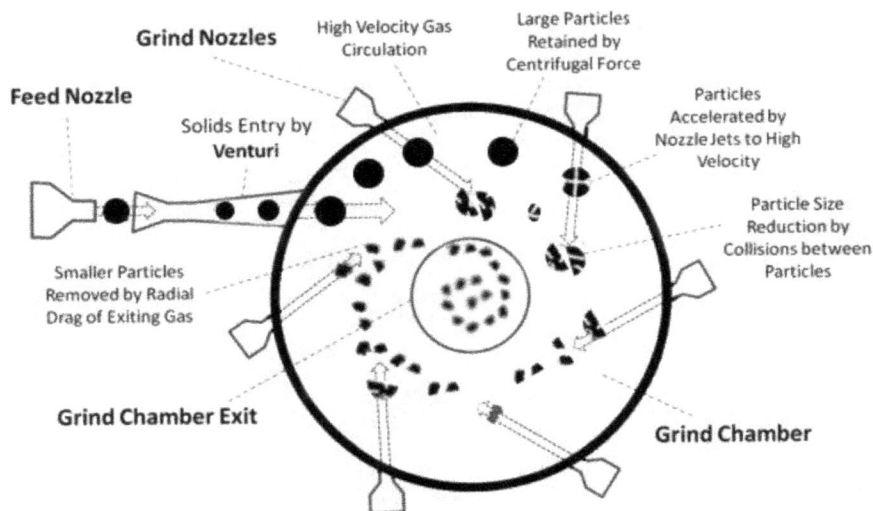

FIGURE 7.7   Schematic diagram of jet milling.

Source: MacDonald et al. (2016)

appetising foods is claimed to be a result of fine flour particles' increased surface area, which increases their water absorption, solubility, and starch digestion (de La Hera et al., 2014; Guan et al., 2020). According to studies on the impacts on wheat bran due to ultrafine grinding, the properties of the bran itself get affected by the particular type of grinding machinery (Hemery et al., 2011). Korolchuk (2005) described a method for creating wheat flour that is ultrafinely processed while retaining the wheat kernels' nutritional content by using a gap mill.

7.3.2 Effect of Milling on Nutrients

Whole wheat grain consists of three parts: germ (lipids proteins), endosperm (protein and carbs), and bran (dietary fibre) (Liu et al., 2015), and milling of whole grains provides the nutritional benefits of all the three constituents of the whole grain. However, the consumption of grains by human requires some kinds of processing. Among various processing techniques, milling is the most prevalent processing technique. When wheat is milled using conventional roller mill, only endosperm fraction is retained and the flour obtained is rich in carbohydrates and proteins (Cubadda et al., 2009) and lacks in valuable nutrients including fibre, phytates, phytochemicals, and minerals (Liu et al., 2015). During milling, germ, bran fraction, and aleurone layer get separated (Clydesdale, 1994; Slavin et al., 2001). Flour obtained is called refined wheat flour, which finds application in various food products. Wheat bran, a rich source of fibre, is generated as the major by-product during conventional milling.

The nutritional content of the resulting wheat flour is determined using the extraction rate of flour, which can be defined as the weight of flour prepared by milling of specific number of wheat grains. The extraction rate of 75% at the maximum presents typical white flour, whereas the extraction rate of greater than 80% reveals the presence of non-endosperm particles in flour and the extraction rate of 100% presents the whole-wheat flour generation (Slavin et al., 2001). The biochemical composition of wheat flours in relation to different extraction rates is listed in Table 7.2.

From Table 7.2, it is evident that different extraction yields followed by milling exhibit a significant negative impact on the fibre profile and nutritional content. It is considered that the starch-rich endosperm of highly processed wheat meals is more sensitive to intestinal hydrolases. The amount of minimally fermentable starch that is available to the colonic microbiota decreases as a result of the rapid and extensive digestion of the starch in the upper gut (Bird and Regina, 2018).

Further, it has been reported that if flour is produced with 100% extraction rate, its nutritional composition may vary using different milling techniques. It is also established that the moisture content of flour is influenced by the milling method's grinding strength (Liu et al., 2015). Milling of wheat cultivars using hammer mill showed

TABLE 7.2

Impact of Extraction Rate on Biochemical Parameters of Resultant Flours after Milling

Biochemical Parameters	Extraction Rates					
	100%	95%	87%	80%	75%	66%
Ash (% db)	1.8	1.5	1.0	0.7	0.6	0.5
Protein (% db)	14.2	13.9	13.8	13.4	13.5	12.7
Fat (% db)	2.7	2.4	2.0	1.6	1.4	1.1
Starch + sugar (% db)	69.9	73.2	77.2	80.8	82.9	84.0
Crude fibre (% db)	2.4	2.1	1.1	0.2	0.3	0.2
Dietary fibre (% db)	12.1	9.4	5.5	3.0	2.8	2.8
Calcium (mg/g)	0.44	0.43	0.33	0.27	0.25	0.23
Phosphorus (mg/g)	3.8	3.3	2.1	1.5	1.3	1.2
Zinc (ppm)	29	25	18	12	8	8
Iron (ppm)	35	33	23	15	13	10
Copper (ppm)	4.0	3.7	2.8	2.4	1.6	1.3
Riboflavin (μg)	0.95	0.79	0.69	0.46	0.39	0.37
Thiamine (μg)	5.8	5.4	4.8	3.4	2.2	1.4
Vitamin B6 (μg)	7.5	6.6	3.4	1.7	1.4	1.3
Folate (μg)	0.57	0.53	0.45	0.11	0.11	0.06
Biotin (μg)	116	108	106	76	46	25
Niacin (μg)	25.2	19.3	10.1	5.9	5.2	3.4

Source: Pedersen et al. (1989)

a greater moisture content in comparison to milled by jet milling which could be due to the evaporation of moisture from the larger surface area of particles produced by jet milling (Kang et al., 2019). When whole-wheat flour from different cultivars was processed with a hammer mill instead of an ultrafine or jet mill, it included more protein and carbohydrates, less total dietary fibre, and less fat (Kang et al., 2019; Liu et al., 2015).

Ash is concentrated in the bran layer, and its concentration rises as you move away from the wheat's centre and towards the husk. Therefore, it is generally believed that ash content reflects how much wheat is crushed up during the milling process and may indicate how refined the white flour is or it is, occasionally, a measure of the flour's quality. The quantity of aleurone that remains in the bran layer after grinding is described as "ash content" of the whole-wheat flour (Keran et al., 2009). However, Pagani et al. (2020) examined stone and roller-milled flour for nutritional and technological properties and did not find any changes in the proximate composition (Table 7.3) and the bioactive compounds of both types of flours.

(Prabhasankar and Rao (2001) prepared whole-wheat flour using stone, plate, and hammer and roller mills and found that method of milling affects the temperature to 90, 85, 55, and 35°C, respectively. Whole-wheat flour produced by stone and plate mills was revealed to have greater levels of protein breakdown and total amino acid content than samples of flour made in hammers and rollers. Similarly, due to higher temperature during processing, there was degradation of essential fatty acid linolenic acid (omega-3) as can be observed from their values, that is, 1.3%, 2.2%, 2.8% and 3.8%, in stone, plate, hammer mill, and roller milled flour, respectively.

The starch granules in the wheat grain are harmed during the milling process. Higher damaged starch content for stone and plate ground flours as compared to flours produced using hammer and roller milling has been reported, which could be due to the severity of grinding in stone milling (Prabhasankar and Rao, 2001). Oladunmoye et al. (2010) have also described that the damage to starch granules is associated with shearing and scrapping activities during milling.

The flours with damaged starch show more water absorption (Manley et al., 2011). Measuring starch degradation is important since it affects the quality of wheat flours in a major way. Starch damage potential is largely hereditary and correlated with kernel hardness. Starch deterioration affects water absorption because it absorbs around four times as much water as native starch. Various researchers have reported that high starch damage is negatively correlated with cookie diameter. Increase in dough stiffness and consistency and decrease in cookie diameter have been observed in sugar-snap cookies due to damaged starch (Faridi et al., 1994). Hard wheat with more severely damaged starch granules is considered more suitable for use because it may absorb additional water and is further prone to amylase digestion. For the duration of the fermentation process, this serves as a substrate for the yeast to feed on (E. Li et al., 2014).

Super fine and ultrafine milling reduces the flour particle size, which results in enhanced surface area of particles (Tóth et al., 2006) and also results in an increase in damaged starch content. There is a need to assess a critical point up to which the reduction in flour particle size is acceptable (Protonotariou et al., 2014). Jet milling has been found to improve the accessibility of biologically active compounds (Sanguansri and Augustin, 2006). Reducing the size of the flour's particles encourages changes in its physicochemical characteristics, which in turn alter the mechanical and starch

TABLE 7.3

Effect of Milling on Proximate Composition, Gluten and Damaged Starch Content of Wheat Flours

Type of Mill Used	Moisture (%)	Protein (%)	Total Fat (%)	Ash (%)	% Composition Dietary Fibre (%)	Carbohydrates (%)	Wet Gluten (%)	Damaged Starch	Reference
Stone	7.3–7.5	9.4–11.3	1.3–1.6	1.7–1.9	-	-	-	12.8–16.3	Prabhasankar and Rao, 2001
	10.4–10.8	13.4–14.8	-	1.1–1.2	-	-	22.5–29.6	-	Liu et al., 2015
Plate	8.5–8.7	9.5–11.0	1.5–1.8	1.5–1.7	-	-	-	11.9–14.5	Prabhasankar and Rao, 2001
Hammer	9.7–11.5	10.1–10.7	2.40–2.55	1.44–1.63	8.81–12.90	73.03–76.48	13.89–16.47	-	Kang et al., 2019
	9.2–9.5	9.5–11.1	1.7–1.9	1.6–1.8	-	-	-	8.0–9.0	Prabhasankar and Rao, 2001
Jet mill	10.9–12.3	13.6–14.6	-	1.1–1.6	-	-	29.4–33.6	-	Liu et al., 2015
	8.9–9.7	9.7–10.3	3.12–4.03	1.61–1.67	12.60–15.03	69.48–72.57	11.30–15.87	-	Kang et al., 2019
	6.9–7.3	13.0–14.4	-	1.5–1.6	-	-	24.2–32.0	-	Liu et al., 2015
Roller	9.5–9.9	9.6–11.3	1.9–2.0	1.5–1.6	-	-	29.4–33.6	5.3–7.0	Prabhasankar and Rao, 2001
	12.9–14.3	10.3–11.8	-	0.59–0.65	-	-	26.9–30.8	-	Liu et al., 2015

gelatinisation characteristics of dough. The particle sizes of flour samples from different cultivars milled with a hammer mill (40.34 μm –188.5 μm) were bigger compared to the one milled using with a jet milling (20.8 μm–41.8 μm) (Kang et al., 2019).

Milling not only affects the wheat nutritional quality but also modifies end product quality. Among the studied milling methods of plate, hammer, pin, and roller mill, flour prepared from the plate mill has been reported to be superior in terms of different characteristics such as texture and pliability of chapati (Patil et al., 2021). The wheat flour samples milled by impact mill exhibited higher trough viscosity, ultimate viscosity, setback values, and less damaged starch compared to wheat flour samples milled using hammer mill (Tian et al., 2020). Ultrafine grinding of whole white wheat resulted in improved colour values of resultant steamed bread (Liu et al., 2015). In another study, the whole meal breads made with roller-milled wheat exhibited high values for sweetness, juiciness, and compactness, whereas those made with stone-milled wheat were distinguished by the qualities of saltiness, deformity, and toasted cereal characteristics (Kihlberg et al., 2004).

7.4 Whole-Wheat Milling: Health Benefits and Challenges

7.4.1 Health Benefits of Whole-Wheat Milling

Whole wheat grains provide a significant source of carbohydrates, proteins, minerals, and dietary fibre. Among them, dietary fibre, a key component of whole grains, is one that is regarded to be the reason for a number of health benefits. Wheat is an ideal substrate for nutritional improvement of overall food supply and enhancing the health of the human population around the globe due to its widespread popularity and dietary importance (Ficco et al., 2014). Wheat has a nutritional profile that is similar to the other whole-grain cereals like rice, maize, and barley. It also contains a number of health-promoting compounds, including choline, betaine, alkylresorcinols, and others (Gupta et al., 2021).

All of these components, alone or in combination, explain the reason behind the protection against non-communicable ailments and the long-term health-related benefits of whole wheat grains. However, the evidence of health-promoting activities of dietary fibre is the highest (Bird and Regina, 2018). Whole wheat is the principal resource of total dietary fibre in developed countries. Higher intake of whole grains and related food products is linked with a reduction (20–30%) in type II diabetes incidence (Gil et al., 2011).

Increased inflammation is also a major factor linked with the progression of weight-related insulin resistance. Previously, a clinical analysis found that using whole wheat-based bread and fibre helped to reduce fundamental exacerbation among diabetic women. Furthermore, whole wheat grain consumption was also reported to improve the liver and inflammatory responses in overweight and obese adults along with slightly elevated cholesterol level in plasma (Hoevenaars et al., 2019). The consumption of whole-wheat bread has also reduced the plasma cholesterol and fatty material while also improving overall steroid discharge in animal studies (Adam et al., 2003). In addition, the whole bran arabinoxylans from wheat were also reported to lower total cholesterol, plasma cholesterol, and LDL-cholesterol fixation levels in hamster models (Tong et al., 2014). Secoisolariciresinol diglucoside (SDG), a wheat lignan, has been linked to antitumor activity in colon cancer SW480 cells (Qu et al., 2005). In animal and human studies, SDG has also presented protection against cancerous growths and various lung and colon diseases (Imran et al., 2015). Apigenin, a type of phytoestrogen found in wheat, was also mentioned as potent antitumor agent for cervical cancer (Sak, 2014). In addition, wheat bran oil with its fractions also showed protective effects against experimental colon carcinogenesis in animal models (Sang et al., 2006).

Recently, researchers have reported the development and biological properties of pigmented wheat varieties containing enhanced content of bioactive compounds which were responsible for antioxidant activity, reduced LDL cholesterol, and neural tube development (Garg et al., 2016; Ficco et al., 2014; Saini et al., 2020). According to a previous report, an anthocyanin-rich diet aids in decreasing eye fatigue, reducing near sightedness, reducing glaucoma, and improving retinal blood flow. It also aids proper cerebrum functioning by moderating age-linked neurodegeneration and psychological degradation (Tsuda, 2012). Coloured wheat was reported to improve the inflammatory profile and glycemia in type-2 diabetic subjects. The phenolic acids, *viz.* gallic acid, syringic acid, ferulic acid, p-coumaric acid, caffeic acid, cinnamic, and sinapic acid in coloured wheat-based diet, were found to exhibit anti-diabetic effects in the animal models (Vinayagam et al., 2016). Blue wheat regulates diabetes by reducing starch absorption catalysed by amylase that in turn is linked to post-prandial blood glucose levels (Tyl and Bunzel, 2014). Recently, it was mentioned that supplementing blue- or purple-coloured wheat protects against paraquat-stimulated oxidative stress in *Drosophila melanogaster*. A coloured wheat-based diet was also reported to enhance life expectancy and prevent cardiac pathologies linked with the consumption of a high-fat diet in flies (Pandey et al., 2022). Another study mentioned the lower glycemic index of coloured wheat compared to conventional wheat. Black wheat was reported to act as a prebiotic and positively modulate the gut microbiome of mice (Kapoor et al., 2022). Recent reviews also mentioned the cardioprotective, anti-cancer, anti-obesity, anti-inflammatory, anti-diabetic, anti-ageing, and large bowel-protective effects of colored wheat attributed to their higher concentration of phytochemicals (Gupta et al., 2021; Dhua et al., 2021; Garg et al., 2022).

It is very well known that the slight variations in the biochemical composition of wheat can significantly influence the nutritional quality of diets across entire populations. That will further help to meet the recommended intakes of essential nutrients and phytochemicals in the target population. In this sense, whole-wheat flour might act as a suitable functional substrate for providing health benefits worldwide through altered grain composition. This strategy may provide a practical and economical means of elevating public well-being by halting or reversing the progression of diet-related chronic disease, as well as reducing the accompanying economic and social burden.

7.4.2 Challenges

Whole-wheat flour is less stable in storage than normal wheat flour. Compared to normal wheat flour, which has expiry of 9 to 15 months after milling, whole-wheat flour has expiry of 3 to 9 months. How long flour really lasts, however, may also depend on variables like storage temperature and humidity (de Almeida et al., 2014).

Lipids are some of the most unstable constituents in whole-wheat flour in comparison to all the other elements. The breakdown of lipids during storage and processing is the fundamental cause of the poor functional, sensory, and nutritional properties of flour (Jia et al., 2021). Lipid degradation involves hydrolytic rancidity followed by oxidative rancidity involving enzymatic and non-enzymatic factors. Briefly, wheat lipids are converted to non-esterified fatty acids in the presence of lipases that reduce the functional properties and sensory acceptability of flour. Further, these non-esterified fatty acids are converted to lipid oxidation products in presence of lipoxygenase and moisture that reduce the overall nutritional, functional, and sensory quality of flours (Doblado-Maldonado et al., 2012).

Glutenin and gliadin, two wheat storage proteins, include intra- and intermolecular disulphide bonds that are essential to flour's functionality. Gliadin only possesses intra-molecular disulphide connections, which are responsible for dough cohesiveness, in contrast to glutenin, which has both intra- and inter-molecular disulphide bonds and is thought to give wheat dough flexibility. But it has been found that, once milled, the flour proteins have a significant amount of sulfhydryl group, which makes them unsuitable for bread baking (Doblado-Maldonado et al., 2012; Poudel and Rose, 2018). The endogenous amylolytic activity also alters the starch of milled wheat flour during storage, which leads to decreased resistance to stretching in the dough. Due to a higher concentration of low molecular weight carbohydrates in whole-wheat flour during storage, the Maillard reaction also increases the colour values of the bread crust. Furthermore, vitamin E activity and thiamine in wheat flour are also reduced to 40% and 7.2–11.5% during the 12-month storage. The oxidation of carotenoids and the degradation of vitamin E in wheat flour are linked with lipid oxidation (Doblado-Maldonado et al., 2012; Di Silvestro et al., 2014).

Doughs are made with more dietary fibres (up to 10%–15%) to boost the nutritious value of noodles, high-fibre bread, and other cereal products. However, adding a lot of fibres slows down the dough's development and increases the water it absorbs, which makes the dough stickier because the dough's tolerances for fermentation and mixing are weakened. Foods supplemented with fibre have been reported to have unpleasant textures, lower brightness, and smaller specific loaf volumes (Chareonthaikij et al., 2016).

The nutritional value and product quality of high-fiber cereal foods clearly trade off with one another. Wheat bran contains a large number of non-starch polysaccharides (fibres), which are essential to gluten's action. Making fiber-enriched cereal products has a noticeable impact on the outcome because adding extra fibres complicates the starch–protein network (Figure 7.8).

Fibres of any size can significantly alter the properties of doughs or cereal items made with gluten. Dietary fibres have a diluting effect on gluten by physically disrupting the gluten network and partially dehydrating gluten due to their contradictory water-binding properties. The lubrication of gluten proteins by water is lessened by naturally occurring water-friendly fibre particles. Large, thick bran particles, in particular, are hypothesised to weaken the gluten matrix and ultimately sterically hinder the formation of the gluten network (Li et al., 2012). The particle size, surface properties, volume fraction, and additional physical fibre features that contribute to the fibres' ability to bind water and their steric hindrance have a significant impact on the physical interactions between dietary gluten protein and fibre (Ortiz de Erive et al., 2020).

Apart from physical interaction, Sivam et al. (2010) explained that the importance of molecular interactions among various major components of dough, that is, fiber, starch, and protein leads to variation in doughs polymeric network. The systematic

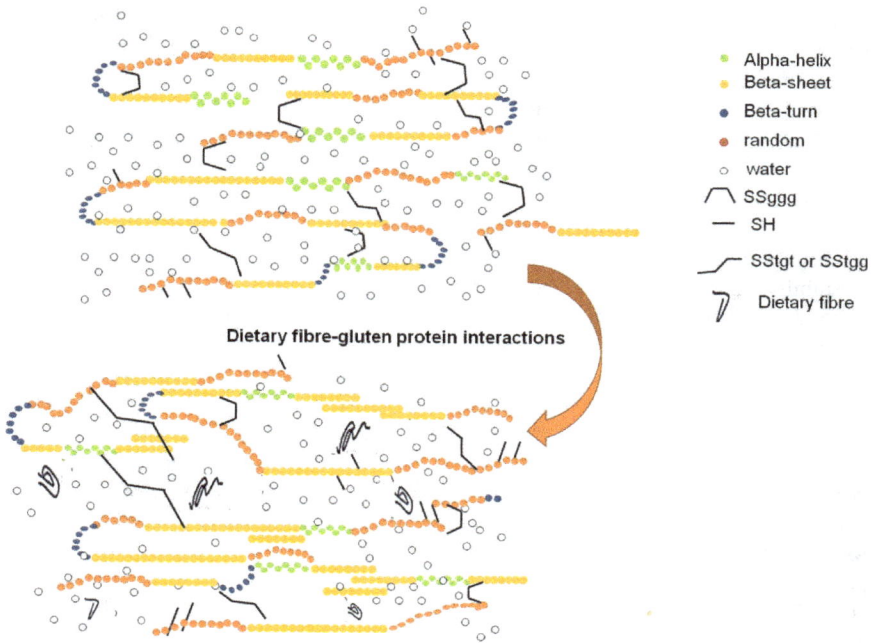

FIGURE 7.8 Dietary fibre–gluten protein interactions in wheat flour dough disrupt the starch–protein network; cause water competition and steric hindrance; and alter protein secondary and tertiary structures, disulphide bridges, and hydrogen-bonding patterns.

Source: Adapted from: Zhou et al. (2021).

interaction between fibres and starches has been discussed by Bemiller (2011). Changes in protein secondary and tertiary structures, disulphide bridges, and hydrogen bonding forms are associated with fibre–gluten interactions (Zhou et al., 2021). Soluble fibres interact with gluten in two non-covalent ways: hydrophobic interactions and hydrogen bonds. In the case of insoluble fibres, the degree of swelling and hydration predominates the interaction. The interactions between fibre and gluten and the functionality of dough are greatly influenced by the molecular weight of the soluble fibres, their capacity to bind water, the competition for water, and steric hindrance.

7.5 Strategies to Retain the Nutritional Quality of Milled Wheat

Various strategies for maintaining the quality characteristics of milled wheat have been reported. During the production of whole-grain flour with all the nutrients of wheat kernel using stone mills, it has been observed that the heat production takes place which causes changes in nutritional composition of whole flour. Since roller milling generates a lesser amount of heat and hence lesser damage to chemical constituents, it can be utilised to avoid losses from heating (Prabhasankar and Rao, 2001). Compared to stone mills, the roller mills are more less expensive and adaptable (Posner and

Hibbs, 2005). Wheat bran and germ could be removed from the endosperm fraction and subjected to post-milling operations such as twin-screw extrusion, heat treatment of the bran and germ (de Kock et al., 1999), or ultrafine grinding for making whole-wheat flours using roller mills (Hemery et al., 2011). These processes can increase the flour's ability to be stored or used in other ways (Posner and Hibbs, 2005).

As whole-wheat flour contains germ and hence is rich in lipids, therefore, it is prone to lipid oxidation. The lipid degradation of wheat flour can be reduced by storage at −20°C. But this method has no practical utility due to its high cost. Furthermore, the modified storage atmosphere and addition of antioxidants are other methods reported to stabilise lipids in whole-wheat flour and resultant foods. However, these methods were also found to be least effective as antioxidants are minimally effective against lipoxygenase-mediated oxidation. In addition, lipid oxidation starts with hydrolytic rancidity in presence of enzymes, and oxygen is not required in this process (Doblado-Maldonado et al., 2012). Additionally, throughout 24 weeks of storage, the oxidative and hydrolytic rancidity of the flour was prevented by conditioning hard red winter wheat with 1% NaCl (flour weight), which reduced lipid degradation (Doblado-Maldonado et al., 2013). The steam treatment (180 s) and superheated steam (155–170°C for 10–90 s and 190 °C for 5 s) treatment of wheat grains before milling have also been mentioned to significantly inactivate lipase, lipoxygenase, polyphenol oxidase, endoxylanase, peroxidase, and α-amylase without gelatinising starch (de Almeida et al., 2014; Jia et al., 2021; Guo et al., 2020; Poudel and Rose, 2018). In another study, ozonation of whole-wheat flour at 5 g/hour for 45 min inactivated lipases in the oil obtained from treated flour, but increased ozonation time enhanced the *p*-anisidine and peroxide values and oxidation of volatile compounds despite enhancing antioxidant activity (Obadi et al., 2018). It has been reported by researchers that the inactivation of lipase enzyme in wheat germ can be done by low-temperature microwave treatment at 60 and 70 °C for 5–10 min (Meriles et al., 2022). Attempts have been made to formulate stabilised whole-wheat flour by adding microwave-treated bran to white flour. The lipase activity also decreased rapidly followed by microwave treatment of wheat bran at 69 °C for 60 s. However, the decline of lipase activity was slow when bran was treated for 90 s at 80 °C (Qu et al., 2021). Rebellato et al. (2018) reported that the fortification of whole-wheat flour with reduced iron and ferric sodium ethylene diamine tetraacetate enhanced the overall stability during one month of storage.

The addition of cellulase reported to increase the paste viscosity and decrease dough mixing properties and bread hardness prepared from whole waxy wheat flour of 48% extraction. While the addition of α-amylase increased final flour viscosity and had no effect on the properties of dough and bread qualities prepared using whole waxy wheat flour, the addition of pentosanase also increased paste viscosity, decreased dough mixing properties, improved loaf volume of bread, and increased the firmness of breadcrumbs made up of whole waxy wheat flour (Hung et al., 2007). It was also mentioned that cold plasma treatment of milled wheat flour affected protein oxidation and encouraged the formation of disulphide bonds between glutenin proteins, which improved dough strength and caused starch to depolymerise and lose its crystallinity. These alterations led to overall improvement in the appearance, porosity structure, and whiteness of bread crumb (Dapčević-Hadnađev et al., 2022). The mixing of *Triticum durum* semolina (50%) with *T. aestivum* refined flour (50%) was reported as an alternate for making pasta having low glycemic index and acceptable physical and sensory properties (Dhiraj and Prabhasankar, 2013). The milled wheat flour obtained after treating with carbohydrase in tempering solution presented improved rheological properties. The use of this treated flour resulted in improved dough proofing and quality of resultant bread (Haros et al., 2002).

Explosion puffing, high-temperature and low-temperature extrusion, and high-temperature jet milling of wheat bran-germ powder have been found to decrease its lipase activities by 31%, 39%, 55%, and 57% and lipoxygenase activities by 17%, 35%, 37%, and 45%, respectively. In addition, the lipoxygenase and lipase activity of all pre-treated powder decreased with increase in storage time up to 28 days (Zhuang et al., 2022).

Another study from Switzerland also explored lipid-stable varieties of wheat ('Velocity' and 'Arina') for developing stable whole-grain flour and whole-grain-based products with enhanced shelf life (Wei et al., 2021). Thus, it is suggested to explore and develop improved wheat varieties along with the application of various physical treatments for maintaining overall quality and enhancing the stability of whole-wheat flour and the final product during storage.

7.6 Conclusion

Wheat is one of the main food cereal grains consumed around the globe in the form of Indian flat breads, breads, cookies, noodles, and other bakery products. Wheat is majorly categorised as soft, mixed, and hard varieties. Wheat quality is determined on the basis of a set of physical and composition parameters which are optimised based on the end use of grains. Milling yield is significantly affected by the grain morphology, size, texture, and hardness. Among the biochemical parameters, starch quality, gluten strength, protein content, and SDS-sedimentation value significantly affect the milling of wheat and quality of the final product. Soft wheat is ground into substantially smaller pieces than hard wheat in order to have a higher 'break flour yield'. Durum (the hardest kind of wheat) generates semolina followed by milling, which is used for making pasta. Milling is an important process to produce flour of wheat grains. Stone milling is one of the oldest methods employed for producing wheat flour, resulting in particle size ranging from more than 363 to less than 85 μm. Roller milling and hammer milling are among the most-used modern milling methods resulting in particle size more than 1,000 to less than 250 μm and from 15 to 50 μm, respectively. Superfine grinding (< 40 μm) μm) and jet milling (<10 μm or 100 nm) are the novel technologies employed for wheat milling. The extraction rate of 100% represents whole-wheat flour, and the extraction rate of 75% at the maximum represents typical white flour. Whole-wheat flour reported to exhibit potent health benefits attributed to its high level of carbohydrates, proteins, dietary fibre, choline, betaine, alkylresorcinols, and several other health-promoting ingredients. However, whole-wheat flour's storage stability is less than normal wheat flour due to significant degradation of lipids, protein, and starch due to milling. Thus, storage at low temperature, modified storage atmosphere, addition of antioxidants, conditioning of wheat with 1% NaCl, ozone treatment, steam, superheated steam, and microwave treatments are recommended for shelf life enhancement. It is vital to promote the regular consumption of whole-grain products for health purposes. The best method for encouraging the consumption of whole-grain foods is to increase their perceived appeal. Therefore, there is a need to undertake more research work on the development of whole-grain-flour-based products.

REFERENCES

Abaye, A. O., Brann, D. E., Alley, M. M., & Griffey, C. A. (1997). Winter durum wheat: Do we have all the answers? In *Virginia Cooperative Extension* (pp. 1–8). www.ext.vt.edu

Adam, A., Lopez, H. W., Leuillet, M., Demigné, C., & Rémésy, C. (2003). Whole wheat flour exerts cholesterol-lowering in rats in its native form and after use in bread-making. *Food Chemistry*, *80*(3), 337–344. https://doi.org/10.1016/S0308-8146(02)00269-8.

Al-Mahasneh, M. A., & Rababah, T. M. (2007). Effect of moisture content on some physical properties of green wheat. *Journal of Food Engineering*, *79*(4), 1467–1473. https://doi.org/10.1016/J.JFOODENG.2006.04.045.

Angelidis, G., Protonotariou, S., Mandala, I., & Rosell, C. M. (2015). Jet milling effect on wheat flour characteristics and starch hydrolysis. *Journal of Food Science and Technology*, *53*(1), 784–791. https://doi.org/10.1007/S13197-015-1990-1.

Babić, L., Babić, M., Turan, J., Matić-Kekić, S., Radojčin, M., Mehandžić-Stanišić, S., Pavkov, I., & Zoranović, M. (2011). Physical and stress-strain properties of wheat (*Triticum aestivum*) kernel. *Journal of the Science of Food and Agriculture*, *91*(7), 1236–1243. https://doi.org/10.1002/JSFA.4305.

Bao, C. L., Xu, X. Y., & Liu, B. W. (2014). The nutrition and comprehensive utilization of wheat bran. *Cereals & Oils*, *27*(8), 58–60.

Bechtel, D. B., Abecassis, J., Shewry, P. R., & Evers, A. D. (2009). Development, structure, and mechanical properties of the wheat grain. In K. Khan & P. R. Shewry (Eds.), *Wheat: Chemistry and Technology* (4th ed., pp. 51–95). AACC International Inc.

Belderok, B., Mesdag, J., & Donner, D. A. (2000). Bread-making quality of wheat. *Bread-Making Quality of Wheat*. https://doi.org/10.1007/978-94-017-0950-7.

Bemiller, J. N. (2011). Pasting, paste, and gel properties of starch–hydrocolloid combinations. *Carbohydrate Polymers*, *86*(2), 386–423. https://doi.org/10.1016/J.CARBPOL.2011.05.064.

Bhise, S. R., Kaur, A., & Manikantan, M. R. (2014). Moisture dependent physical properties of wheat grain (PBW 621). *International Journal of Engineering Practical Research*, *3*(2), 40–45. https://doi.org/10.14355/IJEPR.2014.0302.03.

Bird, A. R., & Regina, A. (2018). High amylose wheat: A platform for delivering human health benefits. *Journal of Cereal Science*, *82*, 99–105. https://doi.org/10.1016/J.JCS.2018.05.011.

Bisht, I. S. (2021). Agri-food system dynamics of small-holder hill farming communities of Uttarakhand in north-western India: socio-economic and policy considerations for sustainable development. *Agroecology and Sustainable Food Systems*, *45*(3), 417–449. https://doi.org/10.1080/21683565.2020.1825585.

Borneo, R., & León, A. E. (2012). Whole grain cereals: functional components and health benefits. *Food & Function*, *3*(2), 110–119. https://doi.org/10.1039/C1FO10165J.

Boz, H., & Murat Karaoğlu, M. (2013). Improving the quality of whole wheat bread by using various plant origin materials. *Czech Journal of Food Sciences*, *31*(5), 457–466.

Brabec, D. L., Pearson, T. C., Maghirang, E. B., & Flinn, P. W. (2015). Detection of fragments from internal insects in wheat samples using a laboratory entoleter. *Cereal Chemistry*, *92*(1), 8–13. https://doi.org/10.1094/CCHEM-08-13-0173-R.

Burmeister, C., Titscher, L., Breitung-Faes, S., & Kwade, A. (2018). Dry grinding in planetary ball mills: Evaluation of a stressing model. *Advanced Powder Technology*, *29*(1), 191–201. https://doi.org/10.1016/j.apt.2017.11.001.

Bushuk, W. (1998). Wheat breeding for end-product use. *Euphytica*, *100*(1), 137–145. https://doi.org/10.1023/A:1018368316547.

Campbell, G. M. (2007). Chapter 7. Roller milling of wheat. *Handbook of Powder Technology*, *12*, 383–419. https://doi.org/10.1016/S0167-3785(07)12010-8.

Campbell, G. M., Webb, C., Owens, G. W., & Scanlon, M. G. (2012). Milling and flour quality. *Breadmaking: Improving Quality*, 188–215. https://doi.org/10.1533/978085 7095695.1.188.

Campbell, J., Hauser, M., & Hill, S. (1991). *Nutritional Characteristics of Organic, Freshly Stone-Ground Sourdough & Conventional Breads*. McGill University.

Carcea, M., Turfani, V., Narducci, V., Melloni, S., Galli, V., & Tullio, V. (2020). Stone milling versus roller milling in soft wheat: influence on products composition. *Foods*, *9*(1), 3. https://doi.org/10.3390/FOODS9010003.

Cauvain, S. P., & Young, L. S. (2006). *Baked Products: Science, Technology and Practice*. Wiley-Blackwell.

Chamayou, A., & Dodds, J. A. (2007). Chapter 8. Air jet milling. *Handbook of Powder Technology*, *12*, 421–435. https://doi.org/10.1016/S0167-3785(07)12011-X.

Chareonthaikij, P., Uan-On, T., & Prinyawiwatkul, W. (2016). Effects of pineapple pomace fibre on physicochemical properties of composite flour and dough, and consumer acceptance of fibre-enriched wheat bread. *International Journal of Food Science & Technology*, *51*(5), 1120–1129. https://doi.org/10.1111/IJFS.13072.

Clydesdale, F. M. (1994). Optimizing the diet with whole grains. *Critical Reviews in Food Science and Nutrition*, *34*(5–6), 453–471. https://doi.org/10.1080/10408399409527675.

Cubadda, F., Aureli, F., Raggi, A., & Carcea, M. (2009). Effect of milling, pasta making and cooking on minerals in durum wheat. *Journal of Cereal Science*, *49*(1), 92–97. https://doi.org/10.1016/J.JCS.2008.07.008.

Curti, E., Carini, E., Bonacini, G., Tribuzio, G., & Vittadini, E. (2013). Effect of the addition of bran fractions on bread properties. *Journal of Cereal Science*, *57*(3), 325–332.

Dapčević-Hadnađev, T., Tomić, J., Škrobot, D., Šarić, B., & Hadnađev, M. (2022). Processing strategies to improve the breadmaking potential of whole-grain wheat and non-wheat flours. *Discover Food*, *2*, 11. https://doi.org/10.1007/s44187-022-00012-w.

Day, L., Bhandari, D. G., Greenwell, P., Leonard, S. A., & Schofield, J. D. (2006). Characterization of wheat puroindoline proteins. *The FEBS Journal*, *273*(23), 5358–5373. https://doi.org/10.1111/J.1742-4658.2006.05528.X.

de Almeida, J. L., Pareyt, B., Gerits, L. R., & Delcour, J. A. (2014). Effect of wheat grain steaming and washing on lipase activity in whole grain flour. *Cereal Chemistry*, *91*(4), 321–326. https://doi.org/10.1094/CCHEM-09-13-0197-CESI.

de Kock, S., Taylor, J., & Taylor, J. R. N. (1999). Effect of heat treatment and particle size of different brans on loaf volume of brown bread. *LWT—Food Science and Technology*, *32*(6), 349–356. https://doi.org/10.1006/FSTL.1999.0564.

de La Hera, E., Gomez, M., & Rosell, C. M. (2013). Particle size distribution of rice flour affecting the starch enzymatic hydrolysis and hydration properties. *Carbohydrate Polymers*, *98*(1), 421–427. https://doi.org/10.1016/J.CARBPOL.2013.06.002.

de La Hera, E., Rosell, C. M., & Gomez, M. (2014). Effect of water content and flour particle size on gluten-free bread quality and digestibility. *Food Chemistry*, *151*, 526–531. https://doi.org/10.1016/J.FOODCHEM.2013.11.115.

B. Dhiraj, P. Prabhasankar, "Influence of Wheat-Milled Products and Their Additive Blends on Pasta Dough Rheological, Microstructure, and Product Quality Characteristics", *International Journal of Food Science*, vol. 2013, Article ID 538070, 11 pages, 2013. https://doi.org/10.1155/2013/538070

Dholakia, B. B., Ammiraju, J. S. S., Singh, H., Lagu, M. D., Röder, M. S., Rao, V. S., Dhaliwal, H. S., Ranjekar, P. K., & Gupta, V. S. (2003). Molecular marker analysis of kernel size and shape in bread wheat. *Plant Breeding*, *122*(5), 392–395. https://doi.org/10.1046/J.1439-0523.2003.00896.X.

Dhua, S., Kumar, K., Kumar, Y., Singh, L., & Sharanagat, V. S. (2021). Composition, characteristics and health promising prospects of black wheat: a review. *Trends in Food Science & Technology, 112*, 780–794. https://doi.org/10.1016/J.TIFS.2021.04.037.

Di Silvestro, R., di Loreto, A., Marotti, I., Bosi, S., Bregola, V., Gianotti, A., Quinn, R., & Dinelli, G. (2014). Effects of flour storage and heat generated during milling on starch, dietary fibre and polyphenols in stoneground flours from two durum-type wheats. *International Journal of Food Science & Technology, 49*(10), 2230–2236. https://doi.org/10.1111/IJFS.12536.

Doblado-Maldonado, A. F., Arndt, E. A., & Rose, D. J. (2013). Effect of salt solutions applied during wheat conditioning on lipase activity and lipid stability of whole wheat flour. *Food Chemistry, 140*(1–2), 204–209. https://doi.org/10.1016/J.FOODCHEM.2013.02.071.

Doblado-Maldonado, A. F., Pike, O. A., Sweley, J. C., & Rose, D. J. (2012). Key issues and challenges in whole wheat flour milling and storage. *Journal of Cereal Science, 56*(2), 119–126. https://doi.org/10.1016/J.JCS.2012.02.015.

Dobraszczyk, B. J., Whitworth, M. B., Vincent, J. F. V., & Khan, A. A. (2002). Single kernel wheat hardness and fracture properties in relation to density and the modelling of fracture in wheat endosperm. *Journal of Cereal Science, 35*(3), 245–263. https://doi.org/10.1006/JCRS.2001.0399.

Dornez, E., Holopainen, U., Cuyvers, S., Poutanen, K., Delcour, J. A., Courtin, C. M., & Nordlund, E. (2011). Study of grain cell wall structures by microscopic analysis with four different staining techniques. *Journal of Cereal Science, 54*(3), 363–373. https://doi.org/10.1016/j.jcs.2011.07.003.

Drakos, A., Kyriakakis, G., Evageliou, V., Protonotariou, S., Mandala, I., & Ritzoulis, C. (2017). Influence of jet milling and particle size on the composition, physicochemical and mechanical properties of barley and rye flours. *Food Chemistry, 215*, 326–332. https://doi.org/10.1016/J.FOODCHEM.2016.07.169.

Du, B., Meenu, M., & Xu, B. (2020). Insights into improvement of physiochemical and biological properties of dietary fibers from different sources via micron technology. *Food Reviews International, 36*, 367–383. https://doi.org/10.1080/87559129.2019.1649690.

Dziki, D., & Laskowski, J. (2000). Investigation of wheat milling properties (in Polish). *InŜynieria Rolnicza, 8*, 63–70.

Dziki, D., & Laskowski, J. (2005). Wheat kernel physical properties and milling process *. Acta Agrophysica, 6*(1), 59–71.

Edmund, J. T., & Perry, K. W. (2008). Soft wheat quality. In G. S. Servet & S. Serpil (Eds.), *Food Engineering Aspects of Baking Sweet Goods* (pp. 1–30). CRC Press.

El-Khayat, G. H., Samaan, J., Manthey, F. A., Fuller, M. P., & Brennan, C. S. (2006). Durum wheat quality I: Some physical and chemical characteristics of Syrian durum wheat genotypes. *International Journal of Food Science & Technology, 41*(Suppl. 2), 22–29. https://doi.org/10.1111/J.1365-2621.2006.01245.X.

El-Naggar, S. M., & Mikhaiel, A. A. (2011). Disinfestation of stored wheat grain and flour using gamma rays and microwave heating. *Journal of Stored Products Research, 47*(3), 191–196. https://doi.org/10.1016/J.JSPR.2010.11.004.

Emanuelson, J., Jørgensen, J. R., Andersen, S. B. F., & Jensen, C. R. (2003). Wheat grain composition and implications for bread quality ministry of food, agriculture and fisheries. *DIAS Report Plant Production, 92*, 3–33. www.agrsci.dk.

Faridi, H., Gaines, C., & Finney, P. (1994). Soft wheat quality in production of cookies and crackers. *Wheat*, 154–168. https://doi.org/10.1007/978-1-4615-2672-8_11.

Feiz, L., Martin, J. M., & Giroux, M. J. (2008). Relationship between wheat (*Triticum aestivum* L.) grain hardness and wet-milling quality. *Cereal Chemistry*, *85*(1), 44–50. https://doi.org/10.1094/CCHEM-85-1-0044.

Ficco, D. B. M., Mastrangelo, A. M., Trono, D., Borrelli, G. M., de Vita, P., Fares, C., Beleggia, R., Platani, C., & Papa, R. (2014). The colours of durum wheat: a review. *Crop & Pasture Science*, *65*, 1–15. https://doi.org/10.1071/CP13293.

Gaines, C. S., Finney, P. L., & Andrews, L. C. (1997). Influence of kernel size and shriveling on soft wheat milling and baking quality. *Cereal Chemistry*, *74*(6), 700–704. https://doi.org/10.1094/CCHEM.1997.74.6.700.

Gao, W., Chen, F., Wang, X., & Meng, Q. (2020). Recent advances in processing food powders by using superfine grinding techniques: a review. *Comprehensive Reviews in Food Science and Food Safety*, *19*(4), 2222–2255. https://doi.org/10.1111/1541-4337.12580.

Garg, M., Chawla, M., Chunduri, V., Kumar, R., Sharma, S., Sharma, N. K., Kaur, N., Kumar, A., Mundey, J. K., Saini, M. K., & Singh, S. P. (2016). Transfer of grain colors to elite wheat cultivars and their characterization. *Journal of Cereal Science*, *71*, 138–144. https://doi.org/10.1016/J.JCS.2016.08.004.

Garg, M., Kaur, S., Sharma, A., Kumari, A., Tiwari, V., Sharma, S., Kapoor, P., Sheoran, B., Goyal, A., & Krishania, M. (2022). Rising demand for healthy foods-anthocyanin biofortified colored wheat is a new research trend. *Frontiers in Nutrition*, 913. https://doi.org/10.3389/FNUT.2022.878221.

Ghafoor, K., Özcan, M. M., AL-Juhaımı, F., Babıker, E. E., Sarker, Z. I., Ahmed, I. A. M., & Ahmed, M. A. (2017). Nutritional composition, extraction, and utilization of wheat germ oil: A review. *European Journal of Lipid Science and Technology*, *119*(7), 1600160.

Gil, A., Ortega, R. M., & Maldonado, J. (2011). Wholegrain cereals and bread: A duet of the Mediterranean diet for the prevention of chronic diseases. *Public Health Nutrition*, *14*(12A), 2316–2322. https://doi.org/10.1017/S1368980011002576.

Graveland, A., & Henderson, M. H. (1991). *Improved flour* (Patent No. EP 0459551 A1). https://patentimages.storage.googleapis.com/cd/ec/c3/9da44de5c68049/EP0459551A1.pdf.

Guan, E., Yang, Y., Pang, J., Zhang, T., Li, M., & Bian, K. (2020). Ultrafine grinding of wheat flour: Effect of flour/starch granule profiles and particle size distribution on falling number and pasting properties. *Food Science & Nutrition*, *8*(6), 2581–2587. https://doi.org/10.1002/FSN3.1431.

Guo, X., Wu, S., & Zhu, K. (2020). Effect of superheated steam treatment on quality characteristics of whole wheat flour and storage stability of semi-dried whole wheat noodle. *Food Chemistry*, *322*, 126738. https://doi.org/10.1016/j.foodchem.2020.126738.

Gupta, R., Meghwal, M., & Prabhakar, P. K. (2021). Bioactive compounds of pigmented wheat (*Triticum aestivum*): Potential benefits in human health. *Trends in Food Science & Technology*, *110*, 240–252. https://doi.org/10.1016/J.TIFS.2021.02.003.

Gürsoy, S., & Güzel, E. (2010). Determination of physical properties of some agricultural grains. *Research Journal of Applied Sciences, Engineering and Technology*, *2*(5), 492–498.

Gutiérrez-García, A. S., Carballo-Carballo, A., Mejía-Contreras, J. A., Vargas-Hernández, M., Trethowan, R., & Villaseñor-Mir, H. E. (2006). Characterization of bread wheat using seed physical and physiological quality parameters. *Technical Agriculture in México*, *32*, 45–55. www.cabdirect.org/cabdirect/abstract/20073255139.

Hammermeister, A. (2008). The Anatomy of cereal seed: Optimizing grain quality involves getting the right proportion within the seed. *Journal of Food Quality*, *20*, 279–289.

Haque, M. A., Shams-Ud-Din, M., & Haque, A. (2002). The effect of aqueous extracted wheat bran on the baking quality of biscuit. *International Journal of Food Science & Technology, 37*(4), 453–462. https://doi.org/10.1046/J.1365-2621.2002.00583.X.

Haros, M., Rosell, C. M., & Benedito, C. (2002). Improvement of flour quality through carbohydrases treatment during wheat tempering. *Journal of Agricultural and Food Chemistry, 50*(14), 4126–4130. https://doi.org/10.1021/jf020059k.

Hemery, Y., Chaurand, M., Holopainen, U., Lampi, A. M., Lehtinen, P., Piironen, V., Sadoudi, A., & Rouau, X. (2011). Potential of dry fractionation of wheat bran for the development of food ingredients, part I: Influence of ultra-fine grinding. *Journal of Cereal Science, 53*(1), 1–8. https://doi.org/10.1016/J.JCS.2010.09.005.

Hoevenaars, F. P. M., Esser, D., Schutte, S., Priebe, M. G., Vonk, R. J., van den Brink, W. J., van der Kamp, J. W., Stroeve, J. H. M., Afman, L. A., & Wopereis, S. (2019). Whole grain wheat consumption affects postprandial inflammatory response in a randomized controlled trial in overweight and obese adults with mild hypercholesterolemia in the Graandioos study. *The Journal of Nutrition, 149*(12), 2133–2144. https://doi.org/10.1093/JN/NXZ177.

Hrušková, M., & Švec, I. (2009). Wheat hardness in relation to other quality factors. *Czech Journal of Food Sciences, 27*(4), 240–248. https://doi.org/10.17221/71/2009-CJFS.

Hung, P. V., Maeda, T., Fujita, M., & Morita, N. (2007). Dough properties and breadmaking qualities of whole waxy wheat flour and effects of additional enzymes. *Journal of the Science of Food and Agriculture, 87*(13), 2538–2543. https://doi.org/10.1002/jsfa.3025.

Imran, M., Ahmad, N., Anjum, F. M., Khan, M. K., Mushtaq, Z., Nadeem, M., & Hussain, S. (2015). Potential protective properties of flax lignan secoisolariciresinol diglucoside. *Nutrition Journal, 14*(1), 1–7. https://doi.org/10.1186/S12937-015-0059-3/PEER-REVIEW.

Işıklı, N. D., Şenol, B., & Çoksöyler, N. (2014). Some physical and mechanical properties of roasted Zerun wheat. *Journal of Food Science and Technology, 51*(9), 1990. https://doi.org/10.1007/S13197-012-0704-1.

Jia, W. T., Yang, Z., Guo, X. N., & Zhu, K. X. (2021). Effect of superheated steam treatment on the lipid stability of whole wheat flour. *Food Chemistry, 363*, 130333. https://doi.org/10.1016/J.FOODCHEM.2021.130333.

Kang, M. J., Kim, M. J., Kwak, H. S., & Kim, S. S. (2019). Effects of milling methods and cultivars on physicochemical properties of whole-wheat flour. *Journal of Food Quality, 2019*. https://doi.org/10.1155/2019/3416905.

Kapoor, P., Kumari, A., Sheoran, B., Sharma, S., Kaur, S., Bhunia, R. K., Rajarammohan, S., Bishnoi, M., Kondepudi, K. K., & Garg, M. (2022). Anthocyanin biofortified colored wheat modifies gut microbiota in mice. *Journal of Cereal Science, 104*, 103433. https://doi.org/10.1016/J.JCS.2022.103433.

Karimi, M., Kheiralipour, K., Tabatabaeefar, A., Khoubakht, G. M., Naderi, M., & Heidarbeigi, K. (2009). The effect of moisture content on physical properties of wheat. *Pakistan Journal of Nutrition, 8*(1), 90–95. https://doi.org/10.3923/PJN.2009.90.95.

Kaukab, S., Mir, N. A., Ritika, & Yadav, D. N. (2022). Interventions in wheat processing quality of end products. In *New Horizons in Wheat and Barley Research* (pp. 789–808). Springer. https://doi.org/10.1007/978-981-16-4449-8_30.

Keran, H., Salkić, M., Odobašić, A., Jašić, M., Ahmetović, N., & Šestan, I. (2009). The importance of determination of some physical–chemical properties of wheat and flour. *Agriculturae Conspectus Scientificus, 74*(3), 197–200.

Kihlberg, I., Johansson, L., Kohler, A., & Risvik, E. (2004). Sensory qualities of whole wheat pan bread—Influence of farming system, milling and baking technique. *Journal of Cereal Science, 39*(1), 67–84. https://doi.org/10.1016/S0733-5210(03)00067-5.

Kim, C. S., & Walker, C. E. (1992). Interactions between starches, sugars and emulsifiers in high ratio cake model systems. *Cereal Chemistry, 69*(2), 206–212.

Korolchuk, T. (2005). *US20060073258A1 — Process for producing an ultrafine-milled whole-grain wheat flour and products thereof—Google Patents.* https://patents. google.com/patent/US20060073258A1/en

Kumari, A., Sharma, S., Sharma, N., Chunduri, V., Kapoor, P., Kaur, S., . . . & Garg, M. (2020). Influence of biofortified colored wheats (purple, blue, black) on physicochemical, antioxidant and sensory characteristics of chapatti (Indian flatbread). *Molecules, 25*(21), 5071.

Létang, C., Samson, M. F., Lasserre, T. M., Chaurand, M., & Abécassis, J. (2002). Production of starch with very low protein content from soft and hard wheat flours by jet milling and air classification. *Cereal Chemistry, 79*(4), 535–543. https://doi. org/10.1094/CCHEM.2002.79.4.535.

Li, E., Dhital, S., & Hasjim, J. (2014). Effects of grain milling on starch structures and flour/starch properties. *Starch-Stärke, 66*(1–2), 15–27.

Li, J., Kang, J., Wang, L., Li, Z., Wang, R., Chen, Z. X., & Hou, G. G. (2012). Effect of water migration between arabinoxylans and gluten on baking quality of whole wheat bread detected by magnetic resonance imaging (MRI). *Journal of Agricultural and Food Chemistry, 60*(26), 6507–6514. https://doi.org/10.1021/JF301195K/ASSET/ IMAGES/MEDIUM/JF-2012-01195K_0004.GIF.

Liu, C., Liu, L., Li, L., Hao, C., Zheng, X., Bian, K., Zhang, J., & Wang, X. (2015). Effects of different milling processes on whole wheat flour quality and performance in steamed bread making. *LWT—Food Science and Technology, 62*(1), 310–318. https://doi.org/10.1016/J.LWT.2014.08.030.

Lyu, H., Gao, B., He, F., Zimmerman, A. R., Ding, C., Huang, H., & Tang, J. (2018). Effects of ball milling on the physicochemical and sorptive properties of biochar: Experimental observations and governing mechanisms. *Environmental Pollution, 233*, 54–63. https://doi.org/10.1016/j.envpol.2017.10.037.

Mabille, F., & Abecassis, J. (2003). Parametric modelling of wheat grain morphology: A new perspective. *Journal of Cereal Science, 37*(1), 43–53. https://doi.org/10.1006/ JCRS.2002.0474.

MacDonald, R., Rowe, D., Martin, E., & Gorringe, L. (2016). The spiral jet mill cut size equation. *Powder Technology, 299*, 26–40.

Manley, M., du Toit, G., & Geladi, P. (2011). Tracking diffusion of conditioning water in single wheat kernels of different hardnesses by near infrared hyperspectral imaging. *Analytica Chimica Acta, 686*(1–2), 64–75. https://doi.org/10.1016/J. ACA.2010.11.042.

Marshall, D. R., Ellison, F. W., & Mares, D. J. (1984). Effects of grain shape and size on milling yields in wheat. I. Theoretical analysis based on simple geometric models. *Australian Journal of Agricultural Research, 35*(5), 619–630. https://doi. org/10.1071/AR9840619.

Meriles, S. P., Steffolani, M. E., Penci, M. C., Curet, S., Boillereaux, L., & Ribotta, P. D. (2022). Effects of low-temperature microwave treatment of wheat germ. *Journal of the Science of Food and Agriculture, 102*(6), 2538–2544. https://doi.org/10.1002/ JSFA.11595.

Meuser, F. (2003). MILLING I types of mill and their uses. *Encyclopedia of Food Sciences and Nutrition*, 3987–3997. https://doi.org/10.1016/B0-12-227055-X/00793-8.

Midoux, N., Hošek, P., Pailleres, L., & Authelin, J. R. (1999). Micronization of pharmaceutical substances in a spiral jet mill. *Powder Technology*, *104*(2), 113–120. https://doi.org/10.1016/S0032-5910(99)00052-2.

Mohsenin, N. N. (1980). *Physical Properties of Plant and Animal Materials. Gordon and Breach Science Publishers* (vol. 1, 2nd ed.). Routledge Taylor & Francis Group. www.routledge.com/Physical-Properties-of-Plant-and-Animal-Materials-v-1-Physical-Characteristics/Mohsenin/p/book/9780677023007

Molenda, M., & Horabik, J. (2005). Part I. Characterization of mechanical properties of particulate solids for storage and handling. In J. Horabik & J. Laskowski (Eds.), *Mechanical Properties of Granular Agro-Materials and Food Powders for Industrial Practice* (pp. 40–41). Institute of Agrophysics PAS.

Morgan, B. C., Dexter, J. E., & Preston, K. R. (2000). Relationship of kernel size to flour water absorption for Canada western red spring wheat. *Cereal Chemistry*, *77*(3), 286–292. https://doi.org/10.1094/CCHEM.2000.77.3.286.

Morris, C. F. (2002). Puroindolines: The molecular genetic basis of wheat grain hardness. *Plant Molecular Biology*, *48*(5–6), 633–647. https://doi.org/10.1023/A:1014837431178.

Muttakin, S., Kim, M. S., & Lee, D. U. (2015). Tailoring physicochemical and sensorial properties of defatted soybean flour using jet-milling technology. *Food Chemistry*, *187*, 106–111. https://doi.org/10.1016/J.FOODCHEM.2015.04.104.

Nakach, M., Authelin, J. R., Corsini, C., & Gianola, G. (2019). Jet milling industrialization of sticky active pharmaceutical ingredient using quality-by-design approach. *Pharmaceutical Development and Technology*, *24*(7), 849–863. https://doi.org/10.1080/10837450.2019.1608449/SUPPL_FILE/IPHD_A_1608449_SM0100.DOCX.

Nesvadba, P., Houška, M., Wolf, W., Gekas, V., Jarvis, D., Sadd, P. A., & Johns, A. I. (2004). Database of physical properties of agro-food materials. *Journal of Food Engineering*, *61*(4), 497–503. https://doi.org/10.1016/S0260-8774(03)00213-9.

Niu, M., Hou, G. G., Wang, L., & Chen, Z. (2014). Effects of superfine grinding on the quality characteristics of whole-wheat flour and its raw noodle product. *Journal of Cereal Science*, *60*(2), 382–388. https://doi.org/10.1016/J.JCS.2014.05.007.

Nuttall, J. G., O'Leary, G. J., Panozzo, J. F., Walker, C. K., Barlow, K. M., & Fitzgerald, G. J. (2017). Models of grain quality in wheat—A review. *Field Crops Research*, *202*, 136–145. https://doi.org/10.1016/j.fcr.2015.12.011.

Obadi, M., Zhu, K. X., Peng, W., Noman, A., Mohammed, K., & Zhou, H. M. (2018). Characterization of oil extracted from whole grain flour treated with ozone gas. *Journal of Cereal Science*, *79*, 527–533. https://doi.org/10.1016/J.JCS.2017.12.007.

Oghbaei, M., & Prakash, J. (2013). Effect of fractional milling of wheat on nutritional quality of milled fractions. *Trends in Carbohydrate Research*, *5*(1).

Oladunmoye, O. O., Akinoso, R., & Olapade, A. A. (2010). Evaluation of some physical-chemical properties of wheat, cassava, maize and cowpea flours for bread making. *Journal of Food Quality*, *33*(6), 693–708. https://doi.org/10.1111/J.1745-4557.2010.00351.X.

Ortiz de Erive, M., He, F., Wang, T., & Chen, G. (2020). Development of β-glucan enriched wheat bread using soluble oat fiber. *Journal of Cereal Science*, *95*, 103051. https://doi.org/10.1016/J.JCS.2020.103051.

Ou, Z. B., & Wang, Q. (2010). Applications on grain and oil processing in micronized technology and its machinery—CNKI. *Guangdong Agricultural Sciences*, *37*(7), 192–194. https://global.cnki.net/kcms/detail/detail.aspx?filename=GDNY201007091&dbcode=CJFQ&dbname=CJFD2010&v=.

Pagani, M. A., Giordano, D., Cardone, G., Pasqualone, A., Casiraghi, M. C., Erba, D., Blandino, M., & Marti, A. (2020). Nutritional features and bread-making performance of wholewheat: Does the milling system matter? *Foods (Basel, Switzerland)*, *9*(8). https://doi.org/10.3390/FOODS9081035.

Pagani, M. A., Marti, A., & Bottega, G. (2014). Wheat milling and flour quality evaluation. *Bakery Products Science and Technology*, 17–53. https://doi.org/10.1002/9781118792001.CH2.

Palaniandy, S., Azizi Mohd Azizli, K., Hussin, H., & Fuad Saiyid Hashim, S. (2008). Mechanochemistry of silica on jet milling. *Journal of Materials Processing Technology*, 205(1–3), 119–127. https://doi.org/10.1016/J.JMATPROTEC.2007.11.086.

Pandey, M., Bansal, S., & Chawla, G. (2022). Evaluation of lifespan promoting effects of biofortified wheat in Drosophila melanogaster. *Experimental Gerontology, 160*. https://doi.org/10.1016/J.EXGER.2022.111697.

Pasha, I., Anjum, F. M., & Morris, C. F. (2010). Grain hardness: A major determinant of wheat quality. *Food Science and Technology International = Ciencia y Tecnologia de Los Alimentos Internacional, 16*(6), 511–522. https://doi.org/10.1177/1082013210379691.

Patil, S., Sonawane, S. K., & Dabade, A. (2021). Recent advances in the technology of chapatti: an Indian traditional unleavened flatbread. *Journal of Food Science and Technology, 58*(9), 3270–3279. https://doi.org/10.1111/jfpp.16529.

Pedersen, B., Knudsen, K. E., & Eggum, B. O. (1989). Nutritive value of cereal products with emphasis on the effect of milling. *World Review of Nutrition and Dietetics, 60*, 1–91. https://doi.org/10.1159/000417519.

Pena, R. J. (2002). Wheat for bread and other foods. In B. C. Curtis, S. Rajaram, & H. G. Macpherson (Eds.), *Bread Wheat—Improvement and Production* (p. 30). FAO Plant Production and Protection Series.

Ponce-García, N., Figueroa, J. D. C., López-Huape, G. A., Martínez, H. E., & Martínez-Peniche, R. (2008). Study of viscoelastic properties of wheat kernels using compression load method. *Cereal Chemistry, 85*(5), 667–672. https://doi.org/10.1094/CCHEM-85-5-0667.

Posner, E. S. (2003). MILLING | principles of milling. In *Encyclopedia of Food Sciences and Nutrition* (pp. 3980–3986). Academic Press. https://doi.org/10.1016/B0-12-227055-X/00792-6.

Posner, E. S. (2009). CHAPTER 5: Wheat flour milling. In *Wheat: Chemistry and Technology* (4th ed., pp. 119–152). AACC International, Inc. https://doi.org/10.1094/9781891127557.005.

Posner, E. S., & Hibbs, A. N. (2005). *Wheat flour milling* (2nd ed.). American Association of Cereal Chemists, Inc. www.cabdirect.org/cabdirect/abstract/20043213266

Poudel, R., & Rose, D. J. (2018). Changes in enzymatic activities and functionality of whole wheat flour due to steaming of wheat kernels. *Food Chemistry, 263*, 315–320. https://doi.org/10.1016/J.FOODCHEM.2018.05.022.

Prabhasankar, P., & Rao, P. H. (2001). Effect of different milling methods on chemical composition of whole wheat flour. *European Food Research and Technology, 213*(6), 465–469. https://doi.org/10.1007/S002170100407.

Protonotariou, S., Batzaki, C., Yanniotis, S., & Mandala, I. (2016). Effect of jet milled whole wheat flour in biscuits properties. *LWT, 74*, 106–113. https://doi.org/10.1016/J.LWT.2016.07.030.

Protonotariou, S., Drakos, A., Evageliou, V., Ritzoulis, C., & Mandala, I. (2014). Sieving fractionation and jet mill micronization affect the functional properties of wheat flour. *Journal of Food Engineering, 134*, 24–29. https://doi.org/10.1016/J.JFOODENG.2014.02.008.

Protonotariou, S., Ritzoulis, C., & Mandala, I. (2021). Jet milling conditions impact on wheat flour particle size. *Journal of Food Engineering, 294*, 110418. https://doi.org/10.1016/J.JFOODENG.2020.110418.

Qu, C., Yang, Q., Ding, L., Wang, X., Liu, S., & Wei, M. (2021). The effect of microwave stabilization on the properties of whole wheat flour and its further interpretation by molecular docking. *BMC Chemistry, 15*(1), 1–9. https://doi.org/10.1186/S13065-021-00782-X/FIGURES/7.

Qu, H., Madl, R. L., Takemoto, D. J., Baybutt, R. C., & Wang, W. (2005). Lignans are involved in the antitumor activity of wheat bran in colon cancer SW480 cells. *The Journal of Nutrition, 135*(3), 598–602. https://doi.org/10.1093/JN/135.3.598.

Rebellato, A. P., Klein, B., Wagner, R., & Azevedo Lima Pallone, J. (2018). Fortification of whole wheat flour with different iron compounds: Effect on quality parameters and stability. *Journal of Food Science and Technology, 55*(9), 3575. https://doi.org/10.1007/S13197-018-3283-Y.

Rosa, N. N., Barron, C., Gaiani, C., Dufour, C., & Micard, V. (2013). Ultra-fine grinding increases the antioxidant capacity of wheat bran. *Journal of Cereal Science, 57*(1), 84–90. https://doi.org/10.1016/J.JCS.2012.10.002.

Rosentrater, K. A., & Evers, A. D. (2017). *Kent's Technology of Cereals* (N. R. Bandeira, Ed., 5th ed.). Elsevier. https://books.google.co.in/books?hl=en&lr=&id=tOZGDgAAQBAJ&oi=fnd&pg=PP1&dq=Kents+technology+of+cereals:+An+introduction+for+students+of+food+science+and+agriculture&ots=rU2gjpuE0C&sig=RFOGmBylZ4J7dSdy5sOL8QtIusU&redir_esc=y#v=onepage&q=Kents%20technology%20of%20cereals%3A%20An%20introduction%20for%20students%20of%20food%20science%20and%20agriculture&f=false

Ross, A. S., & Kongraksawech, T. (2018). Characterizing whole-wheat flours produced using a commercial stone mill, laboratory mills, and household single-stream flour mills. *Cereal Chemistry, 95*(2), 239–252. https://doi.org/10.1002/CCHE.10029.

Saini, P., Kumar, N., Kumar, S., Mwaurah, P. W., Panghal, A., Attkan, A. K., Singh, V. K., Garg, M. K., & Singh, V. (2020). Bioactive compounds, nutritional benefits and food applications of colored wheat: A comprehensive review. *Critical Reviews in Food Science and Nutrition*, 1–14. https://doi.org/10.1080/10408398.2020.1793727.

Sak, K. (2014). Cytotoxicity of dietary flavonoids on different human cancer types. *Pharmacognosy Reviews, 8*(16), 122. https://doi.org/10.4103/0973-7847.134247.

Sakhare, S. D., Inamdar, A. A., Indrani, D., Kiran, M. H. M., & Rao, G. V. (2013). Physicochemical and microstructure analysis of flour mill streams and milled products Effect of ingredients, additives on rheological, physico-sensory and nutritional characteristics of low carbohydrate, high fat bread View project Additives for bakery products view project physicochemical and microstructure analysis of flour mill streams and milled products. *Journal of Food Science and Technology, 52*(1), 407–414. https://doi.org/10.1007/s13197-013-1029-4.

Saldivar, S. O. S. (2010). *Cereal Grains Properties, Processing, and Nutritional Attributes.* CRC Press.

Sang, S., Ju, J., Lambert, J. D., Lin, Y., Hong, J., Bose, M., Wang, S., Bai, N., He, K., Reddy, B. S., Ho, C. T., Li, F., & Yang, C. S. (2006). Wheat bran oil and its fractions inhibit human colon cancer cell growth and intestinal tumorigenesis in Apcmin/+ Mice. *Journal of Agricultural and Food Chemistry, 54*(26), 9792–9797. https://doi.org/10.1021/JF0620665.

Sanguansri, P., & Augustin, A. M. (2006). Nanoscale materials development-a food industry perspective cite this paper related papers. *Trends in Food Science & Technology, 17*, 547–556. https://doi.org/10.1016/j.tifs.2006.04.010.

Shahin, M. A., Symons, S. J., & Hatcher, D. W. (2013). Quantification of mildew damage in soft red winter wheat based on spectral characteristics of bulk samples: A comparison of visible-near-infrared imaging and near-infrared spectroscopy. *Food and Bioprocess Technology, 7*(1), 224–234. https://doi.org/10.1007/S11947-012-1046-8.

Shi, L., Cheng, F., Zhu, P. X., & Lin, Y. (2015). Physicochemical changes of maize starch treated by ball milling with limited water content. *Starch-Stärke, 67*(9–10), 772–779. https://doi.org/10.1002/star.201500026.

Singh, C. B., Jayas, D. S., Paliwal, J., & White, N. D. G. (2010). Identification of insect-damaged wheat kernels using short-wave near-infrared hyperspectral and digital colour imaging. *Computers and Electronics in Agriculture, 73*(2), 118–125. https://doi.org/10.1016/J.COMPAG.2010.06.001.

Sissons, M. (2008). Role of durum wheat composition on the quality of pasta and bread. *Food, 2*(2), 75–90.

Sivam, A. S., Sun-Waterhouse, D., Quek, S. Y., & Perera, C. O. (2010). Properties of bread dough with added fiber polysaccharides and phenolic antioxidants: A review. *Journal of Food Science, 75*(8), R163–R174. https://doi.org/10.1111/J.1750-3841.2010.01815.X.

Slavin, J. L., Jacobs, D., & Marquart, L. (2001). Grain processing and nutrition. *Critical Reviews in Biotechnology, 21*(1), 49–66. https://doi.org/10.1080/20013891081683.

Tabatabaeefar, A. (2003). Moisture-dependent physical properties of wheat. *International Agrophysics, 17*, 207–211.

Thanatuksorn, P., Kawai, K., Kajiwara, K., & Suzuki, T. (2009). Effects of ball-milling on the glass transition of wheat flour constituents. *Journal of the Science of Food and Agriculture, 89*(3), 430–435.

Tian, X., Sun, B., Wang, X., Ma, S., Li, L., & Qian, X. (2020). Effects of milling methods on rheological properties of fermented and non-fermented dough. *Grain & Oil Science and Technology, 3*(3), 77–86. https://doi.org/10.1016/J.GAOST.2020.06.003.

Tong, L. T., Zhong, K., Liu, L., Qiu, J., Guo, L., Zhou, X., Cao, L., & Zhou, S. (2014). Effects of dietary wheat bran arabinoxylans on cholesterol metabolism of hypercholesterolemic hamsters. *Carbohydrate Polymers, 112*, 1–5. https://doi.org/10.1016/J.CARBPOL.2014.05.061.

Török, K., Szentmiklóssy, M., Tremmel-Bede, K., Rakszegi, M., & Tömösközi, S. (2019). Possibilities and barriers in fibre-targeted breeding: Characterisation of arabinoxylans in wheat varieties and their breeding lines. *Journal of Cereal Science, 86*, 117–123.

Tóth, Á., Prokisch, J., Sipos, P., Széles, É., Mars, É., & Gyori, Z. (2006). Effects of particle size on the quality of winter wheat flour, with a special focus on macro- and microelement concentration. *Communications in Soil Science and Plant Analysis, 37*(15–20), 2659–2672. https://doi.org/10.1080/00103620600823117.

Tsuda, T. (2012). Dietary anthocyanin-rich plants: Biochemical basis and recent progress in health benefits studies. *Molecular Nutrition & Food Research, 56*(1), 159–170. https://doi.org/10.1002/MNFR.201100526.

Tyl, C. E., & Bunzel, M. (2014). Activity-guided fractionation to identify blue wheat (UC66049 *Triticum aestivum* L.) constituents capable of inhibiting in vitro starch digestion. *Cereal Chemistry, 91*(2), 152–158. https://doi.org/10.1094/CCHEM-07-13-0138-R.

U.S. Wheat Associates. (2007). *Overview of U.S. Wheat Inspection of Wheat and Flour Testing Methods: A Guide to Understanding Wheat and Flour Quality* (Version 2). www.uswheat.org/

Vinayagam, R., Jayachandran, M., & Xu, B. (2016). Antidiabetic effects of simple phenolic acids: A comprehensive review. *Phytotherapy Research: PTR, 30*(2), 184–199. https://doi.org/10.1002/PTR.5528.

Weaver, G. L. B. S. (2001). A Miller's perspective on the impact of health claims: Nutrition today. *Nutrition Today, 36*(3), 115–118. https://journals.lww.com/nutritiontodayonline/Citation/2001/05000/A_Miller_s_Perspective_on_the_Impact_of_Health.4.aspx.

Wei, C. Y., Zhu, D., & Nyström, L. (2021). Improving wholegrain product quality by selecting lipid-stable wheat varieties. *Food Chemistry, 345*. https://doi.org/10.1016/J.FOODCHEM.2020.128683.

Wieser, H., Koehler, P., & Scherf, K. A. (2020). The two faces of wheat. *Frontiers in Nutrition, 7*, 517313. https://doi.org/10.3389/fnut.2020.517313.

Xing, J., Symons, S., Shahin, M., & Hatcher, D. (2010). Detection of sprout damage in Canada western red spring wheat with multiple wavebands using visible/near-infrared hyperspectral imaging. *Biosystems Engineering, 106*(2), 188–194. https://doi.org/10.1016/J.BIOSYSTEMSENG.2010.03.010.

Xu, Z., Meenu, M., & Xu, B. (2020). Effects of UV-C treatment and ultrafine-grinding on the biotransformation of ergosterol to vitamin D2, physiochemical properties, and antioxidant properties of shiitake and Jew's ear. *Food Chemistry, 309*, 125738. https://doi.org/10.1016/j.foodchem.2019.1257.

Yan, X., Ye, R., & Chen, Y. (2015). Blasting extrusion processing: The increase of soluble dietary fiber content and extraction of soluble-fiber polysaccharides from wheat bran. *Food Chemistry, 180*, 106–115.

Zhang, H., Wang, H., Cao, X., & Wang, J. (2018). Preparation and modification of high dietary fiber flour: A review. *Food Research International, 113*, 24–35. https://doi.org/10.1016/J.FOODRES.2018.06.068.

Zhang, J., Yu, Y., & Xu, G. H. (2006). Technology of superfine grinding and its application in food industry. *Journal of Agricultural Sciences, 27*(2), 88–90.

Zhou, Y., Dhital, S., Zhao, C., Ye, F., Chen, J., & Zhao, G. (2021). Dietary fiber-gluten protein interaction in wheat flour dough: Analysis, consequences and proposed mechanisms. *Food Hydrocolloids, 111*, 106203. https://doi.org/10.1016/J.FOODHYD.2020.106203.

Zhu, K. X., Huang, S., Peng, W., Qian, H. F., & Zhou, H. M. (2010). Effect of ultrafine grinding on hydration and antioxidant properties of wheat bran dietary fiber. *Food Research International, 43*(4), 943–948. https://doi.org/10.1016/J.FOODRES.2010.01.005.

Zhuang, K., Sun, Z., Huang, Y., Lyu, Q., Zhang, W., Chen, X., Wang, G., Ding, W., & Wang, Y. (2022). Influence of different pretreatments on the quality of wheat bran-germ powder, reconstituted whole wheat flour and Chinese steamed bread. *LWT, 161*, 113357. https://doi.org/10.1016/J.LWT.2022.113357.

8

Effect of Storage Conditions on the Nutritional Quality of Wheat

Nitin Kumar Garg, Chirag Maheshwari, Muzaffar Hasan, Amresh Kumar, Jaipraksh Bisen, Rakesh Kumar Prajapat, Nand Lal Meena, and Om Prakash Gupta

CONTENTS

8.1 Introduction

In order to be consumed by farm animals or humans, cereals are stored for various periods of time. Storage periods may extend over many months when stored surplus grain is encouraged to stabilize grain prices during extreme carryovers between crop years. The grain may lose its nutritional value and chemical composition during storage. It is important to keep in mind that the nature and extent of these changes will vary depending on the grain's initial condition, specifically its moisture content and its storage conditions. Due to the industrialization of agriculture, and farmers' ability to grow crops in greater quantities than needed for their immediate use, large quantities of cereals were required to be stored and transported. As biological materials, cereals interact with the environment in which they are placed. It is essential to store and transport seeds, foodstuffs, or raw materials in a manner that maintains their quality. It is possible to store farm products in the farm or in commercial premises outside the farm. The most significant cereal crop grown globally is wheat. When the moisture

DOI: 10.1201/9781003307938-8

level is between 12 and 13%, it can be kept for longer than a year, but it needs special handling and a clean atmosphere to prevent fungus, rat, and insect infection.

It must be preserved properly after harvesting to maintain its nutritional value and purity. Due to poor storage conditions, wheat can deteriorate while being stored. It is required to observe the crucial storage characteristics, including temperature and relative humidity, to keep track of the condition of the wheat. The best possible storage can provide both good quality and ventilation. As a result, following harvesting, the storage spaces must be appropriately set up. This preparation entails a thorough cleaning and then disinfection, derating, and repair of any wall or floor fissures that might eventually turn into nests, ideal for the growth of insect larvae. A significant portion (10–30%) of post-production wheat losses are attributable to insect infestation during storage and are estimated to be worth about $1 trillion annually (Paliwal et al., 2004; Kumar and Kalita, 2017).

Quantitative loss, the excreta, and protein metabolic by-products produced by insects have a foul odour. Wheat also produces a variety of scents while being stored, which over time come to be known as a storage smell. Aliphatic alcohols, amines, ketones, and other carbonyl chemicals are typical that cause odours to develop during storage (Zhang and Wang, 2007). *Tribolium castaneum* and *Rhyzopertha dominica* are responsible for the dominant odour. Medium-polarity aromas similarly rise with increasing storage duration, but low-polarity odours fall off during the same period (Olsson et al., 2000). Wheat seeds can be kept for a variety of times, from short-term drying only storage to longer-term recovery storage and long-term storage for specialized stocks. Storage facilities in farms are often smaller than those in commercial settings. Progressive flour milling takes place between two consecutive harvests. In contrast, flours are typically stored for a shorter amount of time. Milling breaks down the cell walls, putting the various wheat components in direct contact with air and microbes and changing the properties of the flour.

It has frequently been stated that after several weeks of storage in ideal conditions, the quality of flour improves (Baik and Donelson, 2018; Wang and Flores, 1999). However, these advantages are only temporary, and long-term flour storage done in subpar storage conditions can ultimately result in spoiling (Pyler and Gorton, 1973; Wang and Flores, 1999). Several factors can cause the flour to spoil: (1) alteration of chemical origin, such as the deterioration of protein starch and non-enzymatic oxidations; (2) alteration of enzymatic origin; (3) alteration of mechanical origin, primarily caused by shock, leading to cracks or breakage that favour other causes of alteration; and (4) microorganisms' invasion (Hackenberg, Verheyen, Jekle; Becker, 2017; Lancelot et al., 2021).

Temperature, humidity, oxygen, and carbon dioxide, as well as storage duration, determine how flour ages (Wang and Flores, 1999; Warwick, Farrington, and Shearer, 1979). Under specific circumstances, the relative moisture or humidity of the surrounding air can enhance water activity, which can lead to the degradation of organoleptic properties and even the creation of mycotoxin. Moisture can also cause lipids to oxidatively degrade, which increases the amount of free fatty acids and produces the usual rotten smell. The variation in moisture can be brought on by several factors, including poor storage conditions with high air humidity, significant temperature variations, proximity to products that may emit moisture, and pre-harvest weather conditions to increase the moisture content of a product.

Therefore, key factors for flour storage include air temperature and air humidity. In contrast to low air temperatures, higher air temperatures can result in higher specific humidity (kg of water per kilogram of dry air) and more readily available water. Additionally, a rise in temperature quickens the rate of chemical as well as biological and microbiological degradation responses (Vanhanen and Savage, 2006). According to Navarro (2012), oxygen and/or CO_2 may also have an impact on the kind of aerobic and anaerobic metabolism of microbes as well as on the amount of non-enzymatic oxidation and specific enzymatic processes (Zia-Ur-Rehman, 2006). The final component that enhances the modification phenomenon is storage time (Wang and Flores, 1999).

A number of studies have investigated the impact of different packing options, storage settings, storage times, and flour properties (Brandolini, Hidalgo, and Plizzari, 2010; Mis, 2003; Sopiwnyk et al., 2020; Vanhanen and Savage, 2006; Warwick et al., 1979; Zarzycki and Sobota, 2015). The majority of them could only be stored for up to a year. Hrukova and Machova examined two commercial wheat flours stored in jute bags for three months for moisture, wet gluten, acidity, dropping numbers, and alveographs. They demonstrated how, during this brief storage period, the chemical properties and viscoelastic qualities altered drastically. For 32 weeks, Mis (2003) examined the physical characteristics of gluten and reported that the amount of washed-out gluten gradually decreased while the gluten index value rose. In 2015, Zarzycki and Sobota took a six-week storage time into account. They were particularly interested in the diminishing quantity and apparent viscosity of wheat flour gruels kept in tightly covered containers at 20°C, 4°C, and 20°C. It was found that the viscosity of wheat flour rose over the six-week storage period and that the falling number values increased for flours stored at 20°C compared with flours stored at room temperature. Alpha-amylase activity, decreasing number, pasting qualities, and SDS sedimentation volume in overall meal and white flours of einkorn and soft wheat were examined by Brandolini et al. (2010) during 374 days, taking into account storage at temperatures ranging from 20°C to 30°C. They demonstrated that whereas decreasing number, viscosity, and SDS sedimentation rose with age, alpha-amylase activity declined. At high temperatures, the changes were significant; however, at medium and low temperatures, they were either diminished or non-existent. Viscosity and the dropping number of wheat flour were both affected by the storage temperature.

To reduce the nutritious losses of wheat and identify any potential storage issues, it is imperative to consider the losses over the past year caused by poor farming and storage practices. The main concern is to protect the wheat stock after harvest, because, depending on the situation, it might be easily affected by numerous contaminants. Wheat of the desired quality for consumption must be free of germs and other impurities as well as rich in nutrients including minerals, vitamins, and dietary fibres. Humidity, moisture, and temperature are significant elements that affect the quality of wheat when it is being stored. It should be underlined that other elements of wheat grain that promote health, such as proteins, carbs, vitamins, and minerals, are also impacted by inadequate storage conditions. Numerous studies have documented the significant nutritional loss that occurs in wheat during storage (Malaker et al., 2008; Badawi et al., 2017; El-Sisy et al., 2019). Chemical testing has proven that the insects, pests, and other organisms that cause the grain to have an unfavourable taste and smell render it unfit for usage. Unsatisfactory storage practices result in excessive moisture content in wheat stocks, which in turn encourages fungal invasions that directly degrade grain quality (Chattha et al., 2015). Fungal strain contamination of wheat stocks during storage has the potential to result in significant grain quality losses and

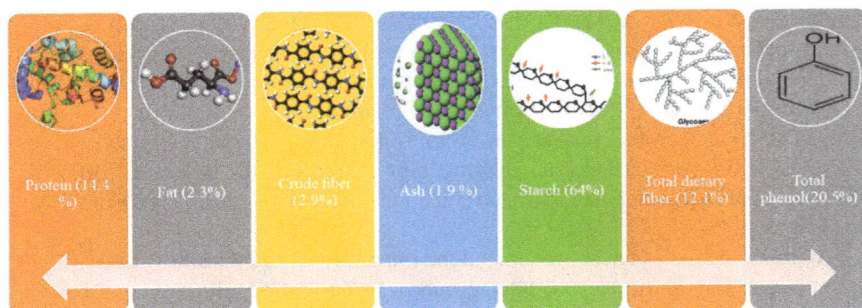

FIGURE 8.1   Nutrient content of wheat grains.

present a further threat to human health (Schmidt et al., 2016). These considerations compelled us to investigate the variables impacted by extended wheat storage. To prevent quality and quantity losses, potential storage options must also be established.

8.2 Nutritional Facts About Wheat

When used in food intake, wheat grain offers nutritional value. Regarding the nourishing component, it contains nearly every type of bioactive compound. Wheat is a food that builds health since it is a great source of protein, vitamins, minerals, and dietary fibres (Kumar et al., 2011). Wheat has 20% calories and 10.8% water by dry weight, respectively. It has dietary fibres (1.5–4.5%), carbohydrates (17%), minerals (4%), crude protein (26.50%), proteins (26–35%), crude fat (8.56%), lipids (10–15%), and ash content (4.18%). It also includes carotenoids, phytosterols, riboflavin, thiamine, tocopherols, and phytosterols. Lysine, leucine, isoleucine, and other important amino acids are incredibly abundant in wheat grains (Mughal et al., 2019).

8.3 Effect of Storage on the Nutritional Quality

Since the dawn of time, man has employed food storage as a requirement for securing the availability of a food supply. Chemical alterations impacting food's nutritional value are influenced by storage conditions. Vitamins are more prone to damage than minerals because they also contain amino acids. However, the right storage conditions help to preserve the original nutritional value of food and also improve the delivery of some nutrients and the whole quality of the product.

8.3.1 Effect on Protein Quality

The majority of the protein in wheat is gluten, which gives dough very good elasticity and extensibility. This characteristic distinguishes wheat from other grain crops. Gluten is primarily made up of the proteins gliadin and glutenin, both of which affect the viscoelastic properties of gluten and cause it to be employed in a variety of baking goods. Wheat that has been stored at high temperatures (greater than 30°C) causes the protein content to degrade, which limits the ability of gluten to function. Lower levels of wet gluten, sedimentation, and poor farinograph stability during wheat grain storage can all be attributed to the decline in wheat gluten proteins (Sisman and Ergin, 2011; Fuertes-Mendizábal et al., 2013; Kibar, 2015; Lukow et al., 1995).

However, the first two months of storage show high sedimentation and gluten content, while the following six months of storage show consecutive declines. In contrast to the initial protein level of 12.6% before preservation, Mhiko (2012) showed that the content of protein reducing to 10.8% after long-duration storage. Lysine, a crucial component of human meals and an important amino acid, was considerably decreased after extended storage. According to Mughal (2019), the lysine content was 10.26 g/100 g and started to decline in proportion to fresh stock after insufficient storage, and it is likely to continue to do so with increased annual storage. A negative impact on protein content was also caused by insect infestation in addition to prolonged storage and high temperatures (Jood et al., 1996).

Crude protein is the storage parameter that is most sensitive to storage time. During storage, the crude protein content decreased from the starting value of 13.48% to 11.37% (Polat, 2013). During the storage of cereal grains at 10°C for six months, the moisture content did not change, but at 25°C and 45°C, the level of moisture gradually decreased. After three months of storage at 25 and 45°C, lowering of the moisture contents started to become noticeable. Freshly harvested wheat grains had a moisture content of 13.75%. These moisture contents dropped to 20.00% at 25°C and 27.27% at 45°C, respectively, after six months of storage. Freshly harvested wheat grains had total accessible lysine concentration that was 2.92%. These contents dropped to varying degrees after storage. During six months of storage at 25°C and 45°C, respectively, the total accessible lysine contents of wheat grains fell by 18.7%. However, for the same periods of storage of wheat grains, the total accessible lysine levels decreased by 6.50% at 10°C. During six months of storage, there was a substantial ($P<0.05$) decrease in the total amount of accessible lysine at 25°C and 45°C, but not at 10°C.

8.3.2 Effect on Gluten Index

The gluten index is determined by using the gluten elasticity test (Raugel et al., 1999), and it depends on how long and how hot wheat flour is stored. For the evaluation of the wheat flour's quality, the gluten content must be known. There are three levels of gluten content: low, medium, and high, which are used in making cookies, biscuits, cakes and bread, respectively, and the gluten level also plays a significant influence in determining the purpose of baking. A high gluten index correlates with a low gluten percentage released by the sieve, which is a reliable sign of high-quality gluten. When the flour is refrigerated for eight weeks, the gluten index declines slowly due to the ageing of the flour. When the gluten index is high (above 95%), the gluten is robust, and when it is low (below 60%), the gluten is weak and unsuitable for the creation of bread (Violeta and Georgeta, 2010).

8.3.3 Effect on Wet Gluten Content

One of the key traits separating the commercial use of flour for bread making, pastries, and pasta from the quality standards of wheat flour is the intensity of the gluten. The storage temperature and relative humidity had an impact on the wet gluten content. After extended storage, the wet gluten content is reduced. Jennifer (2013) observed that the content of wet gluten dropped from 39.5% to 38.1% after eight weeks of storage. Gluten consistency decreases at high temperatures (over 35°C) because the gluten protein loses flexibility and brittleness after more than two weeks of storage. An additional study reported that after 180 days of storage, wet gluten levels had reduced from 30.22% to 25.45% (Karaoglu et al., 2010).

8.3.4 Effect on Dry Gluten Content

The development of dry gluten is identical to that of wet gluten for all storage conditions. The difference between moist gluten and dry gluten is its capacity to bind water. Moist gluten is colloidal in nature, comprises 60–70% water and 75–90% dry protein (gliadins and glutenins), and has high inflammatory qualities (Karaoglu et al., 2010). Dry gluten dropped from 11.40% at the beginning of the maximum storage period to 9.73% at the end, following six months of storage (Karaoglu et al., 2010).

8.3.5 Effect on Carbohydrates

The crucial macro-molecule that maintains the integrity of the grain membrane during dehydration is carbohydrates. The average temperature at which wheat grains are kept higher the concentration of soluble sugar to 25°C. When held at a higher temperature, soluble sugars were shown to decrease due to a non-enzymatic browning reaction (Maillard reaction). Starch makes up the majority of the carbohydrates in the endosperm, the meal portion of the kernel. The amount of starch in the seed decreases by 67.59% after 180 days of storage if it is kept for a long time. Previous investigations have revealed a reduction in carbohydrates after the prolonged storage of wheat (Rehman et al., 2011; Chattha et al., 2015).

The total amount of sugars is shown to greatly drop after eight years of storage (Pixton et al., 1975), while maltose and sucrose values are shown to change relatively little. Six months of cereal grain storage resulted in significant changes in the total soluble sugars. During six months of storage at 45°C, the total soluble sugars in wheat dropped by 37.2%. When cereal grains were stored for six months at 25 and 45°C, respectively, there were increases in total soluble sugars of around 9.30–17.9% and 11.9–31.8%. According to Kramer, Guyer, and Ide (1949), the activity of endogenous amylases may be the cause of the rise in soluble sugars, whereas their participation in Maillard reactions may be the cause of the decline in soluble sugars at 45°C (Glass, Ponte, Christensen, and Gedder, 1959).

8.3.6 Effect on Fatty Acids

The content of fatty acids (FAs) varies with seed variety and storage time, which are closely related to the nature of the grain. The fatty acid content of wheat increases gradually during early storage and then rapidly during storage between 240 and 270 days, according to Tian et al. (2019). Lipase hydrolysis is most likely to increase fatty acid levels and grain acidity during storage (Karaoglu et al., 2010; Pixton et al., 1975). In wheat grains, Pomeranz (1992) found that biological order protects lipids from lipases while lessening oxidation and hydrolysis. In contrast, Lukow et al. (1995)

found that the grain fatty acidity increased linearly over the course of 15 months of storage. As wheat grains were stored for 9 months and 16 years, their total titratable acids increased (Pixton et al., 1975). According to Rehman and Shah (1999), the titratable acidity of wheat grain increased significantly over six months while being stored at two different temperatures (25°C and 45°C), but not at 10°C. It is assumed that lipase enzyme hydrolysed lipids, which mostly take place in the germ and aleurone layers, and this is what causes the significant increase in fatty acid content.

8.3.7 Effect on Vitamins

Wheat grain is a natural best source of vitamins. Various factors, including high temperatures, poor handling, and unfavourable storage conditions, contribute to thiamine loss. Thiamine was identified by several researchers as a crucial component of activities that promote health (Shewry and Hey, 2015). El-Sisy et al. (2019) observed the impact of prolonged storage on the wheat quality and its flour and found that the vitamin content varied greatly depending on the origin of the wheat.

According to Rehman (2006), the amount of thiamine in wheat grain fell by 21.4 and 29.5% at 25°C and 45°C, respectively, throughout six months of storage. The nutritional quality of wheat grains kept at 10°C did not significantly alter. Wheat grains that had just been harvested contained 21.0 mg of thiamine per 100 g. Thiamine content in wheat grains decreased by 21.42% and 29.52%, respectively, after six months of storage at 25°C and 45°C. However, after six months of storage at 10°C, the thiamine content of these cereal grains remained unaltered. These outcomes are in line with those of previous researchers who discovered that storing cowpeas for six months reduced their thiamine concentration by 32% (Onayemi, Osibogun, and Obembe, 1986). Other researchers who stored lima beans, chickpeas, and other cereal at high temperatures also noticed similar thiamine losses (Burr, 1973; Jood and Kapoor, 1994).

8.3.8 Effect on Enzymatic Activity

The main enzyme, amylase, is responsible for the rapid hydrolysis of starch, or the grain food reserves, found in the wheat seeds endosperm after the germination of seed. It produces maltodextrins, which are particles of glucose (Shewry, 2009). Different temperatures and prolonged storage reduce the enzyme's activity to varying degrees. While a greater decrease was seen at 45°C, a slower decline in activity was seen at 10°C (Rehman, 2006).

When stored at 25°C and 45°C, cereal grain's protein and starch digestibilities were significantly impacted, but not at 10°C. Wheat grains' first protein digestibilities were 74.0%. Protein digestibility in wheat grains was reduced by 17.6% after six months of storage at 45°C. However, following six months of storage of wheat grains at 25°C, protein digestibilities barely decreased by 9.45%. Similarly, wheat grains that had just been harvested had starch digestibility values of 62.0%. After six months of storage, wheat grains' starch digestibilities decreased by 11.3% at 25°C and 17.7% at 45°C, respectively.

8.3.9 Effect on Antioxidant Activity

Many of the phytochemicals found in abundant plants have antioxidant properties. The endogenous antioxidant level in the body is insufficient to maintain the redox equilibrium in the body as we age, and we are exposed to a variety of different environmental stimuli. Many degenerative disorders could develop as a result of this imbalance. Therefore, the natural antioxidants found in fruits, vegetables, pulses, and cereals may also provide extra health advantages (Miller et al., 2000; Sinha et al., 2013; Kumar et al., 2014). The majority of studies on antioxidants' activity have discussed how prevalent they are in raw foods. The final product's antioxidant concentration, though, is what matters most. The plant species and type of grains, the percentage of grain, and the processing conditions all affect the antioxidant potential and bioavailability. The number of antioxidants in the finished product can be significantly influenced by storage, various processing methods, formulations, and product creation. Foods' antioxidant activity may change during processing due to interactions between nutrients, antioxidants, and/or oxidants. However, relatively few investigations on the antioxidant activity of foods made from wheat have been published (Angioloni and Collar, 2011). Significant amounts of food antioxidants activity are lost during food preparation, storage, handling at home, and cooking (Nicoli et al., 1997; Sinha and Kumar 2018). Before being ingested, wheat grain and milling products are exposed to some sort of heat processing. This could involve puffing, flaking, extruding, baking, and frying. It is commonly recognised that heat processing has an impact on nutrients like protein and carbohydrates in food, but less research has been done on how it affects the phytochemical composition and antioxidant activity. Results from different food systems demonstrated that thermal treatment greatly lowered the content of natural antioxidants, while the production of Maillard products preserved or even improved the total antioxidant capabilities of food products (Slavin et al., 2000). Although bread-baking quality improves as wheat grain or flour ages with proper storage, the effect of storage on antioxidants has received little attention. From 15 distinct wheat cultivars, whole meals and bran were made and stored for 60 days at four different temperatures (−20°C, 4°C, RT, and 60°C). At regular intervals, samples were examined for their overall antioxidant activity. For 60 days, the overall meal and bran from 15 different types of wheat were kept at four different temperatures (−20°C, 4°C, RT, and 60°C). Samples were collected every 15 days to estimate the antioxidant activity. The activity was reduced by around 25% when the entire meal was kept at −20°C, 4°C, and RT for 60 days. However, at 60°C, the reduction was just 45%. After 60 days of storage, there was a 25% drop in bran at −20°C and RT, only 16% at 4°C, and 64% at 60°C. Within the first 15 days at 60°C, the bran detected a very significant decline (38%) in the activity. However, after an entire dinner, this reduction was lower (22%). Even 15 days of storage at various temperatures revealed a sizable decline. These findings demonstrated that the storage temperature had no impact on antioxidant activity for up to 60 days, except at the higher temperature of 60°C. High storage temperatures may have a negative impact on antioxidant activity. After storage for 60 days at all four temperatures, certain types of wheat showed a lessening in antioxidant activity. However, other cultivars displayed varied patterns of antioxidant activity at various temperatures and storage times.

8.4 Conclusion

Storage of cereal grains at high temperatures had a negative impact on their nutritional quality. After six months of storage at 25°C and 45°C, the protein and starch digestibilities of cereal grains declined to varying degrees. Storage of cereal grains at 25°C and 45°C resulted in significant lysine and thiamine losses. During the six months that cereal grains were stored at 45°C, soluble sugar losses were also noted. Given these details, it is advised that cereal grains should (wheat, rice, and maize) not be stored above 25°C to reduce nutrient losses during storage.

REFERENCES

Angioloni, A., & Collar, C. (2011). Polyphenol composition and "in vitro" antiradical activity of single and multigrain breads. Journal of Cereal Science, 53(1), 90–96. https://doi.org/10.1016/j.jcs.2010.10.002.

Badawi, M. A., Seadh, S. E., Abido, W. A. E., & Hasan, R. M. (2017). Effect of storage treatments on wheat storage. International Journal of Advanced Research in Biological Sciences, 4(1), 78–91. https://doi.org/10.22192/ijarbs.2017.04.01.009.

Baik, B. K., & Donelson, T. (2018). Postharvest and postmilling changes in wheat grain and flour quality characteristics. Cereal Chemistry, 95(1), 141–148. https://doi.org/10.1002/cche.10013.

Brandolini, A., Hidalgo, A., & Plizzari, L. (2010). Storage-induced changes in einkorn (*Triticum monococcum* L.) and breadwheat (*Triticum aestivum* L. ssp. aestivum) flours. Journal of Cereal Science, 51(2), 205–212. https://doi.org/10.1016/j.jcs.2009.11.013.

Burr, H. K. (1973). Effect of storage on cooking qualities, processing and nutritive value of beans. In W. G. Jaffe (Ed.), Nutritional aspects of common beans and other legume seeds as animal and human foods. Riberrao Preto Brazi: Arch Latiaoameri Nutr.

Chattha, S. H., Hasfalina, C. M., Mahadi, M. R., Mirani, B. N., & Lee, T. S. (2015). Quality change of wheat grain during storage in a ferrocement bin. ARPN Journal of Agricultural and Biological Science, 10(8).

El-Sisy, T. T., Abd El Fadel, M. G., Gad, S. S., El-Shibiny, A. A., & Emara, M. F. (2019). Effect of handling, milling process and storage on the quality of wheat and flour in Egypt: Rheological properties of wheat kernels and their flours. Acta Scientific Microbiology, 2(10).

Fuertes-Mendizábal, T., González-Torralba, J., Arregui, L. M., González-Murua, C., González-Moro, M. B., & Estavillo, J. M. (2013). Ammonium as sole N source improves grain quality in wheat. Journal of the Science of Food and Agriculture, 93(9), 2162–2171.

Glass, R. L., Ponte, J. G., Jr., Christensen, C. M., & Gedder, W. F. (1959). Grain storage studies XXVIII. The influence of temperature and moisture level on the behaviour of wheat stored in air or nitrogen. Cereal Chemistry, 36, 341–347.

Hackenberg, S., Verheyen, C., Jekle, M., & Becker, T. (2017). Effect of mechanically modified wheat flour on dough fermentation properties and bread quality. European Food Research and Technology, 243(2), 287–296. https://doi.org/10.1007/s00217-016-2743-8.

Hrušková, M., & Machová, D. (2002). Changes of wheat flour properties during short term storage. Czech Journal of Food Sciences, 20(4), 125–130.

Jennifer, S. J. (2013). Changes in physico-chemical characteristics of wheat flour during storage and the effects on baking quality. Journal of the Science of Food and Agriculture, 210(2), 412–420.

Jood, S., & Kapoor, A. C. (1994). Vitamin contents of cereal grains as affected by storage and insect infestation. Plant Foods for Human Nutrition, 46(3), 237–243. https://doi.org/10.1007/BF01088996.

Jood, S., Kapoor, A. C., & Singh, R. (1996). Chemical composition of cereal grains as affected by storage and insect infestation. Tropical Agriculture, 73, 161–164.

Karaoglu, M. M., Aydeniz, M., Kotancilar, H. G., & Gerçelaslan, K. E. (2010). A comparison of the functional characteristics of wheat stored as grain with wheat stored in spike form. International Journal of Food Science and Technology, 45, 38–47.

Kibar, H. (2015). Influence of storage conditions on the quality properties of wheat varieties. Journal of Stored Products Research, 62, 8–15. https://doi.org/10.1016/j.jspr.2015.03.001.

Kramer, A., Guyer, R. B., & Ide, L. E. (1949). Factors affecting the objective and organoleptic evaluation of quality of sweet corn. Journal of the American Society for Horticultural Science, 54, 342–346.

Kumar, A., Maurya, B. R., & Raghuwanshi, R. (2014). Isolation and characterization of PGPR and their effect on growth, yield and nutrient content in wheat (*Triticum aestivum* L.). Biocatalysis and Agricultural Biotechnology, 3(4), 121–128.

Kumar, D., & Kalita, P. (2017). Reducing postharvest losses during storage of grain crops to strengthen food security in developing countries. Foods, 6(1), 8. https://doi.org/10.3390/foods6010008.

Kumar, P., Yadav, R. K., Gollen, B., Kumar, S., Verma, R. K., & Yadav, S. (2011). Nutritional contents and medicinal properties of wheat: A review. Life Sciences and Medicine Research, 22.

Lancelot, E., Fontaine, J., Grua-Priol, J., & Le-Bail, A. (2021). Effect of long-term storage conditions on wheat flour and bread baking properties. Food Chemistry, 346, 128902. https://doi.org/10.1016/j.foodchem.2020.128902.

Lukow, O. M., White, N. D. G., & Sinha, R. N. (1995). Influence of ambient storage conditions on the breadmaking quality of two hard red spring wheats. Journal of Stored Products Research, 31(4), 279–289. https://doi.org/10.1016/0022-474X(95)00027-5.

Malaker, P. K., Mian, I. H., Bhuiyan, K. A., Akanda, A. M., & Reza, M. M. A. (2008). Effect of storage containers and time on seed quality of wheat. Bangladesh Journal of Agricultural Research, 33(3), 469–477. https://doi.org/10.3329/bjar.v33i3.1606.

Mhiko, T. A. (2012). Determination of the causes and the effects of storage conditions on the quality of silo stored wheat (*Triticum aestivum*) in Zimbabwe. Natural Products and Bioprospecting, 2(1), 21–28. https://doi.org/10.1007/s13659-012-0004-5.

Miller, H. E., Rigelhof, F., Marquart, L., Prakash, A., & Kanter, M. (2000). Antioxidant content of whole grain breakfast cereals, fruits and vegetables. Journal of the American College of Nutrition, 19(3), 312S–319S. https://doi.org/10.1080/07315724.2000.10718966.

Mis, A. (2003). Influence of the storage of wheat flour on the physical properties of gluten. International Agrophysics, 17(2), 71–75.

Mughal, M. H. (2019). Wheat compounds – A comprehensive review. Integrative Food, Nutrition and Metabolism, 6, 1–6.

Navarro, S. (2012). The use of modified and controlled atmospheres for the disinfestation of stored products. Journal of Pest Science, 85(3), 301–322. https://doi.org/10.1007/s10340-012-0424-3.

Nicoli, M. C., Anese, M., Parpinel, M. T., Franceschi, S., & Lerici, C. R. (1997). Loss and/or formation of antioxidants during food processing and storage. Cancer Letters, 114(1–2), 71–74. https://doi.org/10.1016/s0304-3835(97)04628-4.

Olsson, J., Börjesson, T., Lundstedt, T., & Schnürer, J. (2000). Volatiles for mycological quality grading of barley grains: Determinations using gas chromatography–mass spectrometry and electronic nose. International Journal of Food Microbiology, 59(3), 167–178. https://doi.org/10.1016/s0168-1605(00)00355-x.

Onayemi, O., Osibogun, O. A., & Obembe, O. (1986). Effect of different storage and cooking methods on some biochemical, nutritional and sensory characteristics of cowpeas (*V. unguiculata L. Walp.*). Journal of Food Science, 51, 153–156.

Paliwal, J., Wang, W., Symons, S. J., & Karunakaran, C. (2004). Insect species and infestation level determination in stored wheat using near-infrared spectroscopy. Canadian Biosystems Engineering, 46, 7.17–7.24.

Pixton, S. W., Warburton, S., & Hill, S. T. (1975). Long-term storage of wheat-III: Some changes in the quality of wheat observed during 16 years of storage. Journal of Stored Products Research, 11(3–4), 177–185. https://doi.org/10.1016/0022-474X(75)90028-4.

Polat, H. E. (2013). Integration the effects of different storage types on nutritional quality characteristics of some feedstuffs. Journal of Food, Agriculture and Environment, 11, 897–903.

Pomeranz, Y. (1992). Biochemical, functional, and nutritive changes during storage. In D. B. Sauer (Ed.), Storage of cereal grains and their products (4th ed., pp. 55–141). New York: AACC International.

Pyler, E. J., & Gorton, L. A. (1973). Baking science and technology. Chicago, IL: Siebel Publishing Company.

Raugel, P. J., Robertson, P. O., & David, C. T. (1999). Perten instruments. In Rapid food analysis and hygiene monitoring (pp. 439–455). New York: Springer.

Rehman, A., Sultana, K. N., Minhas, M., Gulfraz, G., Raja, K., & Anwar, Z. (2011). Study of most prevalent wheat seed-borne mycoflora and its effect on seed nutritional value. African Journal of Microbiology Research, 5, 4328–4337.

Rehman, Z. U. (2006). Storage effects on nutritional quality of commonly consumed sereals. Food Chemistry, 95, 53–57.

Rehman, Z. U., & Shah, W. H. (1999). Biochemical changes in wheat during storage at three temperatures. Plant Foods for Human Nutrition, 54(2), 109–117. https://doi.org/10.1023/a:1008178101991.

Schmidt, M., Horstmann, S., De Colli, L. D., Danaher, M., Speer, K., Zannini, E., & Arendt, E. K. (2016). Impact of fungal contamination of wheat on grain quality criteria. Journal of Cereal Science, 69, 95–103. https://doi.org/10.1016/j.jcs.2016.02.010.

Shewry, P. R. (2009). Wheat. Journal of Experimental Botany, 60(6), 1537–1553. https://doi.org/10.1093/jxb/erp058.

Shewry, P. R., & Hey, S. J. (2015). The contribution of wheat to human diet and health. Food and Energy Security, 4(3), 178–202. https://doi.org/10.1002/fes3.64.

Sinha, S. K., & Kumar, A. (2018). Condensed tannin: A major anti-nutritional constituent of faba bean (*Vicia faba* L.). Horticulture International Journal, 2(2). https://doi.org/10.15406/hij.2018.02.00022.

Sinha, S. K., Kumar, M., Kumar, A., Bharti, S., & Shahi, V. (2013). Antioxidant activities of different tissue extract of faba bean (*Vicia faba* L.) containing phenolic compounds. Legume Research – An International Journal, 36, 496–504.

Sisman, C. B., & Ergin, A. S. (2011). The effects of different storage buildings on wheat quality. Journal of Applied Sciences, 11(14), 2613–2619. https://doi.org/10.3923/jas.2011.2613.2619.

Slavin, J. L., Jacobs, D., & Marquart, L. (2000). Grain processing and nutrition. Critical Reviews in Food Science and Nutrition, 40(4), 309–326. https://doi.org/10.1080/10408690091189176.

Sopiwnyk, E., Young, G., Frohlich, P., Borsuk, Y., Lagassé, S., Boyd, L., Bourſe, L., Sarkar, A., Dyck, A., & Malcolmson, L. (2020). Effect of pulse flour storage on flour and bread baking properties. LWT, 121, 2019.108971. https://doi.org/10.1016/j.lwt.2019.108971.

Tian, P. P., Lv, Y. Y., Yuan, W. J., Zhang, S. B., & Hu, Y. S. (2019). Effect of artificial aging on wheat quality deterioration during storage. Journal of Stored Products Research, 80, 50–56. https://doi.org/10.1016/j.jspr.2018.11.009.

Vanhanen, L. P., & Savage, G. P. (2006). The use of peroxide value as a measure of quality for walnut flour stored at five different temperatures using three different types of packaging. Food Chemistry, 99(1), 64–69. https://doi.org/10.1016/j.foodchem.2005.07.020.

Violeta, I. C., & Georgeta, S. P. (2010). Comparative evaluation of wet gluten quantity and quality through different methods, the annals of the University Dunarea de Jos of Galati, Fascicle. Food Technology, 34(2), 231–239.

Wang, L., & Flores, R. A. (1999). The effects of storage on flour quality and baking performance. Food Reviews International, 15(2), 215–234. https://doi.org/10.1080/87559129909541187.

Warwick, M. J., Farrington, W. H. H., & Shearer, G. (1979). Changes in total fatty acids and individual lipid classes on prolonged storage of wheat flour. Journal of the Science of Food and Agriculture, 30(12), 1131–1138. https://doi.org/10.1002/jsfa.2740301204.

Zarzycki, P., & Sobota, A. (2015). Effect of storage temperature on falling number and apparent viscosity of gruels from wheat flours. Journal of Food Science and Technology, 52(1), 437–443. https://doi.org/10.1007/s13197-013-0975-1.

Zhang, H., & Wang, J. (2007). Detection of age and insect damage incurred by wheat, with an electronic nose. Journal of Stored Products Research, 43(4), 489–495. https://doi.org/10.1016/j.jspr.2007.01.004.

Zia-Ur-Rehman, Z. U. (2006). Storage effects on nutritional quality of commonly consumed cereals. Food Chemistry, 95(1), 53–57. https://doi.org/10.1016/j.foodchem.2004.12.017.

9

Molecular Mechanisms of Major Bioactive Compounds for Human Health Benefits

Arti Kumari, Nand Lal Meena, Prathap V., Jyoti Prakash Singh,
Muzaffar Hasan, Chirag Maheshwari, Rakesh Kumar Prajapat,
Om Prakash Gupta, and Aruna Tyagi

CONTENTS

9.1 Introduction

9.1.1 Biological Importance of Bioactive Compounds

Plants are the primary source of medicine. The plant produces, which are consumed by us, act as a medium to supply bioactive compounds to our body. Nearly all of phytochemicals are secondary metabolites that are produced in specialized organelles in plants. Secondary bioactive metabolites, 100,000 to 200,000 in number, are ubiquitously distributed in plants (Singh, 2016). They have the ability to affect various physiological processes and actions. These bioactive compounds are used in therapeutics and pharmaceuticals. Among all bioactive components, polyphenols are mostly used for their medicinal purpose. Polyphenols include a group of compounds namely phenols, phenolic acids, phenylpropanoids, and flavonoids. Phenols or polyphenolic compounds belong to the compound containing an aromatic ring with one or more hydroxyl groups substituted to the ring, along with other functional derivatives such as esters, methyl ether, and glycosides (Tsimogiannis and Oreopoulou, 2019). The specific position of the OH group and distinct molecular structure determines the bioactivity of phenols. More than two hydroxyl groups are present in the majority of polyphenolic compounds that provide free radical scavenging activity, therefore polyphenols have a role in reducing oxidative stress. Apart from reducing oxidative stress, they also have antiviral, antibacterial, antithrombotic, anti-allergic, immuno-modulatory, and anti-inflammatory properties (Cory et al., 2018). Therefore, phenols help control dreadful diseases and aging. Nowadays, plant polyphenols are gaining popularity due to their antioxidant property to prevent serious diseases including cancer. Polyphenols are potential neutraceutical compounds. The mechanism involved in different modulatory functions is described as enzymatic detoxification, oxidation, and modulation through the host immune system, apoptosis, and some other mechanisms. Flavonoids also exhibit antioxidant properties as they transfer electrons to free radicals, slow down oxidase activity, stimulate antioxidant enzyme activity, and act as intercalating agents to metal ions (Hajam et al., 2020). The cellular response of flavonoid compounds involves intracellular signaling cascades namely protein kinase signaling cascades, such as protein kinase-C, Akt/protein kinase B, phosphoinositide 3-kinase, etc. (González-Paramás et al., 2019).

9.2 Major Bioactive Compounds in Wheat

The major bioactive compounds in wheat are phenolic acids, phenolic acid oligomers, flavonoids, carotenoids, alkylresorcinols, tocopherols, tocotrienols, benzoxazinoids, etc. (Table 9.1).

TABLE 9.1

List of Bioactive Compounds Present in Wheat

Compounds	Concentration	Structure	Health Benefits	References
Phenolic acids	900 µg/g		Anti-inflammatory, antioxidant, anti-cancer, lower risk of type-II diabetes and CVD	Tian et al., 2022
Flavonoids	214.8–272.9 mg of CE and 17.4 mg of CE per 100 g in bran and endosperm, respectively		Anti-inflammatory, antioxidant, anti-cancer	Adom et al., 2005

TABLE 9.1 (*Continued*)

Compounds	Concentration	Structure	Health Benefits	References
Tocols	20–80 µg/g DM		Anti-inflammatory and antioxidant	Lampi et al., 2008
Carotenoids	130–180 µg/100 g DM		Antioxidant, provitamin A, vision support	Zhai et al., 2016
Alkylresorcinol	200–1489 µg/g	5-Heneicosylresorcinol (AR21)	Anti-cancer, anti-inflammatory	Ross et al., 2003
Phytosterol	670–1187 µg/g		Anti-cancer, cholesterol-lowering effect	Nurmi et al., 2008
Benzoxazinoids	5 µg/g DM		Anti-inflammatory, antioxidant, anticancer, plant defense	Hanhineva et al., 2011

Abbreviations: CE – catechin equivalent, DM – dry matter, CVD – cardiovascular disease

9.2.1 Phenolic Acids

Phenolic acid is the most abundant phytochemical in wheat; it consists of a phenolic ring and carboxylic acid functional group. They are divided into two types: hydroxycinnamic acids and hydroxybenzoic acids. The common phenolic acids found in wheat are caffeic acid, vanillic acid, p-coumaric acid, syringic acid, sinapic acid, 4-hydroxybenzoic acid, cis-(ferulic acid) FA, and trans-FA. Trans-FA is the major portion of nearly 90% of total phenolic acids in wheat (Tian et al., 2022). Phenolic acid concentration ranges from 200 to 900 µg/g in wheat grain. These are present in the bran; therefore, phenolic acid content is higher in whole-grain cereal flours than in refined grain. They act as stabilizers of the cell wall, also involved in defense against microorganisms such as fungi, insects, and pests. They act as bioactive phytochemicals that have a beneficial effect on cardiovascular diseases, obesity, type II diabetes, and cancer. Apart from acting as antioxidants, the FA has a positive effect on vasodilation and increased blood flow through stimulated nitric acid production (Zao et al., 2009). There are two forms of phenolic acids, namely monomeric and oligomeric forms (Bunzel, 2010). Di-ferulic acids (dimers of FA) have been studied among the oligomers (Tian et al., 2022). Majority of DFAs exist in soluble form, whereas some of them exist in an insoluble form (Amić et al., 2020).

9.2.2 Flavonoids

Flavonoids are made of 15-carbon atoms containing two benzene rings separated by a heterocyclic ring (C6–C3–C6 structure). Flavonoids are divided in to six different classes, on the basis of the variation in the basic structure such as anthocyanidins, chalcones flavones, isoflavones, flavanones and flavonols (Singla et al., 2019). Flavonoids exist as anthocyanins, flavones, and flavone C-glycosides in wheat. Flavonoids in a major portion of wheat grain occur in C-glycosidic form. The bran (214.8–272.9 mg) of catechin equivalent (CE)/100 g) has a higher flavonoid content than the endosperm section (17.4 mg of CE/100 g, respectively) (Adom et al., 2005). Total flavonoid content was determined from six different wheat varieties as 20.14–67.69 mg of CE/100 g of grain (Leoncini et al., 2012).

9.2.3 Carotenoids

Carotenoids are terpenoids containing 40 carbon atoms. Approximately 750 carotenoids have been identified to date (Nisar et al., 2015). Common carotenoids found in wheat flour include zeaxanthin, lutein, β-carotene, and β-cryptoxanthin. Luteins are abundant (80–90% of total carotenoids) among all carotenoids. They have a functional role as provitamin-A as well as antioxidants in the diet. Total carotenoid content in winter wheat range from 130 to 180 µg/100 g DM (Zhai et al., 2016). Generally, endosperm has the highest lutein content, whereas bran has the highest content of β-carotene and zeaxanthin (Lu et al., 2014).

9.2.4 Alkylresorcinols

Alkylresorcinols (ARs) are phenolic lipids made up of resorcinol-type phenolic rings with an odd-numbered aliphatic carbon chain. The main food sources of ARs are wheat, barley, and rye (Landberg et al., 2014). AR chemically refers to homologs of 3, 5-dihydroxy-5-n-alkylbenzenes. The unsaturated AR content accounts for less than 5% in wheat. ARs reflect the intake of the bran or whole grain of wheat or rye due to their presence in the outer layer of grain. It is a type of phytochemical that is strongly heritable and is also influenced by environmental conditions such as pesticide application, soil composition, and crop fertilization (Andersson et al., 2014). ARs are absorbed directly in the body in their native form and measured at different sites such as blood plasma, adipose tissue, erythrocytes, and urine (in polar form). AR levels ranged from 200 to 1489 µg/g in 13 different wheat cultivars (Ross et al., 2003)

9.2.5 Tocopherols

Tocopherols are considered as antioxidants naturally found in cereal grains that consist of tocopherols and tocotrienols. Naturally occurring vitamin-E is present as eight isomeric forms (α, β, γ, δ-tocopherols and α, β, γ, δ-tocotrienols). Tocopherols have saturated side chains, whereas tocotrienols have unsaturated side chains. Cereal grains are rich in tocotrienols, whereas animal products are rich in α-tocopherol (Azzi, 2018). Tocopherols protect the biological membranes and act as primary antioxidants by donating hydrogen atoms to peroxy radicals during lipid oxidation, thus inhibiting lipid oxidation. They also quench nitrogen oxide radicals and singlet oxygen. The tocopherol content ranges from 20 to 80 µg/g dry matter in wheat samples (Lampi et al., 2008).

9.2.6 Benzoxazinoids

Earlier, the occurrence of benzoxazinoids (BXs) was believed to be restricted in the root and shoots of some cereals exhibiting a role in the plant defense system; however, recently, mature grains of wheat and rye have been reported to be rich in BXs. Many benefits of BXs have been reported such as anti-inflammatory, anticancer, and antioxidant activities and boosting the central nervous system (Adhikari et al., 2015). Total BX content is very low, i.e. 5 μg/g DM in whole-grain wheat (Hanhineva et al., 2011). BXs act as bioherbicides or allelochemicals in cereal plants. The BXs are present in glycoside form which are hydrolyzed by the enzymes under stress conditions. Main sources of BXs are whole-grain wheat and rye products.

9.2.7 Phytosterols

Phytosterols resemble cholesterol in structure except for the presence of additional methyl and ethyl groups (Zhu and Sang, 2017). The major phytosterol in wheat is sitosterol, covering nearly 50% of total phytosterol. Campesterol, campestanol, and sitostanol are the other phytosterols. These are present in either free form, glycoside, acylated glycosides, or ester form. Phytosterol content from 175 wheat genotypes was reported in the range of 670–1187 μg/g (Nurmi et al., 2008). Cholesterol-lowering properties of phytosterols have been investigated (Plat and Mensink, 2005).

9.3 Molecular Mechanism of the Major Bioactive Compounds Present in Wheat

9.3.1 Flavonoids

The biological activity of flavonoids depends on their bioavailability. The bioavailability is determined by the structure from poorly absorbed anthocyanins to well-absorbed isoflavones. In food, flavonoids are present as glycosides such as quercetin, genistein, daidzein, kaempferol, hesperidin, and naringenin. These flavonoid glycosides are highly stable in cooking methods, gastric enzymes, and gastric pH. They are transported as glycosides in the small intestine and deglycosylated before absorption. Colon microflora enzymes have the capacity to deconjugate flavonoid glycosides. Cancer prevention activity of flavonoids is related to inhibiting inflammation, modulating cell signaling, modulating growth factors and antioxidant activity, enhancing detoxification, and inhibiting the cell cycle. Flavonoids also act as attractants and provide plants with protection from UV light, cold, drought, and insect attacks (Ferreyra et al., 2021). These compounds have antiallergic, antiviral, antiplatelet, antitumor, and anti-inflammatory activity (Perez-Vizcaino and Fraga 2018). The flavonoids are most frequently used in the area of biomedicine to improve human health (Thilakarathna and Rupasinghe 2013). Their free radical scavenging activity and antioxidant activities are helpful in the treatment of various disorders. The acylated derivatives of flavonoids are used in the area of cosmetics, dietic, and therapy (Nakayama et al., 2003). Composition for the esterified flavonoids as the component of pharmaceutical, cosmetic, agri-food, and nutritional foodstuffs has been patented (Fukami et al., 2007). Acylated flavonoids are used to cure arteriosclerosis, angina pectoris, hyperlipidemia, and stroke. Acylated flavanone inhibits cholesterol biosynthesis by inhibiting HMG-CoA reductase enzyme and has no harmful effects on mice (Bok et al., 2002). Some acylated esters of flavonoids showed toxicity against leukemia cell lines, viz. HUT78, MOLT3, HL60, and DAUDI (Demetzos et al., 1997). Flavones isolated from sugarcane have antiproliferative activity in cancer cell lines (Duarte-Almeida et al., 2007). Antitumor and antiangiogenic activities are observed in esterified flavonoid derivatives with polyunsaturated fatty acids (Mellou et al., 2006). Catechins are derivatives of fatty acid esters and have anti-bacterial as well as anti-tumorigenesis properties (Fukami et al., 2007). Acylated quercetagetins with caffeic acid and p-coumaric acid have shown high radical scavenging activity (Parejo et al., 2005).

Flavonoids are used as antioxidant products in cosmetics that have protective properties and a skin cleansing effect and are found beneficial against aging and skin-related discoloration (Malinowska, 2013). Esterified flavonoids safeguard our skin from UV light and are found effective against sunburn, skin wrinkles, and photoaging (Maliar, 2017). The antioxidant property of flavonoids is dependent on the chemical structure of flavonoids and position of hydroxyl groups (Porras et al., 2017). Quercetin, the most abundant dietary flavonol, is a powerful antioxidant due to its right structural orientation that helps in free radical scavenging (Treml and Šmejkal 2016). The mechanism behind the antioxidant activity of flavonoids is associated with their ability to directly scavenge ROS, chelate metals, and control enzyme activity. ROS-scavenging activity of flavonoids is related to the donation of hydrogen atom, and this potential depends on the number of hydroxyl groups, how they are substituted, and how the functional groups are arranged. The catechol group which is present in the B-ring of flavonoids determines its ability to scavenge reactive oxygen and nitrogen species such as superoxide, nitric oxide, nitrogen dioxide, hydroxyl, peroxyl, and lipid peroxyl radicals by donating hydrogen that forms stable flavonoid-derivative radical (aroxyl radical) (Bros et al., 1997). High radical stability is provided by electron delocalization over the three-ring system, which is caused by the existence of a 2,3-double bond coupled to a 4-oxo function in the C-ring and a 3-(and 5-) hydroxyl group (Halliwell and Gutteridge, 1998, Treml et al., 2016, Procházková et al., 2011, Kumar and Pandey 2013). Quercetin and myricetin-derived flavonols are the most effective radical scavengers. Flavonoid aryl radicals are converted to more reactive secondary radicals, namely, quinones or semiquinones that have effects as pro-oxidant or cytotoxic agents (Cotelle, N. 2001). Condensed tannins are formed by the polymerization of flavonoids, which increases the antioxidant capacity due to the presence of many hydroxyl groups (Metodiewa et al., 1999, Dueñas et al., 2010). Reactive oxygen species are created by redox-active metal ions like cobalt, manganese, copper, or iron. Flavonoids form chelating complexes with these metal ions. Fe (II) reacts with H_2O_2 and forms OH, while Cu (II) reacts with H_2O_2 and forms both OH and O_2 (Symonowicz and Kolanek, 2012). They interfere with the enzyme activity of antioxidant enzymes or the enzymes that help in the production of reactive oxygen species. Flavonoids activate antioxidant enzymes, namely superoxide dismutase (SOD), glutathione peroxidase (GPx), and catalase (CAT). SOD acts on superoxide and converts it to H_2O_2. Catalase converts H_2O_2 to water. Peroxidase activity is further resumed through reduction by either glutaredoxin or thioredoxin (TRX). Some detoxification enzymes are activated by flavonoids such as glutathione S-transferase, UDP-glucuronosyl transferase, and NAD(P)H-quinone oxidoreductase that protect from oxidative stress and electrophilic toxicants (Procházková et al., 2011). Flavonoids and polyphenols form toxic quinones that act as a substrate for antioxidant enzymes, in this way activating the antioxidant enzyme system for their detoxification. Cis-acting regulatory elements such as electrophile-responsive element (E) and antioxidant response element (ARE) are present in most of the detoxifying genes (Chen et al., 2000). Flavonoids containing a free hydroxyl group at position 3 are effective inducers of EpRE-mediated gene expression, especially quercetin and myricetin.

Tea polyphenols contain 3-gallate in their structure which induces ARE reporter genes. Flavonoids inhibit ROS production by inhibiting the enzymes such as xanthin oxidase and protein K that are involved in their production (Nijveldt et al., 2001). They also inhibit other enzymes such as cyclooxygenase, NADPH oxidase, lipoxygenase, and microsomal succinimide. NADPH oxidase catalyzes the production of superoxide $O_2^{\cdot-}$. Quercetin, fisetin, or luteolin are flavonoids having a planar benzopyrone ring structure and free hydroxyl groups at positions 3, 4, and 7 that inhibit protein kinase C and ultimately inhibit NADPH oxidase (Ferriola and Cody, 1989). Flavonoids inhibit the activity of nitric oxide synthase (NOS). NOS is involved in the production of NO by the oxidation of L-arginine. NO reacts with free radicals and forms peroxynitrite. Flavonoids also interact with nonenzymatic antioxidants. Alpha-tocopherol is one of the nonenzymatic antioxidants. It scavenges ROS and produces α-tocopheroxyl radical. Flavonoids help in the regeneration of α-tocopherol from α-tocopheroxyl radical. Flavonoids and catechins have been reported to interrupt the oxidation of LDL (Zhu et al., 2000).

9.3.2 Polyphenols

The benefits of polyphenols are associated with the prevention of chronic diseases such as cancers, neurodegeneration, arthritis, and osteoporosis. Mechanisms involve the association with biological activities such as estrogen-like antioxidant activity and the inhibition of tyrosine kinase activity. Inflammatory cells produce oxidants. The first oxidant produced by the action of NADPH oxidase is the superoxide anion radical (O_2^-). Free radical is converted to hydrogen peroxide by superoxide dismutase. Hydrogen peroxide and chloride react in activated neutrophils to produce hypochlorous acid (HOCl) by the action of the enzyme myeloperoxidase. Superoxide ions can also react with nitric oxide and form a powerful nitrating as well as an oxidizing agent, i.e. nitric oxide (NO). At the inflammation site, HOCl and peroxynitrite react with tyrosine residues of the protein. Polyphenols show similarity in structure with tyrosine due to the presence of the phenolic group. Polyphenol gets chlorinated and nitrated when react with HOCl or peroxynitrite. The introduction of chloro- and nitro-group enhances the antioxidant capacity of polyphenols. Chloro- and nitro-polyphenols have structural similarities with estrogen receptors (α and β), and they interfere with the activation of these receptors. Isoflavones, stilbenes, and coumestans have estrogenic activities. Estrogen-like compounds interact with estrogen receptors and activate the same set of genes. Fruits and vegetables are abundant in flavonoids and phenolic acids, which decrease the risk of metabolic diseases. Polyphenolic compounds exhibit antibacterial activity. The antioxidant property of phenolic compounds is related to their free-radical-neutralizing capacity. Polyphenolic compounds contain hydroxyl groups that interact with ROS and reactive nitrogen species and interrupt the cycle of fresh radicals' production. Polyphenolic compounds react with some proteins and decrease the function enzymes involved in radical production, namely, cytochrome P450 isoforms, cyclooxygenases, xanthin oxidases, and cyclooxygenases (Parr and Bolwell 2000).

9.4 Health Benefits of Wheat Phytochemicals and Mechanism Involved

9.4.1 Diabetes Management

Type II diabetes is caused by attenuated insulin sensitivity, elevated blood glucose, and high cholesterol. Balanced diet helps in controlling blood glucose level. Consuming whole-grain foods is linked to a lower incidence of type II diabetes. The antidiabetic effect of phenolic acids has been recognized (Vinayagam et al., 2016). Phytochemicals such as phenolic acids, phytic acids, and ARs are found effective in improving blood sugar levels, insulin sensitivity as well as hyperinsulinemia. Ferulic acid (FA) interferes with sugar digestion and therefore slows down the rise in blood sugar levels after the meal (Adisakwattana et al., 2009). The diet containing 0.5% FA when subjected to C57BL/6 mice with a high-fat diet resulted in decreased blood glucose levels and activity of enzymes, namely, glucose-6-phosphatase (G6pase), phosphoenolpyruvate carboxykinase (PEPCK), and glucokinase (GK). The FA helps in insulin release (Adisakwattana et al., 2008). FA lowered the lipid peroxidation and the level of pro-inflammatory transcription factor such as NF-κB (Ramar et al., 2012). Glucose transporter 2 (GLUT2) is overexpressed in diabetic conditions. FA interferes with GLUT2 expression by impairing the interaction between GLUT2 gene promoter and transcription factors (HNF1α, HNF1β, and SREBP1c) (Narasimhan et al. (2015). To check the effect of esterified FA, the esterified FA-bound oligosaccharides (Fos) were administrated in the rats. Fos was effective in lowering fasting insulin, fasting plasma glucose, lactate dehydrogenase, creatine kinase, and aspartate transaminase which help in diabetes management.

ARs, another phytochemical from wheat, also have a role in diabetes management. Studies reveal that AR21 inhibits the activity of glycerol-3phosphate dehydrogenase and reduces triglyceride accumulation (Luyen et al., 2015). ARs inhibit the activity of hormone-sensitive lipase resulting in decreased level of free fatty acid and thus helping in reducing risk of hypertension and diabetes (Samuel et al., 2010). Wheat ARs are involved in improving insulin sensitivity and glucose tolerance by increasing intestinal cholesterol absorption and hepatic lipid accumulation (Oishi et al., 2015).

9.4.2 Prevention of Cardiovascular Diseases

Cardiovascular diseases (CVDs) involve blood vessel diseases, coronary heart disease (CHD), and hypertensive disease (Sanchis-Gomar et al., 2016). Low-density lipoprotein cholesterol (LDL-C) levels are increased in case of CVDs (Wilson et al., 1998). Wheat phytochemicals have preventive effect against CVDs. Per-day consumption of whole-grain cereal food is associated with a 20–30% reduction in the risk of CVDs (Gil et al., 2011). Flavonoid compounds present in whole grains, vegetables, and fruits are associated with lower risk of CVDs brought on by endothelial dysfunction-related atherosclerosis (Yamagata and Yamori 2020). FA supplementation in the human trial group decreased total cholesterol, triglycerides, and LDL and was also found effective in decreasing oxidative stress biomarkers (MDA) as well as inflammatory biomarkers (TNF-α) (Bumrungpert et al., 2018). ARs from wheat bran were included in the diet (0.4% AR) of the mouse model which resulted in lowered blood cholesterol levels and higher excretion of cholesterol (Oishi et al., 2015). Wheat AR supplementation resulted in reduced plasma concentration in mice (Horikawa et al., 2017). Phytosterols from cereal grains exhibit cholesterol-lowering effects and have a role in the prevention of cardiovascular diseases (CVDs). To analyze the effect of phytosterols in humans, a trial with 25 volunteers was performed. Higher cholesterol level was associated with the group consuming corn oil free from sterol compared to the group supplied with phytosterols containing corn oil (Ostlund et al., 2003).

9.4.3 Prevention of Colorectal Cancer

Whole-grain intake is associated with lower risk of colorectal cancer (Benisi-Kohansal et al., 2016). Phytosterols reduce tumor growth. A study in rats showed that a diet supplemented with 0.2% β-sitosterol resulted from the reduced severity of the tumor condition. DF contributes to good colon health by increasing short-chain fatty acid (SCFAs) production and stool mass. In the case of colorectal cancer, there is a lower abundance of *Lactobacillus spp.* and other SCFA-producing bacteria such as *Clostridium* and *Roseburia spp.* (Costabile et al., 2008; Gong et al., 2019). ARs are unique phytochemicals in wheat and rye. ARs are cytotoxic to the cancer lines (Kruk et al., 2017). The application of ARs resulted in an inhibited growth of colon cancer in human cancer cell models (HCT-116 and HT-29) (Zhu et al., 2011). ARs in wheat bran exhibit a synergistic effect with fiber microbiota metabolites, which include autophagy, apoptosis, and endoplasmic reticulum (ER) stress pathways (Zhao et al., 2019). Insoluble-bound phenolic acids also contribute to colon health. These insoluble-bound phenolic acids resist digestion in the upper part of the gastrointestinal tract and pass to the colon where they are acted upon by microbiota present in the gut, which release FA and thus exhibit anti-inflammatory, antioxidative, and anticancer activity in the colon (Liu, 2007). The effect of FA is compared with the anticancer drug 5-fluorouracil in the rat model, and results suggest that FA significantly upregulates p53 expression, inhibits proliferation, and induces apoptosis (Alazzouni et al., 2021). Poly-FA inhibits colon tumor growth in mice (Alazzouni et al., 2021).

Table 9.2 lists the primary bioactive components of wheat together with their molecular processes and health benefits.

TABLE 9.2

Major Bioactive Compounds of Wheat, Their Health Benefits, and Molecular Mechanisms

Bioactive Compounds	Health Benefits	Molecular Mechanisms	Reference
Phenolic acids and flavonoids	Reduced risk of coronary heart diseases	I. Cytoprotective role in cardiomyocytes against doxorubicin. II. Inhibition of oxidative stress and attenuation of MAP kinases/NF-κB/PI3K/Akt/mTOR signaling.	Sahu et al., 2019

(Continued)

TABLE 9.2 (*Continued*)

Bioactive Compounds	Health Benefits	Molecular Mechanisms	Reference
Phytosterols	I. Reduced risk of cardiovascular diseases II. Reduced LDL-cholesterol	I. Reduction in production of carcinogens II. Reduction of cell growth and cell viability. III. Activate caspases and promote apoptosis IV. Decrease cholesterol absorption V. Inhibit ROS production VI. Inhibit angiogenesis and metastasis VII. Reduction in cholesterol absorption by downregulating the cholesterol transporter protein NPS1L1 VIII. Inhibition of the activity of acyl CoA cholesterol acyltransferase (ACAT2)	Ostlund et al., 2003 Woyengo et al., 2009 Kaur and Myrie 2020.
Alkylresorcinols	I. Lower the risk of heart disease II. Lower the risk of colorectal cancer	ARs act synergistically with short-chain fatty acids, resulting in the induction of autophagy, apoptosis, and ER stress pathways.	Zhao et al., 2019
Ferulic acid	I. Weight control II. Diabetes management III. Reduced risk of cardiovascular disease	Antioxidant potential	Mateo Anson et al., 2008

REFERENCES

Adhikari, K. B., Tanwir, F., Gregersen, P. L., Steffensen, S. K., Jensen, B. M., Poulsen, L. K., . . . & Fomsgaard, I. S. (2015). Benzoxazinoids: Cereal phytochemicals with putative therapeutic and health-protecting properties. *Molecular Nutrition & Food Research, 59*(7), 1324–1338.

Adisakwattana, S., Chantarasinlapin, P., Thammarat, H., & YibchokAnun, S. (2009). A series of cinnamic acid derivatives and their inhibitory activity on intestinal α-glucosidase. *Journal of Enzyme Inhibition and Medicinal Chemistry, 24*(5), 1194–1200. https://doi.org/10.1080/14756360902779326.

Adisakwattana, S., Moonsan, P., & Yibchok-anun, S. (2008). Insulin-releasing properties of a series of cinnamic acid derivatives in vitro and in vivo. *Journal of Agricultural and Food Chemistry, 56*(17), 7838–7844. https://doi.org/10.1021/jf801208t.

Adom, K. K., Sorrells, M. E., & Liu, R. H. (2005). Phytochemicals and antioxidant activity of milled fractions of different wheat varieties. *Journal of Agricultural and Food Chemistry, 53*(6), 2297–2306.

Alazzouni, A. S., Dkhil, M. A., Gadelmawla, M. H. A., Gabri, M. S., Farag, A. H., & Hassan, B. N. (2021). Ferulic acid as an anticarcinogenic agent against 1,2-dimethyl-hydrazine-induced colon cancer in rats. *Journal of King Saud University—Science, 33*(2), 101354. https://doi.org/10.1016/j.jksus.2021.101354.

Amić, A., Marković, Z., Marković, J. M. D., Milenković, D., & Stepanić, V. (2020). Antioxidative potential of ferulic acid phenoxyl radical. *Phytochemistry, 170*, 112218.

Andersson, A. A., Dimberg, L., Åman, P., & Landberg, R. (2014). Recent findings on certain bioactive components in whole grain wheat and rye. *Journal of Cereal Science, 59*(3), 294–311.

Azzi, A. (2018). Many tocopherols, one vitamin E. *Molecular Aspects of Medicine, 61*, 92–103.

Benisi-Kohansal, S., Saneei, P., Salehi-Marzijarani, M., Larijani, B., & Esmaillzadeh, A. (2016). Whole-grain intake and mortality from all causes, cardiovascular disease, and cancer: A systematic review and dose-response meta-analysis of prospective cohort studies. *Advances in Nutrition, 7*(6), 1052–1065. https://doi.org/10.3945/an.115.011635.

Bok, S. H., Jeong, T. S., Lee, S. K., Kim, J. R., Moon, S. S., Choi, M. S., . . . & Choi, Y. K. (2002). *U.S. Patent No. 6,455,577.* Washington, DC: U.S. Patent and Trademark Office.

Bors, W., & Michel, C. (1997). Stettmaier, K. Antioxidant effects of flavonoids. *BioFactors, 6*, 399–402.

Bumrungpert, A., Lilitchan, S., Tuntipopipat, S., Tirawanchai, N., & Komindr, S. (2018). Ferulic acid supplementation improves lipid profiles, oxidative stress, and inflammatory status in hyperlipidemic subjects: A randomized, double-blind, placebo-controlled clinical trial. *Nutrients, 10*(6), 713. https://doi.org/10.3390/nu10060713.

Bunzel, M. (2010). Chemistry and occurrence of hydroxycinnamate oligomers. *Phytochemistry Reviews, 9*(1), 47–64.

Chen, C., Yu, R., Owuor, E. D., & Tony Kong, A. N. (2000). Activation of Antioxidant-Response Element (ARE), Mitogen-Activated Protein Kinases (MAPKs) and caspases by major green tea polyphenol components during cell survival and death. *Archives of Pharmacal Research, 23*, 605–612.

Cory, H., Passarelli, S., Szeto, J., Tamez, M., & Mattei, J. (2018). The role of polyphenols in human health and food systems: Mini-review. *Frontiers in Nutrition, 5*, 87.

Costabile, A., Klinder, A., Fava, F., Napolitano, A., Fogliano, V., Leonard, C., Gibson, G. R., & Tuohy, K. M. (2008). Whole-grain wheat breakfast cereal has a prebiotic effect on the human gut microbiota: A double-blind, placebo-controlled, crossover study. *British Journal of Nutrition, 99*(1), 110–120. https://doi.org/10.1017/S0007114507793923.

Cotelle, N. (2001). Role of flavonoids in oxidative stress. *Current Topics in Medicinal Chemistry, 1*(6), 569–590.

Demetzos, C., Magiatis, P., Typas, M. A., Dimas, K., Sotiriadou, R., Perez, S., & Kokkinopoulos, D. (1997). Biotransformation of the flavonoid salidroside to 7-methyl ether salidroside: bioactivity of this metabolite and its acetylated derivative. *Cellular and Molecular Life Sciences CMLS, 53*(7), 587–592.

Duarte-Almeida, J. M., Negri, G., Salatino, A., de Carvalho, J. E., & Lajolo, F. M. (2007). Antiproliferative and antioxidant activities of a tricin acylated glycoside from sugarcane (*Saccharum officinarum*) juice. *Phytochemistry, 68*(8), 1165–1171.

Dueñas, M., González-Manzano, S., González-Paramás, A., & SantosBuelga, C. (2010). Antioxidant evaluation of O-methylated metabolites of catechin, epicatechin and quercetin. *Journal of Pharmaceutical and Biomedical Analysis, 51*, 443–449.

Ferreyra, M. L. F., Serra, P., & Casati, P. (2021). Recent advances on the roles of flavonoids as plant protective molecules after UV and high light exposure. *Physiologia Plantarum, 173*(3), 736–749.

Ferriola, P. C., Cody, V., & Middleton, E. (1989). Protein kinase C inhibition by plant flavonoids: kinetic mechanisms and structure-activity relationships. *Biochemical Pharmacology, 38*, 1617–1624.

Fukami, H., Nakao, M., Namikawa, K., & Maeda, M. (2007). Esterified catechin, is the process of producing the same, food and drink, or cosmetics containing the same. *Patent EP, 1849779.*

Gil, A., Ortega, R. M., & Maldonado, J. (2011). Wholegrain cereals and bread: a duet of the Mediterranean diet for the prevention of chronic diseases. *Public Health Nutrition,* 14(12A), 2316–2322. https://doi.org/10.1017/S1368980011002576.

Gong, E. S., Gao, N., Li, T., Chen, H., Wang, Y., Si, X., Tian, J., Shu, C., Luo, S., Zhang, J., Zeng, Z., Xia, W., Li, B., Liu, C., & Liu, R. H. (2019). Effect of in vitro digestion on phytochemical profiles and cellular antioxidant activity of whole grains. *Journal of Agricultural and Food Chemistry,* 67(25), 7016–7024. https://doi.org/10.1021/acs. java.9b02245.

González-Paramás, A. M., Ayuda-Durán, B., Martínez, S., González-Manzano, S., & Santos-Buelga, C. (2019). The mechanisms behind the biological activity of flavonoids. *Current Medicinal Chemistry,* 26(39), 6976–6990.

Hajam, Y. A., Rai, S., Kumar, R., Bashir, M., & Malik, J. A. (2020). Phenolic compounds from medicinal herbs: Their role in animal health and diseases–a new approach for sustainable welfare and development. In *Plant Phenolics in Sustainable Agriculture* (pp. 221–239). Singapore: Springer.

Halliwell, B., & Gutteridge, J. M. C. (1998). *Free Radicals in Biology and Medicine.* 3rd ed. Oxford: Oxford University Press, 1998.

Hanhineva, K., Rogachev, I., Aura, A. M., Aharoni, A., Poutanen, K., & Mykkanen, H. (2011). Qualitative characterization of benzoxazinoid derivatives in whole grain rye and wheat by LC-MS metabolite profiling. *Journal of Agricultural and Food Chemistry,* 59(3), 921–927.

Horikawa, K., Hashimoto, C., Kikuchi, Y., Makita, M., Fukudome, S., Okita, K., Wada, N., & Oishi, K. (2017). Wheat alkylresorcinols reduce the micellar solubility of cholesterol in vitro and increase cholesterol excretion in mice. *Natural Product Research,* 31(5), 578–582. https://doi.org/10.1080/14786419.2016.1198347.

Kaur, R., & Myrie, S. B. (2020). Association of dietary phytosterols with cardiovascular disease biomarkers in humans. *Lipids,* 55(6), 569–584.

Kruk, J., Aboul-Enein, B., Bernstein, J., & Marchlewicz, M. (2017). Dietary alkylresorcinols and cancer prevention: A systematic review. *European Food Research and Technology,* 243(10), 1693–1710. https://doi.org/10.1007/s00217-017-2890-6.

Kumar, S., & Pandey, A. K. (2013). Chemistry and biological activities of flavonoids: An overview. *The Scientific World Journal,* 2013, Article ID 162750.

Lampi, A. M., Nurmi, T., Ollilainen, V., & Piironen, V. (2008). Tocopherols and tocotrienols in wheat genotypes in the health grain diversity screen. *Journal of Agricultural and Food Chemistry,* 56(21), 9716–9721.

Landberg, R., Marklund, M., Kamal-Eldin, A., & Åman, P. (2014). An update on alkylresorcinols – Occurrence, bioavailability, bioactivity, and utility as biomarkers. *Journal of Functional Foods,* 7, 77–89.

Leoncini, E., Prata, C., Malaguti, M., Marotti, I., Segura-Carretero, A., et al. (2012). Phytochemical profile and nutraceutical value of old and modern common wheat cultivars. *Plos One,* 7(9), e45997. https://doi.org/10.1371/journal.pone.0045997

Liu, R. H. (2007). Whole grain phytochemicals and health. *Journal of Cereal Science,* 46(3), 207–219. https://doi.org/10.1016/j.jcs.2007.06.010.

Lu, Y., Lv, J., Yu, L., Fletcher, A., Costa, J., Yu, L., & Luthria, D. (2014). Phytochemical composition and antiproliferative activities of bran fraction of ten Maryland-grown soft winter wheat cultivars: Comparison of different radical scavenging assays. *Journal of Food Composition and Analysis,* 36(1–2), 51–58.

Luyen, B. T. T., Thao, N. P., Tai, B. H., Lim, J. Y., Ki, H. H., Kim, D. K., Lee, Y. M., & Kim, Y. H. (2015). Chemical constituents of *Triticum aestivum* and their effects on adipogenic differentiation of 3T3-L1 preadipocytes. *Archives of Pharmacal Research, 38*(6), 1011–1018. https://doi.org/10.1007/s12272-014-0478-2.

Maliar, J. V. T. (2017). from synthesis to biological health effects and application. *Journal of Food and Nutrition Research (ISSN 1336-8672), 56*(3), 232–243.

Malinowska, P. (2013). Effect of flavonoids content on antioxidant activity of commercial cosmetic plant extracts. *Herba Polonica, 59*(3), 63–75.

Mateo Anson, N., van den Berg, R., Havenaar, R., Bast, A., & Haenen, G. (2008). Ferulic acid from aleurone determines the antioxidant potency of wheat grain (*Triticum aestivum* L.). *Journal of Agricultural and Food Chemistry, 56*(14), 5589–5594.

Mellou, F., Loutrari, H., Stamatis, H., Roussos, C., & Kolisis, F. N. (2006). Enzymatic esterification of flavonoids with unsaturated fatty acids: Effect of the novel esters on vascular endothelial growth factor release from K562 cells. *Process Biochemistry, 41*(9), 2029–2034.

Metodiewa, D., Jaiswal, A. K., Cenas, N., Dickancaité, E., & Segura Aguilar, J. (1999). Quercetin may act as a cytotoxic prooxidant after its metabolic activation to semi-quinone and quinoidal product. *Free Radical Biology and Medicine, 26*, 107–116.

Nakayama, T., Suzuki, H., & Nishino, T. (2003). Anthocyanin acyltransferases: Specificities, mechanism, phylogenetics, and applications. *Journal of Molecular Catalysis B: Enzymatic, 23*(2–6), 117–132.

Narasimhan, A., Chinnaiyan, M., & Karundevi, B. (2015). Ferulic acid regulates hepatic GLUT2 gene expression in high fat and fructose-induced type-2 diabetic adult male rat. *European Journal of Pharmacology, 761*, 391–397. https://doi.org/10.1016/j.ejphar. 2015.04.043.

Nijveldt, R. J., van Nood, E., van Hoorn, D. E. C., Boelens, P. G., van Norren, K., & van Leeuwen, P. A. M. (2001). Flavonoids: a review of probable mechanisms of action and potential applications. *The American Journal of Clinical Nutrition, 74*, 418–425.

Nisar, N., Li, L., Lu, S., Khin, N. C., & Pogson, B. J. (2015). Carotenoid metabolism in plants. *Molecular Plant, 8*(1), 68–82.

Nurmi, T., Nystrom, L., Edelmann, M., Lampi, A. M., & Piironen, V. (2008). Phytosterols in wheat genotypes in the HEALTH GRAIN diversity screen. *Journal of Agricultural and Food Chemistry, 56*(21), 9710–9715.

Oishi, K., Yamamoto, S., Itoh, N., Nakao, R., Yasumoto, Y., Tanaka, K., Kikuchi, Y., Fukudome, S., Okita, K., & Takano-Ishikawa, Y. (2015). Wheat alkylresorcinols suppress high-fat, high-sucrose diet-induced obesity and glucose intolerance by increasing insulin sensitivity and cholesterol excretion in male mice. *The Journal of Nutrition, 145*(2), 199–206. https://doi.org/10.3945/jn.114.202754.

Ostlund, R. E., Jr., Racette, S. B., & Stenson, W. F. (2003). Inhibition of cholesterol absorption by phytosterol-replete wheat germ compared with phytosterol-depleted wheat germ. *The American Journal of Clinical Nutrition, 77*(6), 1385–1389. https://doi.org/10.1093/acne/77.6.1385.

Parejo, I., Bastida, J., Viladomat, F., & Codina, C. (2005). Acylated quercetagetin glycosides with antioxidant activity from Tagetes maxima. *Phytochemistry, 66*(19), 2356–2362.

Parr, A. J., & Bolwell, G. P. (2000). Phenols in the plant and man. The potential for possible nutritional enhancement of the diet by modifying the phenols content or profile. *Journal of the Science of Food and Agriculture, 80*(7), 985–1012.

Perez-Vizcaino, F., & Fraga, C. G. (2018). Research trends in flavonoids and health. *Archives of Biochemistry and Biophysics, 646*, 107–112.

Plat, J., & Mensink, R. P. (2005). Plant stanol and sterol esters in the control of blood cholesterol levels: mechanism and safety aspects. *The American Journal of Cardiology, 96*(1), 15–22.

Porras, D., Nistal, E., Martínez-Flórez, S., Pisonero-Vaquero, S., Olcoz, J. L., Jover, R., . . . & Sánchez-Campos, S. (2017). Protective effect of quercetin on high-fat diet-induced non-alcoholic fatty liver disease in mice is mediated by modulating intestinal microbiota imbalance and related gut-liver axis activation. *Free Radical Biology and Medicine, 102*, 188–202.

Procházková, D., Boušová, I., & Wilhelmová, N. (2011). Antioxidant and prooxidant properties of flavonoids. *Fitoterapia, 82*, 513–523.

Ramar, M., Manikandan, B., Raman, T., Priyadarsini, A., Palanisamy, S., Velayudam, M., Munusamy, A., Marimuthu Prabhu, N., & Vaseeharan, B. (2012). Protective effect of ferulic acid and resveratrol against alloxan-induced diabetes in mice. *European Journal of Pharmacology, 690*(1), 226–235. https://doi.org/10.1016/j.ejphar. 2012.05.019.

Ross, A. B., Shepherd, M. J., Schüpphaus, M., Sinclair, V., Alfaro, B., Kamal-Eldin, A., & Åman, P. (2003). Alkylresorcinols in cereals and cereal products. *Journal of Agricultural and Food Chemistry, 51*(14), 4111–4118.

Sahu, R., Dua, T. K., Das, S., De Feo, V., & Dewanjee, S. (2019). Wheat phenolics suppress doxorubicin-induced cardiotoxicity via inhibition of oxidative stress, MAP kinase activation, NF-κB pathway, PI3K/Akt/mTOR impairment, and cardiac apoptosis. *Food and Chemical Toxicology, 125*, 503–519.

Samuel, V. T., Petersen, K. F., & Shulman, G. I. (2010). Lipid-induced insulin resistance: Unravelling the mechanism. *Lancet, 375*(9733), 2267–2277. https://doi.org/10.1016/S0140-6736(10)60408-4.

Sanchis-Gomar, F., Perez-Quilis, C., Leischik, R., & Lucia, A. (2016). Epidemiology of coronary heart disease and acute coronary syndrome. *Annals of Translational Medicine, 4*(13), 256. https://doi.org/10.21037/atm.2016.06.33.

Singh, R. (2016). Chemotaxonomy: A tool for plant classification. *Journal of Medicinal Plants Studies, 4*(2), 90–93.

Singla, R. K., Dubey, A. K., Garg, A., Sharma, R. K., Fiorino, M., Ameen, S. M., . . . & Al-Hiary, M. (2019). Natural polyphenols: Chemical classification, the definition of classes, subcategories, and structures. *Journal of AOAC International, 102*(5), 1397–1400.

Symonowicz, M., & Kolanek, M. (2012). Flavonoids and their properties to form chelate complexes. *Biotechnology and Food Sciences, 76*(1), 35–41.

Thilakarathna, S. H., & Rupasinghe, H. V. (2013). Flavonoid bioavailability and attempts for bioavailability enhancement. *Nutrients, 5*(9), 3367–3387.

Tian, W., Zheng, Y., Wang, W., Wang, D., Tilley, M., Zhang, G., . . . & Li, Y. (2022). A comprehensive review of wheat phytochemicals: From farm to fork and beyond. A *Comprehensive Reviews in Food Science and Food Safety, 21*(3), 2274–2308.

Treml, J., & Šmejkal, K. (2016). Flavonoids as potent scavengers of hydroxyl radicals. *Comprehensive Reviews in Food Science and Food Safety, 15*(4), 720–738.

Tsimogiannis, D., & Oreopoulou, V. (2019). Classification of phenolic compounds in plants. In *Polyphenols in Plants* (pp. 263–284). London: Academic Press.

Vinayagam, R., Jayachandran, M., & Xu, B. (2016). Antidiabetic effects of simple phenolic acids: A comprehensive review. *Phytotherapy Research, 30*(2), 184–199. https://doi.org/10.1002/ptr.5528.

Wilson, P. W. F., D'Agostino, R. B., Levy, D., Belanger, A. M., Silbershatz, H., & Kannel, W. B. (1998). Prediction of coronary heart disease using risk factor categories. *Circulation, 97*(18), 1837–1847. https://doi.org/10.1161/01.CIR.97.18.1837.

Woyengo, T. A., Ramprasath, V. R., & Jones, P. J. H. (2009). Anticancer effects of phytosterols. *European Journal of Clinical Nutrition, 63*(7), 813–820.

Yamagata, K., & Yamori, Y. (2020). Inhibition of endothelial dysfunction by dietary flavonoids and preventive effects against cardiovascular disease. *Journal of Cardiovascular Pharmacology, 75*(1), 1–9. https://doi.org/10.1097/FJC.0000000000000757.

Zhai, S., Xia, X., & He, Z. (2016). Carotenoids in staple cereals: Metabolism, regulation, and genetic manipulation. *Frontiers in Plant Science, 7*, 1197.

Zhao, Y., Shi, L., Hu, C., & Sang, S. (2019). Wheat bran for colon cancer prevention: The synergy between phytochemical alkylresorcinol C21 and intestinal microbial metabolite butyrate. *Journal of Agricultural and Food Chemistry, 67*(46), 12761–12769.

Zhao, Z., Xu, Z., Le, K., Azordegan, N., Riediger, N. D., & Moghadasian, M. H. (2009). Lack of evidence for antiatherogenic effects of wheat bran or corn bran in apolipoprotein E-knockout mice. *Journal of Agricultural and Food Chemistry, 57*(14), 6455–6460.

Zhu, Q. Y., Huang, Y., Chen, Z. Y. (2000). Interaction between flavonoids and α-tocopherol in human low-density lipoprotein. *Journal of Nutritional Biochemistry, 11*, 14–21.

Zhu, Y., Conklin, D. R., Chen, H., Wang, L., & Sang, S. (2011). 5-Alk(en)resorcinol as the major active components in wheat bran inhibit human colon cancer cell growth. *Bioorganic & Medicinal Chemistry, 19*(13), 3973–3982. https://doi.org/10.1016/j.bmc. 2011.05.025.

Zhu, Y., & Sang, S. (2017). Phytochemicals in whole grain wheat and their health-promoting effects. *Molecular Nutrition & Food Research, 61*(7), 1600852. https://doi.org/10.1002/mnfr. 201600852.

10

Micronutrient Biofortification in Wheat

Status and Opportunities

Gopalareddy Krishnappa, H. M. Mamrutha, N. D. Rathan, Hanif Khan, C. N. Mishra, Vishnu Kumar, Krishna Viswanatha Reddy, Vanita Pandey, Rinki Khobra, Charan Singh, K. J. Yashavanthakumar, Suma Biradar, B. S. Tyagi, Gyanendra Singh, and Gyanendra Pratap Singh

CONTENTS

10.1 Introduction

Micronutrient deficiencies is an important global health issue, impacting both physical and mental development, disease vulnerability, mental retardation, blindness, reduced cognitive ability, and general losses in productivity and potential. Unlike protein-energy malnutrition (PEM), the health impacts of micronutrient deficiency may not always be severely visible; hence, they are sometimes termed as "hidden hunger". Globally, more than two billion people suffer from micronutrient deficiency (Gillespie et al., 2016). Iron, zinc, iodine, and vitamin A deficiencies are the most prevalent and calcium and selenium are relatively less common, but they need equal attention, otherwise, they also may become more prevalent. World Health Organization recognized iron, zinc, and vitamin A as the three important limiting micronutrients in the global diet (Ortiz-Monasterio et al., 2007). The iron deficiency causes anemia and affects nearly 25.0% of the global population (de Benoist et al., 2008), leading to 0.12 million deaths and a loss of 48.2 million disability-adjusted life years (DALY) in the year

DOI: 10.1201/9781003307938-10

2010. On the other hand, zinc deficiency affects approximately 17.0% of the global population (Wessells and Brown, 2012), causing the death of 0.09 million people and the loss of 9.1 million DALYs in the year 2010 (Lim et al., 2012). The various interventions to alleviate micronutrient malnutrition are dietary diversification, pharmaceutical supplementation, industrial fortification, and biofortification. The consumption of a diversified diet rich in micronutrients is a simplest and effective strategy, but many people, especially economically weaker sections from developing and undeveloped countries, cannot afford it. The supplementation and fortification are not sustainable over the long term. Additionally, the fortified food is unavailable and unaffordable to the needy people, particularly for the rural poor. Hence, a practice of increasing the nutrient status of agricultural produce through plant breeding, agronomic, and transgenic approaches, known as "biofortification", emerged as a cost-effective and sustainable solution to the problem of micronutrient deficiencies. Micronutrient deficiencies are most common in South Asia, Latin America, and developing countries such as sub-Saharan Africa. Thus, effective policies to address micronutrient deficiencies in food systems such as the use of micronutrient fertilizers and plant breeding to develop nutrient-rich crops for human consumption are being considered (Gregory et al., 2017). In the year 2003, the Consultative Group on International Agricultural Research (CGIAR) launched "HarvestPlus: The Biofortification Challenge Program" with the goal of translating the success of agriculture-based research for food production into a food-based public health program to contain micronutrient deficiency in low- and middle-income populations of developing nations. This program identifies iron, zinc, and vitamin A as limiting micronutrients for human health. The consumption of nutritionally enriched food crops aids in the reduction of micronutrient deficiency and thus improves human health. The development of nutrient-dense staple food crops is currently the scientific community's top research priority.

10.2 Genetic Resources: Valuable Donors for Micronutrient Biofortification

The improvement of grain micronutrient content depends on the availability of sufficient genetic diversity in the gene pool. Ficco et al. (2009) reported a variability in the range of 28.5 to 46.3 mg/kg for grain zinc concentration (GZnC) and 33.6 to 65.6 mg/kg for grain iron concentration (GFeC). Similarly, Zhao et al. (2009) observed a variation between 13.5 and 34.5 mg/kg for GZnC and between 28.8 and 50.8 mg/kg for GFeC among the studied 150 bread wheat genotypes. Velu et al. (2012) evaluated a set of 600 core collection accessions comprising both durum and bread wheat and found that the GZnC and GFeC ranged from 16.85 to 60.77 mg/kg and 26.26 to 68.78 mg/kg, respectively. Gopalareddy et al. (2015) observed variability for GFeC ranging from 27.85 to 54.60 mg/kg and for GZnC it was from 19.30 to 70.55 mg/kg. Pandey et al. (2016) evaluated 150 bread wheat genotypes originated from India and Turkey and observed the variation in the range of 10.7 to 59.4 mg/kg for GZnC and 9.2 to 49.7 mg/kg for GFeC. Rathan et al. (2020) evaluated 94 advanced bread wheat genotypes from CIMMYT's global wheat program and reported the variability of 41.8 to 71.8 mg/kg for GZnC and 33.6 to 51.7 mg/kg for GFeC. The variations found in bread and durum wheat in some of the other studies are similar or lower to the aforementioned studies ranging from 10.0 to 51.0 mg/kg for GFeC and from 10.4 to 61.0 mg/kg for

GZnC (Rawat et al., 2009; Harmankaya et al., 2012; Magallanes-Lopez et al., 2017). Therefore, the genetic variability available in the modern cultivar pool is inadequate, and the inclusion of candidates from secondary and tertiary gene pools would be vital in the breeding program. The variability in wheat landraces for micronutrient concentration was observed between 12.4 to 73.80 mg/kg and 23.7 to 67.67 mg/kg, respectively, for GZnC and GFeC (Badakhshan et al., 2013; Heidari et al., 2016; Goel et al., 2018). They concluded that the landraces have substantially high variation compared to wheat varieties and hence suggested the potentiality of such germplasm in biofortification.

The cultivated wheat varieties have relatively narrow genetic variation for GZnC and GFeC, hence, limiting the genetic gain for higher grain zinc and iron concentration and bioavailability. As a result, wild wheat may act as an important genetic resource for improving grain micronutrient concentrations in cultivated wheat. Cakmak et al. (2000) compared and concluded that the variation for grain micronutrients in modern tetraploid and hexaploid wheat is far lesser than the diploid and tetraploid wheat. The wild wheat, *T. boeoticum*, has shown the highest concentration of zinc and iron with values 178 mg/kg and 159 mg/kg, respectively. Further, *T. dicoccoides* showed up to 159 mg/kg of GZnC, and *T. monococcum* reported as high as 85 mg/kg of GFeC. Also, Cakmak et al. (2004) screened 825 accessions of *Triticum turgidum ssp. dicoccoides* from Fertile Crescent and observed the extensive variation and the highest concentrations of micronutrients ranging from 14.0 to 190.0 mg kg^{-1} for GZnC and from 15.0 to 109.0 mg kg^{-1} for GFeC, significantly higher than those of cultivated wheat. They found that chromosomes 6A, 6B, and 5B are responsible for higher iron and zinc concentration in *T. dicoccoides* using *T. di*coccoides substitution lines. Rawat et al. (2009) analyzed 80 germplasm accessions of 9 species of wild *Triticum* and *Aegilops* along with 15 semi-dwarf cultivars of bread and durum wheat. Around two- to threefolds higher GFeC and GZnC was observed in *Aegilops kotschyi* and *Aegilops longissima* as compared to widely adapted high-yielding wheat cultivar WL 711. They induced homoeologous chromosome pairing between *Aegilops* and wheat genomes using Chinese Spring (*Ph1*) and observed that most interspecific hybrids had higher GZnC and GFeC than parents. Hence, they concluded that the parental *Aegilops* donors possess a more efficient system for uptake and translocation of the micronutrients which could ultimately be utilized for wheat grain biofortification. Thus, they suggested that the parental *Aegilops* donor genotypes possess a more efficient uptake and translocation system for the micronutrients, which could ultimately be used for enhancing the nutrient status of wheat grain through biofortification. Arora et al. (2019) evaluated 167 diverse *Aegilops tauschii* accessions and observed almost twofold (30.3–69.4 mg/kg) and 2.8-fold (17.5–49.8 mg/kg) genetic variation for iron and zinc, respectively. The wild wheat species with increased iron and zinc concentrations are useful to provide an efficient genetic system for enhanced grain micronutrients in modern cultivars. But this is hampered due to crossability barriers and sterility associated with interspecific and intergeneric hybridization. Hence, synthetic hexaploids developed from crossing tetraploid wheat (AB genome) with *Aegilops tauschii* (D genome) generate an extensive variation for the trait of interest and can efficiently avoid problems related to gene introgression. The biofortified varieties like Zinc-Shakti and WB 02 are developed using synthetic hexaploids nested in their pedigree. A total of 47 synthetic hexaploid wheat (SHW) lines developed from crosses between tetraploid wheat cultivar 'Langdon' and 47 *Aegilops tauschii* lines, which were collected from various geographical locations (Gorafi et al., 2018). The GFeC ranged from 22.2 to 78.5 mg/kg with a mean of 45.4 and 20.6–65.8 mg/kg for GZnC with an average of 39.9 mg/kg. Therefore, it is important to utilize natural variability that existed in both cultivated and wild germplasms to develop biofortified cultivars.

10.3 Current Status of Conventional Breeding-Based Biofortified Wheat Varieties

The conventional breeding approach is proved to be the most promising strategy for the development of biofortified wheat varieties. The wide variation for grain micronutrient concentrations present in the wheat gene pool has been utilized to develop modern elite cultivars with enhanced nutrient content. A list of biofortified wheat cultivars released for commercial cultivation is presented in Table 10.1. Some of the biofortified wheat varieties like DBW 187 and WB 02 are widely cultivated.

TABLE 10.1

List of Biofortified Wheat Varieties for Grain Iron and Zinc Concentration and Released for Commercial Cultivation

SN	Variety	Micronutrient (ppm)	Year of Release	Reference	Institute
			India		
1	WB 02	Zn: 42; Fe: 40	2017	Yadava et al. (2022)	ICAR-IIWBR, India
2	DBW 173	Fe: 40.7	2018		
3	DBW 187	Fe: 43.1	2018		
4	DDW 47	Iron: 40.1	2020		
5	DBW 332	Zn: 40.6	2021		
6	DBW 327	Zn: 40.6	2021		
7	HPBW 01	Zn: 40.6; Fe: 40	2017		PAU, India
8	PBW 757	Zn: 42.3	2018		
9	PBW 771	Zinc: 41.4	2020		
10	HI 8759	Zn: 42.8; Fe: 42.1	2017		ICAR-IARI (RS), Indore, India
11	HI 1605	Zn: 35; Fe: 43	2017		
12	HI 8777	Fe: 48.7; Zn: 43.6	2017		
13	HI 8805	Fe: 40.4	2020		
14	HI 1633	Fe: 41.6; Zn: 41.1	2020		
15	HI 1636	Zn: 40.4	2021		
16	HI 8823	Zn: 40.1	2021		
17	HD 3171	Zn: 47.1	2017		ICAR-IARI, New Delhi, India
18	HD 3249	Fe: 42.5	2020		
19	HD 3298	Fe: 43.1	2020		
20	MACS 4028	Zn: 40.3; Fe: 46.1	2018		ARI, India
21	MACS 4058	Fe: 39.5; Zn: 37.8	2020		
22	HUW 838	Zn: 41.8	2021		BHU, India
23	MP (JW) 1358	Fe: 40.6	2021		JNKV, Powarkheda

TABLE 10.1 (*Continued*)

SN	Variety	Micronutrient (ppm)	Year of Release	Reference	Institute
			India		
24	Abhay (Zinc Shakti)	High Zn	2015	Velu et al. (2015)	Nirmal Seeds/ HarvestPlus
25	Zinc Shakti (Chitra)	High Zn	2016	Singh and Velu (2017)	HarvestPlus
			Pakistan		
26	NR-421 (Zincol-16)	High Zn	2015	Singh and Velu (2017)	PARC
27	Akbar-19	High Zn	2019	HarvestPlus, 2020	AARI
			Bangladesh		
28	BARI Gom 33	High Zn	2017	Hossain et al., 2019	BARI/CIMMYT
			Mexico		
29	Nohely-F2018	High Zn	2018	Velu et al., 2019	CIMMYT
			Nepal		
30	Bheri-Ganga, Himganga, Khumal-Shakti, Zinc Gahun 1, Zinc Gahun 2	High Zn	–	Annual Report 2020, CIMMYT	NARC, Nepal, and CIMMYT

Abbreviations: Zn: zinc concentration; Fe: iron concentration; ICAR-IIWBR: Indian Council of Agricultural Research-Indian Institute of Wheat and Barley Research; PAU: Punjab Agricultural University; ICAR-IARI (RS): ICAR-Indian Agricultural Research Institute (Regional Station); ICAR-IARI: ICAR-Indian Agricultural Research Institute; ARI: Agharkar Research Institute; BHU: Banaras Hindu University; JNKVV: Jawaharlal Nehru Krishi Vishwavidyalaya; PARC: Pakistan Agricultural Research Council; AARI: Ayub Agricultural Research Institute; BARI: Bangladesh Agricultural Research Institute; CIMMYT: International Maize and Wheat Improvement Center; NARC: Nepal Agricultural Research Council.

10.4 Molecular Interventions

Understanding the genetics and causal genes of quantitative traits through a molecular mapping approach is required for employing marker-assisted breeding (MAB). The identification of closely linked DNA markers to complex traits is highly valuable and cost-effective for trait improvement, particularly in the post-genomics era where the genotyping cost has decreased significantly. Different studies have shown that the environment and genotype–environment interaction (GEI) have a substantial impact on the expression of cereal grain micronutrients (Velu et al., 2012; Krishnappa et al., 2019).

10.4.1 QTL Mapping

Quantitative trait locus (QTL) mapping is a popular approach to discover genes for polygenic traits including grain micronutrient content, as well as to develop molecular markers to use in breeding programs. The recent developments in next-generation sequencing technologies substantially reduced the genotyping cost and generated thousands of markers faster resulting in the rapid increase in QTL mapping and genomic studies for various traits including grain micronutrients in the last decade. The list of QTL studies that were attempted for the genetic dissection of grain micronutrient concentrations in wheat is presented in Table 10.2. The QTL mapping approaches mainly used bi-parental populations and identified QTL positions at low resolution. Shi et al. (2008) identified four QTLs regulating GZnC on chromosomes 4A, 4D, 5A, and 7A with phenotypic variation (PV) ranging from 5.3 to 11.9%. The QTL on chromosome 4D flanked by markers *Xgwm192–WMC331* explained the highest PV of 11.9%. Tiwari et al. (2009) detected two QTLs for GFeC on chromosomes 2A (*QFe.pau-2A*) and 7A (*QFe.pau-7A*), each explaining the total PV of 12.6% and 11.7%. A QTL for GZnC was detected on chromosome 7A (*QZn. pau-7A*) in the marker interval *Xcfd31–Xcfa2049* with the explained PV of 18.8%. Xu et al. (2012) identified two additive and one epistatic QTL on chromosomes 5A

TABLE 10.2

QTL Mapping for Grain Micronutrient Concentration in Wheat

SN	Micronutrient	Cultivar/Variety	Population	Size	Reference
1	Zn	Hanxuan10 × Lumai 14	DH	119	Shi et al. (2008)
2	Zn	RAC875–2 × Cascades	DH	90	Genc et al. (2009)
3	Zn and Fe	pau5088 × pau14087	RIL	93	Tiwari et al. (2009)
4	Zn and Fe	Langdon × #G18–16	RIL	152	Peleg et al. (2009)
5	Zn and Fe	Xiaoyan 54 × Jing 411	RIL	182	Xu et al. (2012)
6	Zn and Fe	Tabassi × Taifun	RIL	118	Roshanzamir et al. (2013)
7	Zn	PBW343 × Kenya Swara	RIL	177	Hao et al. (2014)
8	Fe and Zn	H+26(PI348449) × HUW234	RIL	185	Srinivasa et al. (2014)
9	Zn, Fe, Se, Cu, Mn	SHW-L1/Chuanmai 32 Chuanmai 42/Chuannong 16	RIL	171 127	Zhi-en et al. (2014)
10	Zn	Saricanak98 × MM 5/4 Adana99 × 70,711	RIL RIL	105 127	Velu et al. (2016)
11	Zn and Fe	Berkut9 × Krichauff	DH	138	Tiwari et al. (2016)
12	Zn and Fe	Seri M82 × CWI76364	RIL	140	Crespo-Herrera et al. (2016)
13	Zn and Fe	WH542 × synthetic derivative	RIL	286	Krishnappa et al. (2017)
14	Zn and Fe	Turtur × Bubo, SHW × Bateleur	RIL	188	Crespo-Herrera et al. (2017)
15	Zn and Fe	Roelfs F2007 × Chinese line	RIL	200	Liu et al. (2019)
16	Zn and Fe	WH542 × synthetic derivative	RIL	163	Krishnappa et al. (2021)
17	Zn and Fe	Kachu' × 'Zinc-Shakti	RIL	190	Rathan et al. (2021)

Abbreviations: DH: doubled haploids; RIL: recombinant inbred lines and 4B for GFeC and GZnC. Hao et al. (2014) identified two novel QTLs with a large effect for enhancing GZnC on chromosomes 2B and 3A. The two QTLs individually explained about 10 to 15% of the total PV. Srinivasa et al. (2014) identified *QZn.bhu-2B* closely linked to the SNP marker *1101425|F|0* and explained the PV greater than 16%, while *QZn.bhu-6A* closely linked to the DArT marker *3026160|F|0* explained the PV of 7.0%. Of the five QTLs detected for iron, three mapped to chromosome 1A were the most consistent and stable, viz., *QFe.bhu-1A.2* (overall PV = 7.5%) and *QFe.bhu-1A.3* (PV = 16.6%), followed by *QFe.bhu-1A.1* (PV = 5.6%). *QFe.bhu-3B* exhibited the highest PV of 26.0%. Crespo-Herrera et al. (2016) identified a major QTL on chromosome 4BS which is pleiotropic to both GZnC and GFeC and explained up to 19.6% of the total PV for GZnC. Tiwari et al. (2016) detected two QTLs for GZnC (1B and 2B) with a QTL (2B) co-localized for GFeC. The QTL located on chromosomes 1B and 2B for GZnC explained up to 23.1 and 35.9% of mean PV respectively, whereas the Fe QTL co-located with the Zn QTL on chromosome 2B explained up to 22.2%. Crespo-Herrera et al. (2017) identified several significant QTLs with a region *QGZn.cimmyt-7B_1P2* on chromosome 7B explaining the largest (32.7%) proportion of PV for GZnC. Krishnappa et al. (2017) identified five QTLs, viz. *QGZn.iari-2A*, *QGZn.iari-4A*, *QGZn.iari-5A*, *QGZn.iari-7A*, and *QGZn.iari-7B* for GZnC and four QTLs, viz. *QGFe.iari-7B*, *QGFe.iari-2A*, *QGFe.iari-5A*, and *QGFe.iari-7A* for GFeC. Velu et al. (2016) identified two major QTLs on chromosomes 1B and 6B for GZnC in both the tetra and hexaploid-mapping populations. Liu et al. (2019) identified ten QTL for GZnC and nine for GFeC. The QTLs *QGFe.co-3B.1* and *QGZn.co-5A* explained the highest PV of 14.56% and 14.22%, respectively. Krishnappa et al. (2021) detected a total of six and three significant QTLs for GFeC and GZnC, respectively.

10.4.2 Genome-Wide Association Studies (GWASs)

Genome-wide association study (GWAS) is another widely used method for analyzing the genetic basis of complex traits. GWAS has been widely used in wheat to investigate the genetic control of complex traits, but it has only been used in a few investigations of the genetic analysis of grain micronutrient concentration which are presented in Table 10.3. Alomari et al. (2018) identified 40 MTAs on 12 chromosomes, viz., 2A, 3A, 3B, 4A, 4D, 5A, 5B, 5D, 6D, 7A, 7B, and 7D which explained 2.5 to 5.2% of PV. The MTA on chromosome 3B was found between 64.5 and 66.8 cM, which is most stably detected in all three years. Bhatta et al. (2018) carried out a GWAS for ten elements, viz., Ca, Cd, Co, Cu, Fe, Li, Mg, Mn, Ni, and Zn in grain and detected 92 MTAs. Gorafi et al. (2018) identified three MTAs each for GFeC and GZnC concentrations. The markers associated with Fe were *Xgwm157*, *Xwmc399*, and *Xwmc357* with the explained PV of 29.0%, 14.0%, and 21.0%, respectively. The markers associated with Zn were *Xgwm102*, *Xwmc357*, and *Xcfd63* explained by respective PVs of 36.0%, 45.0%, and 48.0%. Velu et al. (2018) carried out GWAS on 330 bread wheat HPAM panel, which detected a total of 39 significant MTAs with PVE between 5 and 10.5% for GZnC. The major SNPs, *RAC875_c34757_180* located at 60 cM on chromosome 2A, *IAAV1375* on chromosome 5A, *wsnp_Ex_c5268_9320618* located at 120 cM on chromosome 7B explained PV of 9.0%, 6.0%, and 10.5%, respectively. Arora et al. (2019) performed GWAS on a panel of 114 nonredundant *Aegilops tauschii* accessions and detected a total of nine SNPs (five for Fe, four for Zn) located on all seven chromosomes. The significant MTAs for GFeC were detected on 4D followed by 2D, 1D,

TABLE 10.3

Marker Trait Associations (MTAs) for Grain Micronutrient Concentration through Genome-Wide Association Studies (GWASs)

SN	Micronutrient	Association Mapping Panel	Marker Type	Marker Size	Reference
1	Zn	HPAM panel of 330 genotypes	Illumina iSelect 90 K Infinitum SNP array	14,273	Velu et al. (2018)
2	Fe and Zn	SWRS panel of 246 genotypes	DArT-seq	8,637	Kumar et al. (2018)
3	Ca, Cd, Cu, Co, Fe, Li, Mg, Mn, Ni, Zn	SHW panel of 123 genotypes	GBS-derived SNPs	35,648	Bhatta et al. (2018)
4	Zn	European panel of 369 genotypes	Illumina iSelect 90 K Infinitum and 35k Affymetrix SNP arrays,	15,523	Alomari et al. (2018)
5	Fe and Zn	SHW panel of 47 genotypes	SSR	70	Gorafi et al. (2018)
6	Fe, Zn, Cu, Mn	*Ae. Tauschii* panel of 114 genotypes	GBS-derived SNPs	5249	Arora et al. (2019)
7	Fe, Zn, Cu, Mn	HPAM panel of 330 genotypes	Illumina iSelect 90 K Infinitum SNP Array	17,900	Cu et al. (2020)
8	Zn	Common wheat Panel of 207 genotypes	Wheat Breeders 660 K Axiom® array	244,508	Zhou et al. (2020)
9	Fe, Zn, Mn	161 advanced lines	DArT markers	13,116	Liu et al. (2021a)
10	Cu, Zn	246 wheat varieties	SSR	545	Liu et al. (2021b)
11	Fe and Zn	Bread wheat panel of 280 genotypes	35 K Axiom Array	14,790	Krishnappa et al. (2022)
12	Fe and Zn	184 diverse bread wheat genotypes	35 K Axiom® Wheat Breeder's Array	9,503	Rathan et al. (2022)

7D, and 3D. The significant MTAs for GZnC were identified on 2D, 4D, 6D, and 7D. The GFeC and GZnC MTAs on chromosome 4D were co-localized in the mapping bins of 1.1 cM apart. Cu et al. (2020) identified a total of 72 and 65 MTAs for GZnC and GFeC, respectively. Zhou et al. (2020) identified 29 MTAs for GZnC, of which 7 significant loci located on 1B, 3B, 3D, 4A, 5A, 5B, and 7A chromosomes were nonredundant and found in at least two environments.

10.5 Bioavailability of Mineral Micronutrients

Along with the enrichment of wheat varieties for micronutrients through genetic bio-fortification, a high bioavailability of these nutrients is equally important to contain the hidden hunger. Wheat-based foods are rich in anti-nutrients, particularly phytic acid (Phy), which affects the absorption or utilization of nutrients in the human digestive system (Lonnerdal, 2002). Generally, staple food crops contain low-to-very low bioavailable Fe and Zn (i.e., about 5% and 25% of the total grain Fe and Zn, respectively, are bioavailable). A total of 5–20% increase of Fe bioavailability is roughly comparable to the fourfold increase of total Fe content (Bouis and Welch, 2010). It has been reported that the bioavailability of Fe and Zn can be easily improved through a genetic approach in comparison to the increase of their total micronutrient content in the same magnitude through conventional breeding (Lonnerdal, 2002). The molar ratio of mineral micronutrients with Phy has been widely followed in measuring the bioavailability of minerals in the human diet. The Fe status in the human body is generally regulated through absorption, whereas Zn homeostasis is controlled through absorption as well as gastrointestinal secretion and excretion of endogenous Zn (Oberleas, 1983). Therefore, breeding for low phytic acid concentration has been gaining much importance in the recent past to increase the nutrient bioavailability of micronutrients. To reduce the phytic acid levels, mutation breeding has been widely used in different crop plants and developed low phytate acid mutants (LPA). Studies have shown that Phy inhibits Ca absorption, but Phy's impact on Ca bioavailability appears to be less pronounced than its impact on Fe and Zn bioavailability. Cereals also have promoters that can increase the absorption of micronutrients, such as vitamin C, pro-vitamin A, haemoglobin, and other organic and amino acids. There is considerable genetic diversity in the amount of phytic acid in wheat grain, ranging from 7 to 12 mg/g with Zn fertilization to 8 to 13 mg/g without Zn fertilization (Erdal et al., 2002). Significant genetic variations for phytate concentration have been identified in wheat grain in an intra-species level (Welch et al., 2005; White and Broadley, 2005). Wheat varieties with low levels of phytic acid (LPA) in their grains are preferred due to their favorable impacts on public health. In wheat grain, Zn concentration was correlated with both phytate content (r = 0.37, P < 0.01) and phytase activity (r = 0.28, P < 0.01), although Fe concentration was not correlated with either (Liu et al., 2006). Compared to the uptake, transport, and deposition of Fe and Zn, the biosynthesis and metabolism of inhibitors and promoters involve fewer genes. Therefore, it should be considerably easier to increase the bioavailability of Fe and Zn than to enhance their concentrations in grains (Bouis and Welch, 2010). Micronutrient bioavailability may be increased by enhancing the promoters and reducing the inhibitors (Welch and Graham 2004). However, given the significant functions that phytate and several other anti-nutrients play in plant metabolism and human diets, breeders should exercise caution when making changes to anti-nutrients.

10.6 Genes Involved in Micronutrient Accumulation

Waters and Sankaran (2011) reported several important genes responsible for the uptake and translocation of micronutrient elements from the soil to the grains. In wheat, the *NAM-B1* gene, a NAC transcription factor, influences the levels of grain nutrients like Fe and Zn as well as grain protein content. The remobilization of nutrients from the vegetative tissues and a greater partitioning of resources to grain are the main effects of the NAM genes (Waters et al., 2009). One of the most popular genetic resources to increase the content of zinc in wheat is substitution lines of the 6B chromosome obtained from *Triticum dicoccoides* (Cakmak et al., 2004). The

Gpc-B1 locus, derived from *T. dicoccoides*, is located on the short arm of the 6B chromosome having pleiotropic effects on zinc, iron, and grain protein content in wheat grains (Distelfeld et al., 2007). The increase in zinc and iron levels is induced by a NAC transcription factor (NAM-B1) encoded by *Gpc-B1*, which is possible by way of promoting leaf senescence and the subsequent remobilization of zinc and iron from flag leaves into seeds (Uauy et al., 2006). Higher grain zinc content in synthetic wheat derived from *Ae. tauschii* makes it a valuable genetic resource for increasing grain zinc content in cultivated wheat (Calderini et al., 2003). According to Kohl et al. (2015), several NAC transcription factors were upregulated in the glumes in 14 days following anthesis and were evidently linked with developmental senescence. Proteases are quickly activated during senescence to break down leaf proteins into amino acids (Guitman et al., 1991). In wheat and barley, the major family of serine proteases involved in nitrogen remobilization (NR) during grain filling serves as both a primary regulator and an executor (Hollmann et al., 2014). It has been demonstrated that the particular transcription factors NAC and WRKY, in combination with the hormones, abscisic acid and jasmonic acid, regulate the transition between early grain filling and developmental senescence in wheat and barley (Kohl et al., 2015; Gregersen et al., 2007; Guo et al., 2012).

10.7 Transgenic Biofortified Wheat

Genetic modification (GM) is one of the approaches widely being used for the introduction of desired genes from unrelated species/genera and functional genomics in plants. Many crop varieties have been developed through GM technology over the past two decades. Despite its importance, the progress in wheat GM technology still lags behind other staple crops such as rice, primarily due to the difficulties associated with gene delivery and recovery of transgenic plants. Nevertheless, continued efforts are being made to develop transgenic wheat for various economic traits including micronutrient biofortified traits. The two important bacterial genes (*CrtB* and *CrtI*) were transferred into the common wheat variety Bobwhite to enhance the carotenoid content of wheat grains. Co-expression of both the introduced genes resulted in the increase of roughly eightfold to 4.76 μg per gram of seed dry weight, and a β-carotene content increase of 65-fold to 3.21 μg per gram of seed dry weight. The amount of provitamin A increased similarly, rising 76-fold to 3.82 grams of seed dry weight. Over four generations, the transgenic wheat's high provitamin A concentration was stably inherited (Wang et al., 2014). Similarly, transgenic wheat has been produced by expressing the maize *y1* gene, which encodes phytoene synthase, under the control of the constitutive CaMV 35S promoter in the elite cultivar EM12, along with the bacterial phytoene desaturase *crtI* gene from *Erwinia uredovora*. The total carotenoid content was enhanced up to 10.8-fold as compared with the non-transgenic EM12 cultivar (Cong et al., 2009). Wheat seeds contain an anti-nutritional compound, i.e. phytate, which chelates the metal ions thereby reducing their bioavailability and grains' nutritional value. The bioavailability of mineral nutrients can be increased by the degradation of phytic acid in grains by phytase activity in transgenic plants. The introduction of *Aspergillus japonicus* phytase gene (*phyA*) in wheat endosperm resulted in the expression of transgenic plants with 18.0–99.0% increase in phytase activity and 12.0–76.0% reduction of phytic acid content in wheat seeds (Abid et al., 2017).

Endosperm-targeted intragenic overexpressing of the *TaFer1-A* gene results in a 50.0–85.0% higher iron content in the wheat grain (Borg et al., 2012). Connorton et al. (2017) reported that the overexpression of *TaVIT2* in the wheat endosperm resulted in the twofold increase of iron concentration in white flour.

10.8 Future Perspectives

1. The most vulnerable groups of micronutrient malnutrition are children and pregnant and lactating women, therefore, trials of biofortified cereals should be extended to a wider range of age and gender groups over a prolonged period.

2. More accurate and sensitive biochemical and functional markers, such as plasma Zn content and DNA strand breaks, may be needed to increase the efficacy of trials in order to better understand the effect of biofortified crops on the health of vulnerable populations.

3. National crop breeding programs need to mainstream the biofortification efforts to develop high yielding wheat varieties with enhanced micronutrient levels.

4. Policy interventions like the inclusion of biofortified wheat varieties in mid-day meal scheme, premium price for biofortified produce through market segregation, and enhanced research grant for more bioavailability studies may also be rewarding.

5. Future crop-breeding projects must take into account more in-depth research on chelators, transporters, promoters, and inhibitors in order to enhance nutrition and health.

6. Most of the genetic studies on biofortification concentrated on a whole-grain basis; however, more than 50% of Fe and Zn will be lost during milling; therefore, genetic studies on the identification of genomic regions associated with the accumulation of micronutrients in wheat endosperm may reward more to address hidden hunger.

7. In most of the genetic studies, there is a positive association of flag leaf Fe and/or Zn with grain concentration of these important micronutrients, as they had a common genetic mechanism and are co-localized; therefore, this can effectively be utilized as an indirect selection criteria for high Fe and Zn germplasms.

8. To increase the diffusion of biofortified wheat, there is a need to develop market intelligence for effective seed distribution through seed chain and expand consumer demand for biofortified wheat.

10.9 Conclusion

The enhancement of micronutrient levels in staple food crops like wheat through genetic approaches is both economical and sustainable strategy to contain micronutrient malnutrition. Initial investments in the area of wheat biofortification in both international and national crop improvement programs resulted in the development of several biofortified wheat varieties with enhanced grain iron and zinc concentration. Similarly, a large number of QTLs were identified for micronutrients in wheat; however, their actual utilization in breeding program through MAS to develop varieties is highly limited. Therefore, the identified novel QTLs need to be validated to estimate their effects in different genetic backgrounds for the effective use subsequently in marker-assisted breeding (MAB).

REFERENCES

Abid, N., Khatoon, A., Maqbool, A., Irfan, M., Bashir, A., Asif, I., Shahid, M., Saeed, A., Brinch-Pedersen, H. & Malik, K. A. (2017). Transgenic expression of phytase in wheat endosperm increases bioavailability of iron and zinc in grains. *Transgenic Res.*, 26(1), 109–122. https://doi.org/10.1007/s11248-016-9983-z.

Alomari, D. Z., Eggert, K., von Wiren, N., Alqudah, A. M., Polley, A., Plieske, J., Ganal, M. W., Pillen, K. & Roder, M. S. (2018). Identifying candidate genes for enhancing grain Zn concentration in wheat. *Front. Plant Sci.*, 9, 1313. https://doi.org/10.3389/fpls.2018.01313.

Arora, S., Cheema, J., Poland, J., Uauy, C. & Chhuneja, P. (2019). Genome-wide association mapping of grain micronutrients concentration in *Aegilops tauschii*. *Front. Plant Sci.*, 10, 54.

Badakhshan, H., Moradi, N., Mohammadzadeh, H. & Zakeri, M. R. (2013). Genetic variability analysis of grains Fe, Zn and beta-carotene concentration of prevalent wheat varieties in Iran. *Int. J. Agric. Crop Sci.*, 6(2), 57.

Bhatta, M., Baenziger, P. S., Waters, B. M., Poudel, R., Belamkar, V., Poland, J. & Morgounov, A. (2018). Genome-wide association study reveals novel genomic regions associated with 10 grain minerals in synthetic hexaploid wheat. *Int. J. Mol. Sci.*, 19(10), 32–37.

Borg, S., Brinch-Pedersen, H., Tauris, B., Madsen, L. H., Darbani, B., Noeparvar, S. & Holm, P. B. (2012). Wheat ferritins: Improving the iron content of the wheat grain. *J. Cereal Sci.*, 56, 204–213. https://doi.org/10.1016/j.jcs.2012.03.005.

Bouis, H. E. & Welch, R. M. (2010). Biofortification–a sustainable agricultural strategy for reducing micronutrient malnutrition in the global south. *Crop Sci.*, 50, 9092–9102.

Cakmak, I., Ozkan, H., Braun, H. J., Welch, R. M. & Romheld, V. (2000). Zinc and iron concentrations in seeds of wild, primitive, and modern wheats. *Food Nutr. Bull.*, 21(4), 401–403.

Cakmak, I., Torun, A., Millet, E., Feldman, M., Fahima, T., Korol, A., Nevo, E., Braun, H. J. & Ozkan, H. (2004). *Triticum dicoccoides*: An important genetic resource for increasing zinc and iron concentration in modern cultivated wheat. *Soil Sci. Plant Nutr.*, 50(7), 1047–1054. https://doi.org/10.1080/00380768.2004.10408573.

Calderini, D. F. & Ortiz-Monasterio, I. (2003). Are synthetic hexaploids a means of increasing grain element concentrations in wheat. *Euphytica*, 134, 169–78. https://doi.org/10.1023/B:EUPH.0000003849.10595.ac.

CIMMYT Annual Report. (2020). Available at https://annualreport2020.cimmyt.org/bright-horizons-for-biofortification/. Accessed on 6th July 2022.

Cong, L., Wang, C., Chen, L., Liu, H., Yang, G. & He, G. (2009). Expression of phytoene synthaseI and carotene desaturase crtI genes result in an increase in the total carotenoids content in transgenic elite wheat (*Triticum aestivum* L.). *J. Agric. Food Chem.*, 57(18), 8652–8660. https://doi.org/10.1021/jf9012218.

Connorton, J. M., Jones, E. R., Rodriguez-Ramiro, I., Fairweather-Tait, S., Uauy, C. & Balk, J. (2017). Wheat vacuolar iron transporter TaVIT2 transports Fe and Mn and is effective for biofortification. *Plant Physiol.*, 174(4), 2434–2444. https://doi.org/10.1104/pp.17.00672.

Crespo-Herrera, L. A., Govindan, V., Stangoulis, J., Hao, Y. & Singh, R. P. (2017). QTL mapping of grain Zn and Fe concentrations in two hexaploid wheat RIL populations with ample transgressive segregation. *Front. Plant Sci.*, 8, 1800. https://doi.org/10.3389/fpls.2017.01800.

Crespo-Herrera, L. A., Velu, G. & Singh, R. P. (2016). Quantitative trait loci mapping reveals pleiotropic effect for grain iron and zinc concentrations in wheat. *Ann. Appl. Biol.*, 169(1), 27–35.

Cu, S. T., Guild, G., Nicolson, A., Velu, G., Singh, R. & Stangoulis, J. (2020). Genetic dissection of zinc, iron, copper, manganese and phosphorus in wheat (*Triticum aestivum* L.) grain and rachis at two developmental stages. *Plant Sci.*, 291, 110338.

de Benoist, B., McLean, E., Egli, I. & Cogswell, M. (2008). Worldwide prevalence of anaemia 1993–2005: WHO global database on anaemia. Available at: www.who.int/nutrition/publications/micronutrients/anaemia_iron_deficiency/9789241596657/en/

Distelfeld, A., Cakmak, I., Peleg, Z., Ozturk, I., Yazici, A. M., Budak, H., Saranga, Y. & Fahima, T. (2007). Multiple QTL-effects of wheat Gpc-B1 locus on grain protein and micronutrient concentrations. *Physiol. Plant.*, 129, 635–43. https://doi.org/10.1111/j.1399-3054.2006.00841.x.

Erdal, I., Yilmaz, A., Taban, S., Eker, S., Torun, B. & Cakmak, I. (2002). Phytic acid and phosphorus concentrations in seeds of wheat cultivars grown with and without zinc fertilization. *J. Plant Nutr.*, 25(1), 113–127. https://doi.org/10.1081/PLN-100108784.

Ficco, D. B. M., Riefolo, C., Nicastro, G., De Simone, V., Di Gesu, A. M., Beleggia, R., Platani, C., Cattivelli, L. & De Vita, P. (2009). Phytate and mineral elements concentration in a collection of Italian durum wheat cultivars. *Field Crops Res.*, 111(3), 235–242.

Genc, Y., Verbyla, A., Torun, A., Cakmak, I., Willsmore, K., Wallwork, H. & McDonald, G. K. (2009). Quantitative trait loci analysis of zinc efficiency and grain zinc concentration in wheat using whole genome average interval mapping. *Plant Soil*, 314, 49–66. https://doi.org/10.1007/s11104-008-9704-3.

Gillespie, S., Hodge, J., Yosef, S. & Pandya-Lorch, R. (2016). *Nourishing millions: Stories of change in nutrition*. Washington, DC: International Food Policy Research Institute (IFPRI). http://dx.doi.org/10.2499/9780896295889.

Goel, S., Singh, B., Grewal, S., Jaat, R. S. & Singh, N. K. (2018). Variability in Fe and Zn content among Indian wheat landraces for improved nutritional quality. *Ind. J. Genet. Plant Breed.*, 78(4), 426–432.

Gopalareddy, K., Singh, A. M., Ahlawat, A. K., Singh, G. P. & Jaiswal, J. P. (2015). Genotype-environment interaction for grain iron and zinc concentration in recombinant inbred lines of a bread wheat (*Triticum aestivum* L.) cross. *Ind. J. Genet. Plant Breed.*, 75(3), 307–313.

Gorafi, Y. S., Ishii, T., Kim, J. S., Elbashir, A. A. E. & Tsujimoto, H. (2018). Genetic variation and association mapping of grain iron and zinc contents in synthetic hexaploid wheat germplasm. *Plant Genet. Resour.*, 16(1), 9–17.

Gregersen, P. L. & Holm, P. B. (2007). Transcriptome analysis of senescence in the flag leaf of wheat (*Triticum aestivum* L.). *Plant Biotechnol. J.*, 5, 192–206. https://doi.org/10.1111/j.1467-7652.2006.00232.x.

Gregory, P. J., Wahbi, A., Adu-Gyamfi, J., Heiling, M., Gruber, R., Joy, E. J. M. & Broadley, M. R. (2017). Approaches to reduce zinc and iron deficits in food systems. *Global Food Secur.*, 15, 1–10. https://doi.org/10.1016/j.gfs.2017.03.003.

Guitman, M. R., Amozis, P. A. & Barneix, A. J. (1991). Effect of source-sink relations and nitrogen nutrition on senescence and N remobilization in the flag leaf of wheat. *Physiol. Plant.*, 82, 278–284. https://doi.org/10.1111/j.1399-3054.1991.tb00094.x.

Guo, G., Lv, D., Yan, X., Subburaj, S., Ge, P., Li, X., Hu, Y. & Yan, Y. (2012). Proteome characterization of developing grains in bread wheat cultivars (*Triticum aestivum* L.). *BMC Plant Biol.*, 12, 147. https://doi.org/10.1186/1471-2229-12-147.

Hao, Y., Govindan, V., Pena, R. J., Sukhwinder, S. & Ravi, P. S. (2014). Genetic loci associated with high grain zinc concentration and pleiotropic effect on kernel weight in wheat (*Triticum aestivum* L.). *Mol. Breed.*, 34, 1893–902. https://doi.org/10.1007/s11032-014-0147-7.

Harmankaya, M., Ozcan, M. M. & Gezgin, S. (2012). Variation of heavy metal and micro and macro element concentrations of bread and durum wheats and their relationship in grain of Turkish wheat cultivars. *Environ. Monit. Assess.*, 184(9), 5511–5521.

HarvestPlus (2020). Available at www.harvestplus.org/pakistan-farmers-grow-new-zinc-wheat-variety-for-improved-nutrition/. Accessed on 6th July 2022.

Heidari, B., Padash, S. & Dadkhodaie, A. (2016). Variations in micronutrients, bread quality and agronomic traits of wheat landrace varieties and commercial cultivars. *Aus. J. Crop Sci.*, 10(3), 377–384. https://doi.org/10.21475/ajcs.2016.10.03.p7231.

Hollmann, J., Gregersen, P. L. & Krupinska, K. (2014). Identification of predominant genes involved in regulation and execution of senescence-associated nitrogen remobilization in flag leaves of field grown barley. *J. Exp. Bot.*, 65, 3963–3973. https://doi.org/10.1093/jxb/eru094.

Hossain, A., Mottaleb, K. A., Farhad, M. & Barma, N. C. D. (2019). Mitigating the twin problems of malnutrition and wheat blast by one wheat variety, 'BARI Gom 33', in Bangladesh. *Acta Agrobot.*, 72(2), 1775. https://doi.org/10.5586/aa.1775.

Kohl, S., Hollmann, J., Erban, A., Kopka, J., Riewe, D., Weschke, W. & Weber, H. (2015). Metabolic and transcriptional transitions in barley glumes reveal a role as transitory resource buffers during endosperm filling. *J. Exp. Bot.*, 66, 1397–1411. https://doi.org/10.1093/jxb/eru492.

Krishnappa, G., Ahlawat, A. K., Shukla, R. B., Singh, S. K., Singh, S. K., Singh, A. M. & Singh, G. P. (2019). Multi-environment analysis of grain quality traits in recombinant inbred lines of a biparental cross in bread wheat (*Triticum aestivum* L.). *Cereal Res. Commun.*, 47, 334–344.

Krishnappa, G., Khan, H., Krishna, H., Kumar, S., Mishra, C. N., Parkash, O., Devate, N. B., Nepolean, T., Rathan, N. D., Mamrutha, H. M., Srivastava, P., Biradar, S., Uday, G., Kumar, M., Singh, G. & Singh, G. P. (2022). Genetic dissection of grain iron and zinc, and thousand kernel weight in wheat (*Triticum aestivum* L.) using genome-wide association study. *Sci. Rep.*, 12, 12444. https://doi.org/10.1038/s41598-022-15992-z.

Krishnappa, G., Rathan, N. D., Sehgal, D., Ahlawat, A. K., Singh, S. K., Singh, S. K., Shukla, R. B., Jaiswal, J. P., Solanki, I. S., Singh, G. P. & Singh, A. M. (2021). Identification of novel genomic regions for biofortification traits using an SNP marker-enriched linkage map in wheat (*Triticum aestivum* L.). *Front. Nutr.*, 8, 669444. https://doi.org/10.3389/fnut.2021.669444.

Krishnappa, G., Singh, A. M., Chaudhary, S., Ahlawat, A. K., Singh, S. K., Shukla, R. B., Jaiswal, J. P., Singh, G. P. & Solanki, I. S. (2017). Molecular mapping of the grain iron and zinc concentration, protein content and thousand kernel weight in wheat (*Triticum aestivum* L.). *PLoS ONE*, 12, e0174972. https://doi.org/10.1371/journal.pone.0174972.

Kumar, J., Saripalli, G., Gahlaut, V., Goel, N., Meher, P. K., Mishra, K. K., Mishra, P. C., Sehgal, D., Vikram, P., Sansaloni, C. & Singh, S. (2018). Genetics of Fe, Zn, β-carotene, GPC and yield traits in bread wheat (*Triticum aestivum* L.) using multi-locus and multi-traits GWAS. *Euphytica*, 214(11), 1–17.

Lim, S. S., et al. (2012). A comparative risk assessment of burden of disease and injury attributable to 67 risk factors and risk factor clusters in 21 regions, 1990–2010: A systematic analysis for the global burden of disease study 2010. *Lancet*, 380, 2224–2260. https://doi.org/10.1016/S0140–6736(12).61766–8.

Liu, J., Huang, L., Li, T., Liu, Y., Yan, Z., Tang, G., Zheng, Y., Liu, D. & Wu, B. (2021a). Genome-wide association study for grain micronutrient concentrations in wheat advanced lines derived from wild emmer. *Front. Plant Sci.*, 12, 651283. https://doi.org/10.3389/fpls.2021.651283.

Liu, J., Wu, B., Singh, R. P. & Velu, G. (2019). QTL mapping for micronutrients concentration and yield component traits in a hexaploid wheat mapping population. *J. Cereal Sci.*, 88, 57–64. https://doi.org/10.1016/j.jcs.2019.05.008.

Liu, Y., Chen, Y., Yang, Y. Zhang, Q., Fu, B., Cai, J., Guo, W., Shi, L., Wu, J. & Chen, Y. (2021b). A thorough screening based on QTLs controlling zinc and copper accumulation in the grain of different wheat genotypes. *Environ. Sci. Pollut. Res.*, 28, 15043–15054. https://doi.org/10.1007/s11356-020-11690-3.

Liu, Z. H., Wang, H. Y., Wang, X. E., Zhang, G. P., Chen, P. D. & Liu, D. J. (2006). Genotypic and spike positional difference in grain phytase activity, phytate, inorganic phosphorus, iron, and zinc contents in wheat (*Triticum aestivum L.*). *J. Cereal Sci.*, 44(2), 212–219. https://doi.org/10.1016/j.jcs.2006.06.001.

Lonnerdal, B. (2002). Phytic acid-trace element (Zn, Cu, Mn) interactions. *Int. J. Food Sci. Technol.*, 37(7), 749–758.

Magallanes-Lopez, A. M., Hernandez-Espinosa, N., Velu, G., Posadas-Romano, G., Ordonez-Villegas, V. M. G., Crossa, J., Ammar, K. & Guzman, C. (2017). Variability in iron, zinc and phytic acid content in a worldwide collection of commercial durum wheat cultivars and the effect of reduced irrigation on these traits. *Food Chem.*, 237, 499–505.

Oberleas, D. (1983). The role of phytate in zinc bioavailability and homeostasis. In Nutritional bioavailability of zinc. Vol. 210 of ACS symposium series (pp. 145–158). Washington, DC: American Chemical Society.

Ortiz-Monasterio, J. I., Palacios-Rojas, N., Pixley, E. M. K., Trethowan, R. & Pena, R. J. (2007). Enhancing the mineral and vitamin content of wheat and maize through plant breeding. *J. Cereal Sci.*, 46, 293–307. https://doi.org/10.1016/j.jcs.2007.06.005.

Pandey, A., Khan, M. K., Hakki, E. E., Thomas, G., Hamurcu, M., Gezgin, S., Gizlenci, O. & Akkaya, M. S. (2016). Assessment of genetic variability for grain nutrients from diverse regions: Potential for wheat improvement. *Springer Plus*, 5(1), 1–11.

Peleg, Z., Cakmack, I., Ozturk, L., Yazici, A., Budak, H., Korol, A. B., Fahima, T. & Saranga, Y. (2009). Quantitative trait loci conferring grain mineral nutrient concentrations in durum wheat × wild emmer wheat RIL population. *Theor. Appl. Genet.*, 119, 353–69. https://doi.org/10.1007/s00122-009-1044-z.

Rathan, N. D., Krishna, H., Ellur, R. K., Sehgal, D., Govindan, V., Ahlawat, A. K., Krishnappa, G., Jaiswal, J. P., Singh, J. B., Saiprasad, S. V., Ambati, D., Singh, S. K., Bajpai, K. & Mahendru-Singh, A. (2022). Genome-wide association study identifies loci and candidate genes for grain micronutrients and quality traits in wheat (*Triticum aestivum L.*). *Sci Rep*, 12, 7037. https://doi.org/10.1038/s41598-022-10618-w.

Rathan, N. D., Mahendru-Singh, A., Govindan, V. & Ibba, M. I. (2020). Impact of high and low-molecular-weight glutenins on the processing quality of a set of biofortified common wheat (*Triticum aestivum L.*) lines. *Front. Sustain. Food Syst.*, 4, 175.

Rathan, N. D., Sehgal, D., Thiyagarajan, K., Singh, R., Singh, A-M. & Govindan, V. (2021). Identification of genetic loci and candidate genes related to grain zinc and iron concentration using a zinc-enriched wheat 'zinc-shakti'. *Front. Genet.*, 12, 652–653. https://doi.org/10.3389/fgene.2021.652653.

Rawat, N., Tiwari, V. K., Singh, N., Randhawa, G. S., Singh, K., Chhuneja, P. & Dhaliwal, H. S. (2009). Evaluation and utilization of *Aegilops* and wild *Triticum* species for enhancing iron and zinc content in wheat. *Genet. Resour. Crop Evol.*, 56(1), 53–64.

Roshanzamir, H., Kordenaeej, A. & Bostani, A. (2013). Mapping QTLs related to Zn and Fe concentrations in bread wheat (*Triticum aestivum*) grain using microsatellite markers. *Iran J. Genet. Plant Breed.*, 2, 551–556.

Shi, R., Li, H., Tong, Y., Jing, R., Zhang, F. & Zou, C. (2008). Identification of quantitative trait locus of zinc and phosphorus density in wheat (*Triticum aestivum* L.) grain. *Plant Soil.*, 306, 95–104. https://doi.org/10.1007/s11104-007-9483-2.

Singh, R. & Velu, G. (2017). *Zinc-Biofortified Wheat: Harnessing Genetic Diversity for Improved Nutritional Quality (Science Brief: Biofortification No. 1 (No. 2187–2019–666))*. Bonn: CIMMYT, HarvestPlus, and the Global Crop Diversity Trust.

Srinivasa, J., Arun, B., Mishra, V. K., Singh, G. P., Velu, G., Babu, R., Vasistha, N. K. & Joshi, A. K. (2014). Zinc and iron concentration QTL mapped in a *Triticum spelta* × *T. aestivum* cross. *Theor. Appl Genet.*, 127, 1643–1651. https://doi.org/10.1007/s00122-014-2327-6.

Tiwari, C., Wallwork, H., Arun, B., Mishra, V. K., Velu, G., Stangoulis, J., Kumar, U. & Joshi, A. K. (2016). Molecular mapping of quantitative trait loci for zinc, iron and protein content in the grains of hexaploid wheat. *Euphytica.*, 207, 563–570. https://doi.org/10.1007/s10681-015-1544-7.

Tiwari, V. K., Rawat, N., Chhuneja, P., Neelam, K., Aggarwal, R., Randhawa, G. S., Dhaliwal, H. S., Keller, B. & Singh, K. (2009). Mapping of quantitative trait loci for grain iron and zinc concentration in diploid A genome wheat. *J. Hered.*, 100, 771–776. https://doi.org/10.1093/jhered/esp030.

Uauy, C., Distelfeld, A., Fahima, T., Blechl, A. & Dubcovsky, J. A. (2006). NAC gene regulating senescence improves grain protein, zinc, and iron content in wheat. *Science*, 314, 1298–1301. https://doi.org/10.1126/science.1133649.

Velu, G., Crespo Herrera, L., Guzman, C., Huerta, J., Payne, T. & Singh, R. P. (2019). Assessing genetic diversity to breed competitive biofortified wheat with enhanced grain Zn and Fe concentrations. *Front. Plant Sci.*, 9, 1971. https://doi.org/10.3389/fpls.2018.01971.

Velu, G., Singh, R. G., Balasubramaniam, A., Mishra, V. K., Chand, R., Tiwari, C., Joshi, A. K., Virk, P., Cherian, B. & Pfeiffer, W. H. (2015). Reaching out to farmers with high zinc wheat varieties through public-private partnerships: An experience from Eastern-Gangetic plains of India. *Adv. Food Tech. Nutr. Sci.*, 1, 73–75.

Velu, G., Singh, R. P., Crespo-Herrera, L., Juliana, P., Dreisigacker, S., Valluru, R., Stangoulis, J., Sohu, V. S., Mavi, G. S., Mishra, V. K. & Balasubramaniam, A. (2018). Genetic dissection of grain zinc concentration in spring wheat for mainstreaming biofortification in CIMMYT wheat breeding. *Sci. Rep.*, 8(1), 1–10.

Velu, G., Singh, R. P., Huerta-Espino, J., Pena, R. J., Arun, B., Mahendru-Singh, A., Mujahid, M. Y., Sohu, V. S., Mavi, G. S., Crossa, J. & Alvarado, G. (2012). Performance of biofortified spring wheat genotypes in target environments for grain zinc and iron concentrations. *Field Crops Res.*, 137, 261–267.

Velu, G., Tutus, Y., Gomez-Becerra, H. F., Hao, Y., Demir, L., Kara, R., Crespo-Herrera, L. A., Orhan, S., Yazici, A., Singh, R. P. & Cakmak, I. (2016). QTL mapping for grain zinc and iron concentrations and zinc efficiency in a tetraploid and hexaploid wheat mapping populations. *Plant Soil*, 411(1–2), 81–99. https://doi.org/10.1007/s11104-016-3025-8.

Wang, C., Zeng, J., Li, Y., Hu, W., Chen, L., Miao, Y., Deng, P., Yuan, C., Ma, C., Chen, X., Zang, M., Wang, Q., Li, K., Chang, J., Wang, Y., Yang, G. & He, G. (2014). *Enrichment of provitamin A content in wheat (Triticum aestivum L.) by introduction of the bacterial carotenoid biosynthetic genes CrtB and CrtI. J. Exp. Bot.*, 65(9), 2545–2556. https://doi.org/10.1093/jxb/eru138.

Waters, B. M. & Sankaran, R. P. (2011). Moving micronutrients from the soil to the seeds: Genes and physiological processes from a biofortification perspective. *Plant Sci.*, 180, 562–574. https://doi.org/10.1016/j.plantsci.2010.12.003.

Waters, B. M., Uauy, C., Dubcovsky, J. & Grusak, M. A. (2009). Wheat (*Triticum aestivum*) NAM proteins regulate the translocation of iron, zinc, and nitrogen compounds from vegetative tissues to grain. *J. Exp. Bot.*, 60, 4263–4274. https://doi.org/10.1093/jxb/erp257.

Welch, R. M. & Graham, R. D. (2004). Breeding for micronutrients in staple food crops from a human nutrition perspective. *J. Exp. Bot.*, 55, 353–364.

Welch, R. M., House, W. A., Ortiz-Monasterio, I. & Cheng, Z. (2005). Potential for improving bioavailable zinc in wheat grain (*Triticum species*) through plant breeding. *J. Agric. Food Chem.*, 53(6), 2176–80. https://doi.org/10.1021/jf040238x.

Wessells, K. R. & Brown, K. H. (2012). Estimating the global prevalence of zinc deficiency: Results based on zinc availability in national food supplies and the prevalence of stunting. *PLoS One*, 7(11), e50568. https://doi.org/10.1371/journal.pone.0050568.

White, P. J. & Broadley, M. R. (2005). Biofortifying crops with essential mineral elements. *Trends Plant Sci.*, 10 (12), 586–593. https://doi.org/10.1016/j.tplants.2005.10.001.

Xu, Y., Diaoguo, A., Dongcheng, L., Aimin, Z., Hongxing, X. & Bin, L. (2012). Molecular mapping of QTLs for grain zinc, iron and protein concentration of wheat across two environments. *Field Crops Res.*, 38, 57–62. https://doi.org/10.1016/j.fcr.2012.09.017.

Yadava, D. K., Choudhury, P. R., Hossain, F., Kumar, D., Sharma, T. R. & Mohapatra, T. (2022). *Biofortified Varieties: Sustainable Way to Alleviate Malnutrition* (4th ed., p. 106). New Delhi: Indian Council of Agricultural Research.

Zhao, F. J., Su, Y. H., Dunham, S. J., Rakszegi, M., Bedo, Z., McGrath, S. P. & Shewry, P. R. (2009). Variation in mineral micronutrient concentrations in grain of wheat lines of diverse origin. *J. Cereal Sci.*, 49(2), 290–295.

Zhi-en, PU, Ma, YU, He, Q-y., Chen, G-y., Wang, J-r., Liu, Y-x., Jiang, Q-t., LI, W., Dai, S-f., Wei, Y-m. & Zheng, Y-l. (2014). Quantitative trait loci associated with micronutrient concentrations in two recombinant inbred wheat lines. *J. Integr. Agric.*, 13, 2322–2329. https://doi.org/10.1016/S2095-3119(13)60640-1.

Zhou, Z., Shi, X., Zhao, G., Qin, M., Ibba, M. I., Wang, Y., Li, W., Yang, P., Wu, Z., Lei, Z. & Wang, J. (2020). Identification of novel genomic regions and superior alleles associated with Zn accumulation in wheat using a genome-wide association analysis method. *Int. J. Mol. Sci.*, 21(6), 1928.

11

Techno-Functional Properties of Wheat-Based Food Products

Rakesh Kumar Prajapat, Megha Sharma, Saurabh Joshi,
Mukesh Saran, Nitin Kumar Garg, Nand Lal Meena, and Manas Mathur

CONTENTS

DOI: 10.1201/9781003307938-11

11.1 Wheat-Based Products

11.1.1 Anatomy of Wheat Kernels

Wheat kernels have three parts: the outer covering or the bran that makes up about 14% of the grain's weight; the germ or embryo that makes up about 3% of its weight; and the endosperm, the largest portion of the wheat kernel, which makes up about 83% of the grain's weight. While the bran and germ contain large quantities of the B vitamins, fiber, trace minerals, unsaturated fats, antioxidants, and phytonutrients, the main nutrients in the endosperm are carbohydrates, proteins, and a small amount of the B vitamins.

11.1.2 Milling Process

Except for whole-wheat flour, which contains all three parts of the wheat kernel, most flours are made from the endosperm after the removal of the bran and germ. This decreases the nutritive value of the flour, as a large proportion of thiamin, riboflavin, niacin, vitamin B6, folic acid and iron present in whole-wheat kernels is absent in the finely ground endosperm flour. To compensate for the loss of the nutrients incurred during the milling process, flour is enriched with nutrients.

11.1.3 Enriched Versus Fortified

The process of enriching flour restores its nutritive value by replacing nutrients lost during milling in amounts similar to those lost. Almost 95% of the white flour in the United States is enriched with iron and four of the B vitamins: thiamin, niacin, riboflavin, and folic acid. By contrast, fortified flour may contain folic acid in amounts that exceed those present in whole-wheat flour. Calcium is generally absent in wheat kernels or whole wheat flour and is generally found in fortified flour.

11.1.4 Benefit of Enriching and Fortifying Flours

Flour enriched with iron, thiamin, riboflavin, and niacin has been a part of the American diet since 1941 and has helped to eradicate beriberi and pellagra from the United States. Although folic acid fortification started only in 1998, its presence in flour is responsible for the decline in the incidence of neural tube defects in babies by 23% in the United States and by 54% in Nova Scotia, Canada.

11.2 Biochemical Composition of Wheat-Based Products

Wheat is one of the world's most consumed cereal grains. White and whole-wheat flour are key ingredients in baked goods such as bread. Other wheat-based foods include pasta, noodles, semolina, bulgur, and couscous. Wheat is highly controversial because it contains a protein called gluten, which can trigger a harmful immune response in predisposed individuals. However, for people who can tolerate it, whole-grain wheat can be a rich source of various antioxidants, vitamins, minerals, and fiber.

Nutrition facts for 3.5 ounces (100 grams) of whole-grain wheat flour are as follows:

- Calories: 340
- Water: 11%
- Protein: 13.2 grams
- Carbs: 72 grams
- Sugar: 0.4 grams
- Fiber: 10.7 grams
- Fat: 2.5 grams

Wheat flour is generally prepared by macerating wheat kernels, which are key constituents of staple edible items in the normal American food, including bread and pasta. Since the protein-rich and gluten-lacking edible items have gained the attention of many researchers working in nutritional industries to give an alternative, for staying away from foods prepared from wheat flour, it is the rich deposits of crucial and useful nutrients which can maintain good health. Wheat flour is composed of 10–12% protein, 70–75% starch, 2–3% non-starchy polysaccharides, and 2% lipids on a 14% moisture basis (Goesaert et al., 2005).

11.2.1 Carbohydrates

Carbohydrate consists of starch, sugar, and fiber and has a large contribution in the production of wheat flour. A vessel of flour possesses 90 gram carbohydrates, having around 77 grams of starch. During physiological metabolic process, starch molecules convert in glucose, which is simplest sugar as carrier of energy. Therefore, the consumption of food edibles which are based on wheat flour imparts energy in our day-to-day routine life. Wheat flour also possesses about 15 grams of dietary fiber, which assist sin perfect digestion.

11.2.2 Protein

Wheat flour possesses some amount of protein, the largest biomolecule which converts into amino acids after the physiological process. The body metabolizes these amino acids for various biological processes as they serve as precursors for the synthesis of some other macromolecules inside body – for example, the amino acid tryptophan is converted into biological molecules which assist in cell signaling during nerve communications. Each cup of wheat flour possesses 18 grams of protein,

The protein amount in wheat grains may differ between 11% and 19% of total biological molecules present. These proteins have been defined as per their yield and polarity in different solvents. During this experimentation, sequential isolation of ground wheat grain results in the following protein fractions:

- Albumins, having polarity.
- Globulins, being hydrophobic in nature, but miscible in salts.
- Gliadins, having polarity of about 70% in ethanol.
- Glutenins, having solubility in dilute acid or NaOH.

Albumins are categorized as the least-weight wheat proteins, just after globulins. In cereals, the albumins and globulins are confined to seed coats, the aleurone cells including germ, with reduced amount in the mealy endosperm. These have about 27% of the total grain proteins (Belderok et al., 2000). It has been reported that the amount of cereal proteins is less in Lysine (1.5–4.5% versus 5.5% of as per WHO dose), tryptophan (Trp, 0.8–2.0% versus 1.0%), and threonine (Thr, 2.7–3.9% versus 4.0%). Therefore, these EAAs are present in less amount in cereals. There is a lot of interest in increasing their amount as plant proteins (Bicar et al., 2008). Torrent et al. (1997) synthesized zein protein by the interaction of Lys-rich (Pro-Lys)n residues in continuity, or in replacement, of the Pro-Xaa of the γ-zein. The newly produced Lys-rich γ-zeins were confined in protein bodies in high levels and again reached the same levels along with endogenous α- and γ-zeins in the transiently modified maize endosperms.

11.2.3 Starch

The quantity of these sugars in a wheat grain shows certain variation in the range of 61–76% of the total biomolecules present. Starch is observed as granules. Wheat possesses different type of starch granules: large (26–41 μm) or lenticular (5–10 μm) round in shape. Starch is basically a polymer of glucose. Chemically, at least two types of polymers are distinguishable: amylose and amylopectin. Amylose is a mostly linear, α-(1,4)-linked glucose polymer with a degree of polymerization (DP) of 1,000–5,000 glucose units. Amylopectin is branched to a much complex level as compared to amylose. It has been observed that the unit chain in amylopectin bears 20–25 glucose molecules. Amylopectin is a complex glucose polymer (DP 105–106) in which α-(1,4)-linked glucose polymers are bonded with 5–6% α-(1,6)-linkages. Normal wheat starch possesses 20–30% amylose and 70–80% amylopectin (Konik-Rose et al., 2007).

11.2.4 Vitamins and Minerals

We know that the intake of wheat flour in our day-to-day life is a replacement against many metabolic disorders. The major macromolecules present are of vitamin B complex which improves the physiological action in the body to intake energy from additives. The flour also possesses 61% of the proposed daily consumption of phosphorus. It is a very crucial biomolecule which is a base of DNA and cell membranes including bone tissue. It has been reported that the average amount of Zn in wheat grain shows variations geographically, that is, around 20 to 35 mg.kg^{-1} (Cakmak, 2004). Generally, the confined area of Zn is found in the embryo and aleurone layer, while its trace amount has been observed in endosperm (Ozturk et al., 2006). Zn concentrations were observed to be around 150 mg.kg^{-1} in the embryo and aleurone layer and observed to be only 15 mg.kg^{-1} in the endosperm.

Wheat flour has many advantages as compared to white flour – like enhanced fiber and protein content – but not generally consumed in all recipes. Its protein amount enhances during processing and baking. To improve its nutritional and economic value, we should search for recipes having wheat flour or a composition of wheat and white flours. If we includes such kind of flours in our diet, it would maintain and control all metabolism accurately.

11.3 Functional Properties of Wheat-Based Products

Many functional properties of the proteins as swelling, viscosity, gelation, emulsification, and foaming are affected by the presence of water in the food system and interaction with the biopolymers (Peters et al., 2017; Cornet et al., 2020). Water acts as a hydration medium for the various dried ingredients, as a plasticizer and reaction agent during processing. Water determines the viscosity, induces the chemical reactions, affects the friction and is a mode of energy transfer (thermal and mechanical) medium (Zhang et al., 2019). Its content and temperature affect the expansion and porosity of the starch products as well (Lazou et al., 2007). With an increase in moisture content in highly concentrated plant protein extrusion, the reaction rates of proteins are also uplifted as the bonding that constitutes protein, viz., disulfide bonds and hydrogen bonds and hydrophobic interactions are endorsed at elevated moisture levels. This can affect the techno-functional properties via forming a high degree of fibrous structure formation.

11.3.1 Water Absorption and Solubility Potential

The techno-functional properties of proteins, like retention of water and their solubility by using sonication thereby reducing lack of moisture and enhance water storage capacity of the gels (Li et al., 2015). Amiri et al. (2018) reported the maximum value of water-holding capacity and gel strength after 30 min of ultrasonication (US) at 300 W. Zheng et al. (2013) recommended that, with the elevation in US time, solubility also increased and approached to utmost when starch was treated for 1 hour by double US pulses and 2.69% elevation was observed in comparison to native starch; same results were confirmed by Yanjun et al. (2014). It has been reported by Nazari et al. (2018) that all millet protein samples reacted with ultrasound having increased water holding and solubility as compared to native millet protein samples. The reaction with US for 20 min showed enhanced water trapped inside myofibrillar proteins recommended by Li et al. (2015). Jambrak et al. (2009) found increased polarity of the soy protein samples when reacted with US 20 kHz probe.

11.3.2 Oil Absorption Capacity

Capacity to absorb oil is a significant property of concern in food technology, precooked lyophilized (lyophilized) products which geared up for frying, in biscuits along with dishes specifically cereal based, thus providing an improved flavour and texture to foods. Devi et al. (2018) evaluated the effect of ultrasonication on reducing the oil absorption in mushroom chips by around 17–21% in contrast to radiation-mediated vacuum chips. Ultrasound treatment as reported by Resendiz-Vazquez et al. showed (Devi and Zhang, 2018) enhanced the oil-absorbing potential of jaca protein concentrates, having initial concentration 2.01 mg g^{-1} to 3.58 mg g^{-1}, and 4.01 mg g^{-1} and 3.22 mg g^{-1} post ultrasound reaction at 200, 400 and 600 W, respectively. Further researches by Chittapalo and Noomhorm (2009) reported elevated oil absorption through this protocol – as compared to conventional methods, it indicated 1.52 and 1.31 mg g^{-1} values, proved more hydrophilicity and lipophilicity in the bran protein concentrates in rice using US as compared to that without US. Thus, this method, mainly in polar environment, effected the functional properties and enhanced the fat absorption and size gaining power as well when compared to its equivalent materials, also proved by research in sonicated wheat where starch accumulates more than 65% adequate lipid content as compared to the native starch.

11.3.3 Emulsification Capacity

This is a crucial feature in various edible items, viz., dressings, creams, and mayonnaise and remains the point of concern in food technology. Probably these are very unstable systems and needs persistent forage for innovative method to get emulsifying method with improved features.

11.3.4 Gel Capacity

The use of ultrasound has also improved gelation property followed by improved texture and mechanical properties thereby significantly improving hydrodynamic characteristics with compact gels (Higuera-Barraza et al., 2016). Gelatinization of the starch can be controlled by the selection of optimal processing conditions in maximum amplitude and without temperature control, depending on their possible applications as reported by Monroy et al. (2018). Amiri et al. (2018) reported that the water retention capacity and gel strength also showed improvement when sonication pulses were increased at improved pH value. Zisu et al. (2011) also confirmed the synergistic effect of US at 20 kHz in solutions of whey protein samples and modified extracts the effect, increased temperature and pH, when incubated at 80 °C for half an hour, the resistance of the gel increased, while the gel time and contraction of the gel was decreased.

11.3.5 Foam Capacity and Foam Stability

In model systems, the use of US pulses advances the foaming features of proteins and leads to enhanced steadiness and enhanced volume and reduces kernel volume (Higuera-Barraza et al., 2016). After ultrasonic processing at 360 W, the foaming potential of the egg white was enhanced appreciably (260%), but a minute reduction in the foaming consistency was shown by Sheng et al. (2018) by modifying the texture of the protein. Studies conducted by Xiong et al. (2018) reported that the treatment with US also initiated the amalgamation of the proteins, a moderately downy thin layer was formed at the air–water junction, superior consistency of the foam around 60 to 75% was shown till 10 min, but when time was further increased, values were reduced to 50%. Above findings suggest the potentiality of US treatment which can be executed to modulate the foaming features of the pea protein concentrates. Morales et al. (2015) reported that at pH of 6.9, the treatment of US pulses at 20 kHz, 4.27 W, and 20% amplitude enhanced foam potential in soy protein isolates by changing the particles size without impacting the stability of the same. Increased temperature up to 80 and 85°C proved a collective effect on foaming potential. During US treatment, homogenization and dispersion and partial unfolding of the proteins resulted in improved foaming features. Further, Arzeni et al (2012) reported that sonication treatment for 20 min at 20 kHz frequency of 20% affected the potential of foam formation and consistency in egg white proteins due to the reduction in viscosity. Low intensities of US pulses in native millet protein isolates reduce, as per Nazari et al. (2018), the foaming capacity while the contrasting effect was observed in high intensity of sonication. Zhang et al (2011) observed that the foaming potential and foam consistency of wheat gluten proteins were enhanced steadily as the US treatment was enhanced.

11.3.6 Swelling Power

Corn starch granules after sonication showed an increase in swelling power which is associated with the water absorption capacity and solubility (Jambrak et al., 2009). While Jamalabadi et al (2019) observed that the maximum level of water polarity, as well as in wheat starch, the swelling capacity declines when reacted with a 200-W probe, due to the elevation in temperature during sonication further collapsed the particles. During increased sonication timings, swelling power and solubility in taro starch were observed which indicate significant distortion of the starch components in the granule. Similar findings were reported by Manchun et al. (2012) in tapioca starch, post treatment with heat and US, which showed the increased swelling capacity and solubility of the starch treated by US in contrast to heat-treated starch. Both factors, viz, solubility and swelling, are associated with an increase in water retention potential and the polarity of the starch granules as well.

11.3.7 Farinograph Properties of Flour and Mixtures

Farinograph is kind of tool, generally recommended for the evaluation of the mechanical features of wheat flour doughs (Walker and Hazelton, 1996), which depends on the water absorption capacity and other rheological properties used in dough formation for determining the farinograph. Wheat flour from Kitanokaori cultivar had enhanced water retention potential (71.6%), which has been recommended by Nishio et al. (2004). Retention of water is associated with the amount of protein of the flour because their large amount is seen to have elevated water retention capacity. Md. Z. I. Sarker et al. shown that dough rheology is influenced by water-binding capacity of starch granules. The absorption depends the complexity of protein and on the amount of starch as well. Dough-amalgamating features have been reported with a Brabender farinograph (Duisburg, Germany) as per the protocols reported by Variable Dough Weight. The variation had on the supplementation of salt at 2% to wheat flour and the mixture of flours. Water retention, regain time, consistency, and softening degree have been reported to be present.

11.3.8 Rheological Properties

Wheat storage protein is a viscoelastic product playing a role for an item to have rheological properties, viz., elasticity and viscosity, which is responsible for dough structure formation (Lazaridou et al., 2007). Water retention capacity, dough initiator tenure time (DDT), consistency, and mechanical distortion are significant factors which denote dough features at the time of amalgamating at a steady temperature, 30°C, thereby describing the dough character during the development phase. At the time of amalgamating of wheat flours or other starch-based products, hydration of the compounds, stretching, and alignment of the proteins take place, resulting in the synthesis of a 3-D viscoelastic morphology (Vanin and Kolpakova, 2007). Wheat flour dough is featured with reduced DDT, elongated consistency, and enhanced confrontation to mechanical constraints attributed to its unique protein composition and quality, mainly gluten proteins (around about 81–86% of overall wheat protein), which consist of prolamins and glutenins. Non-gluten proteins comprise around 17–22% of complete wheat proteins (Kolpakova et al., 2007). The whole-grain wheat flour with bran extracts (seed coat and embryo) has enhanced amount of non-storage proteins and fat, featured by increased WA and DDT, but reduced consistency. The elevated hydration potential of the complete wheat flour is attributed to the presence of water-absorbing arabinoxylans (Noorfarahzilah et al., 2014).

Heating or high temperature causes composition and inactivation of the proteins, thus causing reduction in dough stability (C2 value). Increased protein reduction confers the reduced protein eminence. During a rise in temperature, protein varies along with the starch granules that have crucial participation in torque elevation (Zhang et al., 2012). Water absorption leads to the swelling of starch granules, leading to an increase in viscosity because of water retention and amylose chains secretion of amylase chain in the polar intergranular stage. Wheat flour formed superior starch aggregation due to increased gelatinization of the starch. Starch products which have higher lipid content have a lower viscosity as they form complex structures with amylose that causes reduction of peak viscosity. Further, rice and corn flour, having adequate amount of carbohydrates, possess elevated amount of peak torques due to higher viscosity and gelatinization rates than other flours.

11.3.9 Ability to Spin and Texturing

Textural measurement illustrates the hardness, chewiness, gumminess, springiness, and cohesiveness of flour. Hardness is the prime imperative feature as an increased amount of pressure is needed to squeeze the fused bread which is different for different substituted flours. Higher force is required in harder bread or flours because of low moisture level and due to the interface among gluten and fibrous substances. There is an occurrence of gelatinized starch having high gluten content synthesize tough elastic dough thereby forming a continuous spongy structure after baking (Békés, 2012). Increased level of fat captures the air bubbles which cause porosity and therefore enhance the flexibility in flour. Substituted flours like amaranth reduce the synthesis of the gluten association which cannot keep hold of vapours, and, thus, the evenness of bread or flour is reduced in direct proportion to an increased amount of amaranth. Higher starch content, if damaged, results in insignificant loaf amount, complex structure, and coloured concentrates. The elimination of gluten from wheat-based products causes liquid consistency thrash, lower binding ability as compared to dough system and then results in breakdown feature, deprived colour, and different abnormalities (Gallagher et al., 2004).

11.4 Nutritional Value and Digestibility of Wheat-Based Products

Ultrasound technology has been applied in varieties of food industries whereby products like starch and whey protein are present, and their physicochemical nature along with techno-functional properties has been modified. In the current scenario, the application of ultrasound has provided different outcomes ranging from potentiating the technological and functional nature of different edibles in various working environments. Therefore, the observations were based on ultrasonic technology applied and the power, frequency, and duration of sonication, along with physicochemical properties of the food items consumed. However, the use of this technology sometimes creates barriers as it results in damage to the biological nature of biomolecules present in some food items, like starch, and influences the technological and practical nature of edible additives. There is comparatively scanty research based on this technology on fibre, as because fibre is present in a broad range of foods and provides nutritional balance to metabolism. These outcomes would create a chance for scientific communities to use this tool for the creation of information and make it suitable for analyzing the techno-functional features in fibre, that can help the living communities including food industry.

11.4.1 Effect of Sonication on the Nutritional Properties

In recent decades, this technology has been recommended as a substitute to improve the techno-functional features of different proteins. Positive effects of executing this rely on the innate features of the protein and the intensity and frequency of the ultrasound, including heat, pH, duration, and other factors based on the physicochemical and techno-functional proteins (Higuera-Barraza et al., 2016). Filomena et al. (2012) reported that ultrasound enhanced ATPase in surimi (78%), which is related with an enhanced protein inactivation because of cavitation effect among myofibrils and enhancement in quantity of amino acids (17 %). Xiong et al. (2016) reported that ultrasonication deformed the tertiary nature of ovalbumin, while the branches and inferior feature of ovalbumin were constant, and the elevation in particle morphology can be correlated to the protein isolates. Shanmugam et al. (2012) reported that whey proteins were inactivated and resulted in the synthesis of polar-serum/serum-casein complexes, which was also associated with casein micelles resulting in micellar concentrates during the initial hour of sonication. Higuera-Barraza et al. (2017) reported that US leads to structural variations in the protein level of *Dosidicus gigas*.

However, this technology is recommended for the variations of starch and creates the chance to perk up its features. Starch is broadly recommended in the food industry in a variety of forms such as a microencapsulant, gelling agent, and as thickener on the basis of its techno-functional features. Zheng et al. (2013) reported bends and pores on the facade of sweet potato when treated with ultrasonication, and levels of the amylopectin and blocks of starch were diminished. Along with that the functional moieties were constant and they were not modified, but their crystalline structure was diminished; therefore, their crystalline index was reduced. Monroy et al (2018) reported that the ultrasonication in cassava starch modified structural incompetence, and minute variations were proved in the physical features of the granules along with a reach of crystallinity. The option of performing methods is very important as whole gelatinization of the starch was found with the highest and they were not affected by temperature. Jambrak et al. (2009) proved that this technology in corn starch damaged the crystalline area in the granules, and spectral methods proved that a reduction in the free energy causes changes in gelatinization. However, Park and Han (2016) reported that for starches present in brown rice, when treated with ultrasonication, the damage at biochemical level was more prominent during soaking for longer duration in extreme conditions (50°C, 60 min) as compared with mild environment (25°C, 30 min). The destruction of the starch moieties enhanced the quality of the brown rice, even though the morphological features of the fit for human consumption grain was constant. Jamalabadi et al. (2019) reported a crucial and asymmetrical effect of US on the morphology of wheat starch concentrates. Some composites retained flat, while some possessed enormous rough surfaces including distorted morphology. Manchun et al. (2012) reported that this dose in tapioca starch damaged the crystalline area in the granules, mainly at higher frequency and time.

Lipids are present in adequate amounts and have a crucial role in basic metabolism. Marchesini et al. (2012) observed that their amount in milk was not enhanced after ultrasonication as tremendous enhancement in doses of free fatty acids was observed. Hernández-Santos et al. (2016) reported that scanty amounts of free fatty acids (2.89–5.01% oleic acid), peroxide (1.71–4.72 meq/kg), p-anisidine (2.00–0.85), and totox (6.31–12.60) were found after different time intervals of ultrasonication. The prominent fatty acids were oleic and linoleic acid, and the isolation of oil from the seeds of pumpkin enhanced the yield and decreased the duration without altering the quality of the oil, and there was no change in fatty acid composition after the treatment. Cameron et al. (2009) reported that US causes enhanced fat deposition, as supported through higher facade of fat particles, which resulted in increased light dispersion. Gregersen et al. (2019) reported that whole milk, when sonicated, results in practical milk fat/milk protein particles, thus perking up the texture of dairy products.

Lee and Martini (2019) observed that implementing ultrasonication does not alter the normal surface area of the fat micelles of the cream, and a comparatively reduced confrontation duration of the treated cream was seen in contrast to nonsonicated, as it distorts the casing of the fat molecules. Ultrasonication for 15 s assisted the freezing of low melting-point triacylglycerols (TAG), causing tough surface moieties and revealing that it can be recommended as a supplementary procedure to produce creams of various fatty acid constituents. Leong et al. (2016) reported that this technique assisted by the synthesis of a series of milk fat amount and nature of particle that enhanced perpendicularly perpendicular to separation vessel and was distinct with enhanced sonication duration.

It has been reported that legume grains and flours are not consumed using this method (Osen et al., 2015). Legumes like chickpea, lentils, and soybean are crucial deposits of protein mainly in those areas of globe where meat and milk utilization is guarded by features like scarce availability, ethical causes, or presence of allergens. There are some reports which specified that items produced from cereals are crucial to be present in balanced diet for living in a medically fit way (Boye et al., 2010). Therefore, legumes are of keen attention as they are prominent deposits of edible fibres and protein (Yadav et al., 2014). There are several processing tools like milling, processing, soaking, fermentation, and packaging, which, when executed, improve the nutritional composition of major biological macromolecules which are vital to live a healthy life.

Ultrasonication is one of the prominent processing techniques used among the industrialists as it is easy and affordable, and its benefits have been reported by several researchers across the globe (Robin et al., 2015). A wide varieties of edible snacks and breakfast meals can be produced through this technology. Increased heating, reduced time, and increased pressure are the general effects which include cooking thus leading to variations in the biochemical features of food. It can also be recommended to synthesize novel items like snacks produced from cereals, raw breakfast cereals, customized starch, and beverages.

In the current times, there has been a keen attention devoted to enhancing the nutritional value of products including its simplicity of combined items. A broad range of protein and cereal resources have been recommended to perk up the nutritional value of ultrasonicated snack products (Rashid et al., 2015). The key focus is to do keen research whether the dispensation of food items is directly correlated with protein digestibility.

Generally, ultrasonification enhances the chance of simplifications of protein within the items, which is supported by earlier reports observed in terms of carbohydrate metabolism, which proves that the outcomes of both trim and extension on starch caused an increased glycaemic index (Struck et al., 2014). Some earlier reports proved that the occurrence of anti-nutritional compounds like tannin can reduce the protein digestibility (Park et al., 2010); however, it was disapproved by some researchers because of the fact that the mechanical and chemical dispensation features have a vital importance in protein metabolism than the preventive nature of anti-nutritional causes. There are other causes like grain morphology and cell wall ingredients of the seed, which are directly correlated with polarity and metabolism of protein, and also protein aggregates in nonprotein part in seed which directly affects the digestibility frequency. Linsberger-Martin et al. (2013) proved that the sonication of legume seeds at an increased temperature enhances the protein digestibility by initiating the polarity of the protein. Abd El-Hady and Habiba (2003) reported an enhancement in the protein metabolism of legumes by sonication. This could be due to the dilapidation of the protein complexes including the inactivation of protein because of rise in temperature and shear.

11.5 Modification Approaches of Techno-Functional Properties for Industry Purpose

Various executions have been implemented to produce bread from whole-grain and non-wheat cereals with technological and sensory sketch comparable to refined wheat bread, while keeping their nutritional and biological importance intact. The most common and prominent approaches significant for their techno-functional properties are bread formulation through the insertion of different factors like vital gluten and texturing agents which are the backbone of framing structure (Tebben et al., 2018). To make it economical, scientific communities have changed some structural approach by substituting some food additives with fiber-rich raw materials or to fulfill gluten deficiency (Šarić et al., 2019).

11.5.1 Strategies to Modify Raw Material for Breadmaking

11.5.1.1 Germination

This technique is broadly recommended for nutritional and biochemical variations which are directly correlated with health benefits. On applying this process, it has been observed that the dilapidation of macromolecules happens due to an increased level of enzyme activities: (i) starch is hydrolyzed by amylolytic enzymes to maltose, glucose, dextrins, and oligosaccharides; therefore, it results in an increased digestibility (Yang et al., 2021; Bhinder et al., 2022; Lemmens et al., 2019; Benincasa et al.,

2019). Further, germination yields more production of phenolic acids, which is directly correlated with increased free radical scavenging activity. It is also a technique to synthesize crucial phytochemicals like γ-aminobutyric acid (GABA) (Baranzelli et al., 2018), which have been used to control neurological disorders.

Though an elevated enzymatic activity has a drastic effect on the breadmaking potential of cereals, with an appropriate modification of the germination factors, it has been recommended as a crucial tool for enhancing both the nutritional and technological properties of cereal-based food. Further, germination creates soft and fragile grain as a result of enzyme action which resists distortion in starch content while milling (Liu et al., 2017). Further, partial protein hydrolysis and the reduction in hydrophobic fiber assist to reduce the water absorption of flour. The germination also affects dough rheological features including the deterioration of the gluten capacity to form viscoelastic set-up because of a reduction in the concentration of high-molecular-weight glutenin macropolymers, which reduces its tenacity, starch gelatinization, and retrogradation ability due to hydrolysis (Žilić et al., 2016).

11.5.2 Flour Modification Approaches

11.5.2.1 Particle Size Reduction (Micronization)

Flour particle size plays a major role in altering bread functionality and technological quality. If various techniques like jet milling are executed to synthesize fine wheat flour with tremendously minute particle size, flour with enhanced digestible starch is received (Protonotariou et al., 2015). During breadmaking, jet-milled flour somewhat reduced bread glycemic index. Similar correlation between flour mean particle size and technological outcome was received for gluten-free flours. The flours possessing fine particle size are the best for making gluten-free maize bread. de la Hera et al. (2013) reported that the coarser maize flours (>150 μm) rendered breads with more specific volume and lower morsel firmness as compared with finer flour (<106 μm) due to the more ease of use of dough to keep hold of the gas initiated during fermentation. Further, de la Hera et al. (2014) reported that the uncouth portion amalgamated with a high dough hydration was most proficient amalgamation for producing rice bread when keeping factors like the bread volume and crumb texture intact.

11.5.2.2 Heat Treatment

This treatment, when applied to different flours, has different outcomes; for example, dry-heat treatment or hydrothermal treatments are gaining lot of attention to improve the functionality of substituted cereal flour. It has been reported that dry-heat treated sorghum flour synthesized breads with enhanced specific volume and more cells. This is attributed to an elevated viscosity of sorghum flour dough due to starch granule puffiness because of heat-induced partial gelatinization along with the denaturation of both proteins and enzymes (Marston et al., 2016). Further, protein denaturation and the partial gelatinization of starch granules cause an increase in gas retention ability and dough expansion; thus all are responsible for improvements in textures, strength, and volume of dry-heated sorghum-containing bread (Collar et al., 2019). Since sorghum-based additives are featured with pungent off-notes, dry heat treatment can be executed to perk up sorghum bread sensory features (Cappelli et al., 2020). Mann et al. (2014) have reported that the heat treatment of flour results in the formation of gluten and starch aggregates and alters aggregations between gluten and starch. The effects appeared to be more drastically altered in heat-treated flours with elevated moisture content where the higher mobility of the molecules is enabled. It has been observed that gluten-free flours' (maize or rice) blanching creates doughs with higher steadiness, glueyness, springiness, and tackiness due to the partial gelatinisation of the starch, which finally increased the quality of bread (Brites et al., 2010)

11.6 Conclusion and Future Prospects

Agriculture transformation is of utmost importance for regional development. Cutting-edge research involving multidiscipline is the need of the hour and is expected to develop superior genotypes breaking the yield barrier. Despite being cost-intensive, development is mandatory which warrants for higher public and private investments in R&D. Higher protein solubility and colloidal stability of the fine fractions compared to the raw material suggest the potential use of these fractions in various food systems. To reveal the full potential of the fractions for food applications, they should be studied in more detail in terms of functional properties and nutritional quality. For example, the concentration of phytic acid in the protein-enriched fractions should be considered in the further experiments as it may have a negative impact on the protein digestibility in the human digestive tract due to the binding of proteins. However, the results are promising for the use of dry fractionation in the production of functional and nutritionally enhancing rice bran ingredients for food applications.

REFERENCES

Abd El-Hady, E.A., & Habiba, R.A. Effect of soaking and extrusion conditions on antinutrients and protein digestibility of legume seeds. LWT-Food Sci. Technol. 36 (2003), 285–293.

Amiri, P., & Sharifian, N. Soltanizadeh, application of ultrasound treatment for improving the physicochemical, functional and rheological properties of myofibrillar proteins. Int. J. Biol. Macromol. 111 (2018), 139–147.

Arzeni, C., Pérez, O.E., & Pilosof, A.M. Functionality of egg white proteins as affected by high intensity ultrasound. Food Hydrocoll. 29 (2012), 308–316.

Baranzelli, J., Kringel, D.H., Colussi, R., Paiva, F.F., Aranha, B.C., de Miranda, M.Z., da Rosa, Z.E., & Dias, A.R. Changes in enzymatic activity, technological quality and gamma-aminobutyric acid (GABA) content of wheat flour as affected by germination. LWT. 90 (2018), 483–490.

Békés, F. New aspects in quality related wheat research: Challenges and achievements. Cereal Research Communications 40(2) (2012), 159–184.

Belderok, B., Mesdag, H., & Donner, DA. Bread-Making Quality of Wheat. Springer, New York, 2000.

Benincasa, P., Falcinelli, B., Lutts, S., Stagnari, F., & Galieni, A. Sprouted grains: A comprehensive review. Nutrients. 11(2) (2019), 421.

Bhinder, S., Singh, N., & Kaur, A. Impact of germination on nutraceutical, functional and gluten free muffin making properties of tartary buckwheat (*Fagopyrum tataricum*). Food Hydrocoll. 124 (2022), 107268.

Bicar, E.H., Woodman-Clikeman, W., Sangtong, V., Peterson, J.M., Yang, S.S., Lee, M., & Scott, M.P. Transgenic maize endosperm containing a milk protein has improved amino acid balance. Transgenic Res. 17 (2008), 59–71.

Boye, J., Zare, F., & Pletch, A. Pulse proteins: Processing, characterization, functional properties and applications in food and feed. Food Res. Int. 43 (2010), 414–431.

Brites, C., Trigo, M.J., Santos, C., Collar, C., & Rosell, C.M. Maize-based gluten-free bread: Influence of processing parameters on sensory and instrumental quality. Food Bioprocess Technol. 3(5) (2010), 707–715.

Cakmak, I. IFS Proceedings No. 552, pp. 1–28. International Fertiliser Society, York, 2004.

Cameron, M., McMaster, L.D., & Britz, T.J. Impact of ultrasound on dairy spoilage microbes and milk components. Dairy Sci. Technol. 89 (2009), 83–98.

Cappelli, A., Oliva, N., & Cini, E. A systematic review of gluten-free dough and bread: Dough rheology, bread characteristics, and improvement strategies. Appl Sci. 10(18) 2020, 6559.

Chittapalo, T., & Noomhorm, A. Ultrasonic assisted alkali extraction of protein from defatted rice bran and properties of the protein concentrates. Int. J. Food Sci. Technol. 44 (2009), 1843–1849.

Collar, C. Gluten-free dough-based foods and technologies. In Taylor, J.R.N., & Duodu, K.G. (eds.), Sorghum and Millets, pp. 331–354. AACC International Press, Cambridge, 2019.

Cornet, S.H.V., van der Goot, A.J., & van der Sman, R.G.M. Effect of mechanical interaction on the hydration of mixed soy protein and gluten gels. Curr. Res. Food Sci. 3 (2020), 134–145.

De la Hera, E., Rosell, C.M., & Gomez, M. Effect of water content and flour particle size on gluten-free bread quality and digestibility. Food Chem. 151 (2014), 526–531.

De la Hera, E., Talegón, M., Caballero, P., & Gómez, M. Influence of maize flour particle size on gluten-free breadmaking. J Sci Food Agric. 93(4) (2013), 924–932.

Devi, S., Zhang, M., & Law, C.L. Effect of ultrasound and microwave assisted vacuum frying on mushroom (*Agaricusbisporus*) chips quality. Food Biosci. 25 (2018), 111–117.

Filomena, L., Díaz, A., & Puig, I. Sotelo, Efecto de ultrasonidosobre la actividadATPasa y propiedadesfuncionalesen surimi de tilapia (*Oreochromis niloticus*). Vitae 19 (2012), S379–S381.

Gallagher, E., Gormley, T.R., & Arendt, E.K. Crust and crumb characteristics of gluten-free breads. J. Food Eng. 56 (2004), 153–161.

Goesaert, H., Brijs, K., Veraverbeke, W.S., Courtin, C.M., Gebruers, K., & Delcour, J.A. Wheat flour constituents: How they impact bread quality, and how to impact their functionality. Trends in Food Science & Technology 16 (2005), 12–30.

Gregersen, S.B., Wiking, L., & Hammershøj, M. Acceleration of acid gel formation by high intensity ultrasound is linked to whey protein denaturation and formation of functional milk fat globule-protein complexes. J. Food Eng. 254 (2019), 17–24.

Hernández-Santos, B., Rodríguez-Miranda, J., Herman-Lara, E., Torruco-Uco, J.G., Carmona-García, R., Juárez-Barrientos, J.M., Chávez-Zamudio, R., & Martínez-Sánchez, C.E. Effect of oil extraction assisted by ultrasound on the physicochemical properties and fatty acid profile of pumpkin seed oil (*Cucurbita pepo*). Ultrason. Sonochem. 31 (2016), 429–436.

Higuera-Barraza, O.A., Del Toro-Sanchez, C.L., Ruiz-Cruz, S., & Márquez-Ríos, E. Effects of high-energy ultrasound on the functional properties of proteins, Ultrason. Sonochem. 31 (2016), 558–562.

Higuera-Barraza, O.A., Torres-Arreola, W., Ezquerra-Brauer, J.M., Cinco-Moroyoqui, F.J., Figueroa, J.R., & Marquez-Ríos, E. Effect of pulsed ultrasound on the physicochemical characteristics and emulsifying properties of squid (*Dosidicus gigas*) mantle proteins, Ultrason. Sonochem. 38 (2017), 829–834.

Jamalabadi, M., Saremnezhad, S., Bahrami, A., & Jafari, S.M. The influence of bath and probe sonication on the physicochemical and microstructural properties of wheat starch. Food Sci. Nutr. 7(7) (2019), 2427–2435.

Jambrak, A.R., Herceg, Z., Šubarić, D., Babić, J., Brnčić, M., Brnčić, S.R., Bosiljkov, T., Čvek, D., Tripalo, B., & Gelo, J. Ultrasound effect on physical properties of corn starch. Carbohydr. Polym. 79 (2009), 91–100.

Kolpakova, V., Yudina, T., Vanin, S., & Lomakin, A. Dry wheat gluten is an effective flour improver with short-glutinous gluten. Bread Products. 2 (2007), 50–52.

Konik-Rose, Ch., Thistleton, J., Chanvrier, H., Tan, I., Halley, P., Gidley, M., Kosar-Hashemi, B., Wang, H., Larroque, O., Ikea, J., McMaugh, S., Regina, A., Rahman, S., Morell, M., & Li, Z. Effects of starch synthase IIa gene dosage on grain, protein and starch in endosperm of wheat. Theor. Appl. Genet. 115 (2007), 1053–1065.

Lazaridou, A., Duta, D., Papageorgiou, M., Belc, N., & Biliaderis, CG. Effects of hydrocolloids on dough rheology and bread quality parameters in gluten free formulations. Journal of Food Engineering (2007), 1033–1047.

Lazou, A.E., Michailidis, P.A., Thymi, S., Krokida, M.K., & Bisharat, G.I. Structural Properties of corn-legume based extrudates as a function of processing conditions and raw material characteristics. Int. J. Food Prop. 10 (2007), 721–738.

Lee, J., & Martini, S. Modifying the physical properties of butter using high-intensity ultrasound. J. Dairy Sci. 102 (2019), 1918–1926.

Lemmens, E., Moroni, A.V., Pagand, J., Heirbaut, P., Ritala, A., Karlen, Y., Lê, K.A., Van den Broeck, H.C., Brouns, F.J., De Brier, N., & Delcour, J.A. Impact of cereal seed sprouting on its nutritional and technological properties: a critical review. Compr Rev Food Sci Food Saf. 18(1) (2019), 305–328.

Leong, T., Johansson, L., Mawson, R., McArthur, S.L., Manasseh, R., & Juliano, P. Ultrasonically enhanced fractionation of milk fat in a litre-scale prototype vessel. Ultrason. Sonochem. 28 (2016), 118–129.

Li, K., Kang, Z.L., Zou, Y.F., Xu, X.L., & Zhou, G.H. Effect of ultrasound treatment on functional properties of reduced-salt chicken breast meat batter. J. Food Sci. Technol. 52 (2015), 2622–2633.

Linsberger-Martin, G., Weiglhofer, G., Phuong, K.T., & Berghofer, T.P. High hydrostatic pressure influences antinutritional factors and in vitro protein digestibility of split peas and whole white beans. LWT-Food Sci. Technol. 51 (2013), 331–336.

Liu, T., Hou, G.G., Cardin, M., Marquart, L., & Dubat, A. Quality attributes of whole-wheat flour tortillas with sprouted whole-wheat flour substitution. LWT. 77 (2017), 1–7.

Manchun, S., Nunthanid, J., Limmatvapirat, S., & Sriamornsak, P. Effect of ultrasonic treatment on physical properties of tapioca starch. Adv. Mater. Res. 506 (2012), 294–297.

Mann, J., Schiedt, B., Baumann, A., Conde-Petit, B., & Vilgis, T.A. Effect of heat treatment on wheat dough rheology and wheat protein solubility. Food Sci Technol Int. 20(5) (2014), 341–351.

Marchesini, G., Balzan, S., Montemurro, F., Fasolato, L., Andrighetto, I., Segato, S., & Novelli, E. Effect of ultrasound alone or ultrasound coupled with CO2 on the chemical composition, cheese-making properties and sensory traits of raw milk. Innov. Food Sci. Emerg. Technol. 16 (2012), 391–397.

Marston, K., Khouryieh, H., & Aramouni, F. Effect of heat treatment of sorghum flour on the functional properties of gluten-free bread and cake. LWT Food Sci Technol. 65 (2016), 637–644.

Monroy, Y., Rivero, S., & García, M.A. Microstructural and techno-functional properties of cassava starch modified by ultrasound. Ultrason. Sonochem. 42 (2018), 795–804.

Morales, R., Martínez, K.D., Ruiz-Henestrosa, V.M.P., & Pilosof, A.M. Modification of foaming properties of soy protein isolate by high ultrasound intensity: Particle size effect, Ultrason. Sonochem. 26 (2015), 48–55.

Nazari, B., Mohammadifar, M.A., Shojaee-Aliabadi, S., Feizollahi, E., & Mirmoghtadaie, L. Effect of ultrasound treatments on functional properties and structure of millet protein concentrate. Ultrason. Sonochem. 41 (2018), 382–388.

Nishio, Z., Takata, K., Ito, M., Tabiki, T., Iriki, N., Funatsuki, W., & Yamauchi, H. Relationship between physical dough properties and the improvement of bread-making quality during flour aging. Food Sci. Technol. Res. 10 (2004), 208–213.

Noorfarahzilah, M., Lee, J.S., Sharifudin, M.S., Mohd, F.A., & Hasmadi, M. Applications of composite flour in development of food products. International Food Research J. 21(6) (2014), 2061–2074.

Osen, R., Toelstede, S., Eisner, P., & Schweiggert-Weisz, U. Effect of high moisture extrusion cooking on protein–protein interactions of pea (*Pisum sativum* L.) protein isolates. Int. J. Food Sci. Technol. 50 (2015), 1390–1396.

Ozturk, L., Yazici, M.A., Yucel, C., Torun, A., Cekic, C., Bagci, A., Ozkan, H., Braun, H.J., Sayers, Z., & Cakmak, I. Concentration and localization of zinc during seed development and germination in wheat. Physiol. Plant. 128 (2006), 144–152.

Park, D.J., & Han, J.A. Quality controlling of brown rice by ultrasound treatment and its effect on isolated starch. Carbohydr. Polym. 137 (2016), 30–38.

Park, S.J., Kim, T.W., & Baik, B.K. Relationship between proportion and composition of albumins, and in vitro protein digestibility of raw and cooked pea seeds (*Pisum sativum* L.). J. Sci. Food Agric. 90 (2010), 1719–1725.

Peters, J.P.C.M., Vergeldt, F.J., Boom, R.M., & van der Goot, A.J. Water-binding capacity of protein-rich particles and their pellets. Food Hydrocoll. 65 (2017), 144–156.

Protonotariou, S., Mandala, I., & Rosell, C.M. Jet milling effect on functionality, quality and in vitro digestibility of whole wheat flour and bread. Food Bioprocess Technol. 8(6) (2015), 1319–1329.

Rashid, S., Rakha, A., Anjum, F.M., Ahmed, W., & Sohail, M. Effects of extrusion cooking on the dietary fibre content and Water Solubility Index of wheat bran extrudates. Int. J. Food Sci. Technol. 50 (2015), 1533–1537.

Robin, F., Théoduloz, C., & Srichuwong, S. Properties of extruded whole grain cereals and pseudocereals flours. Int. J. Food Sci. Technol. 50 (2015), 2152–2159.

Šarić, B., Dapčević-Hadnađev, T., Hadnađev, M., Sakač, M., Mandić, A., Mišan, A., & Škrobot, D. Fiber concentrates from raspberry and blueberry pomace in gluten-free cookie formulation: Effect on dough rheology and cookie baking properties. J Texture Stud. 50(2) (2019), 124–130.

Shanmugam, A., Chandrapala, J., & Ashokkumar, M. The effect of ultrasound on the physical and functional properties of skim milk. Innov. Food Sci. Emerg. Technol. 16 (2012), 251–258.

Sheng, L., Wang, Y., Chen, J., Zou, J., Wang, Q., & Ma, M. Influence of high-intensity ultrasound on foaming and structural properties of egg white. Food Res. Int. 108 (2018), 604–610.

Struck, S., Jaros, D., Brennan, C.S., & Rohm, H. Sugar replacement in sweetened bakery goods. Int. J. Food Sci. Technol. 49 (2014), 1963–1976.

Tebben, L., Shen, Y., & Li, Y. Improvers and functional ingredients in whole wheat bread: a review of their effects on dough properties and bread quality. Trends Food Sci Technol. 81 (2018), 10–24.

Torrent, M., Alvarez, I., Geli, M.I., Dalcol, I., & Ludevid, D. Lysine-rich modified γ-zeins accumulate in protein bodies of transiently transformed maize endosperms. Plant Mol. Biol. 34 (1997), 139–149.

Vanin, S.V., & Kolpakova, V.V. Functional properties of dry wheat gluten of different quality. IzvestiyaVuzov. Food Technology. 1 (2007), 21–24.

Walker, C.E., & Hazelton, J.L. Dough rheology tests. Cereal Food World 42 (1996), 23–28.

Xiong, T., Xiong, W., Ge, M., Xia, J., Li, B., & Chen, Y. Effect of high intensity ultrasound on structure and foaming properties of pea protein isolate, Food Res. Int. 109 (2018), 260–267.

Xiong, W., Wang, Y., Zhang, C., Wan, J., Shah, B.R., Pei, Y., Zhou, B., Jin, L., & Li, B. High intensity ultrasound modified ovalbumin: Structure, interface and gelation properties. Ultrason. Sonochem. 31 (2016) 302–309.

Yadav, D.N., Anand, T., & Navnidhi Singh, A.K. Co-extrusion of pearl millet-whey protein concentrate for expanded snacks. Int. J. Food Sci. Technol. 49 (2014), 840–846.

Yang, Q., Luo, Y., Wang, H., Li, J., Gao, X., Gao, J., & Feng, B. Effects of germination on the physicochemical, nutritional and in vitro digestion characteristics of flours from waxy and nonwaxy proso millet, common buckwheat and pea. Innov Food Sci Emerg Technol. 67 (2021), 102586.

Yanjun, S., Jianhang, C., Shuwen, Z., Hongjuan, L., Jing, L., Lu, L., Uluko, H., Yanling, S., Wenming, C., Wupeng, G., & Jiaping, L. Effect of power ultrasound pre-treatment on the physical and functional properties of reconstituted milk protein concentrate. J. Food Eng. 124 (2014), 11–18.

Zaidul Islam, S., Yamauchi, H., Sun-Ju, K., Matsuura-endo, C., Shigenobu, T., Hashimoto, N., & Noda, T. A farinograph study on dough characteristics of mixtures of wheat flour and potato starches from different cultivars. Food Sci. Technol. Res. 14 (2) (2008), 211–216.

Zhang, H., Claver, I.P., Zhu, K.X., & Zhou, H. The effect of ultrasound on the functional properties of wheat gluten. Molecules 16 (2011), 4231–4240.

Zhang, H., Claver, J.P., Li, Q., Zhu, K., Peng, W., & Zhou, H. Structural modification of wheat gluten by dry heat – enhanced enzymatic hydrolysis. Food Technology and Biotechnology. 50 (2012), 53–58.

Zhang, J., Liu, L., Liu, H., Yoon, A., Rizvi, S.S.H., & Wang, Q. Changes in conformation and quality of vegetable protein during texturization process by extrusion. Crit. Rev. Food Sci. Nutr. 59 (2019), 3267–3280.

Zheng, J., Li, Q., Hu, A., Yang, L., Lu, J., Zhang, X., & Lin, Q. Dual-frequency ultrasound effect on structure and properties of sweet potato starch. Starch-Stärke. 65 (2013), 621–627.

Žilić, S., Janković, M., Barać, M., Pešić, M., Konić-Ristić, A., & Šukalović, VH. Effects of enzyme activities during steeping and sprouting on the solubility and composition of proteins, their bioactivity and relationship with the bread making quality of wheat flour. Food Funct. 7(10) (2016), 4323–4331.

Zisu, B., Lee, J., Chandrapala, J., Bhaskaracharya, R., Palmer, M., Kentish, S., & Ashokkumar, M. Effect of ultrasound on the physical and functional properties of reconstituted whey protein powders. J. Dairy Res. 78 (2011), 226–232.

12

Gluten-Related Disorders

Current Understanding, Myths, and Facts

**Sunil Kumar, Ajeet Singh, Ankush, Akhlash P. Singh,
Sewa Ram, Om Prakash Gupta, Vanita Pandey,
Hanif Khan, Ramesh Soni, and Gyanendra Pratap Singh**

CONTENTS

12.1 Introduction

Cereal crops contribute as a major food source for *Homo sapiens*. Maize, wheat, and rice constitute almost 89% of total calorific necessity of human population (www.downtoearth.org.in/). Wheat, an annual herb of the Gramineae or Poaceae family, is a widely consumed crop for food purpose in nearly 100 countries mainly attributed to the adoption of Western-style diets (Cummins and Roberts-Thomson,

2009; Shewry, 2009). Worldwide, wheat cultivation is quite extensive ranging from 67° N in Scandinavia and Russia to 45° S in Argentina. The cultivation is extended to the elevated regions of tropics and subtropics (Feldman, 1995; Shewry, 2009). Wheat, the second-most consumed staple after rice, has played a pivotal role in ensuring global food security and continues to do so (Mottaleb et al., 2021). It contributes as an important food source for 30% of the human population; contributing to approximately 20 and 25% of our energy needs (calories) and dietary protein, respectively (Borisjuk et al., 2019; IWGSC, 2014; Li et al., 2020). It also serves as a base material for bread and baking industry products such as bread, biscuit, chapatti, cookies, pasta, and noodles (Li et al., 2021). Currently, about 95% of the wheat grown worldwide is hexaploid bread wheat (*Triticum aestivum* L., 2n = 6x = 42, AABBDD), a major staple worldwide, while the remainder 5% being tetraploid durum wheat (*T. turgidum* var durum) (Sharma et al., 2020). Wheat grains are called caryopses (dry one-seeded fruits), where fruit and seed coats are linked tightly. The grain can be differentiated into five core compartments each with diverse constituents and biological functions: Fruit coat (pericarp) (4–5% of the grain weight) and seed coat (testa) (~1%) are the outermost layers of the whole grain. The endosperm carries the aleurone layer (6–9%) and the starch (80–85%). The germ (3%) is the future embryo consisting of a storage cotyledon and embryonic axis (Mondal et al., 2016; Wieser et al., 2020).

Wheat seed storage proteins (8–15% of total flour weight) can be classified into four prominent fractions, *viz.* albumins, globulins, gliadins (prolamines), and glutenins based on their solubility. Of these, gliadins (soluble in 70% alcohol) and glutenins (insoluble in alcohol) constitute the gluten (literal meaning in Latin: Glue) polymer/protein stored together with starch in endosperm portion of the seed (Schofield, 1994). While in wheat, the prolamins are represented by gliadins; prolamins of rye and barley are termed secalins and hordeins, respectively. Wheat gluten consists of a complex mixture of α/β-, γ- and ω-gliadins, and high and low molecular weight (HMW; LMW) glutenins, all of which are encoded by the medium to large multigene families (Shewry, 2019). Due to the binding and shape-forming characters conferred by its viscoelasticity and extensibility, gluten is the protein of choice in the processed food products industry (Shewry, 2015). Gluten, a structural protein in wheat endosperm and essential for dough-making properties, provides chewy and palatable constituency to baked and processed foods but is linked with certain disorders and allergies among some genetically predisposed individuals. Many people cannot tolerate wheat and its products post-ingestion due to harmful immune response against wheat glutens (Cabanillas, 2020). These gluten-related disorders include celiac disease (CD), wheat allergy, and non-celiac wheat/gluten-sensitivity (NCWGS), which result in compromised quality of life and associated morbidity in affected persons (Stamnaes and Sollid, 2015). Most discussed among these is CD, an autoimmune disorder directly associated with gluten proteins. Wheat allergy (WA) is a condition which arises from contact, inhalation, or ingestion of wheat. However, in addition to gluten, other wheat proteins and carbohydrates of wheat, particularly, fermentable, oligo, di, monosaccharides, and polyols (FODMAPs) also make wheat allergy to prevail (Juhász et al., 2018). The chapter entails various gluten-related disorders, their etiology, and symptoms.

12.2 Gluten: Classification and Significance, Dough Formation Properties

12.2.1 Significance

The use of wheat for making food products is owed to the presence of glutens, major grain storage proteins lodged in the starchy endosperm cells for future germination and seedling growth. Total grain proteins account for between 10 and 15% of the dry weight of grain, of which the gluten proteins (GPs) account for up to 80%. Glutens form a continuous matrix all around the starch granules in the mature starchy endosperm cells, which provides a distinctive amalgamation of elasticity and viscosity thereby enabling the dough to be processed into the range of products, *viz.* chapati, bread, biscuit, muffins, pasta etc., to name a few. Although other temperate cereals (barley, rye, and oats) also possess the related proteins, the techno-functional properties differ altogether among them, and it becomes imperative to blend them with wheat flour to make food products with an enhanced acceptability to most consumers (Shewry and Tatham, 2016). While gliadins impart cohesiveness, viscosity, and dough extensibility, glutenins deliver elasticity to the dough. Thus, more the glutenin, more will be the gluten strength (Barak et al., 2014; Sapone et al., 2012). Such permutations and combinations help accomplish the desired product development. Both gliadins and glutenins thus bet for the well-known techno-functional viscoelasticity to wheat doughs destined for an array of processed product formulations.

12.2.2 Classification

Based on their solubility, wheat GPs are classified into gliadins and glutenins. The gliadins (predominantly present as monomers with minor amounts of polymeric components) are readily extracted from flour with 60% (v/v) ethanol or 50% (v/v) propan-1-ol, while the glutenins (high molecular mass polymers stabilized predominantly by inter-chain disulfide bonds) are extractable using dilute acid or alkali. It has been well documented that both fractions contain related proteins, and their solubility differs due to their presence as monomer or polymers. The monomeric glutenin subunits, obtained after the reduction of disulfide bonds by any means, resemble that of gliadins in structure as well as solubility (Osborne, 1924). Various individual components of GP fractions are having a high level of allelic variation in their composition between different cultivars and can be classified into three families: Sulfur-rich, sulfur-poor (S-poor), and HMW prolamins (Shewry et al., 1986): Electrophoresis of the gliadins at low pH separates four groups of bands, called α-gliadins, β-gliadins, γ-gliadins, and ω-gliadins, with respect to their decreasing mobility. However, sequence comparison of amino acids has deduced that the α- and β-gliadins form a single group, which is sometimes referred to as α-type gliadins. The variability in gluten proteins arises from the presence of multigene families and the high level of polymorphism between genotypes and post-translational modifications (Shewry, 2019). The involvement of multigene families of gliadin and glutenin has been differentiated in Table 12.1.

TABLE 12.1

Gliadin and Glutenin Loci in *Triticum aestivum* [(AABBDD), 2n = 6x = 42] Along With Alleles

Gluten fraction	Component	Located at chromosome number	Arm	Alleles present
Gliadin	α-Gliadin	6	Short arm	*Gli-A2, Gli-B2, Gli-D2*
	γ-Gliadin	1	Short arm	*Gli-A1, Gli-B1, Gli-D1*
	ω-Gliadin	1	Short arm	
Glutenin	LMWGS	1	Short arm	*Glu-A3, Glu-B3, Glu-D3*
	HMWGS	1	Long arm	*Glu-A1, Glu-B1, Glu-D1*

Abbreviations: LMWGS, Low-molecular-weight glutenin subunit; HMWGS, High molecular weight glutenin subunit; Gli, Gliadin; Glu, Glutenin

Source: Sharma et al. (2020) and Shewry and Tatham, (2016)

12.3 Gluten-Related Disorders (GRDs)

The term "gluten-related disorders" is a broader term for encompassing all ailments associated with the ingestion of gluten-containing food. Primarily among these include: CD, dermatitis herpetiformis, gluten ataxia, wheat allergy, NCWGS, and irritable bowel syndrome. Although these GRDs and CD are treated similarly – that is, elimination of gluten from the diet – they are not alike condition-wise. Patients and health-care practitioners should imperatively differentiate between these disorders (Pietzak and Kerner Jr, 2012; Ludvigsson et al., 2013). Wheat/gluten-related diseases can be classified into three different categories of disorders based on a combination of biological, genetic, clinical, and histological data: Autoimmune, allergic, and neither autoimmune nor allergic (Table 12.2) (Sabença et al., 2021; Sapone et al., 2012).

TABLE 12.2

An Overview of Gluten-Related Disorders

Type	Disease	Prevalence	Other Remarks	Environmental Component Triggering Disease
Autoimmune	Celiac disease	1–2%	Mainly affects intestine	Gliadins, glutenins
	Gluten ataxia	Up to 6%	Affects various organs	
	Dermatitis herpetiformis	0.4–2.6 per 1 lac people	Affects skin	
Allergic	IgE-mediated wheat allergy [Baker's asthma (BA), Wheat-dependent exercise-induced anaphylaxis (WDEIA)]	0.5–1.0%	Allergic reaction happens upon inhalation/uptake/ contact of wheat-containing foods	Gliadins, glutenins, albumins, globulins
	Non-IgE mediated wheat allergy			

TABLE 12.2 *(Continued)*

Type	Disease	Prevalence	Other Remarks	Environmental Component Triggering Disease
Non-autoimmune and non-allergic	Non-celiac wheat/ gluten sensitivity	0.6–13.0%	Symptom wise identical to CD; but do not test positive for CD	Gliadins, glutenins, α-amylase/ trypsin inhibitors, fructans, wheat-germ agglutinins

Source: Sabença et al. (2021)

12.4 Pathophysiology of Celiac Disease (CD)

12.4.1 Etiology

There is ample research literature available pertaining to the genetics, environment, and the consumption of wheat gluten among population, using which an etiological model can be suggested that includes the contributions of genes, immune cells, and metabolic pathways in the development of CD. Next-generation sequencing (NGS) has provided the fine mapping of CD-related genes present in the human leukocyte antigen (HLA) and non-HLA regions, while genome-wide association studies (GWAS) have highlighted the potential roles of various loci and haplotypes in a variety of immunological pathways (Abadie et al., 2011). Celiac disease (CD) is a unique autoimmune disease (prevalent in 0.5–1% of population) with well-defined role of its key genetic elements (HLA-DQ2 and HLA-DQ8), the auto-antigen tissue transglutaminase (tTG), and the environmental trigger (gluten) (Caio et al., 2019). The pathological symptoms appear after the interaction of wheat gluten with genetic and environmental factors, leading to immune response activation inside the body (Sharma et al., 2020). Celiac occurs in genetically predisposed individuals carrying specific MHC haplotype DQ2 or DQ8 in HLA (Kuja-Halkola et al., 2016; Lindfors et al., 2019; Sharma et al., 2020). Exposure to gliadin peptide via gluten ingestion leads to an adaptive as well as innate immune response (as evidenced by the presence of intraepithelial lymphocytes) that causes damage to lamina propria (a type of connective tissue found under the thin layer of tissues covering a mucous membrane) (Abadie and Jabri, 2014; Schumann et al., 2017). The intestinal normal microbiota is also considered another factor in the pathogenesis of CD (Tarar et al., 2021).

The CD has been found to be hereditary (having high familial recurrence ~10–15%); also, it has been present in great harmony among monozygotic twins (75–80%) (Lundin and Wijmenga, 2015). As per Caio et al. (2019), the gluten-loaded wheat cereal is considered a latest addition to the human diet as the crop has been introduced 10,000 years ago during the onset of agriculture. Gliadin proteins (of gluten) are bizarrely rich in amino acids [proline (P) and glutamine (Q)] and pose high resistance to protease enzymes (mammalian and microbial)-mediated proteolytic degradation in the human digestive system (Shan et al., 2002). Gluten is consumed chronically in substantive quantity which further crops up multiple immunogenic peptides capable of triggering host responses, *viz.* increased gut permeability, innate and adaptive immune responses, a promulgation toward CD (Caio et al., 2019). So, this can be considered one end of the story where gluten acts as an environmental trigger. Though gliadins are considered responsible for CD, both gliadins and glutenins are liable for CD development. All three variants of gliadins (α/β, ω, and γ-gliadins) pose immunogenicity toward CD; however, the 33-mer pro- and gln-rich peptide sequence (LQLQPFPQPQLPYPQPQLPYPQPQLPYPQPQPF) of α-gliadins has been reported to be the most immunogenic and a powerful stimulator of T-cells (Camarca et al., 2009; Sharma et al., 2020). This 33-mer peptide fragment of α2-gliadin consists of six partially overlapping copies of three highly immunogenic T-cell epitopes. These include PFPQPQLPY (glia-α1a, one copy), PYPQPQLPY (glia-α1b, two copies), and PQPQLPYPQ (glia-α2, three copies) (Anderson et al., 2000; Arentz-Hensen et al., 2000; Sharma et al., 2020; Sollid et al., 2012). The gln and pro rich 33-mer sequence acts as stimulator of T-cells as such and after tissue type 2 transglutaminase (TG2) enzyme-mediated deamidation in intestine (Qiao et al., 2004; Sharma et al., 2020). D-genome (Gli-D2)-encoded α-gliadins are the most immunogenic while B-genome (Gli-B2)-expressed ones exhibit least immunogenicity (within α-gliadins). After α-gliadins (25 to 150 copies per haploid genome), γ-gliadins are second-most immunogenic (copy number 5 to 40) (Anderson et al., 2001). Both components of glutenins, however, pose very less immunogenicity (Shewry and Tatham, 2016).

12.5 Role of Human MHC (Major Histocompatibility Complex) Class II Genes Toward Genetic Predisposition of Individual Toward CD

12.5.1 The Genetics of Human Major Histocompatibility Complex (MHC) Genes

Another aspect in CD development comes from genetic predisposition of individuals. The root for this predisposition lies in MHC. A series of cloning, genomic, as well as immunological experiments, has disclosed the molecular structure and emphatically proposed the link between human MHC and many inflammatory, infectious, and autoimmune diseases such as CD and type I diabetes (Lindfors et al., 2019). The human genome consists of gene-dense and extremely polymorphic region of about 260 genes along with many pseudogenes (human MHC genes) spanning over a 4-Mb supralocus of the 6p21.3 located at short arm of chromosome 6. This polymorphic region (MHC) is also recognized as the human leukocyte antigen (HLA) system in humans. Molecules encoded by this region are involved in various immunity-related processes like antigen presentation, inflammation regulation, the immune response (innate and adaptive), and the complement system, indicating the varying importance of MHC in immune-mediated, autoimmune, and infectious diseases (Shiina et al., 2009). The total genes of MHC can be divided into three major classes: MHC I, MHC II, and MHC III, expressing various antigen-presenting protein complexes which are implicated in a highly intricate immunological response against pathogens and autoimmune ailments in humans and animals. But, sometimes, the human immune system goes berserk and starts to act against its own antigens, causing autoimmune diseases, best characterized here by CD (Houlston and Ford, 1996; Kaslow et al., 1996). In the case of CD, human MHC class I and II molecules impart imperative roles in producing innate and adaptive immune responses against the partially digested gluten peptide (Trynka et al., 2010).

Encoded by MHC genomic region (I and II, respectively), both MHC I and II membranous protein complexes/molecules form a similar type of fold or antigen-binding site with ample space to accommodate a 13–25 amino acid residues-long peptide chain. Both molecules act as receptors on the surface of an antigen-presenting cell (APC), which present the bonded peptide after cellular processing in their binding site (in an adaptive immune response to human diseases), which in turn is further identified by the T-cell receptors (TCRs) of CD8+ and CD4+ T-cells or killer cells (Abadie et al., 2011; Frick et al., 2021). Peptide (antigen)-MHC I (cell receptor) (pMHCI) complexes appear on the nucleated cells (activated natural killer cells and intraepithelial lymphocytes) recognized by the cytotoxic CD8+ T-cells, leading to the killing of intestinal epithelial cells. That is why MHC class I molecules play a significant role in villous atrophy, that is, flattening and disappearance of the finger-like absorptive villi/processes of the small intestine. On the other hand, peptide-MHC II (pMHC-II) complex is processed and presented by APC (*e.g.*, dendritic cells, macrophages, and/or B cells) to interact with CD4+ T-cells which, in turn, coordinates and regulates the function of effector cells via cell-to-cell interactions and their activation process (Lindfors et al., 2019). Gluten and MHC Class II-encoded HLA-DQ2 or DQ8 molecules/proteins at the HLA-DQ locus and certain non-HLA region are responsible for genetic susceptibility and prepare a holistic immunological response that is elicited by gluten peptide in the direction of autoimmunity (Mearin et al., 1983; Wieczorek et al., 2017). Both HLA-DQ2 and HLA-DQ8 genes-encoded proteins are dimeric ones (an α-chain and β-chain governed by the expressions of HLA-DQA1 and HLA-DQB1-specific allelic variants) expressed by MHC Class II. These two genes, HLA-DQ2 and HLA-DQ8, contribute 95 and 5% cases respectively, for CD (Parzanese et al., 2017; Sharma et al., 2020). In addition to the HLA locus, this supra-locus comprises 42 nonconventional HLA genes that encode for non-HLA regions which play a key role in the development of nearly 15% CD-related metabolic pathways by encoding important immunological molecules (Farina et al., 2019; Gutierrez-Achury et al., 2011). Other non-HLA genes (not yet identified) account for 60% of the inherited constituent of the disease (Caio et al., 2019).

12.6 Immune System and Modalities While Disease Development

The development of CD requires both gluten ingestion and a genetic predisposition in terms of the expression of various allelic forms in an affected individual (DQ2/8). In people affected with CD, partially resistant gluten peptides turn on both the adaptive and innate immune systems. In the case of CD, adaptive immune response plays a significant role as evidenced by the generation of CD4+ T-cells inside the small intestinal mucosa. Second, cells produce antibodies against both wheat gliadin and the ubiquitous tissue enzyme TG2 (the CD autoantigen). Structurally, the native gluten peptide contains distinctive amino acid sequences, particularly of proline and glutamine residues (especially in gln-X-pro sequences; X: any amino acid). Thus, gluten peptides contain an ample amount of proline residues which pose great resistance toward proteolytic enzymes of the human digestive system (Withoff et el., 2016). Partially digested gluten (environmental trigger) peptides (protease enzyme mediated) formed in the gut lumen pass via transcellular or passive paracellular routes and enter finally into the lamina propria (Caio et al., 2019). Here, they are acted upon by TG2 enzyme resulting in the deamination of certain glutamine residues into glutamic acid. This deamination imparts negative charge on the glu-residues, which enhances the binding of gluten/gliadin peptides (epitopes) inside the groove of the MHC protein dimeric complex of HLA-DQ2/DQ8 present as the receptor on the surface of APC in the intestinal environment (Kumar et al., 2012) though HLA-DQ2 and/or DQ8 genes are necessary but not absolute for disease development (Caio et al., 2019). In the case of CD, pro-inflammatory dendritic cells are the main APCs which present the HLA-DQ protein molecules bonded with gluten that are recognized by CD4+ T-cells via T-cell receptors (TCRs) which contain a gliadin-specific epitome reorganization site (Kumar et al., 2012). Activated CD4+ T-cells further signal to both gluten-specific and TG2-specific B cells, consequently promoting the activation and differentiation of B cells in the plasma. Next, both gluten and TG2-specific B cells start functioning as antigen-presenting cells and present processed gluten antigens on the surface of B cell receptors (BCRs) and deliver them to gluten-specific CD4+ cells. In nut shell, HLA-DQ2/8-mediated interaction triggers both T and B (including plasma B cells). These activated B cells and CD4+ T-cells produce antibodies against the TG2 enzyme and partially digested gliadin peptides in the intestinal epithelium, and the overall immune response causes villous atrophy (Lindfors et al., 2019). In a short span, the number of antibodies gets proliferated in the circulated blood and mucosal blood vessels of the sub-epithelial basement membrane.

After the exact compatibility between CD4+ T-cells and APC, the cytokines (such as IFNγ and IL-21) generated by CD4+ T-cells (activated during adaptive immune response) further indulge in the innate immune response (including inflammatory response). The cytokines are involved in the triggering of inflammation pathways in small intestine's lamina propria, which leads to tissue damage (Withoff et el., 2016). Intestinal epithelial cells and/or dendritic cells produce IL-15, IL-18, and type I interferons. The resulting IL-15 blocks the regulatory effects of CD4+ regulatory cells, intensifying CD in effect (Gujral et al., 2012).

12.6.1 Symptoms of CD

Celiac disease, by far, the most prominent autoimmune gluten-related disorder, has roughly 1% prevalence with female predominance in the general population. Serological screening data has indicated the actual female: male ratio of 1.5:1. Although the disease can occur at any stage of life irrespective of age, with multifarious symptoms (Caio et al., 2019, Choung et al., 2015), yet specifically can be seen in early childhood (<6 years old) and within 4th to 5th decade (Matthias et al., 2011). While classical CD affects pediatrics (6–24 months), atypical CD ensues at a later age (>5-year-olds) and in adults (Rodrigo-Sáez et al., 2011). The symptoms can be divided into intestinal or extraintestinal as given in Table 12.3. Diagnosis of CD can be ascertained by a combination of serological testing (i.e., IgA-tTG coupled with total IgA levels) along with duodenal biopsies (to verify intestinal damage) (Cabanillas, 2020; Sabença et al., 2021). Marsh (1992) reported that CD can be confirmed in patients showing villous atrophy, increased cell production in a normal tissue or organ, and increased intraepithelial lymphocytes in the duodenal biopsy. HLA typing can be used when the results of the serological and duodenal biopsies are dubious (Ludvigsson et al., 2014). The treatment includes life-long diet devoid of gluten and gluten-related proteins of rye, barley, *etc.* (Rubio-Tapia et al., 2013); although gluten-free diet/GFD (being typically fiber poor) may lead to digestive problems (constipation) (Allen et al.,

TABLE 12.3

Intestinal and Extraintestinal Symptoms of CD

Type	Symptoms	Prevalent in
Intestinal	Diarrhea, abdominal distention, loss of appetite, failure to thrive	Pediatric population and children (aged less than three years)
	Diarrhea, bloating, constipation, abdominal pain, or weight loss; other symptoms include irritable bowel syndrome like symptoms coupled with constipation/nausea/vomiting	Older children and adults
Extraintestinal symptoms	Microcytic anemia, changes in bone mineral density (along with osteopenia or osteoporosis), headache, paresthesia, neuro-inflammation, anxiety and depression, etc.	Both children and adult

Source: Caio et al. (2019); Reilly et al. (2011); Vivas et al. (2008); Volta et al. (2014)

2014; Sabença et al., 2021). The number of individuals adopting a GFD hugely outnumbers the projected number of celiac disease patients. According to an estimate, this adoption of gluten-free products created a global market of $2.5 billion (the United States) in global sales in 2010 (Sapone et al., 2012).

12.6.2 Various Subtypes of CD

Various clinical subtypes of CD are given in Table 12.4 having specific symptoms and prevalent age groups. Nearly 49 peptide sequences of prolamins forming wheat,

TABLE 12.4

Clinical Presentation or Clinical Subtypes of CD

CD Form	Symptoms	Susceptible Class
Atypical	Mostly extraintestinal; no malabsorption reported	Older children and adults
Silent or asymptomatic	Serological and histological abnormalities without any clinical indications	Prevalent in patients with a family history or with associated autoimmune (type I diabetes)/genetic disorders
Latent	Previously asymptomatic with positive serology but without villous atrophy or others related tissue abnormalities	Patients with elevated endomysial antibodies (abs)
Potential	Positive serology associated with normal or mildly abnormal histology but never diagnosed of CD	
Refractory	Malabsorptive symptoms and villous atrophy persisting even one year after a strict GFD (Freeman et al., 2011).	Roughly, 5–30% patients never respond to a GFD (Ho-Yen, 2009), rest initially respond but recurrence of symptoms and intestinal damage afterwards

Source: Lionetti and Catassi (2011).

barley, rye, etc., have been identified as being celiac toxic, collectively referred to as celiac "epitopes", located in repetitive prolamins' domain. High proline levels obviate protease activity and when these epitopes' specific glutamine residues get deamidated to glutamic acid by a tTG2 which is recognized by prolamin-reactive T-cells of CD patients, the situation is further aggravated. This binding forms a stable peptide (epitope)–MHC complex (Sollid et al., 2012). Tye-Din et al. (2010) identified a well-characterized 33 amino acid residues' peptide from α-gliadin resistant to gastrointestinal digestion (with pepsin and trypsin) thought of as the major celiac toxic peptide in the gliadins.

12.7 Other Gluten-Related Disorders (GRDs)

12.7.1 Wheat allergy

Wheat is one of the five most frequent foods causing allergy. IgE-mediated wheat allergens are widely distributed in wheat's different protein fractions. Currently, 28 allergens have been identified in wheat, according to WHO/IUIS Allergen Nomenclature Sub-Committee. WA can be further differentiated into with and without IgE as given in Table 12.5.

IgE-mediated WA can further be differentiated into wheat-dependent exercise-induced anaphylaxis (WDEIA) and Baker's asthma, a brief detail of which has been given in Table 12.6.

TABLE 12.5

Differentiation of Wheat Allergy

Attribute	IgE-Mediated	Non-Ig-Mediated
Prompted by	Allergen ingestion/inhalation/skin contact respectively attributing to food/respiratory/dermal allergy. Dendritic cells present the Ag to CD4+ T-cells which differentiate into T-helper type 2 cells which produce various cytokines like IL-4, 5, and 13. The cytokines stimulate B-cells to finally produce IgE (Sharma et al., 2020); has further two subtypes.	Associated with eosinophilic esophagitis (EoE) or eosinophilic gastritis (EG), which occurs when eosinophils infiltrate the gastrointestinal tract (Cianferoni, 2016).
Symptoms	Nausea, abdominal pain, vomiting, diarrhea, itching, eczema, respiratory issues, rhinitis, asthma, etc. Others include flushing, angioedema, disturbed thinking, headache, and dizziness (Cianferoni, 2016; Ortiz et al., 2017).	Indigestion, diarrhea, vomiting, arthralgia, and headache (Taraghikhah et al., 2020).
Appearance of symptoms	Minutes to hours after the exposure	Hours or days after the consumption of allergens
Treatment	Complete elimination of wheat from the diet	Corticosteroid use and allergen avoidance

Source: Sabença et al. (2021)

TABLE 12.6

Difference Between Types of IgE-Mediated Wheat Allergy

IgE-Mediated WA	Remarks	Occurs When	Symptoms	Diagnosis
Wheat-dependent exercise-induced anaphylaxis (WDEIA)	Exercise can be done only six hours after the consumption of wheat or wheat-containing products	Severe anaphylactic reactions erupt when intense exercise is practiced soon after wheat consumption	Angioedema, chest pain, diarrhea, dysphagia, dyspnea, flushing, headache, hoarseness, nausea, pruritus, and syncope	Skin prick test and IgE against wheat allergens in wheat extracts as well as in blood serum. The double-blind placebo-controlled wheat challenge is more specific
Baker's asthma	0.03–0.24% of pastry factory workers, cereal handlers, confectioners, and bakery workers are more prone to this. Restriction of exposure to wheat flours is endorsed.	After allergen (cereal flour dust) inhalation specifically present in work premises	Consumption of products contaminated with raw wheat flour. Cooked wheat products are harmless	Presence of IgE and bronchial challenge

Source: Armentia et al. (2009); Brant (2007); Christensen (2019); Quirce and Diaz-Perales (2013); Sabença et al. (2021); Scherf et al. (2016)

12.7.2 Non-celiac wheat/gluten sensitivity (NCWGS)

NCWGS is remarkably different from celiac disease and demarcated as a nonallergic, non-autoimmune condition arising from gluten-containing grains, which subsides if the latter are not a part of the diet (Sapone et al., 2012). The disease affects 3–6% of the general population with higher incidence in females (Cascella et al., 2011). The disease has been thought to involve only the activation of innate immunity component, while adaptive immunity participation has not been recorded (cf. CD) (Sapone et al., 2010, 2011). The symptoms characterized by intestinal and/or extraintestinal repercussions (like CD and WA) develop in a slot of few hours to days post wheat/gluten consumption. No specific IgE has been detected against wheat proteins or IgA anti-TG2 autoantibodies in NCWGS patients. Various symptoms encountered for NCWGS include intestinal (abdominal pain, stomach ache, diarrhea, bloating, gas) as well as extraintestinal (headache, depression, rash, anxiety, fatigue, eczema, joint and muscle pain, and anemia) issues (Fasano et al., 2015; Leonard et al., 2017; Sapone et al., 2012; Serena et al., 2020). Recent studies suggest that oligosaccharides/fructans, α-amylase/trypsin inhibitors, and wheat-germ agglutinin (other wheat constituents) may contribute to the development of NCWGS (Sabença et al., 2021). The increased expression of toll-like-receptors (TLRs TLR1, TLR2, and TLR4), a protein class having a vivacious role in innate immunity inside the small intestine, indicates the involvement of activated innate immunity in cases of NCWGS. The patients with NCWGS had less permeable small intestines when compared to CD patients and controls (Sapone et al., 2011). With no specific biomarker, NCWGS diagnosis relies on clinical symptom assessment and ruling out of CD and WA (Carroccio et al., 2012; Sabença et al., 2021). A low FODMAP (fermentable oligosaccharides, disaccharides, monosaccharides, and polyols) diet coupled with the complete elimination of gluten can enhance the improvement of NCWGS symptoms considerably (Biesiekierski et al., 2013). Conversely to CD, NCWGS has no strong genetic predisposition, and only 50% patients express the HLA-haplotype (DQ2 and/or HLA-DQ8) (Sapone et al., 2012).

12.7.3 Gluten Ataxia

Gluten ataxia (GA) is a form of cerebellar ataxia which affects mainly Purkinje cells (present in cerebellum part of brain). It occurs when antibodies (Abs) against gluten mistakenly attack part in the brain of genetically predisposed individuals (Hadjivassiliou et al., 2015). Like other ataxia, GA has symptoms such as gait/lower limb/upper limb ataxia and gaze-evoked nystagmus. Other symptoms include ocular signs and other movement disorders (Hadjivassiliou et al., 2003). Presence of anti-tTG, anti-gliadin, and anti-TG6 antibodies in the serum indicates GA (Hadjivassiliou et al., 2003). Magnetic resonance imaging (MRI) of GA-affected patients shows moderate cerebellar atrophy (60% cases) (Ghezzi et al., 1997). Treatment for the disease is rigorous GFD. Immunotherapy coupled with steroid and intravenous immunoglobulins can be an efficient treatment (Nanri et al., 2014).

12.7.4 Dermatitis herpetiformis (DH)

It is an autoimmune, chronic, and recurrent disorder impacting skin and intestine (Sabença et al., 2021). Exposure to gluten can elicit the generation of anti-tTG antibodies which recognize epidermal transglutaminase (eTG). IgA gets deposited in dermal papillae which ensues pruritic, vesiculobullous, and localized lesions on knees, buttocks, elbows, and scapular areas (Clarindo et al., 2014; Mendes et al., 2013). For diagnosis, direct immunofluorescence (DIF) tests on perilesional skin are conducted. If DIH negative, further confirmation can be done using dosage of anti-tTG (Antiga and Caproni, 2015).

12.8 Concluding Remarks

Gluten is a dietary protein of wheat that is widely used in the global food industry (especially in bread making and baking). Genetic (HLA) and environmental factors (gluten) predispose individuals to a wide range of CD. Currently, there is a need to understand the basic facts regarding the incidence of CD and its difference with NCWGS and other wheat allergies. The CD and other gluten-related disorders have different etiologies and clinical repercussions that, to some extent, can overlap. The CD represents T-cell-mediated chronic inflammatory condition impacting small intestine and jejunum, resulting in villous atrophy in intestine. This results in reduced surface area for nutrient absorption leading to malnutrition and multifarious pathological symptoms. As of now, roughly 1% of world's population is afflicted with this physiological disorder. Also, the genetic predisposition of susceptible individuals contributes more toward CD development rather than gluten alone. Interestingly, other wheat-related disorders such as NCWGS and wheat allergy are not solely dependent upon gluten, but other wheat components play a pivotal role in causing these diseases which otherwise are alleged upon to gluten.

REFERENCES

Abadie, V., & Jabri, B. (2014). IL-15: a central regulator of celiac disease immunopathology. *Immunological Reviews, 260*(1), 221–234.

Abadie, V., Sollid, L. M., Barreiro, L. B., & Jabri, B. (2011). Integration of genetic and immunological insights into a model of celiac disease pathogenesis. *Annual Review of Immunology, 29*, 493–525.

Allen, K. J., Turner, P. J., Pawankar, R., Taylor, S., Sicherer, S., Lack, G., & Sampson, H. A. (2014). Precautionary labelling of foods for allergen content: are we ready for a global framework? *World Allergy Organization Journal, 7*(1), 1–14.

Anderson, O. D., & Hsia, C. C. (2001). The wheat γ-gliadin genes: characterization of ten new sequences and further understanding of γ-gliadin gene family structure. *Theoretical and Applied Genetics, 103*(2), 323–330.

Anderson, R. P., Degano, P., Godkin, A. J., Jewell, D. P., & Hill, A. V. (2000). In vivo antigen challenge in celiac disease identifies a single transglutaminase-modified peptide as the dominant A-gliadin T-cell epitope. *Nature Medicine, 6*, 337–342. https://doi.org/10.1038/73200.

Antiga, E., & Caproni, M. (2015). The diagnosis and treatment of dermatitis herpetiformis. *Clinical, Cosmetic and Investigational Dermatology, 8*, 257–265. doi: 10.2147/CCID.S69127

Arentz-Hansen, H., Körner, R., Molberg, Ø., Quarsten, H., Vader, W., Kooy, Y. M., & McAdam, S. N. (2000). The intestinal T cell response to α-gliadin in adult celiac disease is focused on a single deamidated glutamine targeted by tissue transglutaminase. *The Journal of Experimental Medicine, 191*(4), 603–612.

Armentia, A., Díaz-Perales, A., Castrodeza, J., Duenas-Laita, A., Palacin, A., & Fernández, S. (2009). Why can patients with baker's asthma tolerate wheat flour ingestion? Is wheat pollen allergy relevant? *Allergologia et Immunopathologia, 37*(4), 203–204.

Barak, S., Mudgil, D., & Khatkar, B. S. (2014). Influence of gliadin and glutenin fractions on rheological, pasting, and textural properties of dough. *International Journal of Food Properties, 17*(7), 1428–1438.

Biesiekierski, J. R., Peters, S. L., Newnham, E. D., Rosella, O., Muir, J. G., & Gibson, P. R. (2013). No effects of gluten in patients with self-reported non-celiac gluten sensitivity after dietary reduction of fermentable, poorly absorbed, short-chain carbohydrates. *Gastroenterology, 145*(2), 320–328.

Borisjuk, N., Kishchenko, O., Eliby, S., Schramm, C., Anderson, P., Jatayev, S., & Shavrukov, Y. (2019). Genetic modification for wheat improvement: from transgenesis to genome editing. *BioMed Research International, 2019*. doi.org/10.1155/2019/6216304.

Brant, A. (2007). Baker's asthma. *Current Opinion in Allergy and Clinical Immunology, 7*(2), 152–155.

Cabanillas, B. (2020). Gluten-related disorders: Celiac disease, wheat allergy, and non-celiac gluten sensitivity. *Critical Reviews in Food Science and Nutrition, 60*(15), 2606–2621.

Caio, G., Volta, U., Sapone, A., Leffler, D. A., De Giorgio, R., Catassi, C., & Fasano, A. (2019). Celiac disease: a comprehensive current review. *BMC Medicine*, *17*(1), 1–20.

Camarca, A., Anderson, R. P., Mamone, G., Fierro, O., Facchiano, A., Costantini, S., & Gianfrani, C. (2009). Intestinal T cell responses to gluten peptides are largely heterogeneous: implications for a peptide-based therapy in celiac disease. *The Journal of Immunology*, *182*(7), 4158–4166.

Carroccio, A., Mansueto, P., Iacono, G., Soresi, M., D'Alcamo, A., Cavataio, F., . . . & Rini, G. B. (2012). Non-celiac wheat sensitivity diagnosed by double-blind placebo-controlled challenge: exploring a new clinical entity. *Official journal of the American College of Gastroenterology| ACG*, *107*(12), 1898–1906.

Cascella, N. G., Kryszak, D., Bhatti, B., Gregory, P., Kelly, D. L., Mc Evoy, J. P., & Eaton, W. W. (2011). Prevalence of celiac disease and gluten sensitivity in the United States clinical antipsychotic trials of intervention effectiveness study population. *Schizophrenia Bulletin*, *37*(1), 94–100.

Choung, R. S., Ditah, I., Nadeau, A., Rubio-Tapia, A., Marietta, E., Brantner, T., Camilleri, M., Rajkumar, S., Landgren, O., Everhart, J., & Murray, J. (2015). Trends and racial/ethnic disparities in gluten-sensitive problems in the United States: findings from the National Health and nutrition examination surveys from 1988 to 2012. *American Journal of Gastroenterology*, *110*, 455–61.

Christensen, M. J., Eller, E., Mortz, C. G., Brockow, K., & Bindslev-Jensen, C. (2019). Wheat-dependent cofactor-augmented anaphylaxis: a prospective study of exercise, aspirin, and alcohol efficacy as cofactors. *The Journal of Allergy and Clinical Immunology: In Practice*, *7*(1), 114–121.

Cianferoni, A. (2016). Wheat allergy: diagnosis and management. *Journal of Asthma and Allergy*, *9*, 13.

Clarindo, M. V., Possebon, A. T., Soligo, E. M., Uyeda, H., Ruaro, R. T., & Empinotti, J. C. (2014). Dermatitis herpetiformis: pathophysiology, clinical presentation, diagnosis and treatment. *Anais Brasileiros de Dermatologia*, *89*, 865–877.

Cummins, A. G., & Roberts-Thomson, I. C. (2009). Prevalence of celiac disease in the Asia–Pacific region. *Journal of Gastroenterology and Hepatology*, *24*(8), 1347–1351.

Farina, F., Picascia, S., Pisapia, L., Barba, P., Vitale, S., Franzese, A. & Del Pozzo, G. G. (2019). HLA-DQA1 and HLA-DQB1 Alleles, conferring susceptibility to celiac disease and type 1 diabetes, are more expressed than non-predisposing alleles and are coordinately regulated. *Cells*, *8*(7). https://doi.org/10.3390/cells8070751.

Fasano, A., Sapone, A., Zevallos, V., & Schuppan, D. (2015). Nonceliac gluten sensitivity. *Gastroenterology*, *148*, 1195–204. https://doi.org/10.1053/j.gastro.2014.12.049.

Feldman, M. (1995). Wheats. In Smartt, J., & Simmonds, N. W. (eds.), *Evolution of Crop Plants* (pp. 185–192). Harlow: Longman Scientific and Technical.

Freeman, H. J., Chopra, A., Clandinin, M. T., & Thomson, A. B. (2011). Recent advances in celiac disease. *World Journal of Gastroenterology: WJG*, *17*(18), 2259.

Frick, R., Høydahl, L. S., Petersen, J., du Pré, M. F., Kumari, S., Berntsen, G. & Løset, G. Å. (2021). A high-affinity human TCR-like antibody detects celiac disease gluten peptide-MHC complexes and inhibits T cell activation. *Science Immunology*, *6*(62). https://doi.org/10.1126/sciimmunol.abg4925.

Ghezzi, A., Filippi, M., Falini, A., & Zaffaroni, M. (1997). Cerebral involvement in celiac disease: a serial MRI study in a patient with brainstem and cerebellar symptoms. *Neurology*, *49*(5), 1447–1450.

Gujral, N., Freeman, H. J., & Thomson, A. B. R. (2012). Celiac disease: prevalence, diagnosis, pathogenesis and treatment. *World Journal of Gastroenterology*, *18*(42), 6036–6059. https://doi.org/10.3748/wjg.v18.i42.6036.

Gutierrez-Achury, J., Coutinho de Almeida, R., & Wijmenga, C. (2011). Shared genetics in coeliac disease and other immune-mediated diseases. *Journal of Internal Medicine*, *269*(6), 591–603.

Hadjivassiliou, M., Grünewald, R., Sharrack, B., Sanders, D., Lobo, A., Williamson, C., & Davies-Jones, A. (2003). Gluten ataxia in perspective: epidemiology, genetic susceptibility, and clinical characteristics. *Brain*, *126*(3), 685–691.

Hadjivassiliou, M., Sanders, D. D., & Aeschlimann, D. P. (2015). Gluten-related disorders: gluten ataxia. *Digestive Diseases*, *33*(2), 264–268.

Houlston, R. S., & Ford, D. (1996). Genetics of coeliac disease. *QJM: Monthly Journal of the Association of Physicians*, *89*(10), 737–743. https://doi.org/10.1093/qjmed/89.10.737.

Ho-Yen, C., Chang, F., Van Der Walt, J., Mitchell, T., & Ciclitira, P. (2009). Recent advances in refractory coeliac disease: a review. *Histopathology*, *54*(7), 783–795. www.downtoearth.org.in/news/food/three-crops-rule-the-world-what-it-means-for-the-planet-s-wildlife-81781.

International Wheat Genome Sequencing Consortium (IWGSC), Mayer, K. F., Rogers, J., Doležel, J., Pozniak, C., Eversole, K., & Praud, S. (2014). A chromosome-based draft sequence of the hexaploid bread wheat (*Triticum aestivum*) genome. *Science*, *345*(6194), 1251788.

Juhász, A., Belova, T., Florides, C. G., Maulis, C., Fischer, I., Gell, G., Birinyi, Z., Ong, J., Keeble-Gagnère, G., Maharajan, A., Ma, W., Gibson, P., Jia, J., Lang, D., Mayer, K. F. X., Spannagl, M., Tye-Din, J. A., Appels, R., & Olson, O. A. International Wheat Genome Sequencing Consortium. (2018). Genome mapping of seed-borne allergens and immunoresponsive proteins in wheat. *Science Advances*, *4*. https://doi.org/10.1126/sciadv.aar8602.

Kaslow, R. A., Carrington, M., Apple, R., Park, L., Muñoz, A., Saah, A. J. & Mann, D. L. (1996). Influence of combinations of human major histocompatibility complex genes on the course of HIV-1 infection. *Nature Medicine*, *2*(4), 405–411. https://doi.org/10.1038/nm0496-405.

Kuja-Halkola, R., Lebwohl, B., Halfvarson, J., Wijmenga, C., Magnusson, P. K. & Ludvigsson, J. F. (2016). Heritability of non-HLA genetics in coeliac disease: a population-based study in 107000 twins. *Gut*, *65*(11), 1793–8. https://doi.org/10.1136/gutjnl-2016-311713 108.

Kumar, V., Wijmenga, C., & Withoff, S. (2012). From genome-wide association studies to disease mechanisms: celiac disease as a model for autoimmune diseases. *Seminars in Immunopathology*, *34*(4), 567–580. https://doi.org/10.1007/s00281-012-0312-1.

Leonard, M. M., Sapone, A., Catassi, C., & Fasano, A. (2017). Celiac disease and nonceliac gluten sensitivity: a review. *Jama*, *318*(7), 647–656.

Li, S., Liu, Z., Zhou, Q., Feng, Y., & Chai, S. (2020). Molecular marker assisted gene stacking for disease resistance and quality genes in the dwarf mutant of an elite common wheat cultivar Xiaoyan, 22. *BMC Genetics*, *21*, 45. https://doi.org/10.1186/s12863-020-00854-2.

Li, S., Zhang, C., Li, J., Yan, L., Wang, N., & Xia, L. (2021). Present and future prospects for wheat improvement through genome editing and advanced technologies. *Plant Communications*, *2*(4), 100211.

Lindfors, K., Ciacci, C., Kurppa, K., Lundin, K. E. A., Makharia, G. K., Mearin, M. L. & Kaukinen, K. (2019). Coeliac disease. *Nature Reviews Disease Primers*, *5*(1), 3. https://doi.org/10.1038/s41572-018-0054-z.

Lionetti, E., & Catassi, C. (2011). New clues in celiac disease epidemiology, pathogenesis, clinical manifestations, and treatment. *International Reviews of Immunology*, *30*(4), 219–231.

Ludvigsson, J. F., Bai, J. C., Biagi, F., Card, T. R., Ciacci, C., Ciclitira, P. J., et al. (2014). Diagnosis and management of adult coeliac disease: Guidelines from the British society of gastroenterology. *Gut*, *63*, 1210–1228.

Ludvigsson, J. F., Leffler, D. A., Bai, J. C., Biagi, F., Fasano, A., Green, P. H., & Ciacci, C. (2013). The Oslo definitions for coeliac disease and related terms. *Gut*, *62*(1), 43–52.

Lundin, K. E., & Wijmenga, C. (2015). Coeliac disease and autoimmune disease—genetic overlap and screening. *Nature Reviews Gastroenterology & Hepatology*, *12*(9), 507–515.

Marsh, M. N. (1992). Gluten, major histocompatibility complex, and the small intestine: a molecular and immunobiologic approach to the spectrum of gluten sensitivity ('celiac sprue'). *Gastroenterology*, *102*(1), 330–354.

Matthias, T., Neidhöfer, S., Pfeiffer, S., Prager, K., Reuter, S., & Gershwin, M. E. (2011). Novel trends in celiac disease. *Cellular & Molecular Immunology*, *8*(2), 121–125.

Mearin, M. L., Biemond, I., Peña, A. S., Polanco, I., Vazquez, C., Schreuder, G. T., . . . van Rood, J. J. (1983). HLA-DR phenotypes in Spanish coeliac children: their contribution to the understanding of the genetics of the disease. *Gut*, *24*(6), 532–537. https://doi.org/10.1136/gut.24.6.532.

Mendes, F. B. R., Hissa-Elian, A., Abreu, M. A. M. M. D., & Gonçalves, V. S. (2013). Dermatitis herpetiformis. *Anais Brasileiros de Dermatologia*, *88*, 594–599.

Mondal, S., Rutkoski, J. E., Velu, G., Singh, P. K., Crespo-Herrera, L. A., Guzman, C., & Singh, R. P. (2016). Harnessing diversity in wheat to enhance grain yield, climate resilience, disease and insect pest resistance and nutrition through conventional and modern breeding approaches. *Frontiers in Plant Science*, *7*, 991.

Mottaleb, K. A., Sonder, K., Lopez-Ridaura, S., & Frija, A. (2021). *Wheat Consumption Dynamics in Selected Countries in Asia and Africa: Implications for Wheat Supply by 2030 and 2050* (Integrated Development Program Discussion Paper no. 2, p. 32). El Batan and Texcoco: International Maize and Wheat Improvement Center CIMMYT.

Nanri, K., Mitoma, H., Ihara, M., Tanaka, N., Taguchi, T., Takeguchi, M., & Mizusawa, H. (2014). Gluten ataxia in Japan. *The Cerebellum*, *13*(5), 623–627.

Ortiz, C., Valenzuela, R., & Lucero A, Y. (2017). Enfermedad celíaca, sensibilidad no celíaca al gluten y alergia al trigo: comparación de patologías diferentes gatilladas por un mismo alimento. *Revista Chilena de Pediatría*, *88*(3), 417–423.

Osborne, T. B. (1924). *The Vegetable Proteins* (p. 154). Second ed. London: Longmans Green & Co.

Parzanese, I., Qehajaj, D., Patrinicola, F., Aralica, M., Chiriva-Internati, M., Stifter, S., & Grizzi, F. (2017). Celiac disease: From pathophysiology to treatment. *World Journal of Gastrointestinal Pathophysiology*, *8*(2), 27.

Pietzak, M., & Kerner Jr, J. A. (2012). Celiac disease, wheat allergy, and gluten sensitivity: when gluten free is not a fad. *Journal of Parenteral and Enteral Nutrition*, *36*, 68S–75S.

Qiao, S. W., Bergseng, E., Molberg, Ø., Xia, J., Fleckenstein, B., Khosla, C., & Sollid, L. M. (2004). Antigen presentation to celiac lesion-derived T cells of a 33-mer gliadin peptide naturally formed by gastrointestinal digestion. *The Journal of Immunology*, *173*(3), 1757–1762.

Quirce, S., & Diaz-Perales, A. (2013). Diagnosis and management of grain-induced asthma. *Allergy, Asthma & Immunology Research*, *5*(6), 348–356.

Reilly, N. R., Aguilar, K., Hassid, B. G., Cheng, J. Defelice, A. R., Kazlow, P., Bhagat, G. & Green, P. H. (2011). Celiac disease in normal-weight and overweight children: Clinical features and growth outcomes following a gluten-free diet. *Journal of Pediatrics and Gastroenterology Nutrition*, *53*, 528–531. https://doi.org/10.1097/MPG.0b013e3182276d5e.

Rodrigo-Sáez, L., Fuentes-Álvarez, D., Pérez-Martínez, I., Álvarez-Mieres, N., Niño-García, P., de-Francisco-García, R., & López-Vázquez, A. (2011). Differences between pediatric and adult celiac disease. *Revista Espanola de Enfermedades Digestivas, 103*(5), 238.

Rubio-Tapia, A., Hill, I. D., Kelly, C. P., Calderwood, A. H., & Murray, J. A. (2013). ACG clinical guidelines: diagnosis and management of celiac disease. *Official Journal of the American College of Gastroenterology| ACG, 108*(5), 656–676.

Sabença, C. Ribeiro, M., de Sousa, T., Poeta, P., Bagulho, A. S. & Igrejas, G. (2021). Wheat/gluten-related disorders and gluten-free diet misconceptions: A review. *Foods, 10*, 1765. https://doi.org/10.3390/foods10081765.

Sapone, A., Bai, J. C., Ciacci, C., Dolinsek, J., Green, P. H., Hadjivassiliou, M., & Fasano, A. (2012). Spectrum of gluten-related disorders: consensus on new nomenclature and classification. *BMC Medicine, 10*(1), 1–12.

Sapone, A., Lammers, K. M., Casolaro, V., Cammarota, M., Giuliano, M. T., De Rosa, M., & Fasano, A. (2011). Divergence of gut permeability and mucosal immune gene expression in two gluten-associated conditions: celiac disease and gluten sensitivity. *BMC Medicine, 9*(1), 1–11.

Sapone, A., Lammers, K. M., Mazzarella, G., Mikhailenko, I., Cartenì, M., Casolaro, V., & Fasano, A. (2010). Differential mucosal IL-17 expression in two gliadin-induced disorders: gluten sensitivity and the autoimmune enteropathy celiac disease. *International Archives of Allergy and Immunology, 152*(1), 75–80.

Scherf, K. A., Brockow, K., Biedermann, T., Koehler, P., & Wieser, H. (2016). Wheat-dependent exercise-induced anaphylaxis. *Clinical & Experimental Allergy, 46*(1), 10–20.

Schofield, J. D. (1994). Wheat proteins: structure and functionality in milling and bread-making. In *Wheat* (pp. 73–106). Boston, MA: Springer.

Schumann, M., Siegmund, B., Schulzke, J. D., & Fromm, M. (2017). Celiac disease: role of the epithelial barrier. *Cell Molecular Gastroenterology Hepatology, 3*(2), 150–162.

Serena, G., D'Avino, P., & Fasano, A. (2020). Celiac disease and non-celiac wheat sensitivity: State of art of non-dietary therapies. *Frontiers in Nutrition, 7*, 152. https://doi.org/10.3389/fnut.2020.00152.

Shan, L., Molberg, Ø., Parrot, I., Hausch, F., Filiz, F., Gray, G. M., & Khosla, C. (2002). Structural basis for gluten intolerance in celiac sprue. *Science, 297*(5590), 2275–2279.

Sharma, N., Bhatia, S., Chunduri, V., Kaur, S., Sharma, S., Kapoor, P., Kumari, A. & Garg, M. (2020). Pathogenesis of celiac disease and other gluten related disorders in wheat and strategies for mitigating them. *Frontiers in Nutrition, 7*, 6.

Shewry, P. (2009). Wheat. *Journal of Experimental Botany, 60*(6), 1537–1553. https://doi.org/10.1093/jxb/erp058.

Shewry, P. (2019). What is gluten—why is it special? *Frontiers in Nutrition, 6*, 101.

Shewry, P. R., & Hey, S. J. (2015). The contribution of wheat to human diet and health. *Food and Energy Security, 4*(3), 178–202.

Shewry, P. R., & Tatham, A. S. (2016). Improving wheat to remove coeliac epitopes but retain functionality. *Journal of Cereal Science, 67*, 12–21.

Shewry, P. R., Tatham, A. S., Forde, J., Kreis, M., & Miflin, B. J. (1986). The classification and nomenclature of wheat gluten proteins: a reassessment. *Journal of Cereal Science, 4*(2), 97–106.

Shiina, T., Hosomichi, K., Inoko, H., & Kulski, J. K. (2009). The HLA genomic loci map: expression, interaction, diversity and disease. *Journal of Human Genetics, 54*(1), 15–39.

Sollid, L. M., Qiao, S. W., Anderson, R. P., Gianfrani, C., & Koning, F. (2012). Nomenclature and listing of celiac disease relevant gluten T-cell epitopes restricted by HLA-DQ molecules. *Immunogenetics, 64*(6), 455–460.

Stamnaes, J., & Sollid, L. M. (2015). Celiac disease: autoimmunity in response to food antigen. In *Seminars in Immunology*, 27(5), 343–352.

Taraghikhah, N., Ashtari, S., Asri, N., Shahbazkhani, B., Al-Dulaimi, D., Rostami-Nejad, M., & Zali, M. R. (2020). An updated overview of spectrum of gluten-related disorders: clinical and diagnostic aspects. *BMC Gastroenterology*, 20(1), 1–12.

Tarar, Z. I., Zafar, M. U., Farooq, U., Basar, O., Tahan, V., & Daglilar, E. (2021). The Progression of Celiac Disease, Diagnostic Modalities, and Treatment Options. *Journal of Investigative Medicine High Impact Case Reports*, 9, 23247096211053702.

Trynka, G., Wijmenga, C., & van Heel, D. A. (2010). A genetic perspective on coeliac disease. *Trends in Molecular Medicine*, 16(11), 537–550. https://doi.org/10.1016/j.molmed.2010.09.003.

Tye-Din, J. A., Stewart, J. A., Dromey, J. A., Beissbarth, T., van Heel, D. A., Tatham, A., & Anderson, R. P. (2010). Comprehensive, quantitative mapping of T cell epitopes in gluten in celiac disease. *Science Translational Medicine*, 2(41), 41ra51–41ra51.

Vivas, S., De Morales, J. M. R., Fernandez, M., Hernando, M., Herrero, B., Casqueiro, J., & Gutierrez, S. (2008). Age-related clinical, serological, and histopathological features of celiac disease. *Official Journal of the American College of Gastroenterology| ACG*, 103(9), 2360–2365.

Volta, U., Caio, G., Stanghellini, V., & De Giorgio, R. (2014). The changing clinical profile of celiac disease: a 15-year experience (1998–2012) in an Italian referral center. *BMC Gastroenterology*, 14(1), 1–8.

Wieczorek, M., Abualrous, E. T., Sticht, J., Álvaro-Benito, M., Stolzenberg, S., Noé, F., & Freund, C. (2017). Major histocompatibility complex (MHC) class I and MHC class II proteins: conformational plasticity in antigen presentation. *Frontiers in Immunology*, 8, 292. https://doi.org/10.3389/fimmu.2017.00292.

Wieser, H., Koehler, P., & Scherf, K. A. (2020). The two faces of wheat. *Frontiers in Nutrition*, 7, 517313.

Withoff, S., Li, Y., Jonkers, I., & Wijmenga, C. (2016). Understanding celiac disease by genomics. *Trends in Genetics*, 32(5), 295–308. https://doi.org/10.1016/j.tig.2016.02.003.

www.downtoearth.org.in/. (2022). Three crops rule the world: What it means for the planet's wildlife. Published 3 March 2022. https://www.downtoearth.org.in/news/food/three-crops-rule-the-world-what-it-means-for-the-planet-s-wildlife-81781 (Assessed 10 April 2022).

13

Application of Wheat and Its Constituents in Diverse Functional Food Products

Hanuman Bobade, Harmandeep Kaur, and Deep Narayan Yadav

CONTENTS

13.1 Introduction

The cereals are of utmost importance to the human beings for their daily calorie intake as well as other nutrients' requirements. The cereals like wheat, rice, maize, and others

DOI: 10.1201/9781003307938-13

that include 70–75% carbohydrate and 6–15% protein offer roughly 50% of human energy consumption and are considered a staple diet since ancient times with wheat-based foods providing almost 20% of the world's energy requirement from human nutrition considerations (Curti et al., 2013). The whole wheat meal or flour had been recommended by Hippocrates some 400 years BC for bowel health. Wheat (*Triticum aestivum* L.) is among the most popular grains in the world belonging to grass family in the Triticeae group (Solah et al., 2015).

The demand for functional foods is rising steadily due to increased health consciousness and awareness of consumers about the benefits and risks of various diets (Chen et al., 2013). This has fostered the interest of consumers in foods with multiple health-promoting properties or functional foods for a healthy lifestyle. The functional food refers to foods or dietary components that serve purposes other than basic nutritional needs, such as improving host health and/or lowering the risk of chronic diseases. Such foods are being explored for many years. The functional foods were initially launched in Japan in mid-1980s. These foods are fortified with beneficial substances that provide additional health benefits to the human body in addition to the nutritional benefits they provide to humans (Korhonen, 2002). The functional foods can be made by technical or bioengineering methods to enhance, add, remove, or adjust the concentration of a certain ingredient, which may have a functional effect (Roberfroid, 1999). For instance, the cereal grains could be fermented to enhance its content of bioactive compounds with biological effects and health benefits, thereby transforming the cereals in the functional food.

Wheat, along with starch and protein, also contains several phytochemical and bioactive compounds. These phytochemicals and bioactive compounds, which are distributed in various structural parts of the wheat, have defined physiological activities and demonstrated health benefits (Fardet, 2010). Depending on the milling conditions, these parts can be separated partly or entirely or can be together ground into the whole wheat meal. Therefore, the composition of and the amounts of phytochemicals in the milled flour vary on the basis of contents of different parts of wheat grain in it. The wheat grain can be ground into different type of flours using several milling techniques. The selection of particular milling method depends on the end-use of the wheat flour and its desired characteristics. For example, the refined wheat flour, which is devoid of bran and germ fraction, is commonly preferred for the preparation of the most of bakery products owing to its easy processability. The utilization of whole-wheat flour, which is more nutritious than the refined flour, or the inclusion of bran and germ fractions results in functional food product with inferior physical and sensory characteristics. This necessitates the technological interventions to develop the whole-wheat flour-based or bran and germ-added functional food products with desired and acceptable quality characteristics.

Wheat and its different parts have been effectively utilized in the development of many functional food products including bakery, dairy, meat, and extruded products. These wheat-based foods form an abundant source of many essential nutrients. This chapter briefly outlines the structure and composition of wheat and its parts and milling systems with their principal products and emphasizes the applications of wheat and its parts in the development of a variety of functional food products.

13.2 Structure and Composition

The knowledge and understanding about the structure of wheat grain in its native form are important for its utilization in different functional food products since the structure principally governs the behavior of grain during its processing (Shewry et al., 2009; Solah et al., 2015). For instance, during milling, which is a basic requirement for wheat utilization in added value products, the grain is mechanically separated into its different components like endosperm, bran, and germ, on the basis of how it is composed structurally. The endosperm (which is 80–85% of the grain) of wheat is mostly made up of the starch granules embedded in the protein matrix (Delcour et al., 2012). Along with starch and protein, the endosperm also contains monosaccharides in little amount. The starchy endosperm's primary monosaccharide (96%) is glucose, while arabinoxylans (AX) make up only 2% (Barron et al., 2007; Gebruers et al., 2008). The embryo and scutellum comprise the germ (3% of the grain), which is high in lipids, proteins, neutral carbohydrates, minerals, vitamins, and sterols. The aleurone layer, hyaline layer, seed coat, inner pericarp, and outer pericarp are the peripheral layers that surround the endosperm in order from the inner to the outer sections (Evers and Bechtel, 1988).

13.2.1 Whole Wheat

The majority of wheat grains are oval; however, depending on the wheat type, they can also be almost spherical or long, narrow, and flattened. The cross-sectional shape of the wheat grain ranges from triangular to evenly rounded. The grain typically has a marked longitudinal furrow or crease down in one side where it was once joined to the wheat flower. The dimensions of well-filled caryopsis vary greatly; however, its length ranges from 4 to 10 mm and width from 2.5 to 4.5 mm (Shewry et al., 2009) and weighs between 35 and 50 mg. Wheat has three distinctive structural parts – endosperm, bran, and germ. The endosperm constitutes 80–85%, bran forms 13–17%, and germ fraction makes up 2–3% of the total weight of the grain (Belderok et al., 2000). The structural parts of wheat and their composition are presented in Table 13.1.

TABLE 13.1

Nutrient Content of Whole Wheat and Its Different Parts

Nutrient	Whole Wheat	Wheat Endosperm	What Bran	Wheat Germ
Moisture (%)	10.5	11.9	9.9	11.1
Protein (%)	13.4	10.3	15.6	23.2
Total lipid (%)	2.1	0.98	4.3	9.7
Ash (%)	1.6	0.47	5.8	4.2
Total carbohydrates (%)	72.3	76.3	64.5	51.8
Total dietary fiber (%)	11.5	2.7	47.8	13.2
Sugars (%)	0.41	0.27	0.41	37.6
Total bioactive phytochemicals (%)	1.2	-	6.2	-

Source: USDA National Nutrient Database for Standard Reference, US Department of Agriculture, 2015

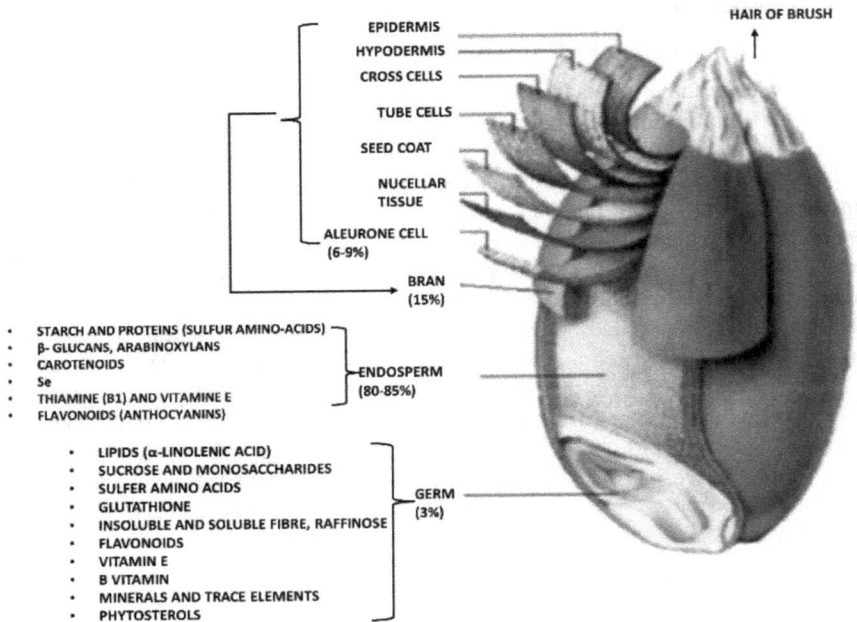

FIGURE 13.1 Structural parts of whole wheat and prominent phytochemicals present.

Source: Anderssen and Haraszi (2009)

Whole wheat is rich in several nutritional compounds, including dietary fibres and phytochemicals, which are well recognized for their health benefits. Starch, followed by protein, is the most abundant biopolymer present in the wheat grain. Starch and proteins are majorly concentrated in the endosperm fraction of the wheat grain. Whole wheat contains many bioactive and phytochemical compounds like phenolic acids, choline, betaine, sulfur amino acids, phytic acid, alkylresorcinol, minerals, and vitamins, besides fibres. The distribution of these bioactive compounds in different parts of the wheat grain is presented in Figure 13.1, and their average/range content is given in Table 13.2. Whole wheat contains about 2% bioactive compounds in addition to approximately 13% dietary fibres (Fardet, 2010). These phytochemicals are surrounded by the complex cellular and molecular structures in the wheat grain (Rosa-Sibakov et al., 2015).

13.2.2 Bran

Wheat bran is the hull of wheat grains and is usually removed during wheat milling. The testa, aleurone, and pericarp layers of wheat bran are separated into three different layers. The bran fraction is generally considered as a by-product of milling and has been used for food as well as nonfood applications (Galanakis, 2018). Wheat bran is rich in dietary fiber and other noncaloric nutrients. Wheat bran has a nutritional fiber content of roughly 53% (xylans, lignin, cellulose, galactan, fructans). Wheat bran is also rich in minerals, B vitamins, and bioactive compounds

TABLE 13.2

Phytochemical Content of Whole Wheat and Its Parts

Nutrient	Whole Wheat	Wheat Bran	Wheat Germ	References
α-Linoleic acid (mg/100 g)	-	0.16	0.53	Fardet (2010)
Sulfuric compounds (mg/100 g)	0.5	0.7	1.2	
Fiber (mg/100 g)	13.2	44.6	17.7	
Lignins (mg/100 g)	1.9	5.6	1.5	
Oligosaccharides (mg/100 g)	1.9	3.7	10.1	
Minerals and trace elements (mg/100 g)	1.12	3.39	2.51	
Carotenoids (μg/100 g)	0.34	0.72	-	
Polyphenols (mg/100 g)	0.15	1.10	>0.37	
Phenolic acids (mg/100 g)	0.11	1.07	>0.07	
Flavonoids (mg/100 g)	0.037	0.028	0.300	
Phytosterols (mg/100 g)	0.08	0.16	0.43	
Lignan (mg/100 g)	0.4	4.75	-	
Betaine (mg/100 g)	156	868	-	
Choline (mg/100 g)	111	172	-	
Phytic acid (mg/100 g)	910–1930	2180–5220	-	Fardet et al. (2008); García-Estepa et al. (1999)
Ferulic acid (mg/100 g)	10–200	500–1500	-	Adom et al. (2003); Barron et al. (2007); Fardet et al. (2008); Zhao and Moghadasian (2008)
Alkylresorcinols (mg/100 g)	28–140	220–400	-	Chen et al. (2004); Fardet et al. (2008); Landberg et al. (2008)
B vitamins, total (mg/100 g)	9.02	33.5	-	Fardet (2010)
B1, thiamine (mg/100 g)	0.56	0.65	-	
B2, riboflavin (mg/100 g)	0.18	0.51	-	
B3, niacin (mg/100 g)	6.5	28	-	
B5, pantothenic acid (mg/100 g)	1.35	3.15	-	
B6, pyridoxine (mg/100 g)	0.38	1.0	-	
B9, folate (mg/100 g)	0.05	0.23	-	
Vitamin E (mg/100 g)	1.4–2.2	1.4mg	-	Cho and Pratt (2008); Fardet et al. (2008)
Folate (μg/100 g)	20–87	79–200	-	Cho and Pratt (2008); Fardet et al. (2008); Patring et al. (2009)
Glutathione (μg/100 g)	82–670	-	-	Sarwin et al. (1992), Schofield and Chen (1995)
Iron (mg/100 g)	3.2	11	-	Cho and Pratt (2008); Fardet et al. (2008)
Manganese (mg/100 g)	3.1	12	-	Cho and Pratt (2008); Fardet et al. (2008)
Zinc (mg/100 g)	2.6	7.3	-	Cho and Pratt (2008); Fardet et al. (2008)
Selenium (μg/100 g)	0.5–75	78	-	Cho and Pratt (2008); Fardet et al. (2008)

including alkylresorcinols, ferulic acid, flavonoids, carotenoids, lignans, and sterols that are known to have health-promoting properties (Apprich et al., 2014; Andersson et al., 2014; De Brier et al., 2014). Among the minerals, the wheat bran is particularly rich in iron (Fe), zinc (Zn), manganese (Mn), magnesium (Mg), and phosphorus (P). However, more than 80% of the phosphorus is stored as phytates, which form complexes with other minerals like Fe, Zn, and Mg. The bioavailability of these minerals is reduced due to the complex formation (Sandberg et al., 1982). Hydrothermal treatment, size reduction, enzymatic treatment, malting, soaking, and fermentation have all been successful attempts to minimize phytic acid in wheat bran over the years (Aivaz and Mosharraf, 2013; Coda et al., 2014). Exogenous phytase from yeast and sourdough fermentation, as well as endogenous phytase in the grain, can release phytic-acid-complexed minerals (Brouns et al., 2012). Incorporating bran into cereal-based dishes generally leads to reduced final product quality and sensory acceptance. Milling, thermal treatments, extraction, extrusion, enzymatic treatment, and fermentation can improve the nutritional, sensory, and physical aspects of wheat bran for food applications (Prueckler et al., 2014).

13.2.2.1 Pericarp

The pericarp, often known as hull, is the outermost coat of the kernel and accounts for 5–7% of the dry weight of the kernel. The pericarp is separated into outer and inner layers, with the outer layer containing insoluble dietary fiber and the inner layer containing bound phenolic acids. (Apprich et al., 2014). The nucellar tissue, also known as the hyaline layer, is found in the intermediate layer of the bran. Alkylresorcinol is mostly found in the testicular layer (Rebolleda et al., 2013). The epidermis, mesocarp, cross cells, tube cells, and seed coat are the tissues of the pericarp from the surface inward. Except for the seed coat, which is amorphous, they are all made up of empty dead cells. The thickness of the pericarp varies throughout the kernel, most likely due to changes in compression degrees rather than the number of cell layers (Brewbaker et al., 1996).

13.2.2.2 Aleurone Layer

The aleurone layer is the deepest layer of bran, and it is partly shared with the endosperm. It has a high concentration of lignans and proteins, as well as bioactive substances, phytic acid, antioxidants, vitamins, and minerals (Muhammad et al., 2012). Aleurone has found a tremendous interest in the scientific community and is still being explored as a beneficial element in cereals. The highly nutritious contents of aleurone contribute to the healthful added value of whole-grain cereals. As a result of this nutritional profile, wheat bran is being extensively used in the development of variety of functional foods (Meziani et al., 2021). A survey conducted at Polish households found that the cereal consumption provides 20–30% of the average protein, thiamin, phosphorus, and zinc intake, as well as 10–20% of a variety of nutrients such as polyunsaturated fatty acids, sodium, potassium, calcium, riboflavin, niacin, vitamin B6, and other micronutrients (Laskowski et al., 2019). The majority lipids of aleurone layer are nonpolar lipids, while polar lipids are concentrated in the starchy endosperm (Marion et al., 2003). During milling, the aleurone layer is treated as an integral component of the wheat bran. When examined in cross or longitudinal sections, aleurone cells surrounding the starchy endosperm are often block-shaped. The cells in the paranormal portion are polygonal with rounded corners and no intercellular gaps. The walls are bilayered (Fulcher et al., 1972), with the outer layer staining more heavily with periodic acid Schiff's reaction or toluidine blue than the inner layer.

13.2.3 Germ

Wheat germ, like an egg yolk, is the embryo of the wheat kernel. The germ is removed during the milling and refining of whole-wheat flour to white flour because it contains lipids that can degrade the quality of milled wheat flour. The wheat germ, which accounts for 2.5–3.8% of the total weight of wheat, is another important milling by-product that is separated from the endosperm during milling due to its susceptibility to oxidation and poor baking properties (Brandolini and Hidalgo, 2012; Kumar and Krishna, 2015). The wheat germ is the nutrient-dense embryo of the wheat kernel or seed and contains about 52% carbohydrates, 23% protein, 10% lipids, and 11% water. The wheat germ is high in minerals like potassium, iron, and zinc, and it is also a good source of vital amino acids and vitamins like thiamin, riboflavin, and niacin (Brandolini and Hidalgo, 2012). The wheat germ proteins consist of about 34.5% albumin as the largest fraction, followed by globulins (15%), gluten (10.6%), and prolamine (4.6%) and have a well-balanced amino acid make-up similar to that of egg proteins (Zhu et al., 2006). Antioxidants such as flavonoids, polyphenols, tocopherols, tocotrienol (vitamin E; recognized for its health advantages), carotenoids, and plant sterols, as well as biogenic amines, particularly polyamines, make up the majority of the bioactive chemicals in WG. Likes et al. (2007) reported the presence of betaine and choline in germ twice the amount in bran. The germ might be a potential low-cost source of raw material for the food and oleochemical industries, as well as a stable oil source for a variety of uses and applications, such as shampoos, soaps, and by-products, margarine, and salad and cooking oils. The wheat germ, with its intrinsic nutritional value, may thus be an excellent substitute for refined vegetable oils in food products.

The oil extracted from germ is rich in triglycerides (57% of total lipids), mainly linoleic acid (18:2), palmitic acid (16:0), and oleic acid (18:1). The wheat germ oil contains fair amount of total tocopherols (288 mg/100 g) consisting of α-tocopherols, γ-tocopherols, and tocotrienols, about 57%, 30%, and 11% of the total tocopherols, respectively. Phospholipids (14–17% of germ oil), policosanols (docosanol, hexacosanol, octacosanol, triacontanol), and phytosterols (mainly sitosterol 60–70%, campesterol 20–30%) are present with monoglycerides and diglycerides of wheat germ oil (Kumar and Krishna, 2015). Wheat germ oil is especially rich in phosphorus (1.4 g/kg). The wheat germ oil is used for its nutritional benefits, notably for its high vitamin-E content, as opposed to commodity oils like soybean and canola oils, which are mostly used for their heat-transfer capabilities during cooking and for producing a pleasant mouth feel in salad dressings. The wheat germ oil is used in a variety of products, including cosmetics, toiletries, medications, and health foods, as well as nutritional supplements (Dunford, 2009).

13.3 Biological Activities of Wheat and Its Parts

Epidemiological studies have demonstrated that consuming whole-grain foods lowers the chances of developing cancer, type II diabetes, and heart disease and possibly even helps people control their weight (Giacco et al., 2011; Mellen et al., 2008). The most significant bioactive elements with proven health advantages in wheat are polyphenols (especially phenolic acids), sulfur amino acids, betaine, total choline, phytic acid, alkylresorcinols, minerals, and vitamins B and E (Fardet, 2010). The total amount of bioactive compounds in food determines how much is consumed, but the microstructure of the food matrix also affects how much is released during digestion and absorbed into the bloodstream (Palzer, 2009). Both the aleurone layer and wheat bran are extremely complex matrices. These complex matrices contain additional bioactive compounds that are chemically linked at the molecular level or encased in the cells and/or cell-wall structures. For instance, ferulic acid, the main phenolic acid in wheat bran, is predominantly ester-linked with arabinoxylan and included in the cell walls of bran and aleurone (Zúñiga and Troncoso, 2012). Similarly, the dietary fibers and other bioactive compounds in various parts of the wheat grain are primarily present in bound forms as opposed to free elements at the molecular level. Although the physiological and health effects of wheat grain and its fractions have been studied, scientific literature rarely acknowledges or takes into account the significance of their structure at the macro, micro, and molecular levels. Through the right assembly of hierarchical structures, foods with higher nutritious characteristics and therefore imparting improved health benefits could be obtained (Zúñiga and Troncoso, 2012).

13.3.1 Anti-Diabetic and Hypoglycemic Activity

The diabetes mellitus (DM), a type of diabetes, is one of the most prevalent chronic diseases in the world and a major contributor to morbidity and mortality (American Diabetes Association, 2010). The diabetes is prevalent in more than 347 million individuals worldwide (Danaei et al., 2011). According to the International Diabetes Federation, China had 114 million diabetic patients aged 20 to 79 in 2017, with the number anticipated to climb to 1.19 billion by 2045. In China, T2DM accounts for more than 90% of cases and is tightly linked to dietary habits. Type I diabetes and decreased responsiveness of peripheral insulin receptors in peripheral tissues both are considered as forms of diabetes, which are characterised by diminished insulin output (type II diabetes) (Poitout, 2008). Hyperglycemia, glucosuria, polyphagia, polydipsia, and polyuria are all related to diabetes (Moussa, 2008). Diabetes causes irreparable tissue damage such as retinopathy, nephropathy, arteriosclerosis, and vascular damage due to persistent hyperglycemia (Luitse et al., 2012).

The consumption of whole wheat foods has been associated with reduced risk of type II diabetes (Jones et al., 2020). The intake of whole grains and their products like whole-wheat bread reduces the risk of type II diabetes by 20–30% (Gil et al., 2011). It has been found that the antidiabetic effect of whole wheat is attributed to its fiber content; however, the fibers from fruits and vegetables have no impact on diabetes. The antidiabetic activity of wheat fibers could be due to their interaction with gut microbiota that alters the body energy metabolism (Davison and Temple, 2018). Alkylresorcinols, which have been suggested as a biomarker of whole-grain intake due to their abundant presence in the bran sections of wheat, are considered to have the ability to lower blood glucose in male mice suffering from the type II diabetes (Biskup et al., 2016). It is suggested that the eating of foods containing wheat germ agglutinin has been linked to a considerable reduction in type II diabetes and associated cardiovascular diseases (van De Sande et al., 2014). The alkylresorcinols from wheat have the ability to prevent the insulin resistance and glucose intolerance induced by high-fat, high-sucrose diet in mice by maintaining the hepatic glucose homeostasis.

13.3.2 Anti-Obesity Activity

Obesity is a medical condition defined by an excess of body fat that further leads to a variety of health issues. According to the World Health Organization (WHO) regulations, the individuals in Asian countries with a body mass index of 25 kg/m^2 or more are considered as obese. Obesity is associated with metabolic syndrome, which comprises type II diabetes mellitus, cardiovascular disease, hypertension, and dyslipidemia. Each of these condition lowers the quality of life and raises the risk of mortality, and they are all linked to obesity. Obesity and overweight have become highly concerning issues in modern culture (Elagizi et al., 2018). The effect of wheat bran consumption on human blood lipid levels and weight has been studied extensively. It has been observed that the consumption of wheat dietary fiber aids in weight loss (Astrup et al., 2010). Whole-grain cereals or cereal brans have been shown to protect against obesity, metabolic syndrome, diabetes mellitus, cardiovascular disease, and cancer in several epidemiological and intervention studies (Fardet, 2010). The ingestion of wheat bran is reported to significantly reduce the obesity in mice (Han et al., 2015). The foods developed with the incorporation of wheat germ agglutinin are considered to aid in long-term weight management.

13.3.3 Anti-Lipidemic Activity

The excess blood lipids or cholesterols are known to cause variety of problems, including atherosclerosis, cardiovascular disease, and chronic renal disease. The wheat germ has been shown to have several possible health and therapeutic advantages, including reducing cholesterol levels in human subjects. The wheat bran is reported to have a favorable effect on the blood lipids and therefore can effect improvements in triglycerides (Cara et al., 1992) and cholesterol (Williams et al., 1999). The rats fed with wheat bran and sucrose showed considerable reduction in liver triglycerides, plasma triglycerides, and liver cholesterol (Chen and Anderson, 1979). The wheat fiber intake by patients suffering from dysmetabolic cardiovascular syndrome causes a decrease in the blood pressure and improvement in the blood glucose and lipid profile (Šabovič et al., 2004). Further, rats supplemented with the wheat bran have shown a significant reduction in their plasma and cholesterol levels (Owen et al., 1975). The wheat bran consumption has remarkably lowered the serum triglyceride levels (Heaton and Pomare, 1974) and serum cholesterol levels (Persson et al., 1976) in human subjects. However, the effect of wheat and its components on lipid metabolism has been contrasting. Many studies have reported that the supplementation of wheat bran has no reducing effect on plasma cholesterol (Mathe et al., 1977), plasma triglycerides (Malinow et al., 1976), and on liver triglycerides (Arvanitakis et al., 1977). The studies related to anti-lipidemic activity of wheat bran are inconsistent as it has been observed that the cholesterol levels slightly increased in a volunteer who took fiber supplement (36 g wheat fiber) for two weeks (Jenkins et al., 1975). Further, it has been observed that the intake of wheat bran by human subjects has not influenced the serum cholesterol and serum triglyceride levels (Jenkins et al., 1975; Persson et al.,

1976). Therefore, further comprehensive research is required to establish the effect of whole wheat and its components on lipid metabolism.

13.3.4 Antitumor Activity

The whole wheat has the potential to reduce the risk of developing colorectal cancer due to its content of fibers, antioxidants, and phytochemicals (Song et al., 2015). The dietary studies in mice employing wheat-containing fibers showed that the mice exhibit considerable protection against the chemically caused colon cancer (Roberfroid, 1993). The anticancer properties of grains have also been attributed to some nonfibrous components of grains. Several phytochemicals present in the cereal bran may have anti-colon cancer capabilities. Orthophenolics, a potent antioxidant family that can scavenge free radicals and chelate metals, are present in significant amounts in whole grains. A decreased risk of developing several cancers has been associated with an increased phenolic consumption. Many such phenolic compounds including ferulic acid, caffeic acid, chlorogenic acid, and diferulic acid are present in the wheat bran (Garcia-Conesa et al., 1997). It has been found that the foods supplemented with 0.1–0.5% plant-derived, pure orthophenolics, including caffeic acid, prevent the gut and colon malignancies in the animal model (Mahmoud et al., 2000). The consumption of foods containing wheat germ agglutinin has been linked to a considerable reduction several types of cancers (van De Sande et al., 2014).

The wheat bran has been linked to reducing the gastrointestinal disease, diverticulitis, diverticulitis-related diabetes, colon cancer, cardiovascular disease, constipation, irritable bowel syndrome, and other conditions (Shahid et al., 2020). However, it is still unclear how it affects hyperlipidemia, hepatotoxicity, obesity, and hyperglycemia. After consuming a high-fat diet, the levels of malondialdehyde, blood sugar, serum and liver lipids, and body weight all considerably rose. However, incorporating coarse and ultrafine wheat bran into a high-fat diet prevented weight gain; decreased triglycerides, malondialdehyde, serum low-density lipoprotein, and glucose levels; and raised serum high-density lipoprotein levels. The wheat bran with the larger particle size is shown to have a larger impact than the wheat bran with smaller particle size.

13.4 Applications in Functional Food Products

13.4.1 Bakery Products

Although bran from various cereals and non-cereals is the most common source of dietary fiber in bakery products, using wheat flour with a higher extraction rate rather than white flour has positive effects on antioxidant activity of the product. The biscuits prepared with high extraction rate flour exhibited higher total phenolics and antioxidant activity than the biscuits prepared from low extraction rate flour. The biscuits prepared from flour with extraction rate up to 90% were found to be acceptable for sensory attributes (Žuljević et al., 2021).

Wheat bran is one of the most common raw ingredients used to increase the amount of nutritional fiber in baked goods. Bran, being an excellent source of phytochemicals (Kamal-Eldin et al., 2009), is considered to be one of the potential ingredients for the production of nutritionally enhanced cereal foods and new components.

However, due to the detrimental effects of bran on the gluten network and subsequent textural qualities of bread, using native bran in wheat baking is a technological challenge (Noort et al., 2010).

13.4.1.1 Bread

Bread is one of the important staple foods and is the most widely consumed bakery product. Generally, white or refined wheat flour, due to its easy processability, is preferred for the bread making. However, such breads have reduced nutrient content compared to breads made from whole-grain cereals (Haruna et al., 2011). Recently, the consumer awareness about functional foods has increased which led the manufacturers to develop the food products with ingredients like wheat bran, wheat germ, and use of whole wheat, having functional properties. However, the incorporation of such ingredients may result in lowering the physical and sensory properties of the food products. Therefore, it necessitates precise process optimization and/other technical intervention to produce the functional foods with desirable properties and consumer acceptance. The breads prepared with the addition of 10–40% whole-wheat flour in bread formulation result in decreased bread volume, specific volume, and increased weight of bread loaf as compared to breads prepared without the addition of whole-wheat flour. The decreased loaf volume has been attributed to the dilution of gluten and subsequently low protein network formation in the dough (Rosell et al., 2001). The fibre addition also results in high water absorption, mixing tolerance and tenacity, and reduced extensibility of the dough (Gómez et al., 2003), which may have caused a reduction in specific volume and increase in bread loaf. The results, however, indicated that the addition of whole-wheat flour up to 20% level produces acceptable breads that have sensory attributes comparable to the breads prepared without the addition of whole-wheat flour (Ngozi, 2014).

The breads prepared using whole-wheat flour contain more phytochemicals like phenolic compounds and other nutrients like vitamin, minerals, and fibers. Therefore, the breads prepared with whole-wheat flour are more nutritious as compared to those prepared from the refined flour. For instance, the whole-wheat flour bread had total phenolic content of 150–165 mg/100 g compared to the 79–103 mg/100 g total phenolics in refined wheat flour bread. Further, the whole-wheat flour bread showed comparatively less reduction (28%) in total phenolics during baking process against 33% losses in phenolics of refined wheat flour bread. The whole-wheat flour breads also have superior in vitro antioxidant activity than the breads prepared from refined flour (Yu et al., 2013). Ndife et al. (2011) prepared the functional breads from whole wheat and soy bean flour blends and observed that the inclusion of soy bean flour in whole-wheat flour decreased the dough expansion and bread volume by 13% and 64.5%, respectively. The functional bread prepared had high protein and fat content and low carbohydrate and calorie content. The sensory properties of the functional bread, however, were not affected by the addition of soy bean flour in whole-wheat flour. Wang et al. (2017) studied the effects of flour particle size on the quality of steamed bread prepared from whole-wheat flour. The results revealed that the steamed bread prepared using whole-wheat flour of smaller particle size had considerably larger specific volume. Further, the use of whole-wheat flour with reduced particle size produced the smaller grain cells having thin cell walls. Therefore, it is implied that the quality of whole-wheat flour steamed bread could be improved by reducing the particle size of

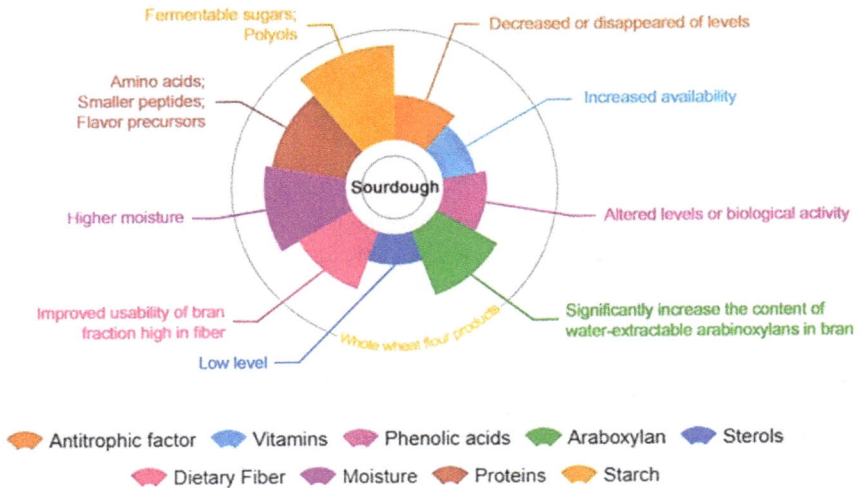

FIGURE 13.2 Effect of sourdough fermentation on various nutrients in whole-wheat flour-based products.

Source: Ma et al. (2021)

the whole-wheat flour. The functional properties of whole-wheat flour bread could be further improved by sourdough. The sourdough fermentation has been increasingly found to be useful in improving the quality of whole-wheat flour products. The sourdough fermentation technology has a great potential in reducing the anti-nutritional factors and toxic and harmful substances in whole-wheat flour products (Ma et al., 2021). The effects of sourdough fermentation on macromolecular nutrient in whole wheat dough systems are represented in Figure 13.2.

The sourdough improved the mineral bioavailability from the reconstituted whole-wheat flour in rats. The phytate contents in yeast and sourdough breads were lowered by 52% and 71%, respectively. The rats fed with the sourdough bread also showed 41% increase in the copper absorption (Lopez et al., 2003). The extraction of particular phytochemical compound, like arabinoxylans, and its use in breadmaking also maintain the properties of such bread close to the control bread or even it results in the improvement of the bread properties besides imparting the functional value (Pietiäinen et al., 2022). The wheat bran arabinoxylans' addition up to 5% level increases the specific volume of bread; higher additions (10–18%), however, negatively impact the bread volume. The crumb structure of bread could also be improved by the addition of arabinoxylans at 2% level. The addition of wheat bran arabinoxylans up to 5% did not have negative impact on the texture of bread; however, the bread texture becomes hard when the level of arabinoxylans is increased to 10% (Zhang et al., 2019). The improvement in the properties of bread as a result of arabinoxylan addition is attributed to the increased porosity, more homogenous cell structure, and, therefore, higher volume and softer texture (Wang et al., 2019). The use of extruded wheat germ also improved the quality of bread. The extruded wheat germ addition has resulted in increased volume and decreased firmness of the bread. The extruded germ addition up to the level of 10 g/100 g of flour was found to produce acceptable breads. The germ extrusion, therefore, improves the dough characteristics and bread quality, and constitutes an appropriate treatment for using wheat germ in bread as a functional ingredient (Gómez et al., 2012). Wheat germ, which in addition to fibers, is rich in many phytochemicals, vitamins, and minerals, is a valuable ingredient for the development of the functional food products (Kevin, 1995). The presence of wheat germ in flour adversely affects the bread-making and storage quality, and, therefore, wheat germ is often separated from the flour fractions during milling. However, by storing the flour and wheat germ separately and mixing them at the time of manufacture, the adverse effect of the wheat germ on bread-making quality can be lessened (Pomeranz, 1970).

13.4.1.2 Biscuits

Biscuits are usually formulated with a high amount of fat and sugar and therefore have high calorie content but less other nutrients (Sozer et al., 2014). The nutritional value of these biscuits could be enhanced with the use of whole-wheat flour or other wheat components in the bran. Leelavathi and Rao (1993) indicated that the wheat flour could be replaced with wheat bran, as a source of dietary fiber, up to the level of 30% without considerably impacting the overall quality of biscuits. The biscuits formulated with highest level of wheat bran have seven times more dietary fiber content than the biscuits prepared without the addition of wheat bran. The use of wheat germ in biscuit making is reported to enhance the its nutritional quality with an increase in protein and mineral content and a decrease in fat, carbohydrate, and calorie content. The sensory characteristics and acceptability of these biscuits also improved with the addition of wheat germ (Al-Marazeeq and Angor, 2017). Protonotariou et al. (2016) prepared the biscuits using whole-wheat flour obtained by jet milling. The biscuits showed increased hardness with increasing the level of whole-wheat flour in formulation. The biscuits also showed significant decrease (23%) in spread factor and increase in density with the substitution of white flour by the whole-wheat flour. The biscuits prepared with 50% jet-milled whole-wheat flour showed good sensory attributes. However, the results of the study indicated that the higher substitution levels are not desirable.

The wheat bran, which is good source of dietary fiber and other phytochemicals, could also be used as a functional ingredient in biscuit making. Haque et al. (2002) also observed that the replacement of flour by wheat bran up to 6% level produced biscuits with quality parameters close to the control. The substitution of flour by bran linearly increased the weight of biscuits. The aqueous extracted wheat bran yielded better-quality biscuits, and, therefore, could provide viable alternative to the wheat bran extracted by alkaline method. The level of bran incorporation with maintaining the biscuit quality could, however, be raised by giving some treatments like roasting, steaming, and others, to the bran. Nandeesh et al. (2011) studied the effect of addition of 30% wheat bran treated with roasting, steaming, and microwave cooking in biscuit making. The biscuits prepared with treated wheat bran showed higher spread ratio than untreated wheat bran biscuits but had less spread ratio than the control (without added bran) sample. Further, the treated bran-added biscuits exhibited lower breaking strength than the biscuits prepared with untreated wheat bran; however, this breaking strength was significantly higher than the control sample. Therefore, it could be implied that the use of treated bran, particularly, roasted bran, increased the quality of biscuits. The quality of biscuits is also affected by the particle size of the wheat bran used. The biscuits prepared by the addition of bran with larger particle size have more hardness. The biscuits formulated with fine bran (68 μm) have more compact structure and have no surface or internal defects compared to the biscuits formulated with coarse bran (450 μm). Further, the breakdown and coarseness in mouth were more significant for the biscuits developed from coarse bran added at highest level (30%). The wheat bran addition in biscuit formulation significantly improved the nutritional profile of the biscuits (Sozer et al., 2014). Inyang et al. (2018) have found that the acceptable quality biscuits having improved nutritional value and high dietary fiber content could be produced from whole-wheat flour supplemented with *acha* and red kidney bean flour.

13.4.1.3 Other Bakery Products

The utilization of whole-wheat flour reduces the quality of saltine crackers. The saltine crackers' quality with respect to breaking strength, stack height, and specific volume has been adversely affected with the addition level of whole-wheat flour. These quality parameters of saltine crackers showed significant correlations with the gluten index, arabinoxylans, and dough extensibility of whole-wheat flour. The inferior baking qualities of whole-wheat flour addition in the preparation of saltine crackers have been linked to the water migration from the gluten network into arabinoxylans matrix in whole wheat dough system (Li et al., 2014).

The cake prepared with the replacement of flour by coarse and fine ground wheat germ at the levels ranging from 10 to 30% has increased nutritional value. The protein content of the cake increased from 7.85 to 11.89% at 30% level of germ enrichment. The addition of wheat germ in cake formulation decreased the volume, volume index, and softness of the cake. The germ addition has resulted in darker, more reddish, and yellowish cake. The sensory characteristics of cake formulated with 10–20% wheat germ are similar to the sensory characteristics of cake without added wheat germ. The cake with 30% wheat germ has reduced appearance, texture, and pore structure (Levent and Bilgiçli, 2013)

The muffins prepared with the addition of 45% waxy whole-wheat flour have shown high overall likeability. The use of waxy whole-wheat flour in the range of 15–30% produces softer and moister muffins. The incremental addition of waxy whole-wheat flour, however, resulted in the production of darker muffins with lower volume (Acosta et al., 2011).

13.4.2 Extruded Products

Extrudates or extruded products are the meals created by strong shearing beneath compression, quick heat treatment, and high-temperature cooking. Low-fat snacks may be made using the extrusion technique, which also causes the synthesis of resistant starch that has no calories and, in addition, functions physiologically like dietary fiber. In order to create snacks with more nutritional value, extrusion is becoming more and more popular, and raw ingredients like protein, starch, and dietary fiber are also used (Igual et al., 2020).

13.4.2.1 Extruded Breakfast Cereals

Breakfast cereal products play a significant role in breakfasts across the world. In affluent nations, the majority of youngsters and nearly half of the population regularly eat breakfast cereal products (Santos et al., 2022). Breakfast cereals cooked via extrusion have high-calorie content. Increasing the amount of dietary fiber in food is one of the strategies used by the food industry to lower the calorie density of meals. Dietary fiber benefits the immune system, heart health, diabetes control, and weight control in addition to its lower calorie content and regulation of digestion (Robin et al., 2012).

The extruded breakfast cereals were developed with the inclusion of wheat bran, and it was observed that the inclusion of wheat bran enhanced the product density, as well as the bulk density of the extruded goods, and changed the pasting characteristics of the raw flour as well as the extruded products. Moreover, the rate and extent of carbohydrate hydrolysis were dramatically decreased by adding wheat bran to the raw bases (Brennan et al., 2008).

The extrudates made from the blend of wheat flour and pinto bean meal with the addition of wheat bran exhibited varying functional properties. The expansion index also increases with increasing the level of wheat bran. Maximum apparent viscosity response surface solution was achieved at 10% bran whereas the response surface solution for the water uptake index was supported to enable at 25% bran. The additional fiber was thought to play a preventive action since the increased rheological behavior and reduced absorption ratio showed very modest harm to the proteins and starch (Hernandez-Diaz et al., 2007). Using a formula based on starch along with two sources of fiber which are wheat bran and oat bran concentrate, extruded cereals were created. It has been shown that with the increase in fiber, there was a decline in crispness, but hardness was increased. The more fiber is added, greater the effect on textural characteristics. While the thickness of the cell wall rises dramatically, only slight changes in porosity and cell size are seen as the whole-grain percentage increases. On the other hand, including fibers reduces the expansion of extruded cereals (Chassagne-Berces et al., 2011). The defatted wheat germ up to a concentration of 15% was used to substitute the yellow corn grits and analyzed for physical, structural, and sensory characteristics after their extrusion with the single-screw extruder. After analysis, the results indicate an increase in bulk density of extrudates as compared to the control sample but a decrease in the level of other properties like water absorption index, water solubility, expansion, and breaking indexes. The conditions optimized for extrudates with 15% defatted wheat bran are 160° extrusion temperature, up to 14% moisture content, and around 240 rpm extrusion speed (Yaseen and Shouk, 2005).

13.4.2.2 Extruded Snacks

The traditional process of boiling and then drying the goods has been superseded by extrusion technology, which is a common approach for creating ready-to-eat foods (Kaur et al., 2015). Increasing the amount of dietary fiber in food, particularly cereal-based goods, is now supported by health and nutritional policies. However, adding fiber to cereals may result in quality problems, which would reduce customer acceptance (Chassagne-Berces et al., 2011). The extruded snacks, which could be inflated with hot air, were prepared by addition of whole-wheat flour and textured soy flour as a partial substitute to refined flour. The findings demonstrated that the amount of whole-wheat flour replacement, particularly at levels of 15%, significantly influenced its organoleptic characteristics measured using the sensory panel. The blend of 15% whole-wheat flour and 5% textured soy flour was shown to have the best expansion index in the prepared snacks (Rodríguez-Vidal et al., 2017).

Extruded crunchy tidbits were prepared by mixing broken rice with wheat bran and examined for their nutritional properties. The results have shown that the content of nutritional components like Ca, P, Fe, Cu, Vitamin B1, Vitamin B2, and lysine have increased along with some anti-nutritional components (Singh et al., 2000). In whole-grain wheat flour extruded products, the impact of jabuticaba (*Myrciaria cauliflora*) peel powder was examined for its technical excellence and sensory acceptability. Rather than only jabuticaba peel powder, the combination of both the whole-grain flour as well as jabuticaba peel powder had an adverse effect on the expansion of products. However, the hardness and crispness of whole-grain wheat-flour-enriched products remain unaffected after the addition of jabuticaba peel powder (Oliveira et al., 2018). Oliveira et al. (2017) optimized the conditions for extruded cereal products with up to 80% of whole-grain wheat flour, 90–110°C extrusion temperature, and up to 14% moisture content by using the Response Surface Methodology tool. Analysis indicates the darkness of products once whole-grain wheat flour was incorporated. However, a better color appearance is produced when high whole-grain wheat flour is combined with greater temperature and lower moisture levels as well as extrudates with a desired low hardness and crispiness can be produced at high whole-grain wheat flour and corn flour ratios as long as the moisture content is under 22%. Whole-grain wheat flour ratio in manufacturing, meanwhile, had no effect on the breakfast cereal's textural characteristics after being soaked in milk.

The whole-wheat flour has been utilized in the development of extruded snacks with value addition from lycopene. The addition of crude lycopene and tomato powder causes an increase in hardness but a considerable reduction in pasting characteristics whereas phenolic content is increased significantly with tomato powder (Bhat et al., 2019). Bobade et al. (2022) also found that the extrusion cooking enhances the antioxidant activity of whole-wheat flour-based honey-added extruded snacks. The desirable-quality extruded snacks from whole-wheat flour are obtainable at the optimized processing conditions comprising 16.50% feed moisture, 151.33°C temperature, and 12.83% honey. The study revealed that the extrusion temperature had a negative impact, while the addition of honey enhanced the total phenolic content and antioxidant activity of extruded snacks. The impact of extrusion process conditions and whole-grain wheat flour addition on the nutritional and technical quality of ready-to-eat cereals was assessed, and it was found that the addition of whole-grain wheat flour resulted in increasing the level of dietary fiber and resistant starch in the final product while it reduced the content of digestible starch. All the extrusion parameters and whole-grain wheat flour content had a substantial impact on the pasting parameters of the prepared product (Oliveira et al., 2015).

The wheat germ can also be added to increase the nutritional value of extruded products. The wheat germ has been utilized to boost the nutritional formulation in the preparation of extruded soryz mix. The results indicate that the structure of the extrudate was significantly impacted in a negative way due to the drop in the expansion coefficient of extruded items by 10%. Other parameters like specific bulk density were increased up to 15%, but the wetting coefficient declined (Iusan, 2022). Fadel et al. (2008) prepared a coffee substitute by extruding the wheat germ with chicory roots. The coffee was then analyzed for its flavor stability and quality parameters. In contrast to actual coffee, this coffee substitute had greater scores for the sweetish/caramel-like note, according to the comparative smell profile analysis, while the other odor quality criteria exhibited a different pattern. The storage analysis of coffee substitutes revealed a striking rise in phenolic compounds and a notable drop in the concentration of the Strecker aldehydes and diketones while pyrazine and furan derivatives did not undergo any significant changes.

13.4.2.3 Pasta

Consuming whole-grain foods, including pasta, has been linked to lower risks of cancer, type II diabetes, and cardiovascular disease; therefore, health-conscious customers demand healthy and health-promoting foods (Vignola et al., 2018). Though pasta is preferentially prepared from the coarsely ground endosperm of durum wheat, whole-wheat flour can also be used to increase the nutritional value of the pasta. However, use of whole wheat and its components may affect the other quality parameters of pasta. Vignola et al. (2018) compared the quality characteristics of refined and whole-wheat flour extruded pasta and found that the whole-grain pasta had shorter cooking times due to gluten matrix instabilities caused by the presence of bran-germ particles in whole-wheat flour. Moreover, the whole-wheat flour pasta had harder texture as compared to refined flour pasta. Whole-grain pasta exhibited a better nutritional profile in terms of increased protein and antioxidant levels as well as other beneficial elements, such as fiber. However, the whole-grain pasta failed to exhibit the same technical excellence as that in refined flour. Kalnina et al. (2015) prepared the pasta using whole wheat and whole triticale flour and analyzed for its rheological characteristics. The results revealed that the addition of whole-grain flour deteriorated the rheological qualities but improved the water absorption characteristics of pasta. Features of pasta made with whole-wheat flour and naturally colored concentrates were examined in terms of technical parameters and antioxidant activity. The results indicated that there is considerable feasibility to produce whole wheat pasta with varied colors and functional bioactive ingredients while keeping the pasta's technical and antioxidant capabilities intact (Wahanik et al., 2021). The antioxidant activity and digestibility properties of cooked pasta developed from whole-wheat flour were studied by Podio et al. (2019). The result of the study found that the polyphenol profile and antioxidant capacity alter as cooked pasta is developed, with the cooking stage being crucial to boost the release of bound polyphenols and improve their antioxidant characteristics.

13.4.3 Dairy Products

The dairy products constitute one of the major categories of food products and cater to the need of diverse group of population. The dairy products are major contributors to the human diet and serve as the source of many nutrients. Though the dairy products are rich sources of several nutrients; however, they lack some phytochemicals like phenolics and flavonoids. The dairy products could be further enriched with these phytochemicals by incorporating various sources. The wheat bran, after its delignification, can be used as a carrier for cell immobilization. Such cells immobilized by wheat bran have been exploited for the production of yogurt. The findings demonstrated that the wheat bran has the capability to promote the viability of microorganisms in yogurt and therefore imparts improved health benefits (Terpou et al., 2017). The wheat bran can be utilized in the preparation of *Gulabjamun*, a traditional Indian sweet prepared by deep-frying khoa dough balls in ghee or vegetable oil followed by dipping in sugar syrup (Rangi et al., 1985). The addition of wheat bran to

Gulabjamun significantly increased the moisture, ash, carbohydrate, and fiber contents, while it reduced the protein and fat content of *Gulabjamun*. An examination of the product's texture profile showed that the addition of wheat bran to *Gulabjamun* gradually reduced its hardness. The *Gulabjamun* prepared with the addition of 20% wheat bran resulted in lowering the ratings for cohesion, adhesiveness, gumminess, and chewiness (Ghube et al., 2015). The wheat bran is reported to protect the beneficial microbial culture used in fermented milk products from the highly acidic environment created as a result of longer fermentation processes. Moreover, the wheat bran usage in fermented milk products has not shown a positive effect on the harmful microorganisms (Terpou et al., 2017). This implied that wheat bran has a selective ability to protect only desirable microorganisms.

The germ portion of wheat can be utilized for enhancing the quality of several dairy products. Majzoobi et al. (2016) have mentioned that the wheat germ can be used in dairy desserts to increase their nutritional value and antioxidant activity. The quality of such desserts is, however, influenced by the amount of wheat germ and its particle size. Therefore, for maintaining the product's quality, it is vital to regulate the particle size and amount of wheat germ. The amount of wheat germ had a significant impact on the physicochemical characteristics of the sweets, whereas the particle size of wheat germ has a moderate effect on the quality of dairy dessert. The increased amount of wheat germ dessert resulted in stiffer, less elastic, denser, and dark-colored dairy dessert. The dairy dessert prepared with the addition of 5% wheat germ was found to be most delectable. Majzoobi et al. (2022) also mentioned that the amount and particle size of wheat germ influence the quality of milk pudding prepared with the addition of wheat germ which can acts as cheap source of plant-based functional component for improving the nutritional quality of the prepared product. The quality of the prepared pudding was more influenced by the amount of added wheat germ than its particle size. The wheat-germ-added puddings have more phenolics and flavonoids than the puddings prepared without any addition of wheat germ. Mohamed et al. (2015) prepared the *labneh* cheese by the addition of wheat germ extract and found that the wheat germ extract addition increased the dry matter content and hardness of the cheese, as well as had an impact on the product's flavour and taste. Çetinkaya and Öz (2020) also found that the cheese made by adding wheat germ increases the solid content of the cheese. The wheat germ addition resulted in balanced amino acid distribution, presence of essential fatty acids, and high protein content in the cheese, and, therefore, it can be stated that the cheese manufacturing can benefit from using wheat germ as an auxiliary raw material to improve nutritional and functional qualities.

The wheat germ has been used to replace the skim milk powder in preparation of goat milk yogurt and camel milk-based beverage. The results indicated a significant decrease in protein, total carbs, and pH levels when skim milk powder was replaced with wheat germ. The replacement of skim milk powder by wheat germ significantly reduced the syneresis in yogurt prepared using goat milk and increased the total bacterial count, *Str. thermophilus*, and *Lb. bulgaricus* counts. The goat's milk yogurt was found to have greater microbial counts than the fermented camel's milk beverage (Seham et al., 2007). The wheat germ powder has shown potential as a substitute to the skim milk powder in the preparation of yogurt-like products. The replacement of skim milk powder by wheat germ powder increased the total ash, fat, and carbohydrate content of the products. It also considerably increased the curd tension of yogurt-like product which was found sensorily acceptable and comparable to the control product without much variations in flavor attributes (Mehanna et al., 2020). The wheat germ has been also used as fat replacer in reduced-fat ice cream having higher nutritional value, more fiber content, and greater antioxidant activity compared to control ice cream (without added wheat germ) (Salem et al., 2016). Soliman et al. (2019) also found that by substituting encapsulated wheat germ oil for 50% of the milk fat, it is possible to create functional *labneh* with strong antioxidant activity and acceptable quality. The encapsulated wheat germ oil improved the oxidative stability and significantly maintained the DPPH radical-scavenging activity of the product. Other quality characteristics and acceptability of the *labneh* prepared with encapsulated wheat germ oil were well comparable to the control. Further, it has been reported that the wheat germ oil has the ability to prevent the oxidation in fluid milk, frozen cream, and powdered whole milk. This oxidation preventive effect has been observed at 0.2% concentration of the wheat germ oil. Even though wheat germ oil is comparatively less efficient than nitrogen gas packing at halting the formation of the oxidized taste in powdered milk, however, the use of wheat germ oil accompanied with nitrogen gas packing increases the shelf life of the whole milk powder more than using either of these (Tracy et al., 1944). The efficacy of wheat germ, due to its content of tocopherols and high antioxidant activity, was used in powdered whole milk for preventing the oxidation. It was observed that the wheat germ oil significantly contributed to the antioxidant activity, which was thought to be attributed to the presence of tocopherols and phosphatides in wheat germ oil (Corbett and Tracy, 1939).

Wheat germ is reported to improve the acid-producing ability of the thermophilic starter in the fermented milk products and can be used up to 20% level in such products. The addition of wheat germ, which is rich source of fibers, vitamin E, and other phytochemicals, in fermented milk products increases the probiotic efficiency as well as the nutritional value of these products (Seleet et al., 2016).

13.4.4 Meat Products

The majority of dietary proteins with a high biological value are derived from the meat. Additionally, meat is a great source of various essential elements including folic acid, iron, and vitamin B12, which either are absent or have a lower bioavailability in other foods. But the consumption of meat and its products in improper amounts might be harmful to human health (Bilek and Turhan, 2009). The wheat and its components could be utilized in variety of ways to improve the quality characteristics of meat and meat products. The fiber-enriched high quality functional nuggets can be prepared by the addition of 8% pre-hydrated wheat bran. The addition of wheat bran results in considerable improvement in fat and moisture retention and few textural properties like hardness, gumminess and chewiness, and color properties (Rindhe et al., 2018). The wheat bran (up to 15% level) can be utilized in the preparation of fiber-rich chicken meat patties, which show an increased efficiency of cooking, dietary fiber content, unsaturated fatty acids, and texture with reduction in the cholesterol level. However, it also results in the reduction of sensory characteristics of the developed patties (Talukder and Sharma, 2010). The use of wheat bran at a level of 5–15% increases the quality characteristics like fat retention, cooking properties, water-holding power as well as moisture retention capacity in meat patties (Tekin et al., 2010). The wheat bran fibers can also be used in the development of reduced calorie meat product. For example, the beef burgers can be prepared by the incorporation of 3.75 g hydrated wheat bran per 80 g of burger having same sensory appeal, but significantly reduced calorie, with the control burger. The hydrated wheat fiber lowers the calorie intake as higher amount of wheat bran can reduce the content of protein and lipid in burgers (Carvalho et al., 2019). Sarıçoban et al. (2009) also mentioned that the use of wheat bran in beef patties reduces the level of proteins and fats and improves the textural properties when used up to certain level.

Wheat bran up to the level of 2% can be used in the preparation of chicken sausages with excellent sensory acceptance. The results, however, indicated that an increment in hardness and shearing force takes place, while cohesiveness and springiness dropped when increasing the fiber content. Additionally, a consistent rise in gumminess and chewiness has been also observed (Yadav et al., 2020). The same experiment was conducted by Yadav et al. (2018) in which the influence of the addition of wheat bran and dried carrot pomace in chicken sausages has been assessed for its different characteristics. The addition of wheat bran along with dried carrot pomace up to 9% was enough to increase the fiber content with moderate acceptance. While adding the wheat bran up to 6%, the chicken sausages can be prepared with better acceptability, greater dietary fiber content as well as a shelf life of around 15 days. Fomenko and Ptichkina (2010) also claimed that the use of wheat bran reduces the bacterial count and doubles the shelf life and improves the digestibility of a minced chicken meat products prepared using wheat bran. The addition of wheat fiber in Chinese-style sausages harden the texture of sausages due to the high amount of dietary fiber. But sausages enriched with 7% wheat fiber scored less six points in terms of overall acceptability than sausages enriched with 3.5% of wheat fiber, which were preferred more (Huang et al., 2011). An experiment was conducted by Choe et al. (2013) for wheat fiber along with pig skin as a fat substitute in frankfurter-type sausages. The findings revealed that the mixture of both wheat fiber and pig skin around 20% can be a good fat replacer for the preparation of frankfurter-type sausages as it can lower the fat by up to 50%, calories up to 32%, less cooking losses, more stability in meat emulsion, and higher moisture and protein content. The wheat bran can also function as a fat replacer in low-fat meatballs. The meatballs prepared with addition of 20% ground wheat bran show increased fiber and ash content but decreased redness value (Yılmaz, 2005).

The wheat germ also has potential applications in the meat and meat products. The beef sausages made with wheat germ flour have been found to have acceptable sensory properties. The beef sausages prepared with 15% replacement level were found to have highest sensory acceptability. The increased level of wheat germ flour also increased the nutritional value of beef sausages with respect to protein, fat, fiber, and ash content. Moreover, wheat germ flour functions as a binder in the manufacturing of beef sausage and is a viable alternative to other plant-based binders that are employed as extenders or meat binders. It is reported that the use of 15% wheat germ flour enhance the quality of comminuted meat products (Elbakheet et al., 2017). Abd EL-Rahman (2015) mentioned the usage of wheat germ as a fat substitute and as a natural antioxidant since it is a most promising source of vitamins, antioxidants, minerals, fiber, and proteins. The wheat germ has also been found to be capable of increasing the shelf life of beef sausages as well as to lower the cholesterol level in rats. The results have shown that increasing the quantities of wheat germ in sausage formulas enhanced the water-holding capacity and cooking loss while decreasing stiffness and cooking output. The results are, however, contrasting after the storage of three months. It has been observed that replacing the sheep fat with 25, 50, and 75% wheat germ can improve the sensory characteristics of a product. Additionally, the presence of total dietary fiber in wheat germ can affect the textural properties of the product. The findings suggested that the replacement of sheep fat with 75% of wheat germ is best suited to decrease the cholesterol level in rats. The use of wheat germ flour in beef patty increases its cooking yield, moisture retention, freshness, softness, pH as well as the color. The addition of wheat germ flour, however, could also negatively impact the cooking loss (Rocha-Garza and Zayas, 1995). The replacement of beef flesh with wheat germ extrudate considerably influences the physical and chemical properties of beef burger (Kadous et al., 2014). The wheat germ extrudate up to the concentration of 10% can be effectively utilized in the preparation of a beef burger. It leads to increase the carbohydrate content; fiber content; and Mg, K, Mn, Zn, and P content of beef burger. Wheat germ protein flour is known to affect the quality characteristics of frankfurters and meat batters. The supplementation of 3.5% wheat germ flour is best suited for this application. In meat batters, the addition of wheat germ protein flour results in the reduction of cooking losses and fat content and increase in processing yields, pH, and adhesiveness. Whereas the viscosity of meat batter remains unaffected by the addition of wheat germ protein flour. Shear force and hardness levels were lower in frankfurters having wheat germ protein flour (Gnanasambandam and Zayas 1994a; Gnanasambandam and Zayas 1994b).

Wheat bran fiber and wheat germ oil can also be used as fat substitutes in low-fat beef patties (Khalid et al., 2021). The use of wheat bran fiber up to 3% level and wheat germ oil up to 4.5% level result in the reduction of TBARS, peroxide, and cholesterol content of low-fat beef patties. However, it leads to reduced sensory acceptance compared to control beef patties despite improving the texture and firmness of the prepared patties.

13.4.5 Beverages

The beverages have a long history and are well-known for their thirst-quenching properties besides delicious flavor and mouthfeel. Several beverages also provide health benefits. The market for practical, natural, and non-alcoholic drinks is continuously expanding globally because of the appealing sensory characteristics and growing consumer awareness of the significance of good nutrition (Ignat et al., 2020). Recently, the demand for non-dairy-based functional beverage alternatives with high acceptability and functionality has increased. The fermented and nonfermented cereal-based beverages offer a great deal of potential to close this gap in the market by serving possible delivery systems for nutritional components such as vitamins, minerals, probiotics, and antioxidants (Waters et al., 2015).

The whole-wheat flour could be utilized in the preparation of the classic Bulgarian drink boza which is characterized by high amount of glucose, low pH, viscosity, the amount of free amino nitrogen, and dry matter. The whole-wheat flour-based boza had viscosity and dry matter content lower than the beverage prepared from the refined wheat flour (Gotcheva et al., 2001). The traditional cereal-based Romanian beverage *borş*, which is made from the fermentation of wheat bran and maize flour, has a huge potential to satisfy the growing desire of consumers for nondairy milk alternatives with high acceptability and functionality, beside its pleasant flavor and antioxidant activity. This product could be also be valued outside of the region of production on the international market with stricter control over processing characteristics (Pasqualone et al., 2018). The nutritional and microbial properties of the fermented wheat bran used to make traditional *borş* beverage in Romania were also examined by Grosu-Tudor et al. (2014). The amounts of readily available amino acids and phenolic compounds increased as a result of the fermentation of wheat bran used to make borş. Additionally, some strains of the *lactobacilli* used in the fermentation of borş might inhibit the growth of other microbes, making this beverage more healthful and microbially safe. The sprouted wheat flour could be exploited in the development of a nutritious beverage using probiotic bacteria and yeast. The addition of sprouted wheat flour resulted in an increased acidity of the prepared beverage. The probiotic beverage containing 10 g sprouted wheat flour had the highest rate of survival of culture (Masoomi et al., 2019).

The wheat bran can be used in the preparation of sour milk drink. The introduction of 2% wheat bran increased the total amino acids and essential amino acid content of the sour milk drink by 15.08% and 10.57%, respectively. This could, therefore, help to improve the nutritional content and the biological value of the proteins in sour milk drink (Nagovska et al., 2018).

The wheat germ powder can be included in the preparation of kefir. The kefir prepared with the addition of 3% germ powder exhibited highest acidity, ethanol, yeast, and total microbial count. The rheological behavior of the samples considerably decreased as wheat germ content increased. The kefir made with up to 1% wheat germ and evaluated within 72 hours of inoculation has scored the highest overall acceptability (Ahmadian Mask et al., 2019). The wheat germ can also be used to prepare the γ-aminobutyric acid-rich beverage. The beverage with desirable quality parameters can be prepared by incorporating 34% wheat germ serum, 54% soy bean milk, 12% syrup, and 0.1% citric acid solution. The wheat-germ-enriched beverage was found pleasant and had milky-white color. The beverage, after sensory analysis, scored 3.82 points out of 5 under ideal circumstances (Chai et al., 2012).

13.4.6 Miscellaneous

Besides the products described before, the whole wheat and its parts have been in use in many diverse food products. The whole-wheat flour has been used in the preparation of nutritional tortilla, and it was found that the tortilla prepared using fine-particle whole-wheat flour (130 μm) has better physical properties than that prepared using coarse particle whole-wheat flour (175 μm). However, the finer particle size of whole-wheat flour decreased the starch retrogradation and stability time (Liu et al., 2016). Niu et al. (2014) also found that the use of fine-particle-size whole-wheat flour enhanced the quality of noodles. The cooking yield of noodles increased significantly on lowering the particle size of whole-wheat flour from 125 μm to 43 μm. The cooking loss was lowest in the noodles prepared from whole-wheat flour with 72 μm particle size. The hardness, cohesiveness, and resilience values of cooked whole-wheat noodles also increased by lowering the particle size of whole-wheat flour.

The whole-wheat flour can be blended with other flours for improving the nutritional properties of the prepared products. Liu et al. (2017) prepared the tortilla from the blend of whole-wheat flour and sprouted whole-wheat flour. The tortillas produced with more sprouted whole-wheat flour have larger diameter and volume, are opaquer and whiter, stronger but more shelf-stable, and have less firmness. It has been implied that the use of sprouted whole-wheat flour might improve tortilla's baking performance, including better appearance, greater customer acceptance, and keeping quality. Doughs and bread prepared from the combination of whole-wheat flour with 5, 10, and 15% of germinated whole-wheat flour exhibited enhanced mixing capabilities, loaf volume, and firmness. When compared to flour from ungerminated wheat, whole-grain flour from germinated wheat had lower levels of phytic acid, thiamine, and dough strength, but higher levels of lysine, asparagine, γ-aminobutyric acid, lipase, esterase, and lipoxygenase activities (Poudel et al., 2019).

The wheat bran can be utilized in the preparation of fried products. The research findings have indicated the use of wheat bran in fried cereal-based foods resulting in significant reduction (2.7–9.4%) in the oil uptake of such products (Kim et al., 2012). The wheat bran, due to its water insolubility, has been shown to have negative influence on the moisture content of Indian deep-fried dough and poori (Yadav and Rajan, 2012). The use of defatted wheat germ has been explored in the preparation of chewing gum. The findings revealed that adding defatted wheat germ had no negative effects on the texture of the chewing gum (Özdoğan et al., 2019). Ge et al. (2001) have also used the defatted wheat germ in the preparation of nutritive noodles. The findings of the research recommended the use of 15% defatted wheat germ for the preparation of noodles with desirable physicochemical and excellent nutritional qualities.

13.5 Conclusion

Wheat is one of the principal crops, and wheat flour constitutes the chief ingredient in the diet of many people. The structural parts of wheat grains include endosperm, bran, and germ. The endosperm mainly contains starch and proteins and little of phytochemicals, whereas the other fractions, bran and germ, are rich in vitamins, minerals, and other bioactive compounds. The refined wheat flour obtained by milling of wheat endosperm and discarding the bran and germ fractions is usually used in the development of the products. However, such products are affluent in starch and carbohydrates and deficient in the nutrients contained in bran and germ fraction of the wheat. Therefore, it is recommended to use the whole-wheat flour containing all the parts of wheat in product development. Alternatively, the functionality of the products can also be increased with the supplementation of bran and germ fractions. The products containing whole-wheat flour, bran, or germ offer many health benefits owing to their content of bioactive compounds with demonstrated biological activities. However, the preparation of such functional products from whole-wheat flour or bran and germ fractions could adversely affect the physical and sensory properties of the developed products. This necessitates precise optimization of the processing methods used in the product development, appropriate pretreatment/conditioning of the ingredients, or use of suitable additives to negate the effects of fibrous components present in whole-wheat flour, bran, and germ fraction.

REFERENCES

Abd EL-Rahman, A. M. (2015) Utilization of Wheat germ as natural antioxidant and fat mimetic to increase shelf-life in beef sausage and as lowering cholesterol in rats. Middle East Journal, 4(3), 555–563.

Acosta, K., Cavender, G., & Kerr, W. L. (2011). Sensory and physical properties of muffins made with waxy whole wheat flour. Journal of Food Quality, 34(5), 343–351.

Adom, K. K., Sorrells, M. E., & Liu, R. H. (2003). Phytochemical profiles and antioxidant activity of wheat varieties. Journal of Agricultural and Food Chemistry, 51(26), 7825–7834.

Ahmadian Mask, S., Tabatabaei Yazdi, F., Mortazavi, S. A., & Koocheki, A. (2019). The effect of wheat germ powder on physicochemical, microbial and sensory properties of kefir beverage. Iranian Food Science and Technology Research Journal, 15(4), 395–406.

Aivaz, M., & Mosharraf, L. (2013). Influence of different treatments and particle size of wheat bran on its mineral and physicochemical characteristics. International Journal of Agricultural Sciences, 3, 608–619.

Al-Marazeeq, K. M., & Angor, M. M. (2017). Chemical characteristic and sensory evaluation of biscuit enriched with wheat germ and the effect of storage time on the sensory properties for this product. Food and Nutrition Sciences, 8(2), 189–195.

American Diabetes Association. (2010). Diagnosis and classification of diabetes mellitus. Diabetes Care, 33(Suppl_1), S62-S69.

Anderssen, R. S., & Haraszi, R. (2009). Characterizing and exploiting the rheology of wheat hardness. European Food Research and Technology, 229(1), 159–174.

Andersson, A. A. M., Dimberg, L., Aman, P., & Landberg, D. (2014). Recent findings on certain bioactive components in whole grain wheat and rye. Journal of Cereal Science, 59, 294–311.

Apprich, S., Tirpanalan, Ö., Hell, J., Reisinger, M., Böhmdorfer, S., Siebenhandl-Ehn, S., Novalin, S., & Kneifel, W. (2014). Wheat bran-based biorefinery 2: Valorization of products. LWT – Food Science and Technology, 56(2), 222–231.

Arvanitakis, C., Stamnes, C. L., Folscroft, J., & Beyer, P. (1977). Failure of bran to alter diet-induced hyperlipidemia in the rat. Proceedings of the Society for Experimental Biology and Medicine, 154(4), 550–552.

Astrup, A., Dyerberg, J., Elwood, P., Hermansen, K., Hu, F. B., Jakobsen, M. U., et al. (2011). The role of reducing intakes of saturated fat in the prevention of cardiovascular disease: where does the evidence stand in 2010? The American Journal of Clinical Nutrition, 93(4), 684–688.

Barron, C., Surget, A., & Rouau, X. (2007). Relative amounts of tissues in mature wheat (*Triticum aestivum* L.) grain and their carbohydrate and phenolic acid composition. Journal of Cereal Science, 45(1), 88–96.

Belderok, B., Mesdag, J., Mesdag, H., & Donner, D. A. (2000). Bread-making quality of wheat: a century of breeding in Europe. Springer Science & Business Media.

Bhat, N. A., Wani, I. A., Hamdani, A. M., & Gani, A. (2019). Effect of extrusion on the physicochemical and antioxidant properties of value added snacks from whole wheat (*Triticum aestivum* L.) flour. Food Chemistry, 276, 22–32.

Bilek, A. E., & Turhan, S. (2009). Enhancement of the nutritional status of beef patties by adding flaxseed flour. Meat Sciences, 82(4), 472–477.

Biskup, I., Kyrø, C., Marklund, M., Olsen, A., van Dam, R. M., Tjønneland, A., Overvad, K., Lindahl, B., Johansson, I., & Landberg, R. (2016). Plasma alkylresorcinols, biomarkers of whole-grain wheat and rye intake, and risk of type 2 diabetes in Scandinavian men and women. The American Journal of Clinical Nutrition, 104(1), 88–96.

Bobade, H., Singh, A., Sharma, S., Gupta, A., & Singh, B. (2022). Effect of extrusion conditions and honey on functionality and bioactive composition of whole wheat flour-based expanded snacks. Journal of Food Processing and Preservation, 46(1), e16132.

Brandolini, A., & Hidalgo, A. (2012). Wheat germ: not only a by-product. International Journal of Food Science and Nutrition, 63(Supp. 1), 71–74.

Brennan, M. A., Merts, I., Monro, J., Woolnough, J., & Brennan, C. S. (2008). Impact of guar and wheat bran on the physical and nutritional quality of extruded breakfast cereals. Starch-Stärke, 60(5), 248–256.

Brewbaker, J. L., Zan, G. H., Larish, H. B., (1996). Pericarp thickness of the indigenous American races of maize. Maydica. 41(2), 105–111.

Brouns, F., Hemery, Y., Price, R., & Anson, N. M. (2012). Wheat aleurone: separation, composition, health aspects, and potential food use. Critical Reviews in Food Sciences and Nutrition, 52, 553–568.

Cara, L., Armand, M., Borel, P., Senft, M., Portugal, H., Pauli, A. M., Lafont, H., & Lairon, D. (1992). Long-term wheat germ intake beneficially affects plasma lipids and lipoproteins in hypercholesterolemic human subjects. Journal of Nutrition, 122(2), 317–326.

Carvalho, L. T., Pires, M. A., Baldin, J. C., Munekata, P. E. S., de Carvalho, F. A. L., Rodrigues, I., Polizer, Y. J., Maloagoli de Mello, J. L., Lapa-Guimarães, J., & Trindade, M. A. (2019). Partial replacement of meat and fat with hydrated wheat fiber in beef burgers decreases caloric value without reducing the feeling of satiety after consumption. Meat Science, 147, 53–59.

Çetinkaya, A., & Öz, F. (2020). The effect of wheat germ on the chemical properties and fatty acids of white cheese during the storage time. Food Science & Nutrition, 8(2), 915–920.

Chai, M., Xu, Q., Yang, T., & Bai, Q. (2012). Preparation of enriched γ-aminobutyric acid wheat germ beverage. Food Research and Development, 33(11), 126–129.

Chassagne-Berces, S., Leitner, M., Melado, A., Barreiro, P., Correa, E. C., Blank, I., Gumy, J., & Chanvrier, H. (2011). Effect of fibers and whole grain content on quality attributes of extruded cereals. Procedia Food Science, 1, 17–23.

Chen, M. F. (2013). Influences of health consciousness on consumers' modern health worries and willingness to use functional foods. Journal of Applied Social Psychology, 43, E1–E12.

Chen, W. J. L., & Anderson, J. W. (1979). Effects of guar gum and wheat bran on lipid metabolism of rats. Journal of Nutrition, 109(6), 1028–1034.

Chen, Y., Ross, A. B., Åman, P., & Kamal-Eldin, A. (2004). Alkylresorcinols as markers of whole grain wheat and rye in cereal products. Journal of Agricultural and Food Chemistry, 52(26), 8242–8246.

Cho, S., & Pratt, C. J. (2008). Active components of whole grain foods. In: L. Marquart, D. R. Jacobs and G. H. McIntosh (Eds.) Whole grains and health. Blackwell Publishing, pp. 137–149.

Choe, J. H., Kim, H. Y., Lee, J. M., Kim, Y. J., & Kim, C. J. (2013). Quality of frankfurter-type sausages with added pig skin and wheat fiber mixture as fat replacers. Meat Sciences, 93(4), 849–854.

Coda, R., Kärki, I., Nordlund, E., Heiniö, R. L., Poutanen, K., & Katina, K. (2014). Influence of particle size on bioprocess induced changes on technological functionality of wheat bran. Food Microbiology, 37, 69–77.

Corbett, W. J., & Tracy, P. H. (1939). Dextrose in commercial ice-cream manufacture. Illinois Agricultural Experiment Station Bulletin, 452.

Curti, E., Carini, E., Bonacini, G., Tribuzio, G., & Vittadini, E. (2013). Effect of the addition of bran fractions on bread properties. Journal of Cereal Science, 57(3), 325–332.

Danaei, G., Finucane, M. M., Lu, Y., Singh, G. M., Cowan, M. J., Paciorek, C. J., et al. (2011). National, regional, and global trends in fasting plasma glucose and diabetes prevalence since 1980: systematic analysis of health examination surveys and epidemiological studies with 370 country-years and 2·7 million participants. The Lancet, 378(9785), 31–40.

Davison, K. M., & Temple, N. J. (2018). Cereal fiber, fruit fiber, and type 2 diabetes: Explaining the paradox. Journal of Diabetes and Complications, 32(2): 240–245.

De Brier, N., Gomand, S. V., Joye, I. J., Pareyt, B., Courtin, C. M., & Delcour, J. A. (2014). The impact of pearling as a treatment prior to wheat roller milling on the texture and structure of bran-rich breakfast flakes. LWT-Food Science and Technology, 62, 668–674.

Delcour, J. A., Joye, I. J., Pareyt, B., Wilderjans, E., Brijs, K., & Lagrain, B. (2012). Wheat gluten functionality as a quality determinant in cereal-based food products. Annual Review of Food Science and Technology, 3(1), 469–492.

Dunford, N. T. (2009). Wheat germ oil. In Gourmet and health-promoting specialty oils (pp. 359–376). AOCS Press.

Elagizi, A., Kachur, S., Lavie, C. J., Carbone, S., Pandey, A., Ortega, F. B., & Milani, R. V. (2018). An overview and update on obesity and the obesity paradox in cardiovascular diseases. Progress in Cardiovascular Diseases, 61(2), 142–150.

Elbakheet, I. S., Elgasim, A. E., & Algadi, M. Z. (2017). Proximate composition of beef sausage processed by wheat germ flour. Journal of Food Processing & Technology, 8(11), 1000704.

Evers, A. D., & Bechtel, D. B. (1988). Microscopic structure of the wheat grain. In: Y. Pomeranz (Ed.) Wheat: Chemistry and technology. American Association of Cereal Chemists, pp. 47–95.

Fadel, H. H. M., Mageed, A., & Lotfy, S. N. (2008). Quality and flavour stability of coffee substitute prepared by extrusion of wheat germ and chicory roots. Amino Acids, 34(2), 307–314.

Fardet, A. (2010). New hypotheses for the health-protective mechanisms of whole-grain cereals: what is beyond fiber? Nutrition Research Review, 23(1), 65–134.

Fardet, A., Rock, E., & Rémésy, C. (2008). Is the in vitro antioxidant potential of whole-grain cereals and cereal products well reflected in vivo? Journal of Cereal Science, 48(2), 258–276.

Fomenko, O. S., & Ptichkina, M. N. (2010). Development of the technology of minced chicken meat products with wheat bran. Meat Industry Magazine, 10, 10–12.

Fulcher, R. G., O'brien, T. P., & Lee, J. W. (1972). Studies on the aleurone layer I. Conventional and fluorescence microscopy of the cell wall with emphasis on phenol-carbohydrate complexes in wheat. Australian Journal of Biological Sciences, 25(1), 23–34.

Galanakis, C. M. (2018). Concluding remarks and future perspectives. In Sustainable recovery and reutilization of cereal processing by-products. Woodhead Publishing, pp. 319–327.

Garcia-Conesa, M. T., Plumb, G. W., Waldron, K. W., Ralph, J., & Williamson, G. (1997). Ferulic acid dehydrodimers from wheat bran: isolation, purification and antioxidant properties of 8-0-4-diferulic acid. Redox Report, 3(5–6), 319–323.

García-Estepa, R. M., Guerra-Hernández, E., & García-Villanova, B. (1999). Phytic acid content in milled cereal products and breads. Food Research International, 32(3), 217–221.

Ge, Y., Sun, A., Ni, Y., & Cai, T. (2001). Study and development of a defatted wheat germ nutritive noodle. European Food Research and Technology, 212(3), 344–348.

Gebruers, K., Dornez, E., Boros, D., Dynkowska, W., Bedő, Z., Rakszegi, M., Delcour, J. A., & Courtin, C. M. (2008). Variation in the content of dietary fiber and components thereof in wheats in the HEALTHGRAIN diversity screen. Journal of Agricultural and Food Chemistry, 56(21), 9740–9749.

Ghube, S. D., Bidwe, K. U., Shelke, R. R., & Shegokar, S. R. (2015). Studies on physico-chemical properties of Gulabjamun prepared from cow milk Khoa blended with wheat bran. Research Journal of Animal Husbandry and Dairy Science, 6(2), 99–104.

Giacco, R., Della Pepa, G., Luongo, D., & Riccardi, G. (2011). Whole grain intake in relation to body weight: from epidemiological evidence to clinical trials. Nutrition, Metabolism and Cardiovascular Diseases, 21(12), 901–908.

Gil, A., Ortega, R. M., & Maldonado, J. (2011). Wholegrain cereals and bread: a duet of the Mediterranean diet for the prevention of chronic diseases. Public Health Nutrition, 14(12A), 2316–2322.

Gnanasambandam, R., & Zayas, J. F. (1994a). Microstructure of frankfurters extended with wheat germ proteins. Journal of Food Science, 59(3), 474–477.

Gnanasambandam, R., & Zayas, J. F. (1994b). Quality characteristics of meat batters and frankfurters containing wheat germ protein flour 1. Journal of Food Quality, 17(2), 129–142.

Gómez, M., González, J., & Oliete, B. (2012). Effect of extruded wheat germ on dough rheology and bread quality. Food and Bioprocessing Technology, 5(6), 2409–2418.

Gómez, M., Ronda, F., Blanco, C. A., Caballero, P. A., & Apesteguía, A. (2003). Effect of dietary fibre on dough rheology and bread quality. European Food Research and Technology, 216(1), 51–56.

Gotcheva, V., Pandiella, S. S., Angelov, A., Roshkova, Z., & Webb, C. (2001). Monitoring the fermentation of the traditional Bulgarian beverage boza. International Journal of Food Sciences, 36(2), 129–134.

Grosu-Tudor, S. S., Stancu, M. M., Pelinescu, D., & Zamfir, M. (2014). Characterization of some bacteriocins produced by lactic acid bacteria isolated from fermented foods. World Journal of Microbiology and Biotechnology, 30(9), 2459–2469.

Han, S., Mao, H., & Dally, W. J. (2015). Deep compression: Compressing deep neural networks with pruning, trained quantization and Huffman coding. arXiv preprint arXiv:1510.00149.

Haque, M. A., Shams-Ud-Din, M., & Haque, A. (2002). The effect of aqueous extracted wheat bran on the baking quality of biscuit. International Journal of Food Sciences, 37(4), 453–462.

Haruna, M., Udobi, C. E., & Ndife, J. (2011). Effect of added brewers dry grain on the physico-chemical, microbial and sensory quality of wheat bread. American Journal of Food and Nutrition, 1(1), 39–43.

Heaton, K. W., & Pomare, E. W. (1974). Effect of bran on blood lipids and calcium. The Lancet, 303(7846), 49–50.

Hernandez-Diaz, J. R., Quintero-Ramos, A., Barnard, J., & Balandran-Quintana, R. R. (2007). Functional properties of extrudates prepared with blends of wheat flour/pinto bean meal with added wheat bran. International Journal of Food Sciences, 13(4), 301–308.

Huang, S. C., Tsai, Y. F., & Chen, C. M. (2011). Effects of wheat fiber, oat fiber, and inulin on sensory and physico-chemical properties of Chinese-style sausages. Asian-Australasian Journal of Animal Sciences, 24(6), 875–880.

Ignat, M. V., Salanță, L. C., Pop, O. L., Pop, C. R., Tofană, M., Mudura, E., Coldea, T. E., Borşa, A., & Pasqualone, A. (2020). Current functionality and potential improvements of non-alcoholic fermented cereal beverages. Foods, 9(8), 1031.

Igual, M., García-Segovia, P., & Martínez-Monzó, J. (2020). Effect of *Acheta domesticus* (house cricket) addition on protein content, colour, texture, and extrusion parameters of extruded products. Journal of Food Engineering, 282, 110032.

Inyang, U. E., Daniel, E. A., & Bello, F. A. (2018). Production and quality evaluation of functional biscuits from whole wheat flour supplemented with acha (fonio) and kidney bean flours. Asian Journal of Agricultural and Food Sciences, 6(6), 193–201.

Iusan, L. (2022). Improving the recipes of extruded soryz mixes using wheat germ. Agricultural Sciences, 94, 86–90.

Jenkins, D. J., Hill, M. S., & Cummings, D. J. (1975). Effect of wheat fiber on blood lipids, fecal steroid excretion and serum iron. The American Journal of Clinical Nutrition, 28(12), 1408–1411.

Jones, J. M., García, C. G., & Braun, H. J. (2020). Perspective: Whole and refined grains and health—Evidence supporting "make half your grains whole". Advances in Nutrition, 11(3), 492–506.

Kadous, M. F., Mahgoub, S. A., & Walid, M. S. (2014). The Utilization of Wheat Germ in Burger Preparation. Annals of Agricultural Science, 52(1), 19–25.

Kalnina, S., Rakcejeva, T., Kunkulberga, D., & Galoburda, R. (2015). Rheological properties of whole wheat and whole triticale flour blends for pasta production. Agronomy Research, 13(4): 948–955.

Kamal-Eldin, A., Lærke, H. N., Knudsen, K. E. B., Lampi, A. M., Piironen, V., Adlercreutz, H., Katina, K., Poutanen, K., & Åman, P. (2009). Physical, microscopic and chemical characterisation of industrial rye and wheat brans from the Nordic countries. Food & Nutrition Research, 53(1), 1912.

Kaur, G. J., Rehal, J., Singh, B., Singh, A. K., & Kaur, A. (2015). Development of multigrain breakfast cereal using extrusion technology. Asian Journal of Dairy and Food Research, 34(3), 219–224.

Kevin, K. (1995). Fascinating phytochemicals. Food Processing, 56(4), 79–81.

Khalid, A., Sohaib, M., Nadeem, M. T., Saeed, F., Imran, A., Imran, M., & Arshad, M. S. (2021) Utilization of wheat germ oil and wheat bran fiber as fat replacer for the development of low-fat beef patties. Food Science and Nutrition, 9(3), 1271–81.

Kim, B. K., Chun, Y. G., Cho, A. R., & Park, D. G. (2012). Reduction in fat uptake of doughnut by microparticulated wheat bran. International Journal of Food Science and Nutrition, 63, 987–995.

Korhonen, H. (2002). Technology options for new nutritional concepts. International Journal of Dairy Technology, 55(2), 79–88.

Kumar, G. S., & Krishna, A. G. (2015). Studies on the nutraceuticals composition of wheat derived oils wheat bran oil and wheat germ oil. Journal of Food Science and Technology, 52(2), 1145–1151.

Landberg, R., Kamal-Eldin, A., Salmenkallio-Marttila, M., Rouau, X., & Åman, P. (2008). Localization of alkylresorcinols in wheat, rye and barley kernels. Journal of Cereal Science, 48(2), 401–406.

Laskowski, W., Górska-Warsewicz, H., Rejman, K., Czeczotko, M., & Zwolińska, J. (2019). How important are cereals and cereal products in the average polish diet?. Nutrients, 11(3), 679.

Leelavathi, K., & Haridas Rao, P. (1993). Development of high fibre biscuits using wheat bran. Journal of Food Science and Technology, 30(3), 187–190.

Levent, H., & Bilgiçli, N. (2013). Quality evaluation of wheat germ cake prepared with different emulsifiers. Journal of Food Quality, 36(5), 334–341.

Li, J., Hou, G. G., Chen, Z., Chung, A. L., & Gehring, K. (2014). Studying the effects of whole-wheat flour on the rheological properties and the quality attributes of whole-wheat saltine cracker using SRC, alveograph, rheometer, and NMR technique. LWT – Food Science and Technology, 55(1), 43–50.

Likes, R., Madla, R. L., Zeisel, S. H., & Craig, S. A. S. (2007). The betaine and choline content of a whole wheat flour compared to other mill streams. Journal of Cereal Science, 46, 93–95.

Liu, T., Hou, G. G., Cardin, M., Marquart, L., & Dubat, A. (2017). Quality attributes of whole-wheat flour tortillas with sprouted whole-wheat flour substitution. LWT – Food Science and Technology, 77, 1–7.

Liu, T., Hou, G. G., Lee, B., Marquart, L., & Dubat, A. (2016). Effects of particle size on the quality attributes of reconstituted whole-wheat flour and tortillas made from it. Journal of Cereal Science, 71, 145–152.

Lopez, H. W., Duclos, V., Coudray, C., Krespine, V., Feillet-Coudray, C., Messager, A., Demigné, C., & Rémésy, C. (2003). Making bread with sourdough improves mineral bioavailability from reconstituted whole wheat flour in rats. Nutrition, 19(6), 524–530.

Luitse, M. J., Biessels, G. J., Rutten, G. E., & Kappelle, L. J. (2012). Diabetes, hyperglycaemia, and acute ischaemic stroke. The Lancet Neurology, 11(3), 261–271.

Ma, S., Wang, Z., Guo, X., Wang, F., Huang, J., Sun, B., & Wang, X. (2021). Sourdough improves the quality of whole-wheat flour products: Mechanisms and challenges-A review. Food Chemistry, 360, 130038.

Mahmoud, N. N., Carothers, A. M., Grunberger, D., Bilinski, R. T., Churchill, M. R., Martucci, C., Newmark, H. L., & Bertagnolli, M. M. (2000). Plant phenolics decrease intestinal tumors in an animal model of familial adenomatous polyposis. Carcinogenesis, 21(5), 921–927.

Majzoobi, M., Ghiasi, F., Eskandari, M. H., & Farahnaky, A. (2022). Roasted wheat germ: a natural plant product in development of nutritious milk pudding; physicochemical and nutritional properties. Foods, 11(12), 1815.

Majzoobi, M., Ghiasi, F., & Farahnaky, A. (2016). Physicochemical assessment of fresh chilled dairy dessert supplemented with wheat germ. International Journal of Food Science & Technology, 51(1), 78–86.

Malinow, M. R., McLaughlin, P., Papworth, L., Naito, H. K., & Lewis, L. A. (1976). Effect of bran and cholestyramine on plasma lipids in monkeys. American Journal of Clinical Nutrition, 29(8), 905–911.

Marion, D., Dubreil, L., & Douliez, J. P. (2003). Functionality of lipids and lipid-protein interactions in cereal-derived food products. Oilseeds Fats Crops Lipids 10(1), 47–56.

Masoomi, M., Sharifan, A., & Yousefi, S. (2019). Formulation of functional beverage with content of folic acid based on wheat sprout flour. Iranian Journal of Food Science And Technology, 16(88), 37–45.

Mathe, D., Lutton, C., Rautureau, J., Coste, T., Gouffier, E., Sulpice, J. C., & Chevallier, F. (1977). Effects of dietary fiber and salt mixtures on the cholesterol metabolism of rats. Journal of Nutrition, 107(3), 466–474.

Mehanna, N., Swelam, S., Almqbil, N., Allah, W. F., & Hafez, Y. (2020). Improvement of the dairy products by wheat germ powder. Fresenius Environmental Bulletin, 29(29), 10954–10959.

Mellen, P. B., Walsh, T. F., & Herrington, D. M. (2008). Whole grain intake and cardiovascular disease: a meta-analysis Nutrition Metabolism and Cardiovascular Diseases, 18(4), 283–290.

Meziani, S., Nadaud, I., Tasleem-Tahir, A., Nurit, E., Benguella, R., & Branlard, G. (2021). Wheat aleurone layer: A site enriched with nutrients and bioactive molecules with potential nutritional opportunities for breeding. Journal of Cereal Science, 100, 103225.

Mohamed, S. H., Seleet, F. L., Bayoumi, A. A., Fathy, F. A. (2015). Effect of wheat germ extract on the viability of probiotic bacteria and properties of Labneh cheese. Research Journal of Pharmaceutical, Biological and Chemical Sciences, 6(4), 674–682.

Moussa, S. A. (2008). Oxidative stress in diabetes mellitus. Romanian Journal of Biophysics 18, 225–236.

Muhammad, M. J., Sana, Z., Sarah, S., Iffat, M., Ambreen, G., Huma, R., Syed, A. I. B., Muhammad, N. A., & Haq, I. (2012). Wheat bran as a brown gold: Nutritious value and its biotechnological applications. African Journal of Microbiology Research, 6(4), 724–733.

Nagovska, V., Hachak, Y., Gutyj, B., Bilyk, O., & Slyvka, N. (2018). Influence of wheat bran on quality indicators of a sour milk beverage. Eastern-European Journal of Enterprise Technologies, 4(11), 28–35.

Nandeesh, K., Jyotsna, R., & Venkateswara Rao, G. (2011). Effect of differently treated wheat bran on rheology, microstructure and quality characteristics of soft dough biscuits. Journal of Food Processing and Preservation, 35(2), 179–200.

Ndife, J., Abdulraheem, L. O., & Zakari, U. M. (2011). Evaluation of the nutritional and sensory quality of functional breads produced from whole wheat and soya bean flour blends. African Journal of Food Science, 5(8), 466–472.

Ngozi, A. A. (2014). Effect of whole wheat flour on the quality of wheat-baked bread. Global Journal of Food Science and Technology, 2(3), 127–133.

Niu, M., Hou, G. G., Wang, L., & Chen, Z. (2014). Effects of superfine grinding on the quality characteristics of whole-wheat flour and its raw noodle product. Journal of Cereal Science, 60(2), 382–388.

Noort, M. W., van Haaster, D., Hemery, Y., Schols, H. A., & Hamer, R. J. (2010). The effect of particle size of wheat bran fractions on bread quality–Evidence for fibre–protein interactions. Journal of Cereal Science, 52(1), 59–64.

Oliveira, L. C., Alencar, N. M., & Steel, C. J. (2018). Improvement of sensorial and technological characteristics of extruded breakfast cereals enriched with whole grain wheat flour and jabuticaba (*Myrciaria cauliflora*) peel. LWT – Food Science and Technology, 90, 207–214.

Oliveira, L. C., Rosell, C. M., & Steel, C. J. (2015). Effect of the addition of whole-grain wheat flour and of extrusion process parameters on dietary fibre content, starch transformation and mechanical properties of a ready-to-eat breakfast cereal. International Journal of Food Science, 50(6), 1504–1514.

Oliveira, L. C., Schmiele, M., & Steel, C. J. (2017). Development of whole grain wheat flour extruded cereal and process impacts on color, expansion, and dry and bowl-life texture. LWT – Food Science and Technology, 75, 261–270.

Owen, D. E., Munday, K. A., Taylor, T. G., & Turner, M. R. (1975). Hypercholesterolemic action of wheat bran and a mould in rats and hamsters. Proceedings of Nutrition Society, 34, 16A.

Özdoğan, A., Gunes, R., & Palabiyik, I. (2019). Investigating release kinetics of phenolics from defatted wheat germ incorporated chewing gums. Journal of the Science of Food and Agriculture, 99(14), 6333–6341.

Palzer, S. (2009). Food structures for nutrition, health and wellness. Trends in Food Science & Technology, 20(5), 194–200.

Pasqualone, A., Summo, C., Laddomada, B., Mudura, E., & Coldea, T. E. (2018). Effect of processing variables on the physico-chemical characteristics and aroma of borș, a traditional beverage derived from wheat bran. Food Chemistry, 265, 242–252.

Patring, J., Wandel, M., Jägerstad, M., & Frølich, W. (2009). Folate content of Norwegian and Swedish flours and bread analysed by use of liquid chromatography–mass spectrometry. Journal of Food Composition and Analysis, 22(7–8), 649–656.

Persson, I., Raby, K., Fønss-Bech, P., & Jensen, E. (1976). Effect of prolonged bran administration on serum levels of cholesterol, ionized calcium and iron in the elderly. Journal of American Geriatrics Society, 24(7), 334–335.

Pietiäinen, S., Moldin, A., Ström, A., Malmberg, C., & Langton, M. (2022). Effect of physicochemical properties, pre-processing, and extraction on the functionality of wheat bran arabinoxylans in breadmaking–A review. Food Chemistry, 132584.

Podio, N. S., Baroni, M. V., Pérez, G. T., & Wunderlin, D. A. (2019). Assessment of bioactive compounds and their in vitro bioaccessibility in whole-wheat flour pasta. Food Chemistry, 293, 408–417.

Poitout, V. (2008). Glucolipotoxicity of the pancreatic β-cell: myth or reality? Biochemical Society Transactions, 36(5), 901–904.

Pomeranz, Y. (1970). Germ bread. Bakers' Digest, 44(6), 30–33.

Poudel, R., Finnie, S., & Rose, D. J. (2019). Effects of wheat kernel germination time and drying temperature on compositional and end-use properties of the resulting whole wheat flour. Journal of Cereal Sciences, 86, 33–40.

Protonotariou, S., Batzaki, C., Yanniotis, S., & Mandala, I. (2016). Effect of jet milled whole wheat flour in biscuits properties. LWT – Food Science and Technology, 74, 106–113.

Prueckler, M., Siebenhandl-Ehn, S., Apprich, S., Hoeltinger, S., Haas, C., Schmid, E., & Kneifel, W. (2014). Wheat bran-based biorefinery 1: Composition of wheat bran and strategies of functionalization. LWT – Food Science and Technology, 56(2), 211–221.

Rangi, A. S., Minhas, K. S., & Sidhu, J. S. (1985). Indigenous milk products. I. Standardization of recipe for Gulabjamun. Journal of Food Science and Technology, 22(3), 191–193.

Rebolleda, S., Beltrán, S., Sanz, M. T., González-Sanjosé, M. L., & Solaesa, Á. G. (2013). Extraction of alkylresorcinols from wheat bran with supercritical CO_2. Journal of Food Engineering, 119(4), 814–821.

Rindhe, S. N., Chatli, M. K., Wagh, R. V., Kumar, P., Malav, O. P., & Mehta, N. (2018). Development and quality of fiber enriched functional spent hen nuggets incorporated with hydrated wheat bran. International Journal of Current Microbiology and Applied Sciences, 7(12), 3331–3345.

Roberfroid, M. B. (1993). Dietary fiber, inulin, and oligofructose: a review comparing their physiological effects. Critical Review in Food Science & Nutrition, 33(2), 103–148.

Roberfroid, M. B. (1999). What is beneficial for health? The concept of functional food. Food and Chemical Toxicology, 37(9–10), 1039–1041.

Robin, F., Schuchmann, H. P., & Palzer, S. (2012). Dietary fiber in extruded cereals: Limitations and opportunities. Trends in Food Science & Technology, 28(1), 23–32.

Rocha-Garza, A. E., & Zayas, J. F. (1995). Effect of wheat germ protein flour on the quality characteristics of beef patties cooked on a griddle 1. Journal of Food Processing and Preservation, 19(5), 341–360.

Rodríguez-Vidal, A., Martínez-Flores, H. E., González Jasso, E., Velázquez de la Cruz, G., Ramírez-Jiménez, A. K., & Morales-Sánchez, E. (2017). Extruded snacks from whole wheat supplemented with textured soy flour: effect on instrumental and sensory textural characteristics. Journal of Texture Studies, 48(3), 249–257.

Rosa-Sibakov, N., Poutanen, K., & Micard, V. (2015). How does wheat grain, bran and aleurone structure impact their nutritional and technological properties? Trends in Food Science & Technology, 41(2), 118–134.

Rosell, C. M., Rojas, J. A., & De Barber, C. B. (2001). Influence of hydrocolloids on dough rheology and bread quality. Food Hydrocolloids, 15(1), 75–81.

Šabovič, M., Lavre, S., & Keber, I. (2004). Supplementation of wheat fibre can improve risk profile in patients with dysmetabolic cardiovascular syndrome. European Journal of Cardiovascular Prevention and Rehabilitation, 11(2), 144–148.

Salem, S. A., Hamad, E. M., & Ashoush, I. S. (2016). Effect of partial fat replacement by whey protein, oat, wheat germ and modified starch on sensory properties, viscosity and antioxidant activity of reduced fat ice cream. Food and Nutrition Sciences, 7(6), 397–404.

Sandberg, A. S., Hasselblad, C., Hasselblad, K., & Hultén, L. (1982). The effect of wheat bran on the absorption of minerals in the small intestine. British Journal of Nutrition, 48(2), 185–191.

Santos, D., Pintado, M., & da Silva, J. A. L. (2022). Potential nutritional and functional improvement of extruded breakfast cereals based on incorporation of fruit and vegetable by-products-A review. Trends in Food Science & Technology, 125, 136–153.

Sarıçoban, C., Yılmaz, M. T., & Karakaya, M. (2009). Response surface methodology study on the optimisation of effects of fat, wheat bran and salt on chemical, textural and sensory properties of patties. Meat Science, 83(4), 610–619.

Sarwin, R., Walther, C., Laskawy, G., Butz, B., & Grosch, W. (1992). Determination of free reduced and total glutathione in wheat flours by an isotope dilution assay. Zeitschrift für Lebensmittel-Untersuchung und Forschung, 195(1), 27–32.

Schofield, J. D., & Chen, X. (1995). Analysis of free reduced and free oxidised glutathione in wheat flour. Journal of Cereal Science, 21(2), 127–136.

Seham, I. F., El-Sonbaty, A. H., Hussein, S. A., Farrag, A. F., & Shahine, A. M. (2007). Effect of substituting added skim milk powder (SMP) with wheat germ (WG) on the quality of goat's milk yoghurt and fermented camel's milk drink. In 10th Egyptian Conference for Dairy Science and Technology, Research Papers, held at The International Agriculture Centre, Cairo, 19–21 November (pp. 315–336). Egyptian Society of Dairy Science.

Seleet, F. L., Assem, F. M., Abd El-Gawad, M. A., Dabiza, N. M., & Abd El-Salam, M. H. (2016). Development of a novel milk-based fermented product fortified with wheat germ. International Journal of Dairy Technology, 69(2), 217–224.

Shahid, Z., Kalayanamitra, R., McClafferty, B., Kepko, D., Ramgobin, D., Patel, R., et al. (2020). COVID-19 and older adults: what we know. Journal of the American Geriatrics Society, 68(5), 926–929.

Shewry, P. R., Evers, A. D., Bechtel, D. B., & Abecassis, J. (2009). Development, structure, and mechanical properties of the wheat grain. In: K. Khan and P. Shewry (Eds.) Wheat chemistry and technology (4th ed.), AACC International Press, pp. 51–96.

Singh, D., Chauhan, G. S., Suresh, I., & Tyagi, S. M. (2000). Nutritional quality of extruded snacks developed from composite of rice brokens and wheat bran. International Journal of Food Properties, 3(3), 421–431.

Solah, V., Fenton, H., & Crosbie, G. (2015). Wheat: Grain structure of wheat and wheat-based products. In: Encyclopedia of food and health, Academic Press, pp. 470–477.

Soliman, T. N., Farrag, A. F., Zahran, H. A. H., & El-Salam, M. E. H. A. (2019). Preparation and properties nano-encapsulated wheat germ oil and its use in the manufacture of functional labneh cheese. Pakistan Journal of Biological Sciences, 22(7), 318–326.

Song, M., Garrett, W. S., & Chan, A. T. (2015). Nutrients, foods, and colorectal cancer prevention. Gastroenterology, 148(6), 1244–1260.

Sozer, N., Cicerelli, L., Heiniö, R. L., & Poutanen, K. (2014). Effect of wheat bran addition on in vitro starch digestibility, physico-mechanical and sensory properties of biscuits. Journal of Cereal Sciences, 60(1), 105–113.

Talukder, S., & Sharma, D. P. (2010). Development of dietary fiber rich chicken meat patties using wheat and oat bran. Journal of Food Science and Technology, 47(2), 224–229.

Tekin, H., Saricoban, C., & Yilmaz, M. T. (2010). Fat, wheat bran and salt effects on cooking properties of meat patties studied by response surface methodology. International Journal of Food Science and Technology, 45(10), 1980–1992.

Terpou, A., Gialleli, A. I., Bekatorou, A., Dimitrellou, D., Ganatsios, V., Barouni, E., Koutinas, A. A., & Kanellaki, M. (2017). Sour milk production by wheat bran supported probiotic biocatalyst as starter culture. Food and Bioproducts Processing, 101, 184–192.

Tracy, P. H., Hoskisson, W. A., & Trimble, J. M. (1944). Wheat germ oil as an antioxidant in dairy products. Journal of Dairy Sciences, 27, 311–318.

US Department of Agriculture, Agricultural Research Service, Nutrient Data Laboratory. (2015). USDA national nutrient database for standard reference, release 28 (Version current). US Department of Agriculture, Agricultural Research Service, Nutrient Data Laboratory.

van De Sande, M. M., van Buul, V. J., & Brouns, F. J. (2014). Autism and nutrition: the role of the gut–brain axis. Nutrition Research Review, 27(2), 199–214.

Vignola, M. B., Bustos, M. C., & Pérez, G. T. (2018). Comparison of quality attributes of refined and whole wheat extruded pasta. LWT – Food Science and Technology, 89, 329–335.

Wahanik, A. L., Neri-Numa, I. A., Pastore, G. M., Felisberto, M. H. F., Campelo, P. H., & Clerici, M. T. P. S. (2021). Technological and antioxidant characteristics of pasta with whole wheat flour and natural colored concentrates. Research, Society and Development, 10(3), e7110312072.

Wang, N., Hou, G. G., & Dubat, A. (2017). Effects of flour particle size on the quality attributes of reconstituted whole-wheat flour and Chinese southern-type steamed bread. LWT – Food Science and Technology, 82, 147–153.

Wang, P., Hou, C., Zhao, X., Tian, M., Gu, Z., & Yang, R. (2019). Molecular characterization of water-extractable arabinoxylan from wheat bran and its effect on the heat-induced polymerization of gluten and steamed bread quality. Food Hydrocolloids, 87, 570–581.

Waters, D. M., Mauch, A., Coffey, A., Arendt, E. K., & Zannini, E. (2015). Lactic acid bacteria as a cell factory for the delivery of functional biomolecules and ingredients in cereal-based beverages: a review. Critical Reviews in Food Science and Nutrition, 55(4), 503–520.

Williams, C. L., Bollella, M. C., Strobino, B. A., Boccia, L., & Campanaro, L. (1999). Plant stanol ester and bran fiber in childhood: effects on lipids, stool weight and stool frequency in preschool children. Journal of the American College Nutrition, 18(6), 572–581.

Yadav, D. N., & Rajan, A. (2012). Fibres as an additive for oil reduction in deep fat fried poori. Journal of Food Science and Technology, 49(6), 767–773.

Yadav, S., Pathera, A. K., Islam, R. U., Malik, A. K., & Sharma, D. P. (2018). Effect of wheat bran and dried carrot pomace addition on quality characteristics of chicken sausage. Asian-Australasian Journal of Animal Sciences, 31(5), 729.

Yadav, S., Pathera, A. K., Islam, R. U., Malik, A. K., Sharma, D. P., & Singh, P. K. (2020). Development of chicken sausage using combination of wheat bran with dried apple pomace or dried carrot pomace. Asian Journal of Dairy and Food Research, 39(1), 79–83.

Yaseen, A. A. E., & Shouk, A. A. (2005). Effect of extrusion variables on physical, structure and sensory properties of wheat germ-corn grits extrudates. Egyptian Journal of Food Science, 33(1), 57–71.

Yılmaz, I. (2005). Physicochemical and sensory characteristics of low fat meatballs with added wheat bran. Journal of Food Engineering, 69(3), 369–373.

Yu, L., Nanguet, A. L., & Beta, T. (2013). Comparison of antioxidant properties of refined and whole wheat flour and bread. Antioxidants, 2(4), 370–383.

Zhang, L., van Boven, A., Mulder, J., Grandia, J., Chen, X. D., Boom, R. M., & Schutyser, M. A. (2019). Arabinoxylans-enriched fractions: From dry fractionation of wheat bran to the investigation on bread baking performance. Journal of Cereal Science, 87, 1–8.

Zhao, Z., & Moghadasian, M. H. (2008). Chemistry, natural sources, dietary intake and pharmacokinetic properties of ferulic acid: A review. Food Chemistry, 109(4), 691–702.

Zhu, K.-X., Zhou, H.-M., & Qian, H.-F. (2006). Proteins extracted from defatted wheat germ: nutritional and structural properties. Cereal Chemistry, 83(1), 69–75.

Žuljević, S. O., & Akagić, A. (2021). Flour-based confectionery as functional food. Functional Foods: Phytochemicals and Health Promoting Potential, 351.

Zúñiga, R. N., & Troncoso, E. (2012). Improving nutrition through the design of food matrices. Scientific, Health and Social Aspects of the Food Industry, 264–320.

14

Trends and Approaches of Wheat Flour and End-Product Fortification Technologies

Rakesh Kumar Prajapat, Manas Mathur, Sarita Kumari, Sumit Kumar, Megha Sharma, Nitin Kumar Garg, Deepak Sharma, and Nand Lal Meena

CONTENTS

14.1 Global Wheat-Based Products, Their Production, and Consumption Scenario

Wheat solely meets the terms of global nutrition demand and supply, with a fifth of food calories (18%) along with proteins (19%) and feed as well (Dixon, 2007; Shiferaw et al., 2013). Wheat is the world's oldest widely used staple food crop and the major component of human diet, domesticated for 10,000 years ago around the time when rice was also domesticated in the Near East's fertile hemisphere and somewhat prior to that of maize (Awika, 2011). These global staple crops, viz., wheat, rice, and maize

DOI: 10.1201/9781003307938-14

account for nearly half of the world's food calorie requirements and two-fifths of protein content. Nearly 120 countries distributed across Europe, Asia, Africa, the Americas, and Oceania cultivate wheat (FAOStat, 2020) and have wheat-based emerging as well as developed economies. From an agronomic point of view, wheat execution is superior in temperate environments.

India is the second-largest producer of wheat after China with the production of 99.7 million tonnes in an area of 29.5 Mha (4th advance estimate, MoA&FW, 2018). Wheat's 752 M tons globally (TE2018) is slightly less but comparable to rice, that is, 768 tons, although both crops are overtaken by maize (1,146 M tons, with some 57% used as feed). Yields of wheat increased steadily globally on average of over 1 ton/ha from 1960s to current 3.5 tons/ha, having nearly four times increase in global wheat production (Figure 14.1). During the wheat-growing season, this staple crop can withstand frost, and around 150 M ha of wheat is grown in areas having freezing temperatures in contrast to other crops which are susceptible to chills.

Three main cultivated species of wheat are *T. aestivum, T. durum,* and *T. dicoccum* (Gupta et al., 2018). Due to a considerable variation in the grain quality of different species, various end products are thus produced (Table 14.1).

Wheat is indeed a vital raw material from baking industry used as household food consumable products. In India, majority of the wheat produced is processed as whole-wheat flour for multiple uses and builds the largest organized food industry in the country. Various forms of wheat prevail depending on their use, viz., specific bread wheats like Hard Red Winter (HRW), Hard Red Spring, Soft Red Winter (SRW), Hard White and Soft White, and Durum wheat used for pasta and couscous. These

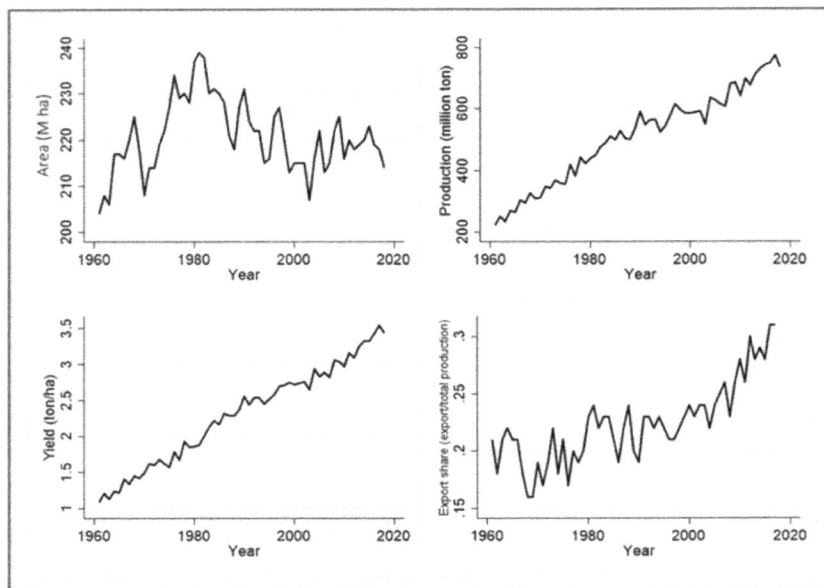

FIGURE 14.1 Dynamics of key wheat indicators 1961–2018: wheat area (M ha), production (M ton), yield (ton/ha), and export share (export/total production).

TABLE 14.1

Wheat-Based Products Prepared From Commonly Grown Species Internationally

	Species	Products
1.	*T. aestivum*	Bread, biscuit, chapati, naan, samosa, *matthi*, tandoori roti, *kachori*, *kulcha*, bhatura, *puri*, pizza, *namakpara*, papad, *balushahi*, *ghewar*, *sattu*, noodles, *pav* buns, cake, pastry, burger, patties, etc.
2.	*T. dicoccum*	*Godihuggi*, sweet pan cake, madel and pasta products etc.
3.	*T. durum*	*Parantha*, *dhebra*, *bhakri*, porridge, *rava idli*, *ravaputtu*, *khichdi*, pasta products, vermicelli etc.

classes of bread wheat differ in their milling characteristics and baking and processing characteristics for flours, pan breads, Asian noodles, hard rolls, cakes, cookies, snack foods, crackers and pastries and for improving blending. Other uses include the industrial production of starch, malt, dextrose, gluten, and alcohol.

Major ingredient forms the basis of biscuit categorization like wheat, oats, millets, and mixed grain. Different biscuits, viz., sweet biscuits, chocolate biscuits, crisp bread, and savory biscuits, possess different taste and nutritional levels on account of increasing concern for health. Dairy-free, gluten-free, and sugar-free fortified products are manufactured as well.

Food use is particularly high in South Asia, sub-Saharan Africa, and Latin America. Protein from wheat, which contains a high amount of gluten, is another frequently used ingredient which has unique film-forming properties resulting in small fibres when applied in meat analogues. The cost-effectiveness of wheat makes it available for industrial use. GDP per capita has increased by 3.5% per annum globally (1961–2019, from US$3.9k pc TE1963 to US$11.1k pc in 2019 at 2010 constant prices) (WorldBank 2020). Urbanization has increased from a little more than 34% of global population in 1961 to nearly 56% off late. In SSA, major wheat consumption is focused among urban populations where the supply of wheat is derived from imports in high ratio (Mason et al., 2015). The use of wheat shows a well-marked deviation between country income status. In L/LM-ICs (low/low and middle income countries), 79% is used as food and 10% as feed with (5%) losses.

On the contrary, in UM/H-ICs (upper- and middle/high-income countries), the percentage of food drops to 60%, while feed counts for 26% increase with lower losses. In addition, feed use is only concentrated in selected geographical areas, for the most part in Australia, trending use by Europe and Eastern Asia. In SSA, much of the wheat consumption is concentrated among urban populations with supplies derived largely from imports (Mason et al., 2015).

14.2 Nutritional Composition in Wheat Flour and Its End Products

Wheat is considered to be the most important staple crop and major source of carbohydrate, energy, protein, minerals, dietary fiber, and B-group vitamins, that is, an excellent health-building food and alone providing 20–25% of the daily intake. Wheat is an important food crop for more than one-third of the globe population and

TABLE 14.2

Contiguous Composition of Whole and Refined Wheat Flours, Wheat Bran, and Wheat Germ

Ingredients	Whole Wheat Flour	Refined Wheat Flour (Unenriched)	Bran	Germ
Carbohydrates (g/100 g)	71.20	76.31	64.51	49.60
Fiber (g/100 g)	10.60	2.70	42.80	15.1
Protein (g/100 g)	15.10	10.33	15.55	29.10
Lipids (g/100 g)	2.73	0.98	4.25	10.70
Calcium (mg/100 g)	38	15	73	45
Iron (mg/100 g)	3.86	1.17	10.75	9.09
Magnesium (mg/100 g)	138	22	611	320
Phosphorus (mg/100 g)	352	108	1013	1182
Potassium (mg/100 g)	372	107	1182	947

Source: Data retrieved from the website Food Data Portal (2020).

contributes more proteins and calories to the world diet than any other cereal crops (Shewry and Hey, 2015). In particular, it is nutritious, easy to transport and store, and can be processed into various types of food. Due to the unique properties of the gluten, wheat has become the principal cereal to produce bread, biscuits, baked goods, pasta, noodles, and a range of processed confectionary food products (Sharma and Gujral, 2014). The nutritional value of wheat is exceptionally imperative as it takes an important place among the few crop species being extensively grown as staple food sources (Shewry, 2009).

Wheat is known to provide a significant part of our diet and contains carbohydrate 78.10%, protein 14.20%, fat 2.10%, fibre 3.20%, minerals 2.10%, and significant proportions of vitamins (thiamine and vitamin B) and minerals (calcium, phosphorus, zinc, iron) (Figure 14.2). The latest edition of the USDA's Dietary Guidelines

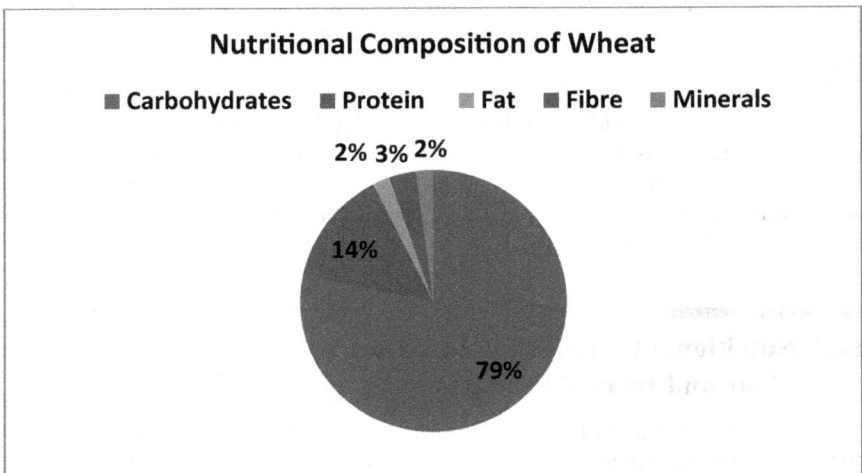

Nutritional Composition of Wheat

■ Carbohydrates ■ Protein ▨ Fat ■ Fibre ▨ Minerals

2% 3% 2%
14%
79%

FIGURE 14.2 Nutritional composition of wheat.

for Americans clearly states that all adults should eat at least three servings of whole grains every day (http://www.dietaryguidelines.gov/sites/default/files/2020-12/Dietary_Guidelines_for_Americans_2020-2025). It helps in preventing both heart diseases and cancer and, therefore, lowers death rates. The protection against heart disease may stem from whole grains, phytochemicals, antioxidants, vitamins, and fiber or trace minerals.

14.3 Issues and Challenges of Nutrient Deficiencies in Wheat Flour and Its End Products

Wheat flour and its processed products (bread, pasta, and bakery food) are considered as the chief source of macronutrients, mainly carbohydrates and protein, and micronutrient minerals (iron, zinc, and potassium) and vitamins, dietary fiber, phenolic acids, and antioxidants and measured as important nutritional diet for humans, which are able to provide health benefits via bioactive compounds (Cappelli et al., 2020). Due to their low quantity in endosperm and interaction with other components, the bioavailability of these nutritional components is low. Deficiencies of vitamins, essential nutrients, and minerals cause irreversible serious physical and cognitive health problems as they affect more than two billion people worldwide, resulting in several illnesses mainly in women and children in developing countries such as decrease in immune competence, growth problems, physical and mental development issues, and poor reproductive outcomes (Lephuthing et al., 2017; Cardoso et al., 2019). Transgenics, food fortification, dietary modifications, and supplementation are the technologically and economically valuable strategies to control micronutrient deficiencies and appear to be most realistic and indelible perspectives that could improve the bioavailability of essential nutritional components.

There are several concerns and challenges experienced in attaining the quality and bioavailability of micronutrients and vitamins in wheat endosperm (Borrill et al., 2014). The major challenge is that complete genome sequence of wheat is not yet available in addition to the uncharacterized process of vitamin, micronutrient, and phenolic acid translocation into the wheat endosperm, which can be useful to understand the biological process of translocation, their biosynthetic pathways, and nutrients' accumulation and assimilation (Borg et al., 2009; Waters and Sankaran, 2011). Thus, it is required to identify the timing at which vitamins, micronutrients and phenolic acids are translocated, assimilated, and biologically synthesized in the endosperm.

To empower the nutritional security of wheat and their end products, the following strategies can be adopted (Pfeiffer and McClafferty, 2007; Cakmak et al., 2010; Carvalho and Vasconcelos, 2013; Sharma et al., 2013; Singh et al., 2013; Goudia and Hash, 2015): (1) The NAC transcription factors can be used to accelerate the identification of the agents for an improved nutritional quality via their remobilization into grain. The NAC gene family is a large group of three transcription factors: NAM (no apical meristem, *Petunia*), ATAF1–2 (*Arabidopsis* thaliana-activating factor), and CUC2 (cup-shaped cotyledon, *Arabidopsis*), which share the same DNA-binding domain. (2) Use of gene-editing tools for improving the numbers of desirable gene of interest. (3) The production of transgenic wheat having *Phytase* for increased calcium, iron, and zinc. (4) Foliar application of nitrogen for improved iron and zinc accumulation. (5) Mineral fortification of wheat.

14.4 Food Fortification Technologies

It is a great challenge to fulfill the increasing demand of the growing population from scarce primary resources, including arable land and fresh water, which are having key roles in global undernourishment reports. They include the disorders caused due to inadequate and unwarranted consumption of nutrients, generally called as undernutrition and overnutrition, respectively. Food fortification is defined as the perfection of the essential micronutrient biomolecules of foods to increase the nutritional and health benefits with least side effects (Dary and Hurrell, 2006). Restoration denotes the replacement of earlier consumed nutrients in food, but deprived or damaged micronutrients because of processing is also a kind of food fortification. The varieties of food fortification comprise market-driven, targeted fortification, and mass fortification, along with microbial fortification and biofortification. Mass fortification is based on a wide range of consumed food within population; targeted fortification is confined to a particular group; market-driven fortification is required by the food industry; and household fortification contains the allocation of additives at community stage (Teshome et al., 2016). Further, microbial fortification involves the secretion of various biomolecules through anerobic respiration, and biofortification has been achieved by various breeding biotechnical applications (Liyanage and Hettiarachchi, 2011).

Although biomolecules and nutrient deficiencies have been reported globally, 80% of the cases are from the Sahara region of Africa. This area is very prone to metabolic diseases, decreased immune response to infections, damage of physical and psychomotor growth, and other diseases related to the deficiency of biologically active molecules (Liyanage and Hettiarachchi, 2011). Fortification of staple crops like rice, maize, and wheat flour has played a crucial role in African continent as a vital source to overcome these metabolic disorders (Saha and Roy, 2020, Swanepoel et al., 2019). Therefore, the profitable food fortification which forced fortified complimentary foods to be easily accessible in developed nations is obviously in demand. The commercial fortified food varieties are much more expensive and are not easily accessible to poor families. The presence of the micronutrients in fortified foods is prejudiced by the liberation vehicle (Shubham et al., 2020). It has been observed that the fortification of the various staple foods including the abandoned and underutilized foods is also useful in controlling the deficiency of major micro and macronutrients. The main objective of food fortification is to make the easy bioavailability of the micronutrients to the society.

14.4.1 Characteristics of Fortified Food Varieties

Most of the food items used in the African continent have a scarcity of certain biological molecules (Khush et al., 2012). Although food fortification has some advantages that it overcomes the barrier of decreasing micronutrient deficiency, the broadly recommended plans often need intricate technologies which are not easily accessible, which enhances the price of food for communities living near to poverty line. Therefore, it is important to accept the easiest protocol and locally keep ready material for food fortification.

Food fortification is termed as a practice of increasing major macromolecules of generally edible food items during processing to enhance their nutritional content. It is a confirmed, protected, and cheap method for humanizing diets and to inhibit micronutrient-deficient disorders. It has been observed that this technology is also useful as a vital tool to combat micronutrient disorders in developed countries since 1920s in Europe and North America when the salt was iodized initially. In earlier decades, food fortification had gained lot of attention due to its advantages like rapid urbanization and enhanced household purchasing power, causing a greater proportion of the population dependence on processed foods (Spohrer et al., 2013). Industrial fortification is based on the concept to increase the level of micronutrients during manufacturing in generally edible items like salt, flours, oil, sugar, and spices. There are certain programs which can be divided as either compulsory, that is, they are executed and synchronized by the government along with private agencies where industrialists add nutrients to their foods on their own decisions but has some limitations. Execution of such fortification programs has gained lot of attention, when focusing on fortified flour. Salt iodization was the first such kind of program in 20th century, thus reaching about 20 to 70% common people (UNICEF, 2008). It was first initiated in 1942, and, at present, 85 nations have adopted this technology. In the United States of America, the addition of folic acid to wheat flour is compulsory to reduce the risk of abnormalities caused during childbirth in females. Edible oils are mostly recommended for fortification, and about 27 nations have made it mandatory for the oils to be enriched with vitamin A; 14 nations have made compulsory milk fortification; and 11 nations have made fortified milk with both vitamin A and D (Osendarp et al., 2018).

Food fortification is also executed by polishing the proficiency of industries to synthesize and distribute fortified foods. Mandatory fortification is the technique by which a food producer adopts to mix various micronutrients in processed foods following all government laws and standards. It has been reported that Olam Ghana Limited in Ghana fortifies long-grain rice with micronutrients including iron, zinc, and B-complex vitamins, thus yielding more than 15% of the minimum recommended dietary allowance per serving. In India and Kenya, such type of programs were drivers of more comprehensive legislation and burly-making conditions (Lalani et al., 2019).

14.4.2 Technologies Executed in Micronutrient Fortification

Nanotechnology is one of the most prominent technologies for the supplementation of micronutrient fortificants. Nano-based micronutrients bear in size of 1–100 nm, which has a direct impact on their biochemical, sensory, and nutritional features as they bear high surface area (Jafari and McClements, 2017). Thus, their encapsulation in nano-based delivery phenomena has been recommended globally in food fortification. It can alleviate sensory variations, provide prominent safety against micronutrient incorporation inhibitors, and protect distortion of the micronutrients, thereby elevating their bioavailability in the gastrointestinal tract (Öztürk, 2017). Therefore, various food items have been fortified with iron through nanotechnology like in cheddar cheese (Siddique and Park, 2019), milk powder (Gupta et al., 2015), yogurt drink (Santillán-Urquiza et al., 2017), fish oil (Choi et al., 2009), soy-based cereals (Kusn and Suyatma, 2017), including dry beverage mixes (Shubham et al., 2020).

Before the development of this technology, it is very crucial to develop a balance among constancy and digestibility of encapsulated materials. Like hydrogenated vegetable oils have been recommended in the coating of iron; in some cases, mono- and diglycerides, maltodextrins, and ethyl cellulose have been recommended. Cheddar cheese fortified with iron by coating gives a prominent superiority in toughness and stickiness due to the proteolysis of the protein matrix as it is embedded with iron during storage (Siddique and Park, 2019). However, milk powder fortified with iron through this technology obtains a fine colour when microcapsules are produced with gum arabic, maltodextrin, and encapsulated sugar moieties by the freeze-drying method. Some reported the application of this technology in an animal model. Hosny et al. (2015) reported that different encapsulation processes were approached after hot mixing and ultrasonication technologies to synthesize iron-embedded lipid nanoparticles. More than 90% encapsulated ratio has been reported during *in vitro* research with a standard size of 25 nm; and a fourfold bioavailability elevation was reported in *in vivo* assessment. Further, the nano-encapsulation tool can be expensive and unreasonable for broad acceptance in areas where micronutrient deficiency is at a large scale. Dary and Hurrell (2006) reported that this is expensive and may cost by three- to five-folds.

14.4.3 Biofortification

A biofortification approach helps to improve the nutritional quality of food crops. This can be gained through agronomic practices, conventional breeding, or biotechnological techniques like genetic engineering and genome editing (Garg et al., 2018). Both these techniques have been recommended to enhance the micronutrient amount in food crops. Plant-breeding-based biofortification has been executed in overcoming global deficiency of micronutrients. The main aim of biofortification is to synthesize safe, stable, adequate, and quality foods (Saltzman et al., 2013). For a broad effect on the uses of biofortification, staple foods play the key role. So products like wheat, rice, maize, sorghum, common bean, lupin, tomato, and potato including sweet potato have been genetically changed using biotechnology and conventional breeding. However, transgenic and agronomic tools are also recommended for the biofortification of soybean, barley, lettuce, canola, mustard, and carrot, while transgenic and breeding techniques are used in cassava, banana and cauliflower, and banana (Garg et al., 2018). Biofortification elevates the level of vitamins in cereals (Brinch-Pedersen et al., 2007). The achievement relies on different issues. The achievement of plant breeding in biofortification depends on the availability of a broad difference within the gene pool for controlling nutrient amounts in selected crop; in that scenario, this transgenic technique is quite crucial (Zhu et al., 2007). Transgenic rice that overexpresses the soy ferritin gene SoyferH2 was found to have higher levels of zinc and iron by 3.4- and 1.3-folds (Aung et al., 2013). So, biofortification is broadly recommended as a vital tool which overcomes micronutrient disorders, mainly in developing countries where it is at a large level. Still biofortification requires a rich skill of plant genetics to productively conduct transgenic variations.

The execution of food fortification policies and various processes has been facing a lot of scientific, ethical, socioeconomic, and political challenges, mainly at a large level clubbed with increasing population. Though it has several advantages, the food regulatory committees still are face with some issues starting from proving the proofs in case of such activities. Therefore, food fortification tools at a broad level of fortification may provide wide varieties of micronutrients to large communities. This creates ethical issues between the autonomy of options and the requirement of individuals with respect to micronutrient consumption, and adequate consumption by groups beyond the communities (Lawrence, 2013). It has been proved that the decision of conflicts in such cases is supreme to consume fortified foods to fight against micronutrient disorders. To overcome this, policymakers often depend on the skills, knowledge, and scientific proofs to validate and illuminate decisions and interventions. Scientific proofs are to execute the public health uses of food fortification on the basis of statistical data and research on population communities. Also, scientific proofs are validated from evaluating the competence of accessible processes. Ethical issues usually occur when food fortification is executed compulsorily on staple foods which are broadly consumed by the population. It has been reported that there was an epidemic of thiamine deficiency in Australia, which, in 1991, encouraged the execution of a mandatory thiamine fortification of bread (Lewis et al., 2003). The overall result was that local people consumed fortified food devoid of initial notification. So, policymakers need to research the benefits to the public health without implementing programs on the community. Similarly, an analysis of voluntary food fortification is required, which is usually not implemented on staple foods, and it is broadly dependent on the industrial aim of the food industry where health benefits are used to introduce the fortified food. It is crucial that consumers, having a chance to opt between fortified and non-fortified foods, should have their decisions based on nutritional properties. Therefore, the mass media has a very important role in the distribution of nutrition knowledge and food fortification awareness (Pambo et al., 2014).

Critical issues in the execution of food fortification program are the socioeconomic factors which are at a large level in developing countries where there is a shortage of food-processing industry. In fact, the scarcity in the development of the industrial sector makes fortified food unreachable and expensive to the people living at the poverty level including local communities who are dependent on them (Bhagwat et al., 2014; Temple and Steyn, 2011). So running such kind of programs in these countries is quite challenging (including the expense of machinery), with the issues of the execution of local scale fortification, money of mixed micronutrients, quality control, stable monitoring (Chadare et al., 2019). However, these economic and technical issues could be overcome by a broader option of the food-to-food fortification tool (Debelo et al., 2020). This approach fortifies staple foods in the form of locally available and available food items with a good quantity of the micronutrients of interest (Uvere et al., 2010). Practically, this requires an amalgamation of various food items in a meal.

14.5 Fortification Approaches of Wheat Flour and Its End Products

Wheat is cultivated and consumed in many parts of the world, and its domestication contributed to the development of farming and human civilization. It was first cultivated 9,000 years ago in the Euphrates Valley of the Middle East. An estimated 65% of the global wheat crop is used for human consumption, 17% is used for animal feed, and 12% is used in industrial applications, including biofuel production (FAO, 2013). Wheat is a staple in many countries due to its agronomic adaptability; ease of grain storage and milling; and suitability for making edible, palatable, acceptable, and satisfying foods (Curtis B, 2002). Isolation of endosperm and sifting into flour are the intermediary processes in wheat flour (VanDer Borght et al., 2005). Most vitamins and minerals in wheat are found in the bran or germ, and flours of 80% or lower extraction rates have significantly reduced micronutrient content. However, high-extraction flour also contains higher levels of phytates, which interfere with intestinal absorption of iron and other minerals as zinc and copper (Field et al., 2021. Fortification is the practice of deliberately increasing the content of one or more micronutrients, that is, vitamins and minerals, in a food to improve the nutritional quality of the food supply and provide a public health benefit with minimal risk to health. Fortification of staple foods is one of the strategies used to safely and effectively prevent vitamin and mineral inadequacies and their associated deficiencies in populations (Allen et al., 2006). Fortification of industrially processed wheat flour, when appropriately designed and implemented, can be an efficient, simple, and inexpensive strategy for supplying vitamins and minerals to the diets of large segments of the population. Industrial fortification of wheat flour has been practiced for many years in several countries where the flour is used in the preparation of different types of bread and national dishes.

Many strategies have been used to combat micronutrient deficiencies, such as food fortification, food diversification, and nutritional supplementation. Particularly, fortification is a method of incorporating nutrients or non-nutritive bioactive components into food products. Food fortification is one of the methods that have been applied increasingly and addressed to all age groups, being widely used to minimize micronutrient deficiency. By placing micronutrients in food products consumed daily, it reaches target populations, for which daily dietary requirements of micronutrients are scarcely satisfied. According to WHO and FAO, food fortification includes different forms (Figure 14.3): (i) mass fortification (fortified foods that are consumed by the general population in good amount), (ii) targeted fortification (fortified food targeted at specific population groups, e.g. small children), and (iii) market-driven fortification (ensuring that food is available in the market). There are also other types of fortification such as household fortification (the addition of micronutrients to homemade foods, namely a fusion of supplements and fortification) and biofortification of staple foods (breeding and genetic modification of plants to improve the content and the absorption of nutrients) (WHO and FAO, 2006).

Mass fortification is often mandatory; targeted fortification is mandatory or voluntary and depends on the importance of the public health; while market-driven fortification is always voluntary, but controlled by regulatory limits (Liyanage and Hettiarachchi, 2011). According to Liyanage and Hettiarachchi (2011), commercial and industrial fortification involves available products, such as flour, rice, cooking

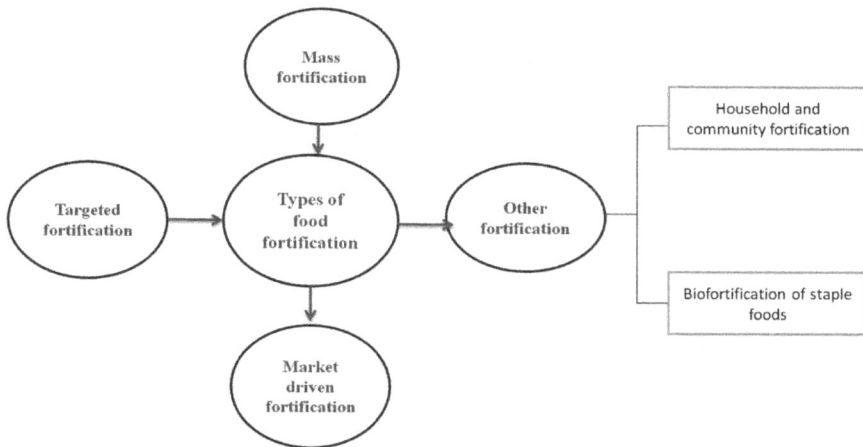

FIGURE 14.3 Different classes of food fortification.

oils, sauces, and butter, and the fortification procedure takes place throughout the manufacturing process. On the contrary, the biofortification encompasses the creation of micronutrient cultures using traditional breeding and/or biotechnology techniques (e.g., transgenic "Golden Rice" containing higher amounts of iron and significant levels of β-carotene. There is also microbial biofortification, which involves the use of probiotic bacteria, which ferment to produce β-carotene in foods or directly in the intestine. For example, the animal feed is enriched with these bacteria (more specific lactic acid bacteria), so that meat, milk, and bioproducts are enriched with vitamin A (Liyanage and Hettiarachchi, 2011). Moreover, there is home fortification, where micronutrients obtained from packages or tablets can be incorporated when cooking and or consumed in a homemade meal to fill micronutrient deficiencies in the populations (Liyanage and Hettiarachchi, 2011).

14.5.1 Fortification of Flour

Worldwide, millions of tons of flours are used for human consumption each year, and they are consumed as noodles, breads, pasta, and other flour products (Serdula, 2010b). In 2016, according to the Food Fortification Initiative, out of the 250 metric tons of industrially milled wheat flour, 26 metric tons of industrially milled maize flour, and 171 metric tons of industrially milled rice, each one of the flours were fortified by 34%, 57%, and 1%, respectively. Between 2016 and 2017, 87 countries decided to fortify at least one of these cereals. Wheat flour can be fortified with several micronutrients, such as iron, folic acid, other B-complex vitamins, vitamin A and zinc (Table 14.3). Some micronutrients are incorporated for the restitution of original nutritional contents of unrefined wheat flour, while others are used for correcting inadequacies and their associated deficiencies of public health significance. The bioavailability of the added micronutrients will depend in part on the grain type and the extraction rate of the flour. The third largest cereal production in the world is of wheat, after maize and rice, and is the second-most consumed by the populations after rice.

TABLE 14.3

Types of Biofortification in Wheat

Sl. No.	Types of Biofortifications		References
1.	Vitamins	B1 (thiamine)	Papathakis et al. (2012); Martorell et al. (2015); Centeno Tablante et al. (2019)
		B2 (riboflavin)	
		B3 (niacin)	
		B6 (pyridoxin)	
		B9 (folic acid)	
		B12 (cobalamine)	
		A (retinol)	Klemm et al. (2010)
		D (calciferol)	Black et al. (2012)
2.	Minerals	Fe	Field et al. (2021)
		Zn	Shah et al. (2016)
		Ca	Cormick et al. (2021)

Wheat flour is considered one of the most appropriate vehicles for multi-micronutrient fortification because of its worldwide consumption and given the high consumption of bread and pasta worldwide (Awika, 2011).

The industrial processed fortification of wheat flour, when properly implemented, is an effective, simple, and inexpensive strategy to provide vitamins and minerals to the world population, thereby improving the nutritional quality of food supply and providing a public health benefit (Cardoso et al., 2019). Globally, the effort to begin fortifying wheat flour was launched during the 1940s as a way to improve the health of populations. Wheat flour has been fortified with different micronutrients such as iron; folic acid; B-complex vitamins; vitamin A, D, and C; zinc; calcium, among others, in different parts of the world. For example, pasta made with white wheat flour, enriched with *Oreochromis niloticus* L. (high levels of essential amino acids and polyunsaturated fatty acids) flour, improves the nutritional composition of the pasta without compromising its sensorial quality (Monteiro et al., 2019).

14.6 Impact of Flour Fortification and Its Health Benefits

Wheat flour fortification is the sustainable approach to combat the malnutrition and hidden hunger among low-income countries. The recommended doses are still the point of discussion between the health and nutrition department. The impact of fortified wheat flour depends upon the amount of fortified nutrition present in the flour readily available for uptake, health competency of the individual for the uptake of a particular nutrient, and doses taken by individuals per days. Wheat is taken as a food crop in major volume. The amount of nutrients added to the flour depends upon the amount of flour taken by the individual per day. The fortification level decreases with the increase of intake amount per day (Serdula, 2010). Wheat flour is fortified with different types of minerals and vitamins to ameliorate the hidden hunger as summarized in Table 14.4.

TABLE 14.4

Wheat Flour Fortification With Micronutrients and Their Health Benefits

S. No.	Micronutrient	Health Benefits
1.	Minerals	Increases the content or bioavailability of specific micronutrients
2.	Vitamins	Prevention/reduction/improvement of the incidence of anemia or specific mineral deficiency
3.	Proteins	Functional food with therapeutic protective effects against diabetes and cardiovascular diseases
4.	Fiber	Effect on the prevention of colon cancer Antioxidant activity
5.	Fatty acids	Prevention of lifestyle diseases (e.g. hypertension, obesity)

14.6.1 Mineral Fortifications

Fe deficiency has been ameliorated with micronutrient powder containing iron alone or in combination with other vitamins and minerals in preschool and school-age children (De-Regil et al., 2017). Fe deficiency in women is a major concern. Fe supplementation with wheat flour leads to the reduction of anemia, but its effect on other aspects is not defined properly (Field et al., 2020, 2021). However, the other health consequences such as mortality, morbidity, and developmental and adverse effects seem to be still scarce due to the lack of data studies with controlled number of populations and small populations. The wheat is fortified with the iron in the form of NaFeEDTA, fumarate, sulfates, and electrolytic powder. Various workers have studied the health impact of these different forms on different age groups. Most of the studies showed positive effect on the health. The health impact of zinc (Zn) and calcium (Ca) fortification studies showed that the fortification with Zn solely improved serum zinc concentration among population, while little or no effect for Zn fortification with other micronutrient was reported. Uptake of Ca was uncertain through flour fortifications (Shah et al., 2016; Cormick et al., 2021).

14.6.2 Vitamin Fortification

Deficiency of vitamin A is a major concern for low- and middle-income countries. It affects more than 190 million children under the age of five years (Imdad et al., 2022). Its deficiency causes illness, night blindness, mortality, and morbidity. Studies reported that vitamin A supplementation in wheat reduced mortality associated with diarrhea, blindness, measles, xerophthalmia, and morbidity, while in few cases vomiting was also observed within 48 hours but that the certainty of vomiting was due to wheat flour is still a matter of skepticism (Mayo-Wilson et al., 2011). Similarly, wheat flour fortified with folic acid either alone or with minerals may reduce the risk of neural tube defects and increase erythrocyte and serum/plasma folate concentrations. The sleep disturbances, gastrointestinal effects, and mental changes are some of the adverse effects that have been observed in few studies. There is no evidence of direct toxicity and reproductive deformities after high doses or long duration of exposure to folic acid. Their effect on the reduction of hemoglobin and anemia still needs research (Tablante et al., 2019). Folic acid fortification has few suspected outcomes such as breast cancer (Castillo-Lancellotti et al., 2012), colorectal cancer, and multiple side effects (Tablante et al., 2019). Vitamin D fortification in wheat increases circulating 25-hydroxyvitamin D [25(OH) D] concentrations in blood and decreases vitamin D deficiency (Black et al., 2012).

The speculated adverse effect of wheat flour fortification is still the point of discussion. Majority of studies have reported the positive effect of wheat flour fortification on the basis of Cuernavaca recommendation (Serdula, 2010). Few reports of certain side effects like vomiting, diarrhoea, and cancers are also there. However, several meta-analyses reported that the reason behind this might be non-following of standard recommended dose as per Cuernavaca recommendation, insufficient data over the large number of populations, lack of long-duration observation, heterogeneity of sample population, and the unavailability of controlled populations. We have to be cautious about the amount of nutrients required for different age groups, amount of nutrients being fortified, amount of nutrients available after fortification, and their short and long-term impacts on the health of different population across the ages and locations.

14.7 Current Challenges in the Quality Control of Wheat Flour and Its End Products

For most of consumers, flour is just a unique and uniformly ground wheat powder. In fact, it is very complex and consists of various elements and qualities. It is a biological material and its protein quality and quantity, amount of moisture and ash, enzymatic activity, color, and other physical properties depend on the sources from where it is obtained. These qualities are analyzed/assessed by various methods using different instruments. Initial tests of grains reaching to a manufacturer or miller are required to get the idea about the type of flour grain, whether it is suitable for whole-grain milling or for white flour or it is suited for specific product or product with regional preferences. Quality control of wheat flour and other similar products is not simple, especially when the results of various qualitative traits are connected to reach a valuable conclusion. Different items of wheat, which are implied for a specific purpose need unequivocal qualities specific to that particular item. Different tests are outlined to check the standard of specific qualities in these products. Quality control aims at gathering the information required for the improvement of the quality of a particular product as per the need and satisfaction of end user. The most important challenge in the quality control of various products is to consider entire quality test results as a whole and improve the quality of the product accordingly.

Various methods are available in the market for controlling flour quality. A comprehensive quality control (QC) consists of three steps.

i. WHAT is in the flour? *I.e.*, quantitative analyses about the composition of the sample.

ii. HOW do the different components behave together? It implies the rheological analyses, behavior of the flour, and the resulting dough after combining with water and during processing.

iii. WHY dough behaves the way it does – that is, functionality analyses.

14.7.1 Composition of Flour

The composition of flour is mainly determined by the parameters like protein, moisture, and ash content. All these factors have definite roles in flavor and behavior of wheat products. Most of the food grains and food products contain moisture within as an important component. However, the presence of moisture beyond 14% in wheat may lead to undesirable microorganisms including molds. Apart from preventing biotic damage, appropriate moisture content should also be maintained as it affects the milling of wheat kernels. For maintaining moisture content within the safe limit, grains are harvested under dry and favourable conditions, which may be followed by drying before packaging and storage. For calculating the amount of different constituents in grain, it is necessary to remove and calculate the moisture content first. Moisture content of the grain is calculated by measuring the loss in kernel weight after drying it. Hot air oven method sufficiently and accurately does this. The sample flour (~2 g) is heated at 266°F (130°C + 1°C) for 1 hour. Alternatively, NIR are being used by various laboratories for this purpose.

Major protein in wheat flour is gluten that affects the behavior of flour after transformation into different products like bread, noodles, and cakes; protein content of the wheat flour may range from 6 to 20 % depending on the product. E.g., cakes and cookies need lesser amount of gluten in contrast with most of the breads. Generally, the protein content of a food product is determined by measuring the nitrogen content in the material and then multiplying it by a factor, which is 6.38 for milk products and 6.25 for most of the cereal grains. In wheat products, it is 5.70. High-protein wheat flours are expensive. However, very low-protein flours that are suitable for cakes also cost high. Kjeldahl procedure is the classic method for estimating nitrogen content, which involves the absorption/mixing of sample in concentrated H_2SO_4, then neutralizing with concentrated NaOH followed by the distillation of ammonia. Nowadays, Dumas combustion procedure has replaced the Kjeldahl process. In this process, sample is mixed with cupric oxide followed by heating in a combustion tube with a stream of CO_2. Organic content of the sample is converted into CO_2, H_2O, and nitrogen. Then, a stream of 50% KOH is led into the gas stream, which absorbs carbon dioxide and sulfur oxides. Now only nitrogen is left as gas whose volume can be determined.

Apart from organic components like starch, protein, fiber, and some fats, some fractions of inorganic minerals like iron and zinc are also present in wheat grain. However, most of these minerals constituting ash content are contained in bran part constituting 1% of wheat. During milling, this part is removed resulting in white flour whereas whole-wheat flours contain large amount of ash. Ash content is determined by using ash furnace. Although the reference method used for determining ash content (NF ISO 2171) has high accuracy, but it is time consuming (about three hours is needed). Some new instruments using NIR can determine ash content within a minute with almost same level of accuracy as ash furnace method. Accuracy in determining ash content is of utmost importance as it determines the milling yield and hence the total amount of the flour (Kumar et., 2011).

14.7.2 Behavior of the Flour

Composition of wheat gives an idea about the behaviour of the flour. For complete assessment about the flour behaviour, various tests have to be performed by the millers or bakers. The starch content of the wheat is broken down to varying extent by enzymes like alpha-amylase. In the brewing industry, it is highly desirable. Nevertheless, for other products, too much activity of this enzyme is not needed. However, these enzymes improve shelf life as well as the structure of bread and smaller sugars because partial breakdown serves as food for the yeast to grow. Therefore, amylase activity during flour processing is a very important quality aspect for various products. Uncontrolled sprouting leads to high amylase activity making dough sticky and low-volume bread and ultimately unfit for human consumption. Therefore, it is necessary to separate the sprout damage as early as possible before flour processing. The most widely accepted and standard method for determining this trait is the Hagberg falling number method developed during 1960s. In this method, the time required for the plunger to fall down the tube and go through the starch gel sample is recorded. A falling number value lesser than 150 s means that the probe has fallen very quickly, which is indicative of higher enzyme activity. Value higher than 400 s shows no meaningful activity of the enzyme. However, few new technologies can record the consistency of dough during the 90 s of shaking, for determining the amylase activity or any amount of sprout damage. This method is called as testogram method, which gives the prediction of α-amylase activity with more accuracy and in lesser time than the traditional falling number (FN) value.

Although composition tells about the constituents of the flour, but the flours having similar composition may have different end-use quality. This necessitates the rheological testing. Different constituents of flour react with water in the process of dough making. Protein and starch of the flour primarily react differently in mixing and perform different functions in baked products. In baking process, gas is evolved which exerts pressure on the dough in a multilinear process which is measured by one of the premiere rheological methods invented in 1920s, that is, the alveograph test. It is the standard and widely accepted method for an analytical evaluation of flour. It measures the amount of pressure and time necessary to create and burst an air bubble in the dough. Alveograph test measures tenacity, extensibility, elasticity, and baking strength. Tenacity (P) refers to the maximum pressure withstood before the formation of bubble and indicates the strength of flour. Extensibility (L) refers to the maximum height of the bubble, which indicates the flexibility of the dough. Elasticity (Ie) refers to the ability of a dough to return to its former shape after the stress is removed. Baking strength or energy (W) is the manifestation of all the above three parameters. Following instruments also measure the dough consistency in similar way.

Amylograph: The wheat flour and a buffer solution is put in a rotating bowl and stirred continuously. This setup is heated by an air bath starting from room temperature to 95°C at a rate of 1.5°C/min. For α-amylase activity, the test may be ended at this temperature. If no amylase activity is there, the consistency of the sample will increase until this temperature is attained. In case of high enzyme activity, the curve will peak at lower temperature and lower viscosity. This peak height serves as a measure of enzyme activity.

Rapid ViscoAnalyzer (RVA): It is a faster version of the amylograph. The temperature control can be programmed in this instrument at varying rates. The load on the stirring motor determines the viscosity in this method. Flexibility in controlling heating/cooling enables it for different uses in laboratories.

Farinograph and Mixograph: These two methods start with the mixing of flour with water for a period and preparing the dough. While mixing, we measure the resistance to mixing and the pattern of changing this resistance. Flours having a high protein quality and containing higher amount of gluten will require more force for a longer period. These methods also let us know the amount of water that can be absorbed by the flour and how quickly it happens.

Alveograph and Extensograph: These instruments help in determining the quality of protein by measuring the extensibility or stretchability. In these methods, dough is prepared as per the protocol. Alveograph blows a bubble through a piece of dough, and the size of the dough indicates the quality of gluten. Similarly, in extensograph approach, the piece of dough is stretched in between the hands until it breaks. These methods look for the elasticity, strength, and extensibility of the flour indirectly.

14.7.3 Use of NIR in Flour Analysis

Rapid analyses of different parameters are the need of the industries dealing with wheat flour and other derived products. Use of NIR for estimating various properties of wheat flour like determination of protein, moisture, and starch can serve this purpose. The wavelength range of NIR lies between 0.75 and 2.5 μm. The first commercial NIR instruments were developed in 1970s, which have been further improved by connecting them with computers. Apart of generating instantaneous, precise, and reproducible results, these instruments can be operated by nontechnical staffs. Absorption bands of NIR range are broad and overlapping, which necessitate the consideration of several bands of the spectrum in order to eliminate the interfering effect of other components. The sample for analysis is fed in the cup, and the radiation is passed through, followed by the measurement of reflectance. Reflectance is then amplified by microcomputer of instrument and converted into numerical results that are displayed on readout screen. This whole process is completed within a minute. Some newer instruments use transmittance instead of reflectance.

Functionality: The last step of a comprehensive quality control process explains the reason behind the varying performances of flours with similar compositions and rheological properties. Properties of dough during dough production process and baking depend on the presence of glutenins (affect extensibility and elasticity), damaged starch (affects stickiness), and pentosans (affect viscosity). The amount of glutenins is measured by solvent retention capacity method using 5% lactic acid in distilled water, whereas 5% sodium carbonate and 50% sucrose in distilled water are used for the quantification of damaged starch and pentosans, respectively.

Free Fatty Acids: This is important for dry products of wheat flour like crackers and cookies. Free fatty acids result from the degradation of lipids due to high temperature and moisture during storage. Rancidity is more frequent in flours having a high level of free fatty acids. For determining its level, the extraction of lipid is done with the help of suitable solvent like petroleum ether. Now, the petroleum ether is evaporated, and lipid is dispersed in a toluene–alcohol mixture. The final amount of free fatty acids can be determined by titration of this mixture with standard potassium hydroxide.

TABLE 14.5

Different Testing Methods or Equipment Used for the Analysis of Various Flour Quality Parameters

Flour Quality	Testing Method/Equipment
A. Basic Qualities	
Moisture	Hot air oven method, NIR-based instruments
Ash content	Ash furnace method, NIR-based methods
Protein content	Kjeldahl method, Dumas combustion process
Dough consistency/α-amylase activity	Falling number test
B. Physical Qualities	
Flour color	Visual examination
Texture characteristics	Single-kernel characterization system
C. Milling Quality	Buhler laboratory flour mill test
D. Wet Gluten	Glutomatic test
E. Dough and Gluten Strength	Farinograph test, Extensograph test, Alveograph test, mixograph test
F. Flour Starch Viscosity	Amylograph test, Rapid ViscoAnalyzer test
G. Others	
Free fatty acid	Titration method
Glutenins	Solvent retention capacity method (5% lactic acid)
Damaged starch	Solvent retention capacity (using 5% sodium carbonate)
Pentosans	Solvent retention capacity (50% sucrose)

Damaged Starch: Some of the starch granules are damaged due to shearing during milling, as these are crystalline. When these damaged granules are exposed to water, these absorb excessive water and are beneficial in bread making but are very negative for cookies and other dry products. These damaged starches are converted into maltose and smaller dextrins due to the action of α- and β-amylase and may interact with gluten.

Flour Color: The color of flour is due to the presence of carotenoid in the endosperm. Endosperm colour is under the genetic control, so the flour colour will vary according to the variety. Flour color affects the color of finished product. The flour color is also affected by the presence of bran. Whole-grain flour is generally brownish creamy white, whereas the milled flour is pure whitish. However, these pigments can be bleached with benzoyl peroxide or by enzyme active soy flour. The determination of color is done by comparing with standard patent flour visually. (Table 14.5)

14.8 Conclusion and Future Prospects

Food fortification is a cost-effective technique with the potential to address malnutrition globally. Studies on the fortification of foods have shown positive results not only in the control and prevention of micronutrient deficiencies among vulnerable populations, especially women and children, but also along social, economic, and environmental dimensions. Sight and life projects offer successful examples to address micronutrient deficiencies through fortification from several LMICs and emphasize the importance of multidimensional partnerships in addressing the many challenges of food fortification strategies. This chapter aimed to demonstrate advantages and disadvantages of food fortification strategies, as well as to provide some key examples of ways in which different programs in LMICs have systematically addressed the issues identified, demonstrating that there is no one-size-fits-all solution.

REFERENCES

Allen L, de Benoist B, Dary O and Hurrell R (2006) *Guidelines on food fortification with micronutrients.* Geneva: World Health Organization and Food and Agriculture Organization of the United Nations.

Aung MS, Masuda H, Kobayashi T, Nakanishi H, Yamakawa T and Nishizawa NK (2013) Iron biofortification of Myanmar rice. *Front. Plant Sci.* 4: 1–14.

Awika J (2011) Major cereal grains production and use around the world. In: Awika J, Piironen V and Bean S (eds) *Advances in cereal science: implications to food processing and health promotion.* Atlantic City: American Chemical Society, pp. 1–13.

Bhagwat S, Gulati D, Sachdeva R and Sankar R (2014) Food fortification as a complementary strategy for the elimination of micronutrient deficiencies: case studies of large scale food fortification in two Indian States. *Asia Pac. J. Clin. Nutr.* 23: S4–S11.

Brinch-Pedersen H, Borg S, Tauris B and Holm PB (2007) Molecular genetic approaches to increasing mineral availability and vitamin content of cereals. *J. Cereal Sci.* 46(3): 308–326.

Black LJ, Seamans KM, Cashman KD and Kiely M (2012) An updated systematic review and meta-analysis of the efficacy of vitamin D food fortification. *J. Nutr.* 142(6): 1102–1108.

Borg S, Brinch-Pedersen H, Tauris B and Holm P (2009). Iron transport, deposition and bioavailability in the wheat and barley grain. *Plant Soil.* 325: 15–24.

Borrill P, Connorton JM, Balk J, Miller AJ, Sanders D and Uauy U (2014) Biofortification of wheat grain with iron and zinc: integrating novel genomic resources and knowledge from model crops. *Front. Plant Sci.* 5: 53.

Cakmak I, Pfeiffer WH and McClafferty B (2010) Biofortification of durum wheat with zinc and iron: a review. *Cereal Chem.* 87: 10–20.

Cappelli A, Oliva N and Cini E (2020) A systematic review of gluten-free dough and bread: Dough rheology, bread characteristics, and improvement strategies. *Appl. Sci.* 10: 6559.

Cardoso RVC, Fernandesa A, Gonzalez-Paramasb AM, Barrosa L and Ferreiraa ICFR (2019) Flour fortification for nutritional and health improvement: a review. *Food Res. Int.* 125: 108576.

Carvalho, SMP and Vasconcelos, MW (2013) Producing more with less: strategies and novel technologies for plant-based food biofortification. *Food Res. Int.* 54: 961–971.

Castillo-Lancellotti C, Tur JA and Uauy R (2012) Folatos y riesgo de cáncer de mama: revisiónsistemática [Folate and breast cancer risk: a systematic review]. *Rev. Med. Chil.* 140(2): 251–260.

Chadare FJ, Idohou R, Nago E, Affonfere M, Agossadou J, Fassinou TK, . . . Hounhouigan DJ (2019) Conventional and food-to-food fortification: an appraisal of past practices and lessons learned. *Food Sci. Nutr.* 7(9): 2781–2795.

Choi SJ, Decker EA and McClements DJ (2009) Impact of iron encapsulation within the interior aqueous phase of water-in-oil-water emulsions on lipid oxidation. *Food Chem.* 116(1): 271–276.

Cormick G, Betran AP, Romero IB, García-Casal MN, Perez SM, Gibbons L and Belizán JM (2021) Impact of flour fortification with calcium-on-calcium intake: a simulation study in seven countries. *Ann. N.Y. Acad. Sci.* 1493(1): 59–74.

Curtis B (2002) Wheat in the world. In: Curtis BC, Rajaram S and Gómez Macpherson H (eds) *Bread wheat– improvement and production* (FAO plant production and protection series. No. 30). Rome: Food and Agriculture Organization of the United Nations.

Dary O and Hurrell R (2006) *Guidelines on Food Fortification With Micronutrients, World Health Organization, Food and Agricultural Organization of the United Nations.* Geneva: FAO.

De-Regil LM, Jefferds MED and Peña-Rosas JP (2017) Point-of-use fortification of foods with micronutrient powders containing iron in children of preschool and school-age. *Cochrane Database Syst Rev.* 11(11): CD009666.

Debelo H, Ndiaye C, Kruger J, Hamaker BR and Ferruzzi MG (2020) African Adansonia digitata fruit pulp (baobab) modifies provitamin A carotenoid bioac cessibility from composite pearl millet porridges. *J. Food Sci. Technol.* 57(4): 1382–1392.

Dixon J (2007) The economics of wheat: research challenges from field to fork. In: Buck H, Nisi J and Salomon N (eds) *Wheat production in stressed environments.* Dordrecht: Springer, pp. 9–22.

FAOStat (2020) FAO Stat. www.fao.org/faostat

FAO Statistical Yearbook (2013) *FAO statistical yearbook.* Rome: Food and Agriculture Organization of the United Nations.

Field MS, Mithra P, Estevez D and Peña-Rosas JP (2020) Wheat flour fortification with iron for reducing anaemia and improving iron status in populations. *Cochrane Database Syst. Rev.* 7: CD011302.

Field MS, Mithra P and Peña-Rosas JP (2021) Wheat flour fortification with iron and other micronutrients for reducing anaemia and improving iron status in populations. *Cochrane Database Syst. Rev.* 1: CD011302.

Food Data Portal. 2020. https://fdc.nal.usda.gov/

Garg M, Sharma N, Sharma S, Kapoor P, Kumar A, Chunduri V and Arora P (2018) Biofortified crops generated by breeding, agronomy, and transgenic approaches are improving lives of millions of people around the world. *Front. Nutr.* 5(2018). https://doi.org/10.3389/fnut.2018.00012.

Goudia BD and Hash CT (2015) Breeding for high grain Fe and Zn levels in cereals. *Int. J. Innov. Appl. Stud.* 12: 342–354.

Gupta OP, Pandey V, Narwal S, Ram S and Singh GP (2018) Production and consumption trends of wheat based products in India. In: *Strengthening value chain in wheat and barley for doubling farmers income.* Karnal: ICAR-IIWBR.

Gupta PC, Arora S, Tomar SK and Singh AK (2015) Iron microencapsulation with blend of gum arabic, maltodextrin and modified starch using modified solvent evaporation method—milk fortification. *Food Hydrocoll.* 43: 622–628.

Hosny KM, Banjar ZM, Hariri AH and Hassan AH (2015) Solid lipid nanoparticles loaded with iron to overcome barriers for treatment of iron deficiency anemia. *Drug Des. Dev. Ther.* 9: 313–320.

Imdad A, Mayo-Wilson E, Haykal MR, Regan A, Sidhu J, Smith A, Bhutta ZA (2022) Vitamin A supplementation for preventing morbidity and mortality in children from six months to five years of age. *Cochrane Database Syst. Rev.* 3(3): CD008524.

Jafari SM and McClements DJ (2017) Nanotechnology approaches for increasing nutrient bioavailability. In: *Advances in food and nutrition research* (1st ed., vol. 81). London: Elsevier Inc.

Khush GS, Lee S, Cho JL and Jeon JS (2012) Biofortification of crops for reducing malnutrition. *Plant Biotechnol. Rep.* 6(3): 195–202.

Klemm RD, West KP Jr, Palmer AC, Johnson Q, Randall P, Ranum P and Northrop-Clewes C (2010) Vitamin A fortification of wheat flour: considerations and current recommendations. *Food Nutr. Bull.* 2010 1(Suppl): S47–S61.

Kumar P, Yadava RK, Gollen B, Kumar S, Verma RK and Yadav S (2011) Nutritional contents and medicinal properties of wheat: a review. *Life Sci. Med. Res.* 22: 1–10.

Kusn F and Suyatma NE (2017) Iron fortification of soya based infant cereal and its stability during storage. *J. Food Technol. Preserv.* 1(2): 1–6.

Lalani B, Bechoff A and Bennett B (2019) Which choice of delivery model(s) works best to deliver fortified foods? *Nutrients* 11: 1594.

Lawrence M (2013) *Food fortification: The evidence, ethics, and politics of adding nutrients to food* (1st ed.). Oxford: Oxford University Press.

Lephuthing MC, Baloyi TA and Tsilo NZSTJ (2017) Progress and challenges in improving nutritional quality in wheat. In: Wanyera R and Owuoche J (eds) *Wheat improvement, management and utilization*. London: IntechOpen.

Lewis J, Broomhead L, Jupp P and Reid J (2003) Nutrition considerations in the development and review of food standards, with particular emphasis on food composition. *Food Control* 14(6): 399–407.

Liyanage C and Hettiarachchi M (2011) Food fortification. *Ceylon Med. J.* 56: 124–127.

Martorell R, Ascencio M, Tacsan L, Alfaro T, Young MF, Addo OY, Dary O and Flores-Ayala R (2015) Effectiveness evaluation of the food fortification program of Costa Rica: impact on anemia prevalence and hemoglobin concentrations in women and children. *Am. J. Clin. Nutr.* 101(1): 210–7.

Mason NM, Jayne TS and Shiferaw B (2015) Africa's rising demand for wheat: trends, drivers, and policy implications. *Dev. Policy Rev.* 33: 581–613.

Mayo-Wilson E, Imdad A, Herzer K, Yakoob MY and Bhutta ZA (2011) Vitamin A supplements for preventing mortality, illness, and blindness in children aged under 5: systematic review and meta-analysis. *BMJ.* 343: d5094.

Osendarp SJM, Martínez H, Garrett GS, Neufeld L, de Regil L, Vossenaar M and Darnton-Hill I (2018) Large-scale food fortification and biofortification in low- and middle-income countries: a review of programs, trends, challenges, and evidence gaps. *Food Nutr. Bull.* 39: 315–331.

Öztürk B (2017) Nanoemulsions for food fortification with lipophilic vitamins: production challenges, stability and bioavailability. *Eur. J. Lipid Sci. Technol.* 119(7): 1500539.

Pambo KO, Otieno DJ and Okello JJ (2014) Consumer awareness of food fortification in Kenya: the case of vitamin-A-fortified sugar. *Int. Food Agribus. Manag. Assoc.* 1–20.

Papathakis PC and Pearson KE (2012) Food fortification improves the intake of all fortified nutrients, but fails to meet the estimated dietary requirements for vitamins A and B6, riboflavin and zinc, in lactating South African women. *Public Health Nutr.* 15(10): 1810–7.

Pfeiffer WH and McClafferty B (2007) HarvestPlus: breeding crops for better nutrition. *Crop. Sci.* 47: 88–105.

Saha S and Roy A (2020) Whole grain rice fortification as a solution to micronutrient deficiency: technologies and need for more viable alternatives. *Food Chem.* 326: 127049.

Saltzman EB, Bouis HE, Boy E, De Moura FF, Islam Y and Pfeiffer WH (2013) Biofortification: progress toward a more nourishing future. *Glob. Food Sec.* 2(1): 9–17.

Santillán-Urquiza E, Méndez-Rojas MA and Vélez-Ruiz JF (2017) Fortification of yogurt with nano and micro sized calcium, iron and zinc, effect on the physicochemical and rheological properties. *LWT Food Sci. Technol.* 80: 462–469.

Serdula M (2010) Second technical workshop on wheat flour fortification. Maximizing the impact of flour fortification to improve vitamin and mineral nutrition in populations. *Food Nutr Bull.* 1(Suppl): S86–S93.

Shah D, Sachdev HS, Gera T, De-Regil LM and Peña-Rosas JP (2016) Fortification of staple foods with zinc for improving zinc status and other health outcomes in the general population. *Cochrane Database Syst. Rev.* 6(6): CD010697.

Sharma A, Patni B and Shankhdhar D (2013) Zinc—An indispensable micronutrient. *Physiol. Mol. Biol. Plants.* 19: 11–20.

Sharma P and Gujral HS (2014). Anti-staling effects of b-glucan and barley flour in wheat flour chapatti. *Food Chem.* 145: 102–108.

Shewry PR (2009). The HEALTHGRAIN programme opens new opportunities for improving wheat for nutrition and health. *Nutr. Bull.* 34(2): 225–231.

Shewry PR and Hey SJ (2015). The contribution of wheat to human diet and health. *Food Energy Secur.* 4(3): 178–202.

Shiferaw B, Smale M, Braun H, Duveiller E, Reynolds MP and Muricho G (2013) Crops that feed the world 10. Past successes and future challenges to the role played by wheat in global food security. *Food Sci.* 5: 291–317.

Shubham K, Anukiruthika T, Dutta S, Kashyap AV, Moses JA and Anandharamakrishnan C (2020) Iron deficiency anemia: a comprehensive review on iron absorption, bioavailability and emerging food fortification approaches. *Trends Food Sci. Technol.* 99: 58–75.

Siddique A and Park YW (2019). Effect of iron fortification on microstructural, textural, and sensory characteristics of caprine milk Cheddar cheeses under different storage treatments. *J. Dairy Sci.* 102(4): 2890–2902.

Singh UM, Sareen P and Sengar RS (2013). Plant ionomics: a newer approach to study mineral transport and its regulation. *Acta Physiol. Plant.* 35: 2641–2653.

Spohrer R, Larson M, Maurin C, Laillou A, Capanzana M and Garrett GS (2013) The growing importance of staple foods and condiments used as ingredients in the food industry and implications for large-scale food fortification programs in Southeast Asia. *Food Nutr. Bull.* 34(Suppl. 2): S50–S61.

Swanepoel E, Havemann-Nel L, Rothman R, Laubscher TM, Matsungo CM and Faber SM (2019) Contribution of commercial infant products and fortified staple foods to nutrient intake at ages 6, 12, and 18 months in a cohort of children from a low socioeconomic community in South Africa. *Matern. Child Nutr.* 15(2): 1–13.

Tablante CE, Pachón H, Guetterman HM and Finkelstein JL (2019) Fortification of wheat and maize flour with folic acid for population health outcomes. *Cochrane Database Syst Rev.* 7(7).

Temple NJ and Steyn NP (2011) The cost of a healthy diet: a South African perspective. *Nutrition* 27(5): 505–508.

Teshome, EM, Otieno W, Terwel SR, Osoti V, Demir AY, Andango PEA, . . . Verhoef H (2017) Comparison of home fortification with two iron formulations among Kenyan children: rationale and design of a placebo-controlled non-inferiority trial. *Contemp. Clin. Trials Commun.* 7: 1–10.

UNICEF. (2008) *Sustainable elimination of iodine deficiency: Progress since the 1990 world summit for children.* New York: UNICEF.

Uvere PO, Onyekwere EU and Ngoddy PO (2010) Production of maize-bambara groundnut complementary foods fortified pre-fermentation with processed foods rich in calcium, iron, zinc and provitamin A. *J. Sci. Food Agric.* 90(4): 566–573.

VanDer Borght A, Goesaert H, Veraverbeke WS and Delcour JA (2005) Fractionation of wheat and wheat flour into starch and gluten: overview of the main processes and the factors involved. *J. Cereal Sci.* 41: 221–237.

Waters BM and Sankaran RP (2011) Moving micronutrients from the soil to the seeds: Genes and physiological processes from a biofortification perspective. *Plant Sci.* 180: 562–574.

World Bank (2020) Data Bank. https://databank.worldbank.org.

Zhu C, Naqvi S, Gomez-Galera S, Pelacho AM, Capell T and Christou P (2007) Transgenic strategies for the nutritional enhancement of plants. *Trends Plant Sci.* 12(12): 548–555.

15

Molecular and Biotechnological Strategies for Improving Nutritional Quality of Wheat

Divya Ambati, Rahul M. Phuke, Uday G. Reddy,
Rahul Gajghate, Singh J. B., and Sai Prasad S. V.

CONTENTS

15.1 Introduction

Over 80% of the food consumed by people comes from plants. Many people, especially in underdeveloped nations, rely on staple foods like wheat, rice, maize, and

DOI: 10.1201/9781003307938-15

potatoes, which do not fully supply all of the needed elements (Rizzo et al., 2021; FAO 2020). A healthy diet is essential for human growth and development. In addition to preserving the body's metabolism for both physical and mental welfare, it aids in illness prevention. Because of the exponential growth in the global population, which is anticipated to reach 9.7 billion by 2050 and 11 billion by 2100, the world is going to experience an extreme resource stress (United Nations News 2019). In response to this massive population strain, food instability and malnutrition have become unprecedented challenges for humanity (WHO 2018; UNICEF 2018).

Around two billion people worldwide suffer from micronutrient deficiencies, sometimes known as hidden hunger, which accounts for around 45% of infant fatalities each year (WHO 2020). The reliable access to enough, safe, inexpensive, and nutritious food is essential for having flourishing societies (Rizzo et al., 2021; Hossain et al., 2022). Our food crops, which are raised on mineral-poor soils, are unable to provide the necessary nutrients, which is the main cause of this nutritional stress. Most developing nations' agricultural production plans don't address the problems associated with hunger; instead, they concentrate on raising grain yield and crop productivity (Gaikwad et al., 2020).

To address the issue of nutritional insufficiency, greater focus should be placed on the availability of healthy foods that counteract nutrient deficit. Cereals contribute the most to daily dietary intake in areas where micronutrient deficiencies are widespread (Cakmak 2010; Bouis et al., 2011). Wheat (*Triticum aestivum* L.) is a major crop in many countries. It is mostly utilized in the making of items that are commonly consumed, including as bread, pasta, cereals, and cakes (Shewry and Hey 2015; Laskowski et al., 2019). As a result, wheat has the ability to reduce malnutrition and diseases caused by nutritional deficiencies by promoting food security and ensuring that people consume the necessary amounts of macro- and micronutrients on a daily basis (Lephuthing et al., 2017). In 2021–22, worldwide wheat output was 779 million metric tonnes. In contrast, India produced 109.5 million metric tonnes in the years 2021–2022 (USDA 2022). Wheat is consumed by 2.5 billion people globally (Listmanand Ordóñez 2019), accounting for around 20% of total dietary calories and proteins ingested (Shiferaw et al., 2013; Gajghate et al., 2020). In order to meet customer demand, wheat has better accessibility, flexibility, and output. Therefore, it is anticipated that the biofortification of wheat will significantly lower micronutrient deficiency in poor nations.

However, wheat and rice (*Oryza sativa*) exhibit suboptimal amounts of micronutrients, particularly Fe and Zn, and milling decreases their concentration even further (Cakmak et al., 2010; Sperotto et al., 2012). Iron deficiency is the most common nutritional condition in the world, impacting 1.6 billion people and contributing to the widespread prevalence of anemia (De Benoist et al., 2008). It can cause a variety of potentially fatal conditions, including chronic kidney and heart failure, as well as inflammatory bowel disease (Lopez et al., 2016). Zn deficiency affects causes a variety of physiological issues, including growth retardation, impaired brain development, increased susceptibility to infectious diseases, diarrhoea, pneumonia, and even an increased risk of infant mortality, childbirth complications, and a variety of other chronic diseases (Terrin et al., 2015).

Among the various strategies have been used to address nutritional qualities of wheat, researchers provide four distinct integrated solutions of dietary diversity, fortification, biofortification, and supplementation to address the micronutrient deficiency

(FAO 2021). Supplementation, a temporary fix for urgent problems, is the direct provision of nutrients in the form of syrup or tablets. Fortification, which involves adding micronutrients to basic foods, is a more sustainable strategy. Iodized salt is the most popular kind, although other wheat and milk products are also fortified with different micronutrients (Müller and Krawinkel 2005). The nutritional value of plant types is raised using traditional, molecular, or transgenic techniques, and genetic biofortification of staple food crops is thought to provide a long-term, sustainable solution to the problem of hidden hunger and nutritional inadequacy (Graham et al., 2007).

Biofortification is a more sustainable solution to eliminate nutritional deficit. Because once the biofortified crops are produced, there are no extra expenditures for purchasing the fortificants and adding them to the food supply during processing. Biofortification is a sustainable, economical, and long-lasting method to battle hidden hunger (Garg et al., 2018). Identifying genetic resources with high essential and useful mineral content from the present germplasm is necessary for breeding biofortified crops. These nutrients are abundant in the wild relatives, landraces, and regional cultivars of the majority of crops, allowing for their efficient use in breeding programs (Gaikwad et al., 2020). Since the quality attributes are polygenic and quantitatively regulated, traditional breeding makes it challenging to enhance them. An efficient way to increase the effectiveness of plant breeding in transmitting these quantitatively inherited traits is to quickly create enhanced genomic tools like molecular markers.

Molecular markers can greatly aid in breeding operations when the precise genomic region/QTL controlling the quality traits could be identified. The detected QTLs can be introgressed into advance breeding materials. Because they are based on bi-parental mapping populations, the value of these markers based on QTL mapping was not very substantial. However, compared to only one or a few meiotic recombination in bi-parental mapping populations, genome-wide association mapping studies (GWAS) provide exceptional opportunity to utilize varied germplasms. The degree of linkage disequilibrium in the panel affects the capacity to resolve marker/trait correlations. The opportunity exists to apply the QTL-linked markers across breeding material for identification and introgression. This chapter discusses the nutritional properties of wheat, various molecular breeding techniques, and advanced biotechnological interventions to improve the nutritional properties of wheat, providing plant breeders with a novel opportunity to learn and use these techniques into breeding program.

15.2 Need for Quality Improvement in Wheat

Wheat is the most important stable food crop for more than one-third of the world population and contributes to most of calories' and protein intake than any other cereal crops. Variety of functional products and end products, viz., bread, chapati (leavened bread), cookies, bakery products, pasta, and noodles, are processed from wheat due to unique protein, that is, gluten in wheat. These unique food products of wheat are gaining global demand because of increasing consumption, industrialization, urbanization, and changing life styles which result in a greater emphasis than ever on the end-use quality of wheat. The preferences of end use quality of wheat varieties by various stack holders of wheat chain, viz., farmers, graders, millers, processing industries, bakers etc., vary accordingly. For farmers, wheat variety which gives highest yield and grain quality that fetches highest price in the market are the need. Well-filled, bold, and large grains are preferred by the millers. Food-processing industries on the other hand are more focused on processing the quality of wheat which determines the wheat's market price and end-use value. The ability of a wheat variety to be processed at low cost, its ability to produce uniform end product, and its acceptance by the consumers are the main criteria for food-processing industry. For both types of quality, grain hardness and gluten quality and quantity are critical. View of end use consumers could vary about the wheat quality according to the product they consume. Color, shelf life of the products, nutritional aspects, and processing conditions are few of some criteria that can determine consumer preference for the wheat end products.

Wheat is a good source of protein, carbohydrates, dietary fiber, and phytochemicals and provides decent amount of vitamins and minerals. In the realm of biofortification, wheat as a staple food crop has the ability to provide nutrients for a complete physical and mental development and a healthy life. Wheat nutritional quality has become one of the priorities for food manufacturers due to the rise in consumer awareness about quality food. The latest edition of the USDA's Dietary Guidelines for Americans clearly states that all adults should eat at least three servings of whole grains every day.

Wheat accounts for about 20% of the food calories and nearly 55% of carbohydrates. It contains carbohydrate around 78%, protein 14%, fat 2.1%, minerals 2.1%, and considerable proportions of vitamins (thiamine and vitamin-B) and minerals (zinc, iron). Wheat is also a good source of trace minerals like selenium and magnesium, nutrients essential to good health (Pawan et al., 2011). Globally, several billions of people rely on wheat for a substantial part of their diet, particularly in less-developed and developing countries. So the nutritional importance of wheat grain, flour, and end products viz., bread, noodles, and other products (e.g., chapati, bulgar, couscous), has to be understood. In these nations, the diets of the rural poor are based predominantly on cereals, which provide 80% of energy and other nutrients. However, among these countries, hidden hunger (micronutrient deficiency) leading to stunted growth in children and reduction in immunity and work efficiency in adults, in particular women, is reported the most. Iron and zinc have been recognized as being the most important among micronutrients needed to combat this hidden hunger. A greater emphasis is given on wheat biofortification by increasing nutrient bioavailability of micronutrients through conventional, molecular, or transgenic approaches. To fulfill all these objectives and to meet all the demands of stakeholders breeding for wheat quality are huge tasks.

15.3 Progress in the Improvement of Wheat Quality

Wheat has become the staple food crop for a long period of time. Bread wheat (*Triticum aestivum*) and durum wheat (*Triticum durum*) have been cultivated in different parts of world, which forms a source of energy and essential nutrients. Presently, China and India are the top two wheat-producing regions, and breeders have continuously improved wheat crop focusing on factors impacting grain yield and, more recently, grain quality. Wheat quality parameters that are ideal for processing and preparing food products have been improved. Wheat quality improvement can be achieved with the effective utilization of genetic resources, allele identification controlling quality traits and dedicated tools to analyze polymer formation, and characterization particularly in response to climatic and other environmental factors. Wheat quality parameters include the nutritional value (the content in fiber, minerals, macro- and micronutrients, vitamins) and the health impact (whether positive or negative). For example, research on the consumption of gluten-based products is need of the hour.

15.3.1 Genetics and Applications in Relation to Wheat Quality Breeding

Wheat quality could be improved by altering the main storage protein genes. From many decades, many effective genes (such as GluD1 (5 + 10) and GluB1 (17 + 18)) have been efficiently utilized for quality improvement (Mohan and Gupta, 2015). HMW-GS, common alleles, have been assigned with quality scores in order to facilitate their application in breeding (Payne et al., 1987). There are six HMW-GS genes in the wheat genome, but most hexaploid wheat varieties possess only three to five HMW glutenin subunits due to the silencing of some genes (Ma et al., 2003), such as the genes encoding the Ay subunit (Yu et al., 2019). It was evidenced that the expression of Ay subunit has positive effects on grain protein content, grain yield, and quality (Roy et al., 2020). A new storage protein family consisting of the avenin-like proteins has also been identified to have a great breeding value for the improvement of wheat quality (Chen et al., 2016). In recent years, important progress has been achieved in research on the regulation of wheat storage protein synthesis (**Table 15.1**).

TABLE 15.1

Identified Transcription Factors Regulate Seed Storage Protein Synthesis in Wheat

Transcription Factor	Function	Target Gene	Sequence	Reference
SPA	Transcriptional activation	Glutenin promoters	ACGTG	Ravel et al. (2014)
SHP	Transcriptional repression	Glutenin promoters	ATGAG/CTCAT	Boudet et al. (2019)
WPBF	Transcriptional activation	Gliadin gene promoters	TGTAAAG	Dong et al. (2007)
TaPBF-D	Transcriptional activation	HMW-GS gene promoters	TGTAAAG	Zhu et al. (2018)
TaGAMyb	Transcriptional activation	HMW-GS gene promoters	C/TAACAAA/CC	Guo et al. (2015)
TaFUSCA3	Transcriptional activation	HMW-GS gene promoters	CATGCA	Sun et al. (2017)
TaNAC019	Transcriptional activation	Glutenin promoters	[AT] NNNNNN[ATC] [CG]A[CA] GN[ACT]A	Gao et al. (2021)
TaNAC100	Transcriptional repression	HMW-GS gene promoters	CATGT	Li et al. (2021)
TaSPR	Transcriptional repression	SSP gene promoters	CANNTG	Shen et al. (2021)

15.3.2 New Approaches and Achievements in Wheat Quality Improvement

Quality wheat grains are the major component of diet for a large proportion of world's population; hence new methodologies in pre-breeding and breeding techniques require innovation to achieve the desired quality. But to sustain an increased productivity along with grain quality, improvement happens to be a major challenge for breeders. This aspect is mainly important as it is estimated that the world population would increase by approximately one third to over nine billion in the next 40 years.

15.3.2.1 Progress in Improving Nutritional Quality

Nutritional quality improvement involves a series of processes to ensure that the nutrients are bioavailable to the human system upon consumption. The major process requires a genotypic and phenotypic characterization of key biological processes or pathways, which are involved in assimilation, accumulation, biosynthesis, translocation, and remobilization of desired nutritional quality components such as Fe, Zn, vitamins, and phenolic acids in the wheat grain (Ma et al., 2016). Genetic biofortification is a technique which involves the screening of germplasm to decipher any genetic variation for grain Fe and Zn levels across different wheat genotypes grown in different environments (Cakmak et al., 2010). Progress has been made to establish genetic variation of Fe and Zn across various wheat species. An important quantitative trait locus (QTL) Gpc-B1 from wild emmer wheat (*Triticum turgidum* ssp. Dicoccoides) was discovered and mapped on chromosome arm 6BS (Joppa et al., 1997). The gene of this locus was then cloned and effectively improved Zn, Fe, and protein concentrations by 12%, 18% and 38%, respectively (Uauy et al., 2006). Besides, the Xuhw89 marker, linked to the Gpc-B1 locus with a 0.1 cM genetic distance, can be used to identify and select lines with improved levels of selected micronutrients in the wheat grain (Distelfeld et al., 2007).

15.3.2.2 Efforts in Utilizing the Genetic Variations for Wheat Quality Improvement

Genetic variation associated with levels of total Fe, Zn, vitamins, and phenolic acids in different wheat genotypes cultivated across different environments can be explained with the aid of analytical instrument (Lampi et al., 2010). Such studies are made using the wholemeal flour, and few reports are available on establishing the genetic variation in grain Fe and Zn concentration in white flour. Velu et al. (2014) reported progresses made in screening more than 7,800 wheat genotypes for their variation in Zn concentration in bread wheat, durum wheat, wheat landraces and their wild relatives from several studies conducted. Amiri et al. (2015) also reported certain level of genetic variation for grain's protein, Fe, and Zn concentration among 80 irrigated bread wheat genotypes. Wheat genotypes possessing the highest levels of Fe and Zn could be selected as donors in order to improve the levels of Fe and Zn in recurrent parents who have lower levels of Fe and Zn. However, it is imperative to ensure that important traits, such as grain yield, protein content, disease resistance, and other agronomic traits, are not compromised upon the development of varieties with improved nutritional quality. Genetic variation was also reported with respect to phenolic acid content of various wheat genotypes (Mpofu et al., 2006). Hence, progress has been made to selectively breed for genotypes to improve phenolic acid content. However, further studies must be carried out to prove that genetic variation exists in phenolic acids among different wheat genotypes through germplasm screening of other wheat genotypes including wild relatives and landraces. Genetic variation on the concentration of vitamins, manganese, magnesium, copper, potassium, as well as concentrations of other anti-nutritional components is also found in the wheat grain. However, more studies are needed in this field. The establishment of genetic variation in minerals has led to the improvement of several wheat germplasms. The selected genotypes were used to improve the levels of Zn by more than twofold in other instances (Velu et al., 2014). However, the conventional breeding sets some drawbacks that it may take several years to develop a new variety with improved nutritional quality. In addition, only the total grain Fe and Zn can be increased. Therefore, breeders have no control on improving the levels of selected nutrients into that of desired grain compartments.

15.3.2.3 Wheat Quality Enhancement Through Nutrient Translocation Into Grain

Nutrient uptake and translocation or remobilizations are complex processes that contribute to the translocation of minerals – mainly Fe and Zn into the wheat grain. The process of moving micronutrients from the soil into the seeds is a complex process, which still requires further characterization. In wheat, Gpc-B1 locus derived from *T. dicoccoides* (Uauy et al., 2006) is responsible for nutrient remobilization. Wherein Distelfeld et al. (2007) showed that recombinant substitution lines (RSLs) carrying the dicoccoides Gpb-B1 allele had 12%, 18%, and 38% more Zn, Fe, and grain protein content (GPC), respectively, than (RSLs) carrying a Gpc-B1 locus acquired from durum wheat. Consequently, there is a great need to trace the origin of nutrients found in different grain compartments.

15.3.2.4 Identification of Genes Involved in Micronutrient and Phenolic Acid Accumulation

Waters and Sankaran (2011) reported that genes are involved in the uptake of Fe but, and no gene(s) involving in Fe uptake has been reported for wheat. Hence, there is a need to characterize and identify the genes involved in the uptake of Fe from soil to the seeds in wheat crop. Giehl et al. (2009) provided a comprehensive overview of genes and pathways involved mainly in Fe uptake from roots to other plant compartments. Waters et al. (2009) conducted a more comprehensive investigation on the role of the NAM-B1 gene, which affects Fe and Zn in wheat. Gpc-B1 locus from *Triticum dicoccoides* was mapped and found to enhance Zn and Fe concentrations and encoded a NAC transcription factor responsible to accelerate senescence. Senescence involves programmed degradation of cell constituents, which makes nutrients available for remobilization from leaves to developing seeds (Distelfeld et al., 2007). Kohl et al. (2015) reported that some NAC transcription factors were upregulated in the glumes in 14 days after anthesis and associated with developmental senescence. During senescence, proteases are rapidly activated to degrade leaf proteins into amino acids. Serine proteases are the most important family of proteases participating in nitrogen remobilization (NR) during grain filling, which act as major regulators and executors in wheat and barley (Hollmann et al., 2014). In wheat and barley, the specific NAC and WRKY transcription factors, in combination with hormones (abscisic acid and jasmonic acid), have been involved in the regulation of transition between early grain filling and developmental senescence (Gregersen et al., 2007). Zhao et al. (2015) identified a novel NAC1-type transcription factor, TaNAC-S, in wheat, with gene expression located primarily in the leaf/sheath tissues. The overexpression of TaNAC-S in transgenic wheat plants resulted in a delayed leaf senescence, which led not only to increased GPC but also to increased grain yields; thus, this result further verified the improved NR from vegetative organs to growing grain in transgenic lines (Zhao et al., 2015).

Ma et al. (2016) reported five key enzymes, namely phenylalanine ammonia lyase (PAL), coumaric acid 3-hydrolase (C3 H), cinnamic acid 4-hydrolase (C4 H), 4-coumarate CoA ligase (4CL), and caffeic acid/5-hydroxyferulic acid O-methyltransferase (COMT), which are essential for the biosynthesis of phenolic acids. Ma et al. (2016) characterized gene expression patterns of nine candidate genes associated with phenolic acid biosynthesis during early and late grain filling stages which are crucial growth stages in polyphenol accumulation. The study also revealed that seven genes (namely TaPAL1, TaPAL2, Ta4CL1, Ta4CL2, TaCOMT1, TaCOMT2, and TaC3H2) were found to be expressed highly during the early stages of grain development among white, red, and purple wheat. However, TaC3H1 was the single gene that was expressed only during the later stage of wheat improvement, management and utilization of grain development. Finally, five genes (namely TaC4H, TaPAL1, TaPAL2, Ta4CL2, and TaCOMT1) showed higher expressions in both early and later grain developmental stages. Hence, there is still a need to conduct studies to further characterize the process of phenolic acid accumulation in seeds.

15.4 Molecular Breeding Approaches for Enhancing Wheat Quality

Molecular breeding helps to hasten the conventional breeding as it is mainly based on the phenotypic selection of genotypes, and genotype × environment (G × E) interaction plays a main role. As wheat is mainly utilized for the production of wide range of products, viz., pasta, chapati, biscuits, cakes, and bread, breeding for wheat quality traits, viz., hardness of grain, grain protein content, composition and content of gluten, and starch properties is given prime importance by the breeders. Detection and determining wheat quality parameters include destructive methods which make analysis of quality parameters not possible in advanced wheat breeding material till the F6 or F7 generations. Determining quality traits in breeding program through conventional methods is cumbersome and time-consuming. So, breeding for wheat quality traits through molecular means would ensure the screening of large amount of breeding material in the early stages.

15.4.1 Grain Hardness

Grain hardness influences milling quality, physical and chemical seed properties, and end-use properties of wheat. Kernel hardness, along with grain protein content (GPC) and seed color, is the one of main criteria involved in the grading of wheat grain. Flour obtained from grain with varying texture has an effect on water absorption and the end product, viz., bread-making industry prefers flour from grain with hard endosperm as it has higher water absorption. For best biscuit quality, flour of soft grain endosperm having 11% protein along with weak and extensible gluten is preferred. Grain hardness is primarily controlled by the *Hardness* locus on chromosome 5DS. This locus consists of three small genes: *Pina-D1, Pinb-D1 (Puroindoline a/b)*, and *grain softness protein-1 (Gsp-1)*. In addition, minor loci other than the *Ha* may also be involved in modifying grain hardness. For example, *QTLs* associated with grain hardness of wheat are located on different chromosomes: *1A, 2A, 5A, 7A, 2B, 6B, 7B, 2D, 6D*, and *7D* (Sourdille et al., 1996; Perretant et al., 2000; Galande et al., 2001; Turner et al., 2004; Tsilo et al., 2011; Geng et al., 2012). Wheat varieties with the wild-type alleles *Pina-D1a* and *Pinb-D1a* normally have soft grain, while deletions or other loss-of-function mutations in one or both *Pin* genes cause harder grain (Morris et al., 2011; Li et al., 2014).

Distribution and diversity of *Pin* genes and their influence on kernel hardness were studied in diverse populations of wheat, and plenty of information is available about puroindoline diversity (Kumar et al., 2015). With a high hardness index [HI] of more than 75, durum wheat kernels are very hard as they lack the D genome and *Ha* locus. In particular, the Ha locus has been transferred from common wheat into durum wheat variety Svevo using homoeologous recombination (Morris et al., 2011). Effects of *Pin* genes on kernel softening was demonstrated in durum wheat, and many durum wheat lines with soft-kernel were developed and their grain characteristics, milling quality, and food-processing qualities being comprehensively studied (Boehm et al., 2017). The development of soft kernel durum wheat lines and their significance have been reviewed (Morris et al., 2019). DBW 14, an Indian wheat variety with hard kernel, was introgressed with puroindoline grain softness gene *Pina-D1a* from an Australian soft-grained variety Barham (Rai et al., 2019). Association analyses of 372 diverse European wheat varieties revealed a strong quantitative genetic nature of GPC and GSC with associations on groups 2, 3, and 6 chromosomes. Based on wheat's reference and pan-genome sequences, the physical characterization of two loci, *viz.*, QGpc. ipk-2B and QGpc.ipk-6A, facilitated the identification of the candidate genes for GPC (Muqaddasi et al., 2020).

15.4.2 Gluten

Gluten is a major storage protein in wheat with complex nature, present in the endosperm of the grain and has profound influence on the quality of end products. Various studies have indicated that strong and extensible flour dough is preferred by pizza and bread-making industries, whereas weak and extensible dough offers cake, cookies, and other short-textured products. In south Asian nations where chapati is the main end product of wheat, the flour which gives medium strong and extensible gluten is preferred. Pelshenke and sodium dodecyl sulfate (SDS) sedimentation test, farinograph, the extensograph are the few lab tests which measure the quantity and quality of gluten (AbuHammad et al., 2012; Singh et al., 2018). HMW-GS, high molecular weight proteins (80–130 kDa), and LMW-GS, low molecular weight (10–70 kDa) proteins, which are polymeric and monomeric gliadins constitute gluten in wheat (Shewry 2009). The genes coding these high- and low-molecular-weight glutenins and gliadins are positioned on six chromosomes (Payne 1987). Different forms/ alleles of corresponding genes give rise to structural changes in each fraction of these polypeptides. The allelic differences for HMW-GS and LMW-GS have been correlated with rheological parameters, and the corresponding molecular markers identified to discern the presence of these alleles have been developed (Wang et al., 2009, 2010). Information about the HMW and LMW-GS genes from more than 8,000 diverse wheat varieties is available in the form of databases, and these can be utilized to predict the end-use potential of the genotype (Békés and Wrigley 2013). Analysis of HMW-GS and technological quality in the wheat cultivars of Slovak breeding programs has led to the development of high bread-making quality wheat varieties. Through genetic transformation, many of these genes were introduced into bread and durum wheat (Tosi et al., 2005; Blechl et al., 2007), viz., the high-molecular-weight glutenin HMW-GS *1Ax1* gene was introduced into the cultivar Bobwhite to have a superior bread-making quality (Altpeter et al., 1996). De Bustos et al. (2001) carried out marker-assisted selection to improve HMW-GS in wheat. Landraces of wheat representing genetic variability for HMW-GS and LMW-GS not available in many modern cultivars can be screened and utilized for the improvement of quality in wheat (Gregová et al., 1999).

15.4.3 Grain Protein Content (GPC)

Grain protein content (GPC) in wheat has been a major trait of interest for breeders since it has enormous end-use potential. Genes attributing high protein content (high GPC) were introgressed into elite wheat cultivars. In 1970s, high GPC gene, that is, Gpc-B1, was identified from a wild emmer wheat, *Triticum turgidum* ssp. *dicoccoides*, Israel accession FA-15–3 (Avivi 1978), using alien chromosome substitution lines and was mapped on 6BS (Joppa et al., 1997, Olmos et al., 2003). Advanced breeding tools, viz., molecular markers and QTL analysis, have allowed genetic dissection of GPC in wheat (Bernardo 2008). Many number of major and minor QTLs were identified controlling GPC in wheat and are well utilized in the breeding programs. In particular, MAS has been successfully used in wheat to increase GPC through the introgression of the Gpc-B1 quantitative trait locus (QTL) for high grain protein content (Uauy et al., 2006; Sherman et al., 2008). Marker-assisted selection (MAS) was performed for the introgression of a major gene for high GPC (*Gpc-B1*) into ten wheat genotypes and seven MAS-derived progenies with significantly higher GPC (14.83–17.85%) without yield penalty (Kumar et al., 2011). Marker-assisted backcrossing (MABC) was successfully used to improve GPC in wheat cultivar HUW468 by crossing with Glu269 and transferred Gpc-B1 within five crop cycles indicating the practical utility of MABC for developing high GPC lines (Viswakarma et al., 2014). Similar efforts were made by Gupta et al. (2008), and Mishra et al. (2015) to introgress GPC genes into elite wheat backgrounds.

In bread wheat, some success stories, but not of commercial success, of the introgression of major gene Gpc-B1 (high protein content) with MAS have been reported by several researchers. In durum wheat, one successful example of a cultivar having high protein content using MAS, namely, "Desert King High Protein", had been developed at the University of California wheat-breeding program (Gaikwad et al., 2020). However, little success has been achieved in the development of nutritionally rich commercial cultivars with MAS in other food crops. The simple reason is that most of these traits are controlled by a large number of genes having a little effect individually on the expression of these traits.

15.4.4 Micronutrients

Micronutrient malnutrition or hidden hunger, observed in around three billion people worldwide, is a major concern (Welch and Graham 2004). Deficiency of key micronutrients, *viz.*, iron (Fe), zinc (Zn), iodine (I), selenium (Se), and vitamin A among women and children of developing countries, has a great effect on health care system and economy of the countries (Ghandilyan et al., 2006). Among the remedies available to combat hidden hunger, the economic and environment-friendly solution which attained huge importance in the recent past is "biofortification" (Singh et al., 2005). Wheat, being a staple crop, becomes ideal crop for biofortification and enrichment of iron (Fe); zinc (Zn) and vitamin A which were limiting micronutrients according to World Health Organization attain importance (Ortiz-Monasterio et al., 2007). Therefore, the development of biofortified wheat varieties with improved concentrations of Fe and Zn can help to control malnutrition among the populations who are mostly dependent on cereal diet. Genetic variability for the micronutrient content was studied among the wheat cultivars and wild relatives, and it was found that wild relatives of wheat may have a greater potential than common wheat cultivars for micronutrient biofortification (Welch and Graham 2004; Ortiz-Monasterio et al., 2007). Many studies have shown strong associations among grain micronutrient content (Fe and Zn) and protein content. High-protein content gene Gpc-B1 was found effective in improving grain Fe and Zn concentrations. This will help wheat breeders to enhance the micronutrient and protein content together in their breeding schema.

These micronutrient content traits are polygenic in nature, and genetic variation for these traits is available in the genetic resources and wild species of important food crops. Quantitative trait locus (QTL) mapping has been a useful tool to tap the variation available for grain Fe and Zn concentrations (Ghandilyan et al., 2006). Many QTLs for micronutrient concentration in wheat grain were mapped in recent years (Distelfeld et al., 2007; Ozkan et al., 2007; Shi et al., 2008; Peleg et al., 2009; Tiwari et al., 2009; Wani et al., 2022) utilizing RILs and doubled haploid populations as the mapping populations. QTLs for high-grain Fe and Zn concentrations were reported on various chromosomes, viz., 1A, 1D, 1B, 2A, 2B, 3B, 3D, 4B, 5A, 6A, 6B, 7A, 7B, and 7D in hexaploid wheat (Srinivasa et al., 2014; Crespo-Herrera et al., 2016; Tiwari et al., 2016; Velu et al., 2017; Liu et al., 2019) in the recent past. GWAS of harvest plus wheat genotypes panel was done by Velu et al. (2018) which was tested across India and Mexico. It showed the presence of two QTL regions on chromosomes 2 and 7 for high-grain Zn content, that is, QZn2A and QZn7B, which were explaining only 11.9% genetic variance.

The development of biofortified wheat varieties is in the forefront being led by various public and private sector organizations in Asia and African countries in collaboration with international organizations like CIMMYT, ICARDA, and Harvest Plus. In India, the development of Fe- and Zn-rich wheat varieties has gained a momentum in last few years and a total of 11 varieties of bread wheat and 5 varieties of durum wheat rich in Fe or Zn or both.

15.5 Advanced Biotechnological Interventions for Quality Improvement in Wheat

The conventional breeding approaches have improved the crop plants to a great extent for economic yield, pest, and disease resistance including quality traits. In the recent decade, the conventional breeding approaches have been accelerated through the identification of genes/QTLs for the favourable traits and their introgression in cultivars through molecular breeding approaches like QTL mapping and marker-assisted backcrossing and selection. However, the projected results are not achieved especially in the case of nutritional traits and are mostly led to undesirable agronomic traits with improved nutrition. The recent biotechnological approaches based on direct genetical intervention like RNAi, miRNA, and CRISPER/CAS showed immense potential to enhance nutritional quality in food crops with the help of recent advanced technologies in genetic engineering.

15.5.1 RNAi (RNA Silencing) for the Improvement of Quality Traits in Wheat

RNAi is the reverse genetic approach, as it downregulates the expression of targeted genes in a sequence-specific manner, and RNAi works as a post-transcriptional process using double-strand RNA (dsRNA) leading to gene silencing (Gil-Humanes et al., 2008). The RNAi approach has a great potential in polyploid species since there are multiple homologous copies of targeted traits, and they can be simultaneously silenced with a single RNAi construct (Fu et al., 2007). The allopolyploid wheat with three homoeologous genomes (A, B, and D) shares 95% sequence similarity, and hence the chemical or insertion mutagenesis targeting mutants with loss of function through phenotypic screening is less effective in wheat; in contrast to this, the RNAi gene silencing strategy is a viable alternative for gene functional analysis through simultaneous knockdown of multiple copies of related genes. In wheat, grain is the most important part as it is widely consumed, and therefore modifying the grain composition to improve the grain quality for processing and human consumption becomes the priority research activity. The RNAi approach is useful to know the functional activities of different genes responsible for grain quality and to lose the function of deleterious genes for improving the overall quality such as loss of function of γ galadin to reduce the celiac disease problem in the gluten intolerance population.

The first attempt to improve wheat grain quality through RNAi approach was carried out for silencing of the waxy gene to reduce the amylose content in wheat (Li et al., 2005), In this study, the gene "granule bound starch synthase 1" (*GBSSI* or waxy protein) was targeted and silenced, which could develop four transgenic plants, among which cultivar "Yangmai 10" showed a significant reduction in amylose content in the endosperm. In another study to improve the wheat grain for making it suitable for human consumption by improving health and reducing the risk of non-infectious diseases, the RNAi strategy was used through the down-regulating of two isoforms of starch-branching enzymes (SBE)II viz., SBE IIa and SBE IIb for improving the resistant starch; the suppression of both these isoforms (SBE IIa and SBE IIb) showed the significant effect on amylose content in comparison to individual isoform supersession, and the finding also confirms that the high amylose wheat through its resistant starch content has a significant potential to improve human health (Regina et al., 2006).

Biofortifying wheat through increased grain iron (Fe) and Zinc (Zn) is one of the sustainable ways of wheat quality improvement and to know the functions of the gene involved in increasing grain Fe and Zn content; the attempt was carried out to silence the (NAM-B1) gene which accelerates the senescence and increases the nutrient mobilization from leaves to developing grains, and the silencing of NAM-D1 through RNA interference leads to a delay in senescence by more than three weeks and increased grain protein – Fe and Zn by more than 30% (Distelfeld et al., 2006). Chronic enteropathy due to the ingestion of gluten protein from wheat, barley, and rye is known as celiac disease (CD) (Trier 1998; Sollid 2002). The ingestion of gluten results in lesion formation in the small intestine, which is characterized by microvilli flattening, hyperplasia of crypt cells, and infiltration of leukocytes (Sollid 2002), and ultimately this results in symptoms of diarrhea and malabsorption of food. The RNAi approach has also been used to manage celiac disease due to wheat consumption, and the control of celiac disease due to the ingestion of gluten from wheat is possible only through the lifelong gluten exclusion diet; these inflammatory reactions in a celiac patient are governed by T-cells that recognized gluten peptides in the context of human leukocytes antigen HLA-DQ2 or HLA-DQ8 molecules. The RNAi approach was used to downregulate the expression of gliadins in transgenic lines, the total gluten protein was extracted from these transgenic lines and tested for the ability to stimulate four different T-cells clones derived from the intestinal lesion of celiac disease patients specific for four different epitopes, the five transgenic lines resulted in 1.5–2 log of reduction in the number of different epitopes. Additionally, these transgenic lines were tested with other T-cell lines that were reactive with ω-gliadin epitopes; among these lines for three transgenic lines the total gluten extracts were unable to elicit T-cells, and six transgenic lines showed the reduction response. This study proves that the downregulation of gliadin through RNAi can develop wheat genotypes with low toxicity for celiac disease patients (Gil-Humanes et al., 2010). The wheat processing quality is governed by the high molecular weight-glutenin subunits (HMW-GS); hence, this is very pivotal in understanding the role of these subunits in flour processing quality of wheat. A study was conducted using RNAi in wheat cultivar Bobwhite which has five actively expressed HMW-GS genes, viz., *1Ax2**, *1Dx5*, *1Bx9*, *1By9*, and *1Dy10*, and the aim was to silence the gene *1Dx5*. Among the six silenced transgenic events characterized, four events resulted in a complete blockage of *1Dx5* expression, and in other two events partial reduction was observed. This silencing of *1Dx5* subunit caused a decrease in the processing quality of wheat studied through faringraph, gluten, and zeleny tests (Yue et al., 2008).

15.5.2 Clustered Regularly Interspaced Short Palindromic Repeats CRISPER/Cas9 for Grain Quality Improvement in Wheat

Another direct gene manipulation method, cluster regulatory interspersed short palindromic repeats (CRISPR), has achieved the directed gene editing and sequence variation (Gilbert et al., 2013; Shan et al., 2013; Soda et al., 2018). Wheat grain quality is governed by many genes, which is also affected by many factors like testing environment and analysis method and further gets complicated due to the complex wheat hexaploid nature. CRISPER/Cas9-based gene editing is a good solution through creating more allelic variations in a much faster and more precise manner. The wheat quality improvement was successfully attempted in the recent past using the CRISPER/Cas9 technology to silence HMW-GS to produce low gluten foodstuff and for the modification of starch composition, structure, and properties. In the case of celiac disease, the 33 mer is the main immunodominant peptide of α-gliadin gene family of wheat responsible for celiac disease in the patients. Sánchez-León et al. (2018) targeted the conserved region adjacent to the coding sequence of 33 mer in α-gliadin gene resulting in the mutation of 35 different genes in one of the lines that result in a reduction of immunoreactivity by 85%; further, low-gluten, transgenic free lines are identified with no off target mutation, and these lines are the potential source for the production of low-gluten wheat products and to use as a donor in the varietal development program.

The end-use quality of wheat is determined by the stability time (ST) and SDS-sedimentation value (SDS), where ST determines the final quality of wheat products such as bread, steamed bread, and noodles (Tsilo et al., 2011), and SDS is an indicator for determining the gluten quality of wheat, which is positively correlated with dough rheological properties (Kaur et al., 2013). The QTL (*Qst/Sv-6A-2851*) on chromosome 6A is known for ST and SDS through testing under multiple environments, and this QTL has the xylanase inhibitor protein (Xip) gene; hence, the study was conducted to functionally validate the *Xip* gene through CRISPER/Cas9 mutagenesis system by Sun et al. (2022). The three homologous gens *TaXip-6A, TaXip-6B*, and *TaXIp-6D* were edited using CRISPER/Cas9 in the wheat variety "Fielder", and from $T_{2:3}$ generation, two mutant types were obtained (aaBBDD and AAbbdd) for SDS – both the mutants showed a significant improvement of 31.77 ml and 27.30 ml over the wild type 20.08 ml, whereas for ST only one mutant aaBBDD (2.60) was significantly higher than that of wild type (2.25). An attempt was also made to increase the amylose and resistant starch content in both spring and winter wheat using the CRISPER/Cas9 approach by Li et al. (2021). The targeted mutagenesis of *TaSBEIIa* in winter wheat cultivar *cv* Zhengmai 7698(ZM) and in spring wheat cultivar cv Bobwhite was done for increasing the amylose content. A number of transgenic mutants, with either partial or triple null *TaSBEIIa*, were generated in both the cultivars. The mutants were tested for starch content, structure, and properties, and it was observed that the changes are dosage dependent, showing that triple null mutant resulted in a more profound impact on starch composition, fine structure of amylopectin, and physiochemical and nutritional properties. It contains significant higher amylose, resistant starch, and soluble pentosane contents which are beneficial for human health. The transgenic free lines were developed with high amylose content in both spring and winter wheat, and the finding confirms the pivotal role of *TaSBEIIa* on end-use quality of wheat. The grain overall qualities such as grain hardness, starch quality, and dough color were also targeted for genome editing using the known four grain quality traits, *viz.*, *pinb, waxy, ppo*, and *psy*. The Hi-Tom sequencing was used to identify the CRISPER-targeted mutants. Although the complex genome of wheat makes it difficult to edit all alleles simultaneously, in T2 and T3 generations, most of the mutants showed 100% editing efficiency. The selected mutants with the same mutation in all three genomes A, B, and D for *pinb-47, waxy-2, ppo-7*, and *Psy-13* were analyzed, which showed a significant decrease in the expression of all targeted four genes compared to wild type (WT) (Zhang et al., 2021)

15.6 Conclusion and Future Perspectives

Various biotechnological approaches like in vitro tissue culture, gene transfer, and the use of DNA markers have emerged as new approaches to complement the conventional methods of breeding. DNA-based markers were identified for numerous quality traits in wheat which can offer various advantages to transfer the quality traits in breeding program, compared to the traditional methods saving time and cost. In near future, molecular approaches including genetic transformation, wheat whole-genome and pan-genome sequencing, GWAS, trait dissection, gene discovery, speed breeding, CRISPR-cas, gene editing, high-throughput genotype and phenotype profiling, and multiplex genome editing will provide massive amounts of genetic information and accelerate wheat improvement. We believe that combining molecular approaches with other state-of-the-art breeding technologies will underpin efforts to breed green super wheat varieties for sustainable agriculture and a better environment to ensure global food security.

REFERENCES

AbuHammad, W. A., Elias, E. M., Manthey, F. A., Alamri, M. S., & Mergoum, M. (2012). A comparison of methods for assessing dough and gluten strength of durum wheat and their relationship to pasta cooking quality. International Journal of Food Science & Technology, 47(12), 2561–2573.

Altpeter, F., Vasil, V., Srivastava, V., & Vasil, I. K. (1996). Integration and expression of the high-molecular-weight glutenin subunit 1Ax1 gene into wheat. Nature Biotechnology, 14(9), 1155–1159.

Amiri, R., Bahraminejad, S., Sasani, S., Jalali-Honarmand, S., & Fakhri, R. (2015). Bread wheat genetic variation for grain's protein, iron and zinc concentrations as uptake by their genetic ability. European Journal of Agronomy, 67, 20–26. https://doi.org/10.1016/j.eja.2015.03.004.

Avivi, L. (1978). High protein content in wild tetraploid Triticum dicoccoides Korn. In Proceedings of the 5th International Wheat Genetics Symposium. New Delhi, India: Indian Society of Genetics and Plant Breeding (ISGPB) (pp. 372–380).

Békés, F., & Wrigley, C. W. (2013). Gluten alleles and predicted dough quality for wheat varieties worldwide: a great resource—free on the AACC International Website. Cereal Foods World, 58(6), 325–328.

Bernardo, R. (2008). Molecular markers and selection for complex traits in plants: learning from the last 20 years. Crop Science, 48(5), 1649–1664.

Blechl, A., Lin, J., Nguyen, S., Chan, R., Anderson, O. D., & Dupont, F. M. (2007). Transgenic wheats with elevated levels of Dx5 and/or Dy10 high-molecular-weight glutenin subunits yield doughs with increased mixing strength and tolerance. Journal of Cereal Science, 45(2), 172–183.

Boehm Jr, J. D., Ibba, M. I., Kiszonas, A. M., & Morris, C. F. (2017). End-use quality of CIMMYT-derived soft-kernel durum wheat germplasm: I. grain, milling, and soft wheat quality. Crop Science, 57(3), 1475–1484.

Boudet, J., Merlino, M., Plessis, A., Gaudin, J. C., Dardevet, M., Perrochon, S., et al. (2019). The bZIP transcription factor SPA Heterodimerizing Protein represses glutenin synthesis in Triticum aestivum. The Plant Journal: For Cell and Molecular Biology, 97(5), 858–871. https://doi.org/10.1111/tpj.14163.

Bouis, H. E., Hotz, C., McClafferty, B., Meenakshi, J. V., & Pfeiffer, W. H. (2011). Biofortification: A new tool to reduce micronutrient malnutrition. Food and Nutrition Bulletin, 32(Suppl_1), S31–S40. https://doi.org/10.1177/15648265110321S105.

Cakmak, I. (2010). Biofortification of cereals with zinc and iron through fertilization strategy. In 19th World Congress of Soil Science, Soil Solutions for a Changing World, August 2010. Brisbane, Australia (pp. 1–6). Published on DVD.

Cakmak, I., Pfeiffer, W. H., & McClafferty, B. (2010). Biofortification of durum wheat with zinc and iron. Cereal Chemistry, 87(1), 10–20. https://doi.org/10.1094/CCHEM-87-1-0010.

Chen, X. Y., Cao, X. Y., Zhang, Y. J., Islam, S., Zhang, J. J., Yang, R. C., et al. (2016). Genetic characterization of cysteine-rich type-b avenin-like protein coding genes in common wheat. Scientific Reports, 6(1), 1–12.

Crespo-Herrera, L. A., Velu, G., & Singh, R. P. (2016). Quantitative trait loci mapping reveals pleiotropic effect for grain iron and zinc concentrations in wheat. Annals of Applied Biology, 169(1), 27–35. https://doi.org/10.1111/aab.12276.

De Benoist, B., McLean, E., Egli, I., & Cogswell, M. (2008). Worldwide Prevalence of Anaemia 1993–2005. Geneva: WHO Global Database of Anaemia.

De Bustos, A., Rubio, P., Soler, C., Garcia, P., & Jouve, N. (2001). Marker assisted selection to improve HMW-glutenins in wheat. In Wheat in a Global Environment (pp. 171–176). Dordrecht: Springer.

Distelfeld, A., Cakmak, I., Peleg, Z., Ozturk, L., Yazici, A. M., Budak, H., Saranga, Y., & Fahima, T. (2007). Multiple QTL-effects of wheat *Gpc-B1* locus on grain protein and micronutrient concentrations. Physiologia Plantarum, 129(3), 635–643. https://doi.org/10.1111/j.1399 3054.2006.00841.x.

Distelfeld, A., Uauy, C., Fahima, T., & Dubcovsky, J. (2006). Physical map of the wheat high-grain protein content gene Gpc-B1 and development of a high-throughput molecular marker. New Phytologist, 169(4), 753–763.

Dong, G., Ni, Z., Yao, Y., Nie, X., & Sun, Q. (2007). Wheat Dof transcription factor WPBF interacts with TaQM and activates transcription of an alpha-gliadin gene during wheat seed development. Plant Molecular Biology, 63(1), 73–84. https://doi.org/10.1007/s11103-006-9073-3.

FAO. (2020). International year of plant health – protecting plants, protecting life. http://www.fao.org/plant-health-2020.

FAO. (2021). The state of food security and nutrition in the world 2021. In building climate resilience for food security and nutrition; Food and Agriculture Org.: Rome, Italy.

Fu, D., Uauy, C., Blechl, A., & Dubcovsky, J. (2007). RNA interference for wheat functional gene analysis. Transgenic Research, 16(6), 689–701.

Gaikwad, K. B., Rani, S., Kumar, M., Gupta, V., Babu, P. H., Bainsla, N. K., & Yadav, R. (2020). Enhancing the nutritional quality of major food crops through conventional and genomics-assisted breeding. Frontiers in Nutrition, 7, 533453.

Gajghate, R., Chourasiya, D., & Harikrishna, & Sharma, R. K. (2020). Plant morphological, physiological traits associated with adaptation against heat stress in wheat and maize. In Plant Stress Biology (pp. 51–81). Singapore: Springer.

Galande, A., Tiwari, R., Ammiraju, J. Santra, D. K., Lagu, M. D., Rao, V. S., Gupta, V. S., Misra, B. K., Nagarajan, S., & Ranjekar, P. K. (2001). Genetic analysis of kernel hardness in bread wheat using PCR-based markers. Theoritical and Applied Genetics, 103, 601–606. https://doi.org/10.1007/PL00002915

Gao, Y., An, K., Guo, W., Chen, Y., Zhang, R., Zhang, X., et al. (2021). The endosperm-specific transcription factor TaNAC019 regulates glutenin and starch accumulation and its elite allele improves wheat grain quality. The Plant Cell, 33(3), 603–622. https://doi.org/10.1093/plcell/koaa040.

Garg, M., Sharma, N., Sharma, S., Kapoor, P., Kumar, A., Chunduri, V., & Arora, P. (2018). Biofortified crops generated by breeding, agronomy, and transgenic approaches are improving lives of millions of people around the world. Frontiers in Nutrition, 12. https://doi.org/10.3389/fnut.2018.00012.

Geng, H., Beecher, B. S., He, Z., & Morris, C. F. (2012). Physical Mapping of Puroindoline b-2 Genes in Wheat using 'Chinese Spring' Chromosome Group 7 Deletion Lines. Crop Science, 52(6), 2674–2678.

Ghandilyan, A., Vreugdenhil, D., & Aarts, M. G. (2006). Progress in the genetic understanding of plant iron and zinc nutrition. Physiologia Plantarum, 126(3), 407–417.

Giehl, R. F., Meda, A. R., & von Wiren, N. (2009). Moving up, down, and everywhere: signaling of micronutrients in plants. Current Opinion in Plant Biology, 12(3), 320–327. https://doi.org/10.1016/j.pbi.2009.04.006.

Gilbert, L. A., Larson, M. H., Morsut, L., Liu, Z., Brar, G. A., Torres, S. E., Stern-Ginossar, N., Brandman, O., Whitehead, E. H., Doudna, J. A., Lim, W. A., Weissman, J. S., & Qi, L. S. (2013). CRISPR-mediated modular RNA-guided regulation of transcription in eukaryotes. Cell, 154(2), 442–451. https://doi.org/10.1016/j.cell.2013.06.044.

Gil-Humanes, J., Pistón, F., Hernando, A., Alvarez, J. B., Shewry, P. R., & Barro, F. (2008). Silencing of γ-gliadins by RNA interference (RNAi) in bread wheat. Journal of Cereal Science, 48(3), 565–568.

Gil-Humanes, J., Pistón, F., Tollefsen, S., Sollid, L. M., & Barro, F. (2010). Effective shutdown in the expression of celiac disease-related wheat gliadin T-cell epitopes by RNA interference. Proceedings of the National Academy of Sciences, 107(39), 17023–17028.

Graham, R. D., Welch, R. M., Saunders, D. A., Ortiz-Monasterio, I., Bouis, H. E., Bonierbale, M., De Haan, S., Burgos, G., Thiele, G., Liria, R., 2007. Nutritious subsistence food systems. Advances in Agronomy, 92, 1–74. https://doi.org/10.1016/S0065-2113(04)92001-9.

Gregersen, P. L., & Holm, P. B. (2007). Transcriptome analysis of senescence in the flag leaf of wheat (*Triticum aestivum* L.). Plant Biotechnology Journal, 5(1), 192–206. https://doi.org/10.1111/j.1467-7652.2006.00232.x.

Gregová, E., Hermuth, J. í., Kraic, J. & Dotlacil, L. (1999). Protein heterogeneity in European wheat landraces and obsolete cultivars. Genetic Resources and Crop Evolution, 46(5), 521–528.

Guo, W., Yang, H., Liu, Y., Gao, Y., Ni, Z., Peng, H., Xin, M., Hu, Z., Sun, Q., Yao, Y. (2015). The wheat transcription factor TaGAMyb recruits histone acetyltransferase and activates the expression of a high-molecular-weight glutenin subunit gene. The Plant Journal, 84(2), 347–359. https://doi.org/10.1111/tpj.13003.

Gupta, P. K., Balyan, H. S., Kumar, J., Kulwal, P. K., Kumar, N., Mir, R. R., Kumar, A., & Prabhu, K. V. (2008). QTL analysis and marker assisted selection for improvement in grain protein content and pre-harvest sprouting tolerance in bread wheat. *11th Wheat Genetics Symposium (IWGS)*, Brisbane, Australia. pp. 1–3.

Hollmann, J., Gregersen, P. L., & Krupinska, K. (2014). Identification of predominant genes involved in regulation and execution of senescence-associated nitrogen remobilization in flag leaves of field grown barley. Journal of Experimental Botany, 65(14), 3963–3973. https://doi.org/10.1093/jxb/eru094.

Hossain, F., Zunjare, R. U., Muthusamy, V., Bhat, J. S., Mehta, B. K., Sharma, D., et al. (2022). Biofortification of maize for nutritional security. In Biofortification of Staple Crops (pp. 147–174). Springer, Singapore. https://doi.org/10.1007/978-981-16-3280-8_6.

Joppa, L. R., Du, C., Hart, G. E., & Hareland, G. A. (1997). Mapping gene (s) for grain protein in tetraploid wheat (*Triticum turgidum* L.) using a population of recombinant inbred chromosome lines. Crop Science, 37(5), 1586–1589. https://doi.org/10.2135/cropsci1997.0011183X003700050030x.

Kaur, A., Singh, N., Ahlawat, A. K., Kaur, S., Singh, A. M., Chauhan, H., & Singh, G. P. (2013). Diversity in grain, flour, dough and gluten properties amongst Indian wheat cultivars varying in high molecular weight subunits (HMW-GS). Food Research International, 53(1), 63–72. https://doi.org/10.1016/j.foodres.2013.03.009.

Kohl, S., Hollmann, J., Erban, A., Kopka, J., Riewe, D., Weschke, W., & Weber, H. (2015). Metabolic and transcriptional transitions in barley glumes reveal a role as transitory resource buffers during endosperm filling. Journal of Experimental Botany, 66(5), 1397–1411. https://doi.org/10.1093/jxb/eru492.

Kumar, J., Jaiswal, V., Kumar, A., Kumar, N., Mir, R. R., Kumar, S., Dhariwala, S., Tyagi, S., Khandelwal, M., Prabhu, K. V., Prasad, R., Balyan, H. S., & Gupta, P. K. (2011). Introgression of a major gene for high grain protein content in some Indian bread wheat cultivars. Field Crops Research, 123(3), 226–233.

Kumar, R., Arora, S., Singh, K., & Garg, M. (2015). Puroindoline allelic diversity in Indian wheat germplasm and identification of new allelic variants. Breeding science, 65(4), 319–326.

Lampi, A. M., Nurmi, T., & Piironen, V. (2010). Effects of the environment and genotype on tocopherols and tocotrienols in wheat in the HEALTHGRAIN diversity screen. Journal of Agricultural and Food Chemistry, 58(17), 9306–9313. https://doi.org/10.1021/jf100253u.

Laskowski, W., Górska-Warsewicz, H., Rejman, K., Czeczotko, M., & Zwolińska, J. (2019). How important are cereals and cereal products in the average polish diet?. Nutrients, 11(3), 679.

Lephuthing, M. C., Baloyi, T. A., Sosibo, N. Z., & Tsilo, T. J. (2017). Progress and challenges in improving nutritional quality in wheat. In Wheat improvement, Management and Utilization. (pp. 77–95). Croatia: IntechOpen.

Li, J. R., Zhao, W., Li, Q. Z., Ye, X. G., An, B. Y., Li, X., & Zhang, X. S. (2005). RNA silencing of Waxy gene results in low levels of amylose in the seeds of transgenic wheat (*Triticum aestivum* L.). Acta Genetica Sinica, 32(8), 846–854.

Li, J., Xie, L., Tian, X., Liu, S., Xu, D., Jin, H., et al. (2021). TaNAC100 acts as an integrator of seed protein and starch synthesis exerting pleiotropic effects on agronomic traits in wheat. The Plant Journal, 108(3), 829–840. https://doi.org/10.1111/tpj.15485.

Li, Y., Mao, X., Wang, Q., Zhang, J., Li, X., Ma, F., Sun, F., Chang, J., Chen, M., Wang, Y., Li, K., Yang, G., & He, G. (2014). Overexpression of Puroindoline a gene in transgenic durum wheat (*Triticum turgidum* ssp. durum) leads to a medium–hard kernel texture. Molecular Breeding, 33(3), 545–554.

Listman, M., & Ordóñez, R. (2019). Ten Things You Should Know about Maize and Wheat. CIMMYT: Batan, Mexico. www.cimmyt.org/news/ten-things-you-should-know-about-maize-and-wheat

Liu, J., Wu, B., Singh, R. P., & Velu, G. (2019). QTL mapping for micronutrients concentration and yield component traits in a hexaploid wheat mapping population. Journal of Cereal Science, 88, 57–64. https://doi.org/10.1016/j.jcs.2019.05.008.

Lopez, A., Cacoub, P., Macdougall, I. C., & Peyrin-Biroulet, L. (2016). Iron deficiency anaemia. The Lancet, 387(10021), 907–916. https://doi.org/10.1016/s0140-6736(15)60865-0.

Ma, D., Li, Y., Zhang, J., Wang, C., Qin, H., Ding, H., Xie, Y., & Guo, T. (2016). Accumulation of phenolic compounds and expression profiles of phenolic acid biosynthesis-related genes in developing grains of white, purple, and red wheat. Frontiers in Plant Science, 7, 528. https://doi.org/10.3389/fpls.2016.00528.

Ma, W., Zhang, W., & Gale, K. R. (2003). Multiplex-PCR typing of high molecular weight glutenin alleles in wheat. Euphytica, 134(1), 51–60. https://doi.org/10.1023/A:1026191918704.

Mishra, V. K., Gupta, P. K., Arun, B., Chand, R., Vasistha, N. K., Vishwakarma, M. K., Yadav, P. S., & Joshi, A. K. (2015). Introgression of a gene for high grain protein content (Gpc-B1) into two leading cultivars of wheat in Eastern Gangetic Plains of India through marker assisted backcross breeding. Journal of Plant Breeding and Crop Science, 7(8), 292–300.

Mohan, D., & Gupta, R. K. (2015). Gluten characteristics imparting bread quality in wheats differing for high molecular weight glutenin subunits at Glu D1 locus. Physiology and molecular biology of plants, 21(3), 447–451. https://doi.org/10.1007/s12298-015-0298-y.

Morris, C. F., Kiszonas, A. M., Murray, J., Boehm Jr, J., Ibba, M. I., Zhang, M., & Cai, X. (2019). Re-evolution of durum wheat by introducing the Hardness and Glu-D1 loci. Frontiers in Sustainable Food Systems, 3, 103.

Morris, C. F., Simeone, M. C., King, G. E., & Lafiandra, D. (2011). Transfer of soft kernel texture from *Triticum aestivum* to durum wheat, *Triticum turgidum* ssp. durum. Crop Science, 51(1), 114–122.

Mpofu, A., Sapirstein, H. D., & Beta, T. (2006). Genotype and environmental variation in phenolic content, phenolic acid composition, and antioxidant activity of hard spring wheat. Journal of Agricultural and Food Chemistry, 54(4), 1265–1270. https://doi.org/10.1021/jf052683d.

Müller, O., & Krawinkel, M. (2005). Malnutrition and health in developing countries. Canadian Medical Association Journal, 173(3), 279–286. https://doi.org/10.1503/cmaj.050342.

Muqaddasi, Q. H., Brassac, J., Ebmeyer, E., Kollers, S., Korzun, V., Argillier, O., Stiewe, G., Plieske, J., Ganal, M. W., & Röder, M. S. (2020). Prospects of GWAS and predictive breeding for European winter wheat's grain protein content, grain starch content, and grain hardness. Scientific Reports, 10(1), 1–17. https://doi.org/10.1038/s41598-020-69381-5.

Olmos, S., Distelfeld, A., Chicaiza, O., Schlatter, A. R., Fahima, T., Echenique, V., & Dubcovsky, J. (2003). Precise mapping of a locus affecting grain protein content in durum wheat. Theoretical and Applied Genetics, 107(7), 1243–1251.

Ortiz-Monasterio, J. I., Palacios-Rojas, N., Meng, E., Pixley, K., Trethowan, R., & Pena, R. J. (2007). Enhancing the mineral and vitamin content of wheat and maize through plant breeding. Journal of Cereal Science, 46(3), 293–307.

Ozkan, H., Brandolini, A., Torun, A., Altintas, S., Eker, S. E. L. İ. M., Kilian, B., Braun, H. J., Salamini, F., & Cakmak, I. (2007). Natural variation and identification of microelements content in seeds of einkorn wheat (*Triticum monococcum*). In Wheat Production in Stressed Environments (pp. 455–462). Dordrecht: Springer.

Pawan, Y., Yadava, R. K., Babita, G., Sandeep, K., Verma, R. K., & Sanjat, Y. (2011). Nutritional contents and medicinal properties of wheat: A review. *Life Sciences and Medicine Research*, LMSR-22.

Payne, P. I. (1987). Genetics of wheat storage proteins and the effect of allelic variation on bread-making quality. Annual Review of Plant Physiology, 38(1), 141–153. https://doi.org/10.1146/annurev.pp. 38.060187.001041.

Peleg, Z., Cakmak, I., Ozturk, L., Yazici, A., Jun, Y., Budak, H., Korol, A. B., Fahima, T., & Saranga, Y. (2009). Quantitative trait loci conferring grain mineral nutrient concentrations in durum wheat x wild emmer wheat RIL population. Theoritical and Applied Genetics, 119(2), 353–369. doi: 10.1007/s00122-009-1044-z. Epub 2009 Apr 30. PMID: 19407982.

Perretant, M. R., Cadalen, T., Charmet, G., Sourdille, P., Nicolas, P., Boeuf, C., Tixier, M. H., Branlard, G., Bernard, S., & Bernard, S. (2000). QTL analysis of bread-making quality in wheat using a doubled haploid population. Theoretical and Applied Genetics, 100(8), 1167–1175.

Rai, A., Mahendru-Singh, A., Raghunandan, K., Kumar, T. P. J., Sharma, P., Ahlawat, A. K., et al. (2019). Marker-assisted transfer of PinaD1a gene to develop soft grain wheat cultivars. 3 Biotech, 9(5), 1–10. https://doi.org/10.1007/s13205-019-1717-5.

Ravel, C., Fiquet, S., Boudet, J., Dardevet, M., Vincent, J., Merlino, M., Michard, R., & Martre, P. (2014). Conserved cis-regulatory modules in promoters of genes encoding wheat high-molecular-weight glutenin subunits. Frontiers in Plant Science, 5, 621. https://doi.org/10.3389/fpls.2014.00621.

Regina, A., Bird, A., Topping, D., Bowden, S., Freeman, J., Barsby, T., Kosar-Hashemi, B., Li, Z., Rahman, S., & Morell, M. (2006). High-amylose wheat generated by RNA interference improves indices of large-bowel health in rats. Proceedings of the National Academy of Sciences, 103(10), 3546–3551.

Rizzo, D. M., Lichtveld, M., Mazet, J. A. K., Togami, E., & Miller, S. A. (2021). Plant health and its effects on food safety and security in a One Health framework: Four case studies. One Health Outlook, 3(1), 1–9. https://doi.org/10.1186/s42522-021-00038-7.

Roy, N., Islam, S., Al-Habbar, Z., Yu, Z., Liu, H., Lafiandra, D., Masci, S., Lu, M., Sultana, N., & Ma, W. (2020). Contribution to breadmaking performance of two different HMW glutenin 1Ay alleles expressed in hexaploid wheat. Journal of Agricultural and Food Chemistry, 69(1), 36–44. https://doi.org/10.1021/acs.jafc.0c03880.

Sánchez-León, S., Gil-Humanes, J., Ozuna, C. V., Giménez, M. J., Sousa, C., Voytas, D. F., & Barro, F. (2018). Low-gluten, nontransgenic wheat engineered with CRISPR/Cas9. Plant Biotechnology Journal, 16(4), 902–910. https://doi.org/10.1111/pbi.12837.

Shan, Q., Wang, Y., Li, J., Zhang, Y., Chen, K., Liang, Z., Zhang, K., Liu, J., Xi, J. J., Qiu, J. L., & Gao, C. (2013). Targeted genome modification of crop plants using a CRISPR-Cas system. Nature Biotechnology, 31(8), 686–688. https://doi.org/10.1038/nbt.2650.

Shen, L., Luo, G., Song, Y., Xu, J., Ji, J., Zhang, C., Gregová, E., Yang, W., Li, X., Sun, J., Zhan, K., Cui, D., Liu, D., & Zhang, A. (2021). A novel NAC family transcription factor SPR suppresses seed storage protein synthesis in wheat. Plant Biotechnology Journal, 19(5), 992–1007. https://doi.org/10.1111/pbi.13524.

Sherman, J. D., Lanning, S. P., Clark, D., & Talbert, L. E. (2008). Registration of Near-Isogenic Hard-Textured Wheat Lines Differing for Presence of a High Grain Protein Gene. Journal of Plant Registrations, 2(2), 162–164.

Shewry, P. R. (2009). Wheat. Journal of experimental botany, 60(6), 1537–1553. https://doi.org/10.1093/jxb/erp058.

Shewry, P. R., & Hey, S. J. (2015). The contribution of wheat to human diet and health. Food and Energy Security, 4(3), 178–202. https://doi.org/10.1002/fes3.64.

Shi, R., Li, H., Tong, Y., Jing, R., Zhang, F., & Zou, C. (2008). Identification of quantitative trait locus of zinc and phosphorus density in wheat (*Triticum aestivum* L.) grain. Plant and Soil, 306(1), 95–104.

Shiferaw, B., Smale, M., Braun, H. J., Duveiller, E., Reynolds, M., & Muricho, G. (2013). Crops that feed the world 10. Past successes and future challenges to the role played by wheat in global food security. Food Security, 5(3), 291–317.

Singh, B., Natesan, S. K. A., Singh, B. K., & Usha, K. (2005). Improving zinc efficiency of cereals under zinc deficiency. Current Science, 36–44.

Singh, N., Katyal, M., Virdi, A. S., Kaur, A., Goyal, A., Ahlawat, A. K., & Mahendru Singh, A. (2018). Effect of grain hardness, fractionation and cultivars on protein, pasting and dough rheological properties of different wheat flours. International Journal of Food Science & Technology, 53(9), 2077–2087. https://doi.org/10.1111/ijfs.13794.

Soda, N., Verma, L., & Giri, J. (2018). CRISPR-Cas9 based plant genome editing: Significance, opportunities and recent advances. Plant Physiology and Biochemistry, 131, 2–11. https://doi.org/10.1016/j.plaphy.2017.10.024.

Sollid, L. M. (2002). Coeliac disease: dissecting a complex inflammatory disorder. Nature Reviews Immunology, 2(9), 647–655. https://doi.org/10.1038/nri885.

Sourdille, P., Perretant, M. R., Charmet, G., Leroy, P., Gautier, M. F., Joudrier, P., Nelson, J. C., Sorrells, M. E., & Bernard, M. (1996). Linkage between RFLP markers and genes affecting kernel hardness in wheat. Theoretical and Applied Genetics, 93(4), 580–586.

Sperotto, R. A., Vasconcelos, M. W., Grusak, M. A., et al. (2012). Effects of different Fe supplies on mineral partitioning and remobilization during the reproductive development of rice (*Oryza sativa* L.). Rice, 5, 27. https://doi.org/10.1186/1939-8433-5-27.

Srinivasa, J., Arun, B., Mishra, V. K., Singh, G. P., Velu, G., Babu, R., et al. (2014). Zinc and iron concentration QTL mapped in a Triticum speltax T. aestivum cross. Theoretical and Applied Genetics, 127(7), 1643–1651. https://doi.org/10.1007/s00122-014-2327-6.

Sun, F., Liu, X., Wei, Q., Liu, J., Yang, T., Jia, L., Wang, Y., Yang, G., & He, G. (2017). Functional Characterization of TaFUSCA3, a B3-Superfamily Transcription Factor Gene in the Wheat. Frontiers in Plant Science, 8, 1133. https://doi.org/10.3389/fpls.2017.01133.

Sun, Z., Zhang, M., An, Y., Han, X., Guo, B., Lv, G., Zhao, Y., Guo, Y., & Li, S. (2022). CRISPR/Cas9-Mediated Disruption of Xylanase inhibitor protein (XIP) Gene Improved the Dough Quality of Common Wheat. Frontiers in Plant Science, 13, 811668. https://doi.org/10.3389/fpls.2022.811668.

Terrin, G., Berni Canani, R., Di Chiara, M., Pietravalle, A., Aleandri, V., Conte, F., & De Curtis, M. (2015). Zinc in early life: A key element in the fetus and preterm neonate. Nutrients, 7(12), 10427–10446. https://doi.org/10.3390/nu7125542.

Tiwari, C., Wallwork, H., Arun, B., Mishra, V. K., Velu, G., Stangoulis, J., et al. (2016). Molecular mapping of quantitative trait loci for zinc, iron and protein content in the grains of hexaploid wheat. Euphytica, 207(3), 563–570. https://doi.org/10.1007/s10681-015-1544-7.

Tiwari, V. K., Rawat, N., Chhuneja, P., Neelam, K., Aggarwal, R., Randhawa, G. S., Dhaliwal, H. S., Keller, B., & Singh, K. (2009). Mapping of quantitative trait loci for grain iron and zinc concentration in diploid A genome wheat. Journal of Heredity, 100(6), 771–776.

Tosi, P., Masci, S., Giovangrossi, A., D'Ovidio, R., Bekes, F., Larroque, O., Napier, J., & Shewry, P. (2005). Modification of the low molecular weight (LMW) glutenin composition of transgenic durum wheat: effects on glutenin polymer size and gluten functionality. Molecular Breeding, 16(2), 113–126.

Trier, J. S. (1998). Diagnosis of celiac sprue. Gastroenterology, 115(1), 211–216.

Tsilo, T. J., Simsek, S., Ohm, J. B., Hareland, G. A., Chao, S., & Anderson, J. A. (2011). Quantitative trait loci influencing endosperm texture, dough-mixing strength, and bread-making properties of the hard red spring wheat breeding lines. Genome, 54(6), 460–470. https://doi.org/10.1139/g11-012.

Turner, A. S., Bradburne, R. P., Fish, L., & Snape, J. W. (2004). New quantitative trait loci influencing grain texture and protein content in bread wheat. Journal of Cereal Science, 40(1), 51–60.

Uauy, C., Distelfeld, A., Fahima, T., Blechl, A., & Dubcovsky, J. (2006). A NAC gene regulating senescence improves grain protein, zinc, and iron content in wheat. Science, 314(5803), 1298–1301. https://doi.org/10.1126/science.1133649.

UNICEF. (2018). UNICEF Annual Report. Available at https://www.unicef.org/reports/annual-report-2018.

United Nations News. (2019). www.un.org/development/desa/en/news/population/world-population-prospects-2019.html.

USDA. (2022). https://apps.fas.usda.gov/psdonline/circulars/production.

Velu, G., Ortiz-Monasterio, I., Cakmak, I., Hao, Y., & Singh, R. Á. (2014). Biofortification strategies to increase grain zinc and iron concentrations in wheat. Journal of Cereal Science, 59(3), 365–372. https://doi.org/10.1016/j.jcs.2013.09.001.

Velu, G., Singh, R. P., Crespo-Herrera, L., Juliana, P., Dreisigacker, S., Valluru, R., et al. (2018). Genetic dissection of grain zinc concentration in spring wheat for mainstreaming biofortification in CIMMYT wheat breeding. Scientific Reports, 8(1), 1–10. https://doi.org/10.1038/s41598-018-31951-z.

Velu, G., Tutus, Y., Gomez-Becerra, H. F., Hao, Y., Demir, L., Kara, R., et al. (2017). QTL mapping for grain zinc and iron concentrations and zinc efficiency in a tetraploid and hexaploid wheat mapping populations. Plant and Soil, 411(1), 81–99. https://doi.org/10.1007/s11104-016-3025-8.

Vishwakarma, M. K., Mishra, V. K., Gupta, P. K., Yadav, P. S., Kumar, H., & Joshi, A. K. (2014). Introgression of the high grain protein gene Gpc-B1 in an elite wheat variety of Indo-Gangetic Plains through marker assisted backcross breeding. Current Plant Biology, 1, 60–67.

Wang, L. H., Zhao, X. L., He, Z. H., Ma, W., Appels, R., Peña, R. J., & Xia, X. C. (2009). Characterization of low-molecular-weight glutenin subunit Glu-B3 genes and development of STS markers in common wheat (*Triticum aestivum* L.). Theoretical and Applied Genetics, 118(3), 525–539. https://doi.org/10.1007/s00122-008-0918-9.

Wang, L., Li, G., Peña, R. J., Xia, X., & He, Z. (2010). Development of STS markers and establishment of multiplex PCR for Glu-A3 alleles in common wheat (*Triticum aestivum* L.). Journal of Cereal Science, 51(3), 305–312.

Wani, S. H., Gaikwad, K., Razzaq, A., Samantara, K., Kumar, M., & Govindan, V. (2022). Improving zinc and iron biofortification in wheat through genomics approaches. Molecular Biology Reports, 1–17. https://doi.org/10.1007/s11033-022-07326-z.

Waters, B. M., & Sankaran, R. P. (2011). Moving micronutrients from the soil to the seeds: genes and physiological processes from a biofortification perspective. Plant Science, 180(4), 562–574. https://doi.org/10.1016/j.plantsci.2010.12.003.

Waters, B. M., Uauy, C., Dubcovsky, J., & Grusak, M. A. (2009). Wheat (*Triticum aestivum*) NAM proteins regulate the translocation of iron, zinc, and nitrogen compounds from vegetative tissues to grain. Journal of Experimental Botany, 60(15), 4263–4274. https://doi.org/10.1093/jxb/erp257.

Welch, R. M., & Graham, R. D. (2004). Breeding for micronutrients in staple food crops from a human nutrition perspective. Journal of Experimental Botany, 55(396), 353–364.

WHO (2018). World Health Organization. www.who.int/news-room/factsheets/

WHO (2020). Children: Improving Survival and Well-being. www.who.int/en/news-room/fact-sheets/detail/children-reducing-mortality

Yu, Z., Peng, Y., Islam, M. S., She, M., Lu, M., Lafiandra, D., Roy, N., Juhasz, A., Yan, G., & Ma, W. (2019). Molecular characterization and phylogenetic analysis of active y-type high molecular weight glutenin subunit genes at Glu-A1 locus in wheat. Journal of Cereal Science, 86, 9–14. https://doi.org/10.1016/j.jcs.2019.01.003.

Yue, S. J., Li, H., Li, Y. W., Zhu, Y. F., Guo, J. K., Liu, Y. J., Chen, Y., & Jia, X. (2008). Generation of transgenic wheat lines with altered expression levels of 1Dx5 high-molecular weight glutenin subunit by RNA interference. Journal of Cereal Science, 47(2), 153–161. https://doi.org/10.1016/j.jcs.2007.03.006.

Zhang, S., Zhang, R., Gao, J., Song, G., Li, J., Li, W., Qi, Y., Li, Y., & Li, G. (2021). CRISPR/Cas9-mediated genome editing for wheat grain quality improvement. Plant Biotechnology Journal, 19(9), 1684–1686. https://doi.org/10.1111/pbi.13647.

Zhao, D., Derkx, A. P., Liu, D. C., Buchner, P., & Hawkesford, M. J. (2015). Overexpression of a NAC transcription factor delays leaf senescence and increases grain nitrogen concentration in wheat. Plant Biology, 17(4), 904–913. https://doi.org/10.1111/plb.12296.

Zhu, J., Fang, L., Yu, J., Zhao, Y., Chen, F., & Xia, G. (2018). 5-Azacytidine treatment and TaPBF-D over-expression increases glutenin accumulation within the wheat grain by hypomethylating the Glu-1 promoters. Theoretical and Applied Genetics, 131(3), 735–746. https://doi.org/10.1007/s00122-017-3032-z.

16

Wheat Biofortification and Mainstreaming Grain Zinc and Iron in CIMMYT Wheat Germplasm

Velu Govindan, Ravi P. Singh, and Maria Itria Ibba

CONTENTS

16.1 Introduction

As one of the world's major staple food crops, wheat is consumed by 35% of the population, contributing almost 20% of protein and daily dietary energy to the human diet (Braun et al., 2010). The development of nutrient-dense wheat has the potential to alleviate micronutrient deficiency issues that affect a significant portion of the world's population, primarily in developing or underdeveloped countries. Wheat is an ideal candidate for biofortification because of its important role in ensuring food security. Biofortification is an intervention that improves the mineral and vitamin content of staple food crops through plant breeding to address mineral and vitamin deficiencies (Bouis, 2000). More than 720 million people faced hunger in 2020, and approximately three billion people lacked access to a healthy diet. All of these issues, exacerbated by the current COVID-19 crisis, have increased the number of people affected by so-called hidden hunger, which is caused by an inadequate intake of essential nutrients in the daily diet. Biofortification, defined as improving the nutritional quality of food crops through conventional breeding, agronomic practices, or modern biotechnologies, represents a long-term, cost-effective, and sustainable approach to addressing micronutrient deficiency. Given their importance in the human diet, staple crops are typically the primary focus of most biofortification studies. Wheat, in particular, contributes approximately 20% of total energy and protein intake, as well as approximately 30% of Fe and Zn intake worldwide. However, the current level of micronutrients in most wheat-derived food products is insufficient to meet the minimum daily intake

DOI: 10.1201/9781003307938-16

of such components, particularly in the world's poorest regions where wheat is the primary source of energy. For these reasons, it is critical to continue working on wheat biofortification in all of its forms in order to ensure the production of nutritious and sustainable food while also contributing to the reduction of micronutrient deficiency. This chapter summarizes some of the most recent advances in wheat biofortification research, with studies ranging from the development of genetic tools to accelerate conventional biofortification breeding to the development of novel agronomic methods to increase the micronutrient content of wheat grain.

16.2 Targeted Breeding for Micronutrient Traits in Wheat

Tremendous progress has been made on wheat biofortification over the past decade to ensure a significant reduction in the incidence of hidden hunger. Targeted breeding for micronutrient content in CIMMYT wheat began in 2006 by crossing high micronutrient-carrying synthetic wheats, *T. spelta*, and landraces with high yielding adapted wheat germplasm and then selecting plants with agronomic and disease resistance traits and in advanced generations phenotyped for grain Zn and Fe in segregating populations. The continued use of conventional breeding methods has resulted in the inclusion of several novel alleles for grain Zn and Fe in elite, high-yielding germplasm (Velu et al., 2012). More than 25 biofortified wheat varieties have been released to date, including 'Zincol 2016' in Pakistan; Zinc Shakti (Chitra), WB 02, PBW 01 Zn, and Ankur Shiva in India; BARI Gom 33 in Bangladesh; and Nohely F2018 in Mexico (Velu et al., 2022). Since 2006, over 5,000 simple and top/back crosses have been made between parents with high micronutrient and yield potential, followed by early generation (F2–F4) selection for agronomic traits and advanced generation selection for grain Zn, grain yield, and yield stability (F5–F7). A high-throughput nondestructive technique based on X-ray fluorescence (XRF) has been developed and is routinely used to screen a large number of breeding lines for grain Zn and Fe (Paltridge et al., 2012). The advancement was made possible by a strong emphasis on phenotyping of materials in Zn-homogenized fields at Ciudad Obregon, Mexico. Wheat lines found to have higher Zn levels in grain also showed high Zn levels in South Asia when grown in trials known as the Harvest Plus Yield Trial (HPYT). The multi-site analysis revealed high heritability and high genetic correlations between locations including Cd. Obregon and target environments (Velu et al., 2012). The increased interest in growing via the HPYT by national partners is shown in Figure 16.1. The HPYT trials grown over the past 12 years period shown yield gain progress of 1.5% per annum with zinc and iron increase of about 1% per annum (Govindan et al., 2022).

Biofortified wheat is expected to make a significant difference among resource-poor wheat consumers in South Asia. In India and Pakistan, where nearly two out of every five children under the age of five do not have enough iron (Fe) and zinc (Zn) in their bodies, biofortified wheat promises to become an important source of these nutrients. The biofortified wheat can meet up to 50% of a consumer's daily Zn requirement (Sazawal et al., 2018) and large share of grain Zn is localized in the outer layer of the grain (Wan et al., 2022). Meanwhile, farmers in South Asia can expect even more benefits from Zn-enriched wheat as more varieties with improved rust resistance and yield potential with tolerance to abiotic stresses become available.

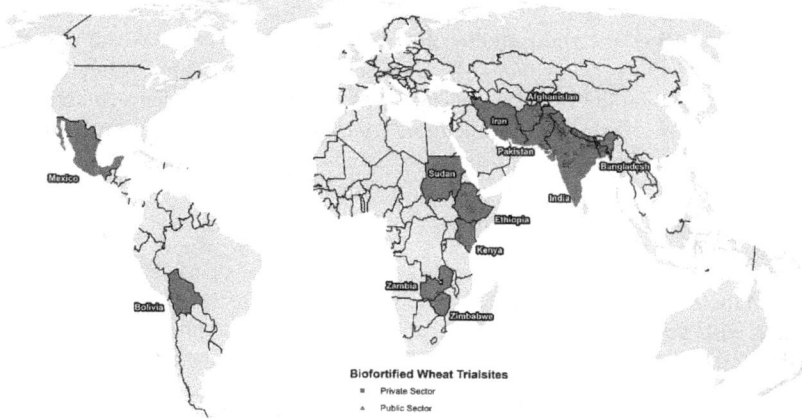

FIGURE 16.1 Distribution of HPYT trial across different countries for phenotyping and variety release.

16.3 Gene Discovery and Candidate Gene Identification

Several studies using bread wheat recombinant inbred line (RIL) populations or association mapping studies to identify different quantitative trait loci (QTL) associated with the variation in the iron (Fe) and zinc (Zn) grain content show several promising hotspot candidate regions for grain Zn and Fe. Traditional QTL mapping and association mapping studies were used to identify "hotspot" genomic regions and associated molecular markers for grain Zn concentration, which will be used in marker-assisted breeding (Srinivasa et al., 2014; Liu et al., 2019; Rathan et al., 2021). According to preliminary genomic prediction analysis, the correlation between observed and predicted values for grain Zn and Fe was approximately 50% across environments, implying that genomic selection models can be used to accelerate breeding efficiency (Velu et al., 2016). Wang et al. (2021) grew an RIL population in nine different environments and identified seven different genomic regions associated with grain Zn content, accounting for 2.2 to 25.1% of the observed variation and four genomic regions associated with grain Fe accumulation, accounting for 2.3 to 30.4% of the variation. Some of the QTL were novel, while others were most likely previously reported in other mapping studies, such as the QTL found on chromosome 4DS and associated with the *Rht2* allele, which is related to plant height. Interestingly, three of the QTL identified in this study appeared to be linked to Fe and Zn accumulation. As a result, these QTL were transformed into high-throughput Kompetitive Allele Specific PCR (KASP) markers that could be easily used to speed up biofortification within a conventional breeding program. Similar to Wang et al. (2021), Krishnappa et al. (2021) used a RIL population grown in various environments to investigate the genetic control of Fe and Zn accumulation in wheat grains. However, in this case, additional traits associated with grain quality (grain protein content and thousand kernel weight) were also studied. Several QTL associated with Fe, Zn, protein content, and thousand kernel weight were identified as a result of the high marker density and D-genome coverage. Several of them were found on the D genome, with chromosome 7D harboring several QTL associated with all of the traits studied. Putative candidate genes for the observed phenotypic variation were also identified, paving the way for future research. The identification of genomic regions associated with higher micronutrient contents and the development of genetic tools for the rapid transfer of high-micronutrient traits are the best approaches to speed up the development of nutrient-rich wheat.

16.3.1 Agronomic Biofortification

Several studies have shown that agronomic biofortification is an efficient and effective short-term method of increasing wheat grain micronutrient content, especially when combined with genetic biofortification. Yu et al. (2021) investigated the potential of Zn foliar application on wheat grown in China's Quzhou County, where more than 90% of the population works in agriculture and 39% of children suffer from Zn deficiency. According to the findings of this study, which was conducted in 16 different locations throughout Quzhou County, wheat treated with Zn foliar application had 97.7% and 68.2% higher Zn content in wheat grain and flour, respectively, when compared to wheat grain and flour from non-treated controls.

Even though agronomic biofortification has been widely demonstrated to be an efficient biofortification approach, the dynamics of absorption and translocation of the applied microelements are complex and influenced by several factors such as the chemical form of the microelement to be increased, the rate at which it is applied, and the method of application. In this regard, Ramkissoon et al. (2021) reported on the time-dependent changes in the absorption, transformation, and distribution of selenium (Se) applied to wheat leaves at two growth stages and with or without the inclusion of urea. When considering agronomic biofortification, it is important to consider the potential negative impact on the environment. Microbial-assisted biofortification may be a solution for combining the short-term effects of traditional agronomic biofortification while minimizing the negative impact that increased fertilizer application may have on both soils and waters. The use of an endophytic strain of *Bacillus altitudinis* (Sun et al., 2021) confirms the technique's potential for wheat Fe biofortification. The authors tested two different inoculation methods and demonstrated that, particularly after spraying the microbe inoculum in the soil, grain Fe accumulation increased significantly and that this *B. altitudinis* strain could efficiently colonize and translocate within wheat.

16.4 Mainstreaming Grain Zinc and Iron in CIMMYT Wheat Germplasm

The wheat biofortification breeding program at CIMMYT has made a steady progress toward the development, distribution, and deployment of Zn-enriched wheat lines with increased yield potential and disease resistance and better processing quality through national program partners. The public–private partnership (PPP) in South Asia enabled to achieve dissemination of high Zn wheat varieties to farmers, and the adoption of Zn-enriched wheat is being accelerated by the new partnership with small and medium enterprise (SME) seed companies in South Asia. For instance, in 2021, five Zn-biofortified wheat varieties released in Nepal, and the area of adoption of these varieties is increasing every year (Thapa et al., 2022).

Scaling up of the adoption of Zn wheat varieties in India and Pakistan is in the early stages, and further testing and scaling out to other South Asian countries (Bangladesh, Nepal, and Afghanistan) and in Ethiopia is underway. Over the next two decades, as the mainstreaming of Zn in the CIMMYT wheat-breeding program is implemented, and as additional high-yielding, high Zn-competitive varieties are released for large scale adoption by farmers, it is likely that a high percentage of total wheat supplies in South Asia will be dense in Zn, a trait which is invisible but improves nutritional uptake of major wheat-dependent consumers. The ratio of public health benefits to the costs of developing and deploying these varieties is estimated to be as high as 100-to-1, if the adoption rates of biofortified varieties reach 75% of total supply.

Scientists continue to refine or adopt methods and techniques to enhance the efficiency of breeding and accuracy of selection. The adoption of X-ray fluorescence (XRF) analysis, for instance, is facilitating the near-accurate screening of thousands of grain samples for Zn and Fe concentration. Gene discovery and mapping techniques are expected to improve the efficiency of breeding programs for the high Zn trait, which facilitate mainstreaming of Zn in CIMMYT wheat-breeding programs as well as the national wheat-breeding programs in South Asia and sub-Saharan Africa.

Most high-Zn lines developed to date yield 5–10% less than the best wheat lines from the main CIMMYT pipeline, whose products serve approximately 70% of the wheat area in the developing world, either directly or as parents of locally bred varieties. Currently, CIMMYT devotes 20% of its wheat-breeding effort to developing high-Zn varieties. In India and Pakistan, several Zn-enriched lines have been released. However, the performance of Zn-enhanced varieties must be superior to that of current elite non-biofortified varieties in order to incentivize smallholders who produce for both home consumption and market adoption. Because yield and Zn content are both polygenic traits, more breeding effort and novel approaches are required to combine them at a high frequency in CIMMYT's elite germplasm. The CIMMYT wheat-breeding program is currently working on mainstreaming grain Zn content as a breeding target, greatly increasing the frequency of elite lines combining high Zn content and high yield. This will be accomplished by implementing Zn screening throughout the program and shortening breeding cycle times, allowing for simultaneous gains in Zn and grain yield. Within ten years, all CIMMYT breeding lines distributed globally will outperform current varieties in yield and meet the Zn biofortification target of 37 ppm, which is approximately 50% higher than the current levels. Achieving this could provide up to 90% of the daily Zn requirement for women and children in vulnerable, high wheat-consuming populations.

16.5 Conclusion

Genetic biofortification breeding in wheat aimed to increase grain Zn and Fe appears to be very promising. To achieve higher Zn and Fe biofortification targets, additional landraces, wild relatives, and other germplasm sources from around the world will need to be screened, and favorable alleles from high-nutrient genetic resources will need to be introgressed. Appropriate testing conditions that mimic farmers' conditions will be critical for biofortification breeding programs due to the relatively high genotype x environment interaction effect for grain Fe and Zn concentrations, as well as the high heritability and moderately significant positive association between environments for grain mineral micronutrient concentrations under diverse target environments. The lack of associations between grain yield and grain Fe and Zn concentrations, combined with favorable associations between grain Fe and Zn, should allow for efficient breeding for nutritious and high-yielding wheat varieties, with the potential to help reduce micronutrient malnutrition among wheat-dependent consumers in the developing world.

REFERENCES

Bouis, H. E. (2000). Enrichment of food staples through plant breeding: A new strategy for fighting micronutrient malnutrition. *Nutrition* https://doi.org/10.1016/S0899-9007(00)00266-5.

Braun, H. J., Atlin, G., and Payne, T. (2010). Multi-location testing as a tool to identify plant response to global climate change. *Climate Change Crop Prod.* https://doi.org/10.1079/9781845936334.0115.

Govindan, V., Atanda, S., Singh, R. P., Huerta-Espino, J., Crespo-Herrera, L. A., Juliana, P., et al. (2022). Breeding increases grain yield, zinc and iron supporting enhanced wheat biofortification. *Crop Sci.* https://doi.org/10.1002/csc2.20759.

Krishnappa, G., Rathan, N. D., Sehgal, D., Ahlawat, A. K., Singh, S. K., Singh, S. K., et al. (2021). Identification of novel genomic regions for biofortification traits using an SNP marker-enriched linkage map in wheat (*Triticum aestivum* L.). *Front. Nutr.* 8, 1–13. https://doi.org/10.3389/fnut.2021.669444.

Liu, J., Wu, B., Singh, R. P., and Velu, G. (2019). QTL mapping for micronutrients concentration and yield component traits in a hexaploid wheat mapping population. *J. Cereal Sci.* 88. https://doi.org/10.1016/j.jcs.2019.05.008.

Paltridge, N. G., Milham, P. J., Ortiz-Monasterio, J. I., Velu, G., Yasmin, Z., Palmer, L. J., et al. (2012). Energy-dispersive X-ray fluorescence spectrometry as a tool for zinc, iron and selenium analysis in whole grain wheat. *Plant Soil* 361. https://doi.org/10.1007/s11104-012-1423-0.

Ramkissoon, C., Degryse, F., Young, S., Bailey, E. H., and McLaughlin, M. J. (2021). Using 77Se-labelled foliar fertilisers to determine how se transfers within wheat over time. *Front. Nutr.* 8, 1–10. https://doi.org/10.3389/fnut.2021.732409.

Rathan, N. D., Sehgal, D., Thiyagarajan, K., Singh, R., Singh, A.-M., and Govindan, V. (2021). Identification of genetic loci and candidate genes related to grain zinc and iron concentration using a zinc-enriched wheat 'Zinc-Shakti.' *Front. Genet.* 12.

Sazawal, S., Dhingra, U., Dhingra, P., Dutta, A., Deb, S., Kumar, J., et al. (2018). Efficacy of high zinc biofortified wheat in improvement of micronutrient status, and prevention of morbidity among preschool children and women—A double masked, randomized, controlled trial. *Nutr. J.* 17, 86. https://doi.org/10.1186/s12937-018-0391-5.

Srinivasa, J., Arun, B., Mishra, V. K., Singh, G. P., Velu, G., Babu, R., et al. (2014). Zinc and iron concentration QTL mapped in a Triticum spelta × T. Aestivum cross. *Theor. Appl. Genet.* 127. https://doi.org/10.1007/s00122-014-2327-6.

Sun, Z., Yue, Z., Liu, H., Ma, K., & Li, C. (2021) Microbial-Assisted Wheat Iron Biofortification Using Endophytic *Bacillus altitudinis* WR10. *Front. Nutr.* 8, 704030. https://doi.org/10.3389/fnut.2021.704030.

Thapa, D. B., Subedi, M., Yadav, R. P., Joshi, B. P., Adhikari, B. N., Shrestha, K. P., et al. (2022). Variation in grain zinc and iron concentrations, grain yield and associated traits of biofortified bread wheat genotypes in Nepal. *Front. Plant Sci.* 13, 1–13. https://doi.org/10.3389/fpls.2022.881965.

Velu, G., Crossa, J., Singh, R. P., Hao, Y., Dreisigacker, S., Perez-Rodriguez, P., et al. (2016). Genomic prediction for grain zinc and iron concentrations in spring wheat. *Theor. Appl. Genet.* 129, 1595–1605. https://doi.org/10.1007/s00122-016-2726-y.

Velu, G., Singh, R. P., Huerta-Espino, J., Peña-Bautista, R. J., Crossa, J., Andersson, M. S., et al. (2012). Breeding progress and genotype × environment interaction for zinc concentration in CIMMYT spring wheat germplasm. In *Wheat: Productivity enhancement under changing climate*. India: Narosa Book Distributors Pvt Ltd, pp. 340–349.

Velu, G., Singh, R. P., Joshi, A. K., and Virk, P. (2022). Advances in wheat biofortification and mainstreaming grain zinc in CIMMYT wheat breeding. In *Biofortification of staple crops*. Singapore: Springer, pp. 105–117.

Wan, Y., Stewart, T., Amrahli, M., Evans, J., Sharp, P., Govindan, V., et al. (2022). Localisation of iron and zinc in grain of biofortified wheat. *J. Cereal Sci.* 105, 103470. https://doi.org/10.1016/j.jcs.2022.103470.

Wang, Y., Xu, X., Hao, Y., Zhang, Y., Liu, Y., Pu, Z., et al. (2021). QTL mapping for grain zinc and iron concentrations in bread wheat. *Front. Nutr.* 8, 1–11. https://doi.org/10.3389/fnut.2021.680391.

Yu, B. G., Liu, Y. M., Chen, X. X., Cao, W. Q., Ding, T. B., and Zou, C. Q. (2021). Foliar zinc application to wheat may lessen the zinc deficiency burden in Rural Quzhou, China. *Front. Nutr.* 8, 1–9. https://doi.org/10.3389/fnut.2021.697817

Index

Note: Page numbers in *italic* indicate a figure and page numbers in **bold** indicate a table on the corresponding page.

For Product Safety Concerns and Information please contact our EU
representative GPSR@taylorandfrancis.com
Taylor & Francis Verlag GmbH, Kaufingerstraße 24, 80331 München, Germany

www.ingramcontent.com/pod-product-compliance
Lightning Source LLC
Chambersburg PA
CBHW060744220326
41598CB00022B/2324